D1385463

Advances in

INORGANIC CHEMISTRY

———

Volume 54

ADVISORY BOARD

Advances in

INORGANIC CHEMISTRY

Including Bioinorganic Studies

Rudi van Eldik and Colin D. Hubbard

Institute for Inorganic Chemistry
University of Erlangen-Nürnberg
91058 Erlangen
Germany

VOLUME 54: Inorganic Reaction Mechanisms

ACADEMIC PRESS

An imprint of Elsevier Science

Amsterdam   Boston   London   New York   Oxford   Paris
San Diego   San Francisco   Singapore   Sydney   Tokyo

Academic Press
An imprint of Elsevier Science
525 B Street, Suite 1900
San Diego, California 92101-4495
USA
http://www.academicpress.com

First edition 2003

ISBN: 0-12-023654-0
ISSN: 0898-8838

♾ The paper used in this publication meets the requirements of ANSI/NISO Z39.48-1992 (Permanence of Paper).

Printed in The Netherlands.

CONTENTS

Reaction Mechanisms of Nitric Oxide with Biologically
Relevant Metal Centers

PETER C. FORD, LEROY E. LAVERMAN AND IVAN M. LORKOVIC

Homogeneous Hydrocarbon C–H Bond Activation and
Functionalization with Platinum

ULRICH FEKL AND KAREN I. GOLDBERG

Density Functional Studies of Iridium Catalyzed Alkane Dehydrogenation

MICHAEL B. HALL AND HUA-JUN FAN

Recent Advances in Electron-Transfer Reactions

DAVID M. STANBURY

Metal Ion Catalyzed Autoxidation Reactions: Kinetics and Mechanisms

ISTVÁN FÁBIÁN AND VIKTOR CSORDÁS

PUBLISHER'S NOTE

This special volume of *Advances in Inorganic Chemistry*, with the thematic title Inorganic Reaction Mechanisms, is the first volume to be published under the auspices of the new editor, Professor Rudi van Eldik.

A Dutch national, Professor van Eldik studied chemistry at Potchefstroom University in South Africa, where he gained his D.Sc. in 1971. After a number of years working abroad, he was appointed Professor of Chemistry at Potchefstroom in 1979. In 1982 he received his Habilitation at the University of Frankfurt where he was Group Leader at the Institute for Physical Chemistry between 1980 and 1986. From 1987 to 1994 he was Professor of Inorganic Chemistry at the University of Witten/Herdecke, Germany and was then appointed to his present position as Professor of Inorganic and Analytical Chemistry at the University of Erlangen-Nürnberg in Germany. In the intervening years he has travelled widely, being a Visiting Professor at the University of Utah in the USA, the University of Canterbury in New Zealand, Ben Gurion University in Israel and at the moment is Wilsmore Visiting Professor at the University of Melbourne in Australia.

A prolific author, Professor van Eldik has been responsible for some 580 papers in refereed journals, and four books as editor or co-editor. His current research intrests are the application of high pressure techniques in mechanistic studies; metal-catalyzed autoxidation processes; and bioinorganic studies. As such he is eminently qualified to edit the prestigious *Advances in Inorganic Chemistry*. We are confident that he is a worthy successor to Professor Geoff Sykes and that he will maintain the high standards for which the series is known.

PREFACE

I am especially honoured to have been appointed the Editor of *Advances in Inorganic Chemistry*, and to be associated with a highly cited, very successful series. The series began in 1959, edited by H.J. Eméleus and A.G. Sharpe, and at that time included reviews on Radiochemistry. With Volume 31, Professor A.G. Sykes assumed the editorship and extended the area of interest to include bioinorganic studies. He prefaced that issue with an outline of his views on the aims of the series, and (paraphrasing) he intended to continue to provide a forum for scholarly and critical reviews by recognized experts, rather than seeking to catalogue each and every event. Those opinions and other comments were very appropriate then and remain so, now and in the future.

Contributions will be solicited by the editor, who will also be guided by the Editorial Advisory Board. Issues on thematic topics will in general involve a co-editor as a specialist in that particular field. Suggestions for subjects for reviews in the future will be welcome at any time. Of particular interest will be reviews of rapidly developing areas that do not necessarily fit into traditional subject sub-areas, thus appealing to newer readers and research colleagues. It is felt that a presentation of diverse topics will assist in creative thinking and help to ensure that the overall subject of Inorganic Chemistry continues to develop and thrive. In this respect, I would like to welcome the new Advisory Board members and will look forward to interacting with them.

To honour the accomplishments of Professor Geoff Sykes, both in his own research, principally in mechanistic studies in Inorganic and Bioinorganic Chemistry, and as editor of this series, the present volume on Inorganic Reaction Mechanisms is dedicated to him. The Publisher's note in Volume 53 referred to the high impact factor as a reflection of the high standards set and the quality of the contributing authors. I echo these comments and personally acknowledge Professor Sykes' significant contribution to the success of the series. I am delighted and feel privileged to succeed illustrious editors and will endeavour to match their high standards.

For my first volume as Editor, I have invited Professor Colin D. Hubbard (University of Erlangen-Nürnberg, Erlangen, Germany and University of New Hampshire, Durham, NH, USA) as co-editor. Professor Hubbard studied chemistry at the University of Sheffield, and obtained his PhD with Ralph G. Wilkins. Following post-doctoral work at MIT, Cornell University and University of California in Berkeley, he joined the academic staff of the University of New Hampshire, Durham, where he became Professor of Chemistry in 1979. His interests cover the areas of high-pressure chemistry, electron transfer reactions, proton tunnelling and enzyme catalysis.

The first chapter by F.A. Dunand, L. Helm and A.E. Merbach is a comprehensive account of the mechanism of solvent exchange processes. Metal complex formation can be controlled by solvent exchange. This topic, as well as ligand substitution in general, form the subject of the second chapter by J. Burgess and C.D. Hubbard. Following this, J.H. Espenson describes 'Oxygen Transfer Reactions: Catalysis by Rhenium Compounds'. The fourth chapter by P.C. Ford, L.E. Laverman and J.M. Lorkovic is an account of the reaction mechanisms of nitric oxide with biologically relevant metal centers. In chapter 5, U. Fekl and K.I. Goldberg discuss 'Platinum Involvement in Homogeneous Hydrocarbon C–H Bond Activation and Functionalization'. Chapter 6 by M.H. Hall and H.-J. Fan is titled 'Density Functional Studies of Iridium Catalyzed Alkane Dehydrogenation'. 'Recent Advances in Electron Transfer Reactions' are reported by D.M. Stanbury. The final chapter by I. Fábián and V. Csordás is on 'The Kinetics and Mechanism of Metal Ion Catalyzed Autoxidation Reactions'. I thoroughly believe that these contributions cover the present advances accomplished in the general area of Inorganic Reaction Mechanisms.

Rudi van Eldik
University of Erlangen-Nürnberg
Germany
December 2002

SOLVENT EXCHANGE ON METAL IONS

FRANK A. DUNAND, LOTHAR HELM and ANDRÉ E. MERBACH

Institut de chimie moléculaire et biologique, Ecole polytechnique fédérale de
Lausanne, EPFL-BCH, CH-1015 Lausanne, Switzerland

I. Introduction

A. GENERAL ASPECTS

Solvent exchange reactions on metal cations are among the most simple chemical reactions: a solvent molecule situated in the first coordination shell of the ion is replaced by another one, normally entering from the second shell. They are generally considered as fundamental reactions for metal ions in solution, since they constitute an important step in complex-formation reactions on metal cations. The reaction is

1

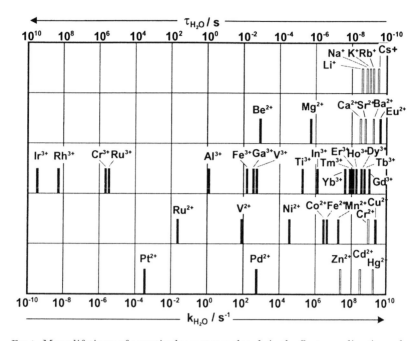

FIG. 1. Mean lifetimes of a particular water molecule in the first coordination sphere of a given metal ion, τ_{H_2O}, and the corresponding water exchange rate constants, k_{H_2O} at 298 K. The filled bars indicate directly determined values, and the empty bars indicate values deduced from ligand substitution studies.

symmetrical: reactants and reaction products are identical, which has important drawbacks in the determination of the rate constants.

The rates of solvent exchange vary widely with the nature of the cation and, to a lesser extent, with that of the solvent. As an example Fig. 1 shows that k_{H_2O}, the exchange rate constant for water molecules, covers nearly 20 orders of magnitude. At the "slow end" of the lability scale the mean life time of a water molecule in the 1st coordination shell of $[Ir(H_2O)_6]^{3+}$, τ_{H_2O} ($=1/k_{H_2O}$) is about 290 years (1), whereas at the other extreme the shortest mean life time is 2×10^{-10} s ($=200$ ps) directly measured on $[Eu(H_2O)_7]^{2+}$ (2). Variation of the rate constant with solvent on the same metal ion is less pronounced and generally below 2 orders of magnitude. Strong back-bonding from the metal to solvent molecules, however, can slow down the exchange process by several orders of magnitude, as observed for example for $[Ru(MeCN)_6]^{3+}$ (3). It is therefore convenient to divide the discussion of solvent exchange into categories of metal ions.

The first category includes the main group ions, which exhibit in general, for a given ionic charge, increasing exchange rate constants, k_{ex}, with increasing ionic radius. This is nicely illustrated by the water exchange on Al^{3+}, Ga^{3+} and In^{3+}, all being six-coordinate, with a rate increase of more than 6 orders of magnitude. Solvent exchange on mono-valent alkali and large divalent alkali-earth ions is very fast and the exchange rate constants can only be deduced from complex-formation reactions (4,5).

The second category is the d-transition metal ions. Their solvent exchange properties are strongly influenced by the electronic occupancy of their d orbitals. This is best illustrated by the 1st row transition metal ions. On the basis of their ionic radii, r_M, they should all show labilities similar to Zn^{2+} for the divalent ions and similar to Ga^{3+} for the trivalent ones. However, the water exchange rates vary by 7 and 15 orders of magnitude, respectively, depending largely on the electronic configuration of the metal ion. Within this category, square-planar complexes constitute a special sub-category.

The third category is the high coordination number lanthanides and actinides. The trivalent lanthanides show a decrease in r_M with the progressive filling of the 4f orbitals, called the lanthanide contraction. Since the 4f orbitals are shielded by the filled 5s and 5p orbitals, the electronic configuration has no remarkable effect and therefore the variation in r_M and an eventual change in coordination number and geometry determine the lability of the 1st coordination shell.

Solvent exchange reactions have been reviewed several times in the last 10 years. A comprehensive review by Lincoln and Merbach was published in this series in 1995 (6). More recent reviews focused more on high pressure techniques for the assignment of reaction mechanisms (7–9) or on water exchange (10). This review is a follow up of the exhaustive Lincoln and Merbach review (6). The main features of solvent exchange on metal ions will be pointed out, taking into account developments and new results from the last 10 years.

B. EXPERIMENTAL METHODS

Only a few experimental techniques are available to measure solvent exchange rate constants directly. Nuclear magnetic resonance (NMR) spectroscopy has shown the widest range of application. Most exchange rate constants given in this review were determined by NMR using a variety of specific methods. A common method is the observation of NMR lineshape (11). If the exchange rate constant is in the range defined by the natural linewidth of the NMR resonance and there is a chemical shift difference between resonance signals of bulk solvent and solvent molecules in the 1st coordination shell (both in $rad\,s^{-1}$), k_{ex} can be directly determined from the NMR spectrum. This means that rates in the order of several s^{-1} can be measured by ^1H or ^{13}C NMR as well as rates of the order of $10^6\,s^{-1}$ if chemical shift differences are large due to paramagnetic effects. Even faster exchange rates can be determined using methods based on relaxation rate measurements. Using these methods, the fastest rate measured by NMR up to now is $\sim 5 \times 10^9\,s^{-1}$ (2). Relatively slow exchange reactions from k_{ex} 0.1 to $10\,s^{-1}$ can be followed using magnetization transfer experiments. By exciting spins on a chemical site, well defined by its NMR signal, and observing how this excitation is transferred by the chemical reaction to another site, reaction rates can be obtained if the reaction proceeds faster than nuclear spin relaxation (12,13).

Very slow reactions can be followed by isotopic labeling techniques. The solvent molecules in the 1st shell of the metal ion or the bulk solvent may be labeled either using stable isotopes as for example ^2H, ^{13}C, ^{15}N or ^{17}O or either radioactive isotopes as ^{14}C or ^3H. The exchange of the labeled molecules can be followed with various techniques like NMR, mass spectroscopy, or radiation counting after precipitation. There is in principle no lowest limit of k_{ex} to be measured by the labeling method; samples can be stored in a thermostated bath for months and transferred from time to time into a NMR spectrometer to record the progress of the reaction. To accelerate the reaction, measurements can be performed at high temperature. Water exchange on the extremely slow $[Ir(H_2O)_6]^{3+}$ was measured at temperatures from 358 to 406 K and the value at 298 K was then extrapolated from these data (1).

Measuring very fast solvent exchange rates is very difficult, especially on diamagnetic ions where the NMR technique described above cannot be applied. Some information on the dynamics of water protons in aqueous ionic solution is available from high-resolution incoherent quasielastic neutron scattering (IQENS). Based on differences of translational diffusion constants for bulk and bound water, this technique allows one to establish a time-scale for ion-water proton binding (14). From the experimental data it can be decided if the binding time τ_{H_2O} is short ($\leq 10^{-10}\,s$) or long ($\geq 5 \times 10^{-9}\,s$) on the IQENS observation timescale (15). Between these two limits binding times can be estimated from the IQENS spectra. Very fast water exchange rates can be estimated from rates of complex formation on the aqua ions, measured by ultra sound absorption or temperature-jump techniques (4,5). If the formation of the inner sphere complex involves the movement of a ligand from the second to the first coordination sphere and this is the rate-determining step in the complex-formation reaction, then this rate can be considered to be close to the water exchange rate.

Measuring the pressure dependence of the exchange rate constant leads to activation volumes, $\Delta V^{\ddagger}$, and this technique has become a major tool for the mechanistic identification of solvent exchange mechanisms (8,16,17). In the last 25 years high-pressure, high-resolution NMR probes were developed which allow the application of all NMR techniques described to pressures up to several hundreds of mega Pascals (18).

C. CLASSIFICATION OF MECHANISMS

The mechanistic classification generally accepted for ligand substitution reactions was proposed by Langford and Gray in 1965 (19). This classification was often discussed in the literature and its principles are only summarized here for convenience.

$$\text{MX}_n \underset{+X}{\overset{-X}{\rightleftharpoons}} \{\text{MX}_{n-1}\} \underset{-Y}{\overset{+Y}{\rightleftharpoons}} \text{MX}_{n-1}\text{Y} \quad (\mathbf{D} = \text{dissociative}) \quad (1)$$

$$\text{MX}_n \underset{-Y}{\overset{+Y}{\rightleftharpoons}} \{\text{MX}_n\text{Y}\} \underset{+X}{\overset{-X}{\rightleftharpoons}} \text{MX}_{n-1}\text{Y} \quad (\mathbf{A} = \text{associative}) \quad (2)$$

$$\text{MX}_n \cdots \text{Y} \rightleftharpoons \text{MX}_{n-1}\text{Y} \cdots \text{X} \quad (\mathbf{I} = \text{interchange}) \quad (3)$$

Langford and Gray divided substitution reactions into three categories of stoichiometric mechanisms: *associative* (**A**) where an intermediate of increased coordination number is inferred, or *dissociative* (**D**) where an intermediate of reduced coordination number is inferred, and *interchange* (**I**) where there is no kinetically detectable intermediate [Eq. (3), $\text{MX}_n \cdots \text{Y}$ represents an outer-sphere complex]. They further distinguish two categories of *intimate mechanisms*: mechanisms with an associative activation mode (**a**) and mechanism with a dissociative activation mode (**d**). In the first case the reaction rate is sensitive to the nature of the entering group whereas in the second case the reaction rate is not sensitive to the variation of the entering group but to the nature of the leaving group.

All **A** mechanisms must be associatively and all **D** mechanisms must be dissociatively activated. The interchange mechanisms (**I**) include a continuous spectrum of transition states where the degree of bond-making between the entering ligand and the complex ranges from very substantial ($\mathbf{I_a}$ mechanism) to negligible ($\mathbf{I_d}$ mechanism) and inversely for bond-breaking (*16*). For a solvent exchange reaction, the forward and backward reaction coordinates must be symmetrical.

How can mechanisms be assigned to solvent exchange reactions? The rate law for solvent exchange reaction can be determined using an inert diluent. Unfortunately such a diluent does not exist for all solvents: there is no inert diluent known for water for example. The variation of the enthalpy and entropy of activation, obtained from variable temperature experiments, within a series of similar ligand substitution systems can give a guide to a mechanistic changeover. Thus **d**-activated reactions tend to have greater $\Delta H^{\ddagger}$ values than do **a**-activated reactions, and $\Delta S^{\ddagger}$ tends to be positive for **d**- and negative for **a**-activated reactions. However, the magnitudes of the contributions to these two parameters arising from interactions that occur beyond the 1st coordination shell can be uncertain; hence the determination of $\Delta S^{\ddagger}$ is often prone to systematic errors. By the choice of adequate conditions, the precision of measurement of $\Delta V^{\ddagger}$ is high and there is a direct relationship between its sign and the increase or decrease of the rate constant with pressure. Therefore, establishment of the dependence of the exchange rate constant on pressure, leading to volumes of activation $\Delta V^{\ddagger}$, provides a major tool for the experimental identification of solvent exchange mechanisms (*20–22*).

D. THE VOLUME OF ACTIVATION

The volume of activation, $\Delta V^{\ddagger}$, is defined as the difference between the partial molar volumes of the transition state and the reactants. It is related to the pressure variation of the rate constant by Eq. (4):

$$\left(\frac{\partial \ln(k)}{\partial P}\right)_T = -\frac{\Delta V^{\ddagger}}{RT} \tag{4}$$

Assuming that $\Delta V^{\ddagger}$ is slightly pressure dependent leads to the approximate Eq. (5),

$$\ln(k_P) = \ln(k_0) - \frac{\Delta V_0^{\ddagger} P}{RT} + \frac{\Delta \beta^{\ddagger} P^2}{2RT} \tag{5}$$

where k_P and k_0 are the rate constants at pressures P and 0, respectively, $\Delta V_0^{\ddagger}$ is the activation volume at zero pressure and $\Delta \beta^{\ddagger}$ is the compressibility coefficient of activation. For solvent exchange the quadratic term in Eq. (5) is very often small compared to the linear one for pressures generally applied in kinetic studies (typically 0–150 MPa) and therefore $\Delta V^{\ddagger} \cong \Delta V_0^{\ddagger}$.

The interpretation of the activation volume for solvent exchange reactions on metal ions is based on application of the transition state theory (23), where no differences due to pressure variation in solvent interactions beyond the first coordination shell are taken into account. The measured $\Delta V^{\ddagger}$ is usually considered to be the combination of an intrinsic and an electrostriction contribution. The intrinsic contribution, $\Delta V_{int}^{\ddagger}$, results from a change in internuclear distances and angles within the reactants during the formation of the transition state, whereas the electrostriction contribution, $\Delta V_{elec}^{\ddagger}$, arises from changes in the electrostriction between the transition state and the reactant (24). For solvent exchange processes, where the charge of the complex remains unchanged, $\Delta V_{elec}^{\ddagger} \approx 0$ and therefore $\Delta V^{\ddagger} \cong \Delta V_{int}^{\ddagger}$. Consequently, the observed activation volume is a direct measure of the degree of bond formation and bond breaking on going to the transition state, assuming no changes in bond length of the non-exchanging water molecule.

The relation between the pressure induced changes of the observed exchange rates and the underlying solvent exchange reaction mechanisms is visualized in Fig. 2. Applying pressure to a **d**-activated exchange

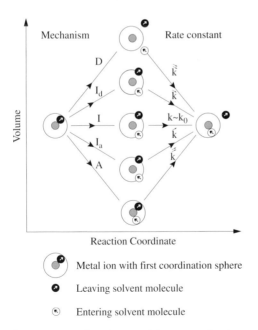

FIG. 2. Volume profiles, connected to the transition states, for the spectrum of solvent exchange processes.

process will result in a decrease in the observed reaction rate, k_P, because the approach to the transition state (mainly bond-breaking) requires an increase in volume. For an **a**-activated exchange process, however, where the approach to the transition state (mainly bond-making) requires a decrease in volume, the observed reaction rate is increased. Briefly, a negative $\Delta V^{\ddagger}$ is indicative of associatively activated processes and a positive $\Delta V^{\ddagger}$ is indicative of dissociatively activated processes. It is, however, more difficult to decide on the basis of activation volumes if the mechanism is a limiting one, **A** or **D**, in the two extreme cases, or an interchange **I** with a relatively strong contribution of the entering solvent molecule ($\mathbf{I_a}$) or with a negligible contribution of the entering solvent molecule ($\mathbf{I_d}$). As a rule of thumb we can say that the larger the absolute value of the activation volume, $|\Delta V^{\ddagger}|$, the closer the mechanism is to a limiting one. There are, however, exceptions to this guideline and every case has to be considered individually.

II. Solvent Exchange on Main Group Metal Ions

A. GENERAL CHARACTERISTICS

There are only very few main group metal ions where solvent exchange reactions can be studied experimentally: Be^{2+}, Mg^{2+}, Al^{3+}, Ga^{3+}, and In^{3+}. All alkali metal ions and the large alkaline earth ions Ca^{2+}, Sr^{2+}, and Ba^{2+} are very labile due to the low surface charge density and the absence of ligand field stabilization effects. The only direct experimental data on water exchange rates of some of these ions come from incoherent neutron scattering (IQENS) (14,25). IQENS has an observation time scale $t_{obs} \sim 1$ ns and allows to give limits for ion to water-proton binding times τ_i (Table I, (26)). Therefore, the rate constants given in Fig. 1 are those deduced from ligand substitution reactions (4,27).

An estimation for the rate of water exchange on Sr^{2+} can be obtained from comparison with Eu^{2+}. This divalent lanthanide behaves similarly to main group ions with the same ionic radius. The characteristic rate constant for H_2O substitution on Sr^{2+} estimated by Eigen (4) ($k_{H_2O} \sim 2 \times 10^8 \, s^{-1}$) using ultra sound absorption is about 25 times slower than that for water exchange on Eu^{2+} measured by ^{17}O NMR ($k_{ex} \sim 5 \times 10^9 \, s^{-1}$, $\tau_M \sim 0.2$ ns). Recent structural investigations using EXAFS have shown that the aquated Sr^{2+} ion is similar to Eu^{2+} although not identical (discussed in Section IV.B).

TABLE I

ION-WATER PROTON BINDING TIMES, τ_i, OBTAINED
FROM IQENS EXPERIMENTS ON CONCENTRATED
AQUEOUS IONIC SOLUTIONS AT $\sim$ 298 K (26)

$\tau_i < 0.1$ ns	$\tau_i > 0.1$ ns	$\tau_i > 5$ ns
Li^+	Zn^{2+}	Mg^{2+}
Cs^+	Nd^{3+}	Ni^{2+}
Ca^{2+}	Dy^{3+}	Al^{3+}
Cu^{2+}		Cr^{3+}
F^-		Fe^{3+}
Cl^-		Ga^{3+}
I^-		
ClO_4^-		

B. DIVALENT MAIN GROUP IONS

The only divalent main group ions where exchange rates and activation parameters are experimentally available are beryllium(II), which is also the smallest metal ion ($r_M = 27$ pm) (28,29),[1] and magnesium(II) ($r_M = 72$ pm). An important consequence of this substantial difference in ionic radii is that Be^{2+} forms predominantly tetrahedral complexes (30) and Mg^{2+} is in general surrounded by six solvent molecules forming octahedral complexes (31).

Beryllium(II) is the only tetrahedral metal ion for which a significant quantity of solvent exchange data is available (Table II, (30,32–36)). Water exchange on $[Be(H_2O)_4]^{2+}$ is characterized by a very negative activation volume $\Delta V^{\ddagger}$ (-12.9 cm^3 mol^{-1}) (30) close to a calculated limiting contribution for an **A** mechanism. This is strong evidence for operation of a limiting **A** mechanism for water exchange on $[Be(H_2O)_4]^{2+}$. In non-aqueous solvents the bulkiness of the solvent molecules becomes an important parameter that determines the exchange mechanism. The data in Table II indicate that solvent exchange on $[BeS_4]^{2+}$ (S = solvent) occurs according to a general rate law as expressed in Eq. (6).

$$k_{obs} = k_1 + k_2 [S] \qquad (6)$$

For DMSO, TMP, DMMP and MMPP the second-order term, k_2, dominates and either an **A** or an $\mathbf{I_a}$ mechanism operates, as it is also indicated by the negative $\Delta V^{\ddagger}$ measured for DMSO and TMP (30). Both terms

---

[1] All ionic radii in this revue are from Refs. (28,29).

TABLE II

RATE CONSTANTS AND ACTIVATION PARAMETERS FOR SOLVENT EXCHANGE ON $[Be(solvent)_4]^{2+}$

Solvent	k_1 (298 K)[a] (s^{-1})	k_2 (298 K)[a] ($dm^3\,mol^{-1}\,s^{-1}$)	$\Delta H^{\ddagger}$ ($kJ\,mol^{-1}$)	$\Delta S^{\ddagger}$ ($J\,K^{-1}\,mol^{-1}$)	$\Delta V^{\ddagger}$ ($cm^3\,mol^{-1}$)	Mechanism	Reference
H_2O[b]	730	13.2[c]	59.2	+8.4	−13.6	A	(30)
DMSO[d]		213	35.0	−83.0	−2.5	A, I_a	(30)
TMP[d]		140	51.1	−32.3	−4.1	A, I_a	(32)
		4.2	43.5	−87.1		A, I_a	(30)
DMMP[d]		1.5	56.0	−54.0		A, I_a	(33)
MMPP[d]		0.81	60.2	−44.4		A, I_a	(33)
DMF[d]		0.22	68.7	−26.1		A, I_a	(34)
		16	52.0	−47.5	−3.1	A, I_a	(30)
	0.2		74.9	−7.3		D	(30)
NMA[d]		8.5	58.1	−32.0		A, I_a	(35)
	0.1		83.6	+16.3		D	(35)
DMA[d]		0.32	76.8	+3.1		A, I_a	(35)
	0.23		71.5	−17.3		D	(35)
DEA[d]		0.34	66.7	−30.1		A, I_a	(35)
	0.38		56.9	−62.1		D	(35)
DMADMP[d]		0.59	68.5	−19.6		A, I_a	(34)
	0.044		76.4	−14.6		D	(36)
TMU[d]	0.0073		89.1	+12.6		D	(33)
	1.0		79.6	+22.3	+10.5	D	(30)
DMPU[d]	1.4		77.1	+16.4		D	(32)
	0.1		92.6	+47.5	+10.3	D	(30)

[a] Rate constant for the exchange of a particular coordinated solvent molecule (6).
[b] Neat solvent.
[c] In $kg\,mol^{-1}\,s^{-1}$, corresponds to 730 s^{-1}/55.5 $mol\,kg^{-1}$.
[d] In d_3-nitromethane diluent.

TABLE III

RATE CONSTANTS AND ACTIVATION PARAMETERS FOR SOLVENT EXCHANGE ON
[Mg(solvent)$_6$]$^{2+}$

Solvent	k_1 (298 K)^a (s^{-1})	$\Delta H^{\ddagger}$ (kJ mol^{-1})	$\Delta S^{\ddagger}$ (J K^{-1} mol^{-1})	$\Delta V^{\ddagger}$ (cm^3 mol^{-1})	Mechanism	Reference
H$_2$O ^{b}	6.7×10^5	49.1	+31.1	+6.7	**D, I$_d$**	(37)
H$_2$O ^{b}	5.3×10^5	42.7	+8		**D, I$_d$**	(38)
MeOH ^{b}	4.7×10^3	69.9	+59		**D, I$_d$**	(39)
EtOH ^{b}	2.8×10^6	74.1	+126		**D, I$_d$**	(40)
DMF^c	4.3×10^5	54.8	+46.8	+8.5	**D, I$_d$**	(41)
DMF^d	6.2×10^6	77.8	+146		**D, I$_d$**	(42)
TMP^d	7.4×10^5	51.3	+39.5		**D, I$_d$**	(43)

^{a} Rate constant for the exchange of a particular coordinated solvent molecule (6).

^{b} Neat solvent.

^{c} In d$_3$-nitromethane diluent.

^{d} In d$_2$-dichloromethane diluent.

apply for DMF, NMA, DMA and DEA which is consistent with a parallel operation of **d**- and **a**-activated mechanisms. For the most bulky solvents, DMADMP, TMU, and DMPU, a **D** mechanism dominates: only the k_1 term applies and the $\Delta V^{\ddagger}$ measured on TMU and DMPU are very positive (Table II). In summary, a crossover from **a**- to **d**-activation is observed as the steric hindrance at the metal center increases with increasing size of the solvent molecules.

The increase in ionic radius from Be^{2+} to Mg^{2+}, which is accompanied by an increase in coordination number from 4 to 6, is responsible for a substantial increase in lability (Table III, (37–43)). The two activation volumes measured are positive as well as all the activation entropies. The rate laws determined for non-aqueous solvents in inert diluent are first order, showing a limiting **D** mechanism for all solvent exchange reactions on [MgS$_6$]$^{2+}$.

C. TRIVALENT MAIN GROUP IONS

Solvent exchange on Al^{3+} ($r_M = 54$ pm), Ga^{3+} ($r_M = 62$ pm), and In^{3+} ($r_M = 80$ pm) are most conveniently discussed together with Sc^{3+} ($r_M = 75$ pm) which has also a closed shell electronic configuration. Together, they permit the assessment of the influence of metal ion size on solvent exchange on octahedral trivalent metal ions. The general trend observed from the data in Table IV (16,44–57) is that the lability of the solvent exchange process increases with increasing size of the [MS$_6$]$^{3+}$. There is, furthermore, a change in the rate law and activation

TABLE IV

RATE CONSTANTS AND ACTIVATION PARAMETERS FOR SOLVENT EXCHANGE ON $[M(\text{solvent})_n]^{3+}$

$[M(\text{solvent})_n]^{3+}$	k_1 (298 K)^a (s^{-1})	k_2 (298 K)^a (dm^3 mol^{-1} s^{-1})	$\Delta H^{\ddagger}$ (kJ mol^{-1})	$\Delta S^{\ddagger}$ (J K^{-1} mol^{-1})	$\Delta V^{\ddagger}$ (cm^3 mol^{-1})	Mechanism	Reference
$[\text{Al}(\text{H}_2\text{O})_6]^{3+\,b}$	1.29		84.7	+41.6	+5.7	I_d	(44)
	2.0		73	+6.0		D	(45)
					+5.6 (ab initio)		(46)
$[\text{Al}(\text{OH})(\text{H}_2\text{O})_5]^{2+}$	3.1×10^4		36.4	−36.4		D	(47)
$[\text{Al}(\text{DMSO})_6]^{3+\,c}$	0.30		82.6	+22.3	+15.6	D	(48,49)
$[\text{Al}(\text{DMF})_6]^{3+\,c}$	0.05		88.3	+28.4	+13.7	D	(48,49)
$[\text{Al}(\text{TMP})_6]^{3+\,c}$	0.78		85.1	+38.2	+22.5	D	(16)
$[\text{Al}(\text{TMP})_6]^{3+\,c}$	0.38		98.3	+76.1		D	(50)
$[\text{Al}(\text{DMMP})_6]^{3+\,c}$	5.1		79.5	+33.0		D	(50)
$[\text{Al}(\text{HMPA})_4]^{3+\,c}$		4800	32.2	−42.7		A	(50)
$[\text{Ga}(\text{H}_2\text{O})_6]^{3+\,b}$	400		67.1	+30.1	+5.0	I_d	(51)
					+4.8 (ab initio)		(46)
$[\text{Ga}(\text{OH})(\text{H}_2\text{O})_5]^{2+}$	1×10^5		58.9		+6.2	I_d	(51)
$[\text{Ga}(\text{DMSO})_6]^{3+\,c}$	1.87		72.5	+3.5	+13.1	D	(48,49)
$[\text{Ga}(\text{DMF})_6]^{3+\,c}$	1.72		85.1	+45.1	+7.9	D	(48,49)
$[\text{Ga}(\text{TMP})_6]^{3+\,c}$	6.4		76.5	+27.0	+20.7	D	(16)
	5.0		87.9	+63.2		D	(52)

[Sc(TMP)$_6$]$^{3+}$ ^{b}	1200	37.4	−60.5	−20.1	A, I$_a$	(53)
[Sc(TMP)$_6$]$^{3+}$ ^{c}	38.4	21.2	−143.5	−18.7	A, I$_a$	(16)
[Sc(TMP)$_6$]$^{3+}$ ^{c}	45.3	26.0	−126		A, I$_a$	(54)
[Sc(DMMP)$_6$]$^{3+}$ ^{c}	13.2	29.7	−124		A, I$_a$	(55)
[Sc(TMU)$_6$]$^{3+}$ ^{c}		91.2	+47.8		D	(56)
[In(H$_2$O)$_6$]$^{3+}$ ^{b}	0.21	19.2	−96		A, I$_a$	(57)
	≥4×10^4		−5.2 (ab initio)		A	(46)
	>3×10^4					
[In(TMP)$_6$]$^{3+}$ ^{c}	7.6	32.8	−118	−21.4	A, I$_a$	(16)
[In(TMP)$_6$]$^{3+}$ ^{c}	7.2	35.6	−109		A, I$_a$	(53)

^{a} Rate constant for the exchange of a particular coordinated solvent molecule (6).

^{b} Neat solvent.

^{c} In d$_3$-nitromethane diluent.

for Al^{3+} and Ga^{3+} (1st order, **d**) versus Sc^{3+} and In^{3+} (2nd order, **a**). This shift is clearly shown by the values of $\Delta V^{\ddagger}$; these were obtained for TMP exchange on all four ions and change from $+22.5/+20.7$ (Al^{3+}/Ga^{3+}) to $-18.7/-21.5 \, cm^3 mol^{-1}$ (Sc^{3+}/In^{3+}) (*16*). Unfortunately, there are no experimental data on water exchange on Sc^{3+} and only lower limits for k_{ex} on $[In(H_2O)_6]^{3+}$ (*46*). Ab-initio theoretical calculations suggest that the mechanisms of solvent exchange at $[Al(H_2O)_6]^{3+}$, $[Ga(H_2O)_6]^{3+}$, and $[In(H_2O)_6]^{3+}$ are **D**, **D**, and **A**, respectively (*46*). The activation volume for water exchange on In^{3+}, not yet available experimentally, has been predicted to be $-5 \pm 1 \, cm^3 mol^{-1}$ (Fig. 3).

Kowall *et al.* (*46*) also gave, within the limits of their computational model, an alternative view of interpretation of $\Delta V^{\ddagger}$ for water substitution mechanisms. Swaddle rationalized the systematic trend in $\Delta V^{\ddagger}$ for H_2O exchange on $[M(H_2O)_6]^{z+}$ through a More-O'Ferral type diagram (Fig. 4a) (*58*). The bond-making and -breaking contributions to $\Delta V^{\ddagger}$ are plotted on the two axes, which are scaled to volumes for limiting **D** and **A** mechanisms, calculated from a semi-empirical equation as $+13.5$ and $-13.5 \, cm^3 mol^{-1}$, respectively (*58–60*). The only measurable point on the curves is the volume change between transition and ground state, $\Delta V^{\ddagger}$, and it is not obvious where to place it on the two-dimensional space. If, as Swaddle argued (*61*), all water exchange reactions had interchange mechanisms spanning a continuum from very dissociative to very associative interchange, the transition state would lie on the dashed diagonal in Fig. 4a. It should be stressed that in this model the **A** and **D** mechanisms are considered as the limiting cases for the interchange mechanistic continuum and are therefore following the edges of the square diagram with "transition states/intermediates" located at the corners of the square diagram.

However, when intermediates can be identified for solvent exchange reactions, a different view of the square diagram has to be defined. For concerted mechanisms **I**, the single transition state lies as before on the $\Delta V^{\ddagger}$ axes diagonal. However, for stepwise mechanisms, the *intermediate* lies on the diagonal (and not necessarily on a corner), while the two symmetric *transition states* lie symmetrically off to the diagonal. Computed structures and volume changes for Al^{3+}, Ga^{3+}, and In^{3+} water exchange can be used to introduce the position of transition states and intermediates in the diagram. Figure 4b shows that the transition states are about half way between the reactant and the intermediate. This is the reason why for a non-concerted **A** or **D** mechanism the volume of activation can be quite far away from the estimated limiting values of -13.5 or $+13.5 \, cm^3 mol^{-1}$, respectively. It is interesting to note that the unfavorable dissociative reaction of In^{3+}, which can nevertheless

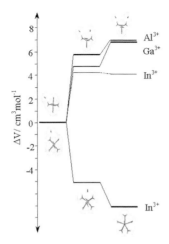

FIG. 3. Volume change (reactants-transition states-intermediates) for a dissociative **D** water exchange (top) and an associative **A** water exchange (bottom) estimated from ab-initio cluster calculations.

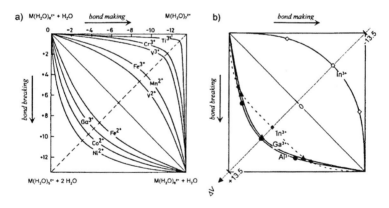

FIG. 4. Interpretation of volumes of activation (in $cm^3 mol^{-1}$) for water exchange on aqueous $M(H_2O)_6^{z+}$ in terms of contributions from bond-making and -breaking: (a) summary of volumes of activation for metal aqua ions; (b) calculated curves for Al(III), Ga(III), and In(III) with use of the Connolly volumes in Table IV.

be computed (Fig. 3), gives rise to a quite peculiar reaction pathway (Fig. 4b).

Water exchange on aluminium(III) complexes has received an increasing interest in recent years due to the geological importance of this ion. Casey and co-workers have studied H_2O exchange on different complexes containing 2, 4 or 5 water molecules (Table V (44,45,47,62–66)). A common

TABLE V

ON MAIN GROUP METAL IONS BY Al^{3+}: EFFECT OF NON-LEAVING LIGANDS ON THE
RATE CONSTANTS AND ACTIVATION PARAMETERS FOR SOLVENT EXCHANGE ON Al^{3+}

[M(solvent)$_6$]$^{3+}$	k_1 (298 K)^a (s^{-1})	$\Delta H^{\ddagger}$ (kJ mol^{-1})	$\Delta S^{\ddagger}$ (J K^{-1} mol^{-1})	Reference
[Al(H$_2$O)$_6$]$^{3+}$	1.29	84.7	+41.6	(44)
	2.0	73	+6.0	(45)
[Al(OH)(H$_2$O)$_5$]$^{2+}$	3.1×10^4	36.4	−36.4	(47)
[Al(F)(H$_2$O)$_5$]$^{2+}$	1.1×10^2	79	+60	(45)
[Al(F)(H$_2$O)$_5$]$^{2+}$	2.3×10^2	65	+19	(62)
[Al(F)$_2$(H$_2$O)$_4$]$^+$	1.71×10^4	66	+57	(62)
[Al(F)$_2$(H$_2$O)$_4$]$^+$	1.96×10^4	69	+70	(45)
[Al(C$_2$O$_4$)(H$_2$O)$_4$]$^+$	1.09×10^2	68.9	+25.3	(63)
[Al(mMal)(H$_2$O)$_4$]$^+$	6.6×10^2	66	+31	(64)
[Al(mMal)$_2$(H$_2$O)$_2$]$^-$	6.9×10^3	55	+12.8	(64)
[Al(ma)(H$_2$O)$_4$]$^{2+}$	3.4×10^2	63	+14	(65)
[Al(ma)$_2$(H$_2$O)$_2$]$^+$	1.95×10^3	49	−19	(65)
[Al(Sal)(H$_2$O)$_4$]$^+$	4.9×10^3	35	−57	(66)
[Al(sSal)(H$_2$O)$_4$]$^+$	3.0×10^3	37	−54	(66)

^{a} Rate constant for the exchange of a particular coordinated solvent molecule (6).

feature of these data is that water exchange is progressively enhanced by several orders of magnitude in substituting the first shell water molecules by fluoride, malonate, methylmalonate or salicylate.

III. Solvent Exchange on d-Transition Metal Ions

A. GENERAL CHARACTERISTICS

The labilities and the solvent exchange mechanisms of the transition metal ions are strongly affected by the electronic occupancy of their d-orbitals. The order of reactivity on 3d transition metal ions (V^{2+} < Ni^{2+} < Co^{2+} < Fe^{2+} < Mn^{2+} < Cu^{2+}) is independent of the reaction mechanism and semi-quantitatively correlates with the ligand field activation energies (LFAE). The variations of ionic radii, r_M, LFAE for **D** and **A** mechanisms, $\Delta G^{\ddagger}_{298.2}$ and $\Delta V^{\ddagger}$ for water exchange on [M(H$_2$O)$_6$]$^{2+/3+}$ are shown in Fig. 5 for the 3d series. The LFAE values are calculated as the differences in ligand field energies between octahedral ground states and square pyramidal and pentagonal bipyramidal transition states for **D** and **A** mechanisms, respectively, using a

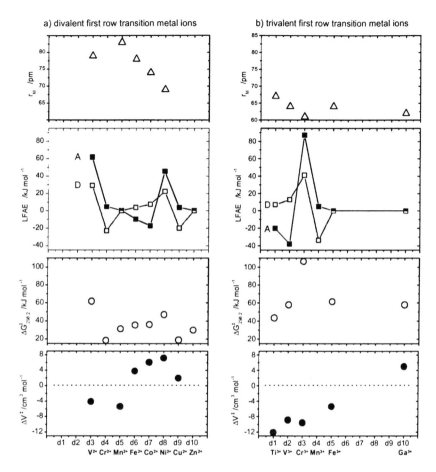

FIG. 5. Variation with successive filling of d-orbitals of ionic radii, r_M, ligand field activation energies, LFAE, calculated for **D** (■) and **A** (□) mechanisms, $\Delta G^{\ddagger}_{298.2}$ and $\Delta V^{\ddagger}$ for water exchange on $[M(H_2O)_6]^{2+/3+}$.

one-electron formula (17,67). The variation of LFAE reflects the variation of $\Delta G^{\ddagger}_{298.2}$: a small LFAE coincides with a high lability and a large LFAE with a smaller lability. The *relative* labilities predicted by LFAE for $[MS_6]^{2+/3+}$ (S=solvent) are similar for all solvents.

The increase of the exchange rate constants for solvent exchange on $[MS_6]^{2+/3+}$ in the sequence MeOH < MeCN < DMF < H$_2$O is independent of the metal ion and cannot be linked to specific properties like dielectric constant, donor number or size. It appears to reflect the "overall characteristics" of the solvent.

B. Divalent First-Row Transition Metals

Solvent exchange on the octahedral first-row transition metal complexes has been extensively studied as can be seen from Table VI (*68–
101*). For the small solvent molecules H_2O, MeOH and MeCN the solvent
exchange mechanism changes from I_a to I_d as the number of d-electrons
increases and the ionic radius decreases. This variation in mechanism
can be explained with the combined effect of: (i) the steric crowding
(decrease in r_M); (ii) increase in t_{2g} electron occupancy which increases
the electronic repulsion between an entering solvent molecule approach-
ing on a trigonal face in an **a**-activated mechanism; and (iii) increase in
e_g^* electron occupancy which facilitates solvent-metal bond breaking in
a d-activated mechanism. Thus, $[VS_6]^{2+}$ and $[MnS_6]^{2+}$ exchange solvent
molecules through **a**-activated mechanism, while their Fe^{2+}, Co^{2+} and
Ni^{2+} analogues show **d**-activated mechanism.

In a series of articles Rotzinger investigated H_2O exchange reactions
on 1st row transition metal ions with ab-initio quantum-chemical calculations (*102–104*). The activation energies for **D**, **A** and I_a, **I**, I_d mechanisms can be computed on the basis of Eqs. (7)–(9). The transition states,
designed by $[\cdots]^{\ddagger}$, are characterized by a single imaginary vibrational
frequency whereas intermediates with lower or higher coordination
number have no imaginary frequency.

$$M(OH_2)_6^{n+} \rightarrow \left[M(OH_2)_5 \cdots OH_2^{n+} \right]^{\ddagger} \rightarrow M(OH_2)_5 \cdot OH_2^{n+} \quad (\mathbf{D} \text{ mechanism})$$

(7)

$$M(OH_2)_6 \cdot OH_2^{n+} \rightarrow \left[M(OH_2)_6 \cdots OH_2^{n+} \right]^{\ddagger} \rightarrow M(OH_2)_7 \quad (\mathbf{A} \text{ mechanism})$$

(8)

$$M(OH_2)_6^{n+} \rightarrow \left[M(OH_2)_5 \cdots (OH_2)_2^{n+} \right]^{\ddagger} \quad (\mathbf{I_a}, \mathbf{I}, \mathbf{I_d} \text{ mechanism}) \quad (9)$$

The I_a and **A** mechanisms are mutually exclusive, since the 7-coordinate
transition state has an imaginary mode that either describes a concerted
entry/leaving of the ligands and the mechanism is I_a (Fig. 6a) or only
the motion of the entering (or leaving ligand) and the mechanism is A
(Fig. 6b) (*104*). The I_d and **D** mechanisms are not mutually exclusive and
up to now no I_d has been computed, most probably due to the omission
of the second coordination sphere (*104*).

Rotzinger's calculations confirmed the mechanistic crossover for
water exchange on the first-row transition metal ions. The calculations
predict I_a mechanisms for V^{2+} Mn^{2+} and **D** (or I_d) mechanisms for
Mn^{2+}, Fe^{2+}, Cu^{2+}, and Zn^{2+}. A **d**-activated mechanism for water exchange

TABLE VI

RATE CONSTANTS AND ACTIVATION PARAMETERS FOR SOLVENT EXCHANGE ON FIRST-ROW DIVALENT TRANSITION-METAL IONS

$[M(solvent)_6]^{2+}$	d-Electronic configuration	k_1 (298 K)[a] (s^{-1})	$\Delta H^{\ddagger}$ $(kJ\,mol^{-1})$	$\Delta S^{\ddagger}$ $(J\,K^{-1}\,mol^{-1})$	$\Delta V^{\ddagger}$ $(cm^3\,mol^{-1})$	Mechanism	Reference
$[V(H_2O)_6]^{2+}$	t_{2g}^3	8.7×10^1	61.8	-0.4	-4.1	I_a	(68)
$[Cr(MeOH)_6]^{2+\,b}$	$t_{2g}^3 e_g$	1.2×10^8	31.6	+16.6		I_d	(69)
$[Mn(H_2O)_6]^{2+}$	$t_{2g}^3 e_g^2$	2.1×10^7	32.9	+5.7	-5.4	I_a	(70)
$[Mn(MeOH)_6]^{2+}$		3.7×10^5	25.9	-50.2	-5.0	I_a	(71,72)
$[Mn(DMF)_6]^{2+}$		2.2×10^6	34.6	-7.4	+2.4	I_d	(73)
		2.7×10^6	35.8	-2	+1.6	I_d	(74)
		6.3×10^5	17.5	-75.1	+4.2	I_d	(75)
$[Mn(MeCN)_6]^{2+}$		1.4×10^7	29.6	-8.9	-7.0	I_a	(76)
		1.3×10^7	28.6	-13	-5.8	I_a	(77)
		1.2×10^7	30.3	-8		I_a	(78)
		3.1×10^7	35.9	+19		I_a	(79)
$[Mn(EtCN)_6]^{2+}$		1.3×10^7	29.6	-10	-2.1	I_a	(77)
$[Mn(PrCN)_6]^{2+}$		9.9×10^6	31.3	-6	-5.0	I_a	(77)
$[Mn(Pr^iCN)_6]^{2+}$		1.1×10^7	40.0	+24	-2.5	I_a	(77)
$[Mn(ButCN)_6]^{2+}$		9.3×10^6	35.6	+8	+0.6	I	(77)
$[Mn(PhCN)_6]^{2+}$		1.2×10^7	36.9	+14			(77)
$[Mn(DMTF)_6]^{2+}$		3.9×10^7			+11.5	I_d	(75)
$[Mn(HOAc)_6]^{2+\,c}$		1.6×10^7	29	-10	+0.4	I_d	(80)
$[Mn(OAc)_2(HOAc)_4]^{\,d}$		4.8×10^7	32	+9	+6.7	I_d	(80)
$[Mn(en)_3]^{2+}$		1.7×10^6	24.7	-43	-0.6	I	(81)
$[Mn(tn)_3]^{2+}$		2.5×10^6	21.9	-50	+0.1	I	(82)
$[Mn(pa)_6]^{2+}$		3.7×10^7	26.2	-13			(82)

(Continued)

TABLE VI
(Continued)

$[M(solvent)_6]^{2+}$	d-Electronic configuration	k_1 (298 K)[a] (s^{-1})	$\Delta H^{\ddagger}$ ($kJ\,mol^{-1}$)	$\Delta S^{\ddagger}$ ($J\,K^{-1}\,mol^{-1}$)	$\Delta V^{\ddagger}$ ($cm^3\,mol^{-1}$)	Mechanism	Reference
$[Fe(H_2O)_6]^{2+}$	$t_{2g}^4 e_g^2$	4.39×10^6	41.4	+21.2	+3.8	I_d	(70)
$[Fe(MeOH)_6]^{2+}$		5.0×10^4	50.2	+12.6	+0.4	I	(71,72)
$[Fe(MeCN)_6]^{2+}$		6.6×10^5	41.4	+5.3	+3.0	I_d	(76)
$[Fe(DMF)_6]^{2+}$		9.7×10^5	43.0	+13.8	+8.5	I_d	(73)
$[Fe(en)_3]^{2+}$		4.3×10^4	46.3	−1	−1.2	I_a	(81)
$[Fe(tn)_3]^{2+}$		3.9×10^5	47.9	+23	+5.8	I_d	(82)
$[Fe(pa)_6]^{2+}$		6.9×10^7	32.4	+14			(82)
$[Co(H_2O)_6]^{2+}$	$t_{2g}^5 e_g^2$	3.18×10^6	46.9	+37.2	+6.1	I_d	(70)
$[Co(MeOH)_6]^{2+}$		1.8×10^4	57.7	+30.1	+8.9	I_d	(83,84)
$[Co(AcOH)_6]^{2+}$		1.3×10^6	37	−6			(85)
$[Co(MeCN)_6]^{2+}$		3.4×10^5	49.5	+27.1	$+8.1^e$	I_d	(84,86,87)
$[Co(DMF)_6]^{2+}$		3.9×10^5	56.9	+52.7	+6.7	I_d	(84,88)
$[Co(DMSO)_6]^{2+}$		4.5×10^5	49	+28			(89)
$[Co(TMU)_6]^{2+}$		1.7×10^7	26	−20			(90)
$[Co(en)_3]^{2+}$		5.4×10^3	56.5	+16	+0.9	I_d	(81)
$[Co(tn)_3]^{2+}$		2.9×10^5	49.3	+25	+6.6	I_d	(91)
$[Co(pa)_6]^{2+}$		2.0×10^8	36.2	+35		I_d	(91)
$[Ni(H_2O)_6]^{2+}$	$t_{2g}^6 e_g^2$	3.15×10^4	56.9	+32.0	+7.2	I_d	(70)
$[Ni(MeOH)_6]^{2+}$		1.0×10^3	66.1	+33.5	+11.4	I_d	(83,84)

$[Ni(MeCN)_6]^{2+}$	2.8×10^3	64.3	+37	+8.5	I_d	(84,87)
$[Ni(EtCN)_6]^{2+}$	6.2×10^3	41.4	−30	+12.0	I_d	(92)
$[Ni(PrCN)_6]^{2+}$	1.3×10^4	42.0	−25	+13.7	I_d	(92)
$[Ni(Pr^iCN)_6]^{2+}$	1.0×10^4	43.3	−23	+13.1	I_d	(92)
$[Ni(BuCN)_6]^{2+}$	1.6×10^4	43.3	−19	+12.4	I_d	(92)
	1.0×10^4	47.1	−10	+14.4	I_d	(92)
$[Ni(PhCN)_6]^{2+}$	9.4×10^3	51.6	+4	+13.1	I_d	(92)
$[Ni(DMF)_6]^{2+}$	3.8×10^3	62.8	+33.5	+9.1	I_d	(83,88)
	6.9×10^3	54.4	+25.0		I_d	(93)
	7.7×10^3	39.3	−37.7		I_d	(94)
	3.7×10^3	59.3	+22.3		I_d	(95)
$[Ni(DMTF)_6]^{2+}$	8.7×10^6	57.4	+80.5	+20.6/+21.8	D	(75)
$[Ni(en)_3]^{2+}$	2.0×10^1	69	+10	+11.4	I_d	(81)
$[Ni(tn)_3]^{2+}$	3.1×10^3	61.3	+28	+7.2		(82)
$[Ni(pa)_6]^{2+}$	1.3×10^7	37.1	+16			(82)
$[Cu(H_2O)_5]^{2+}$	5.7×10^9	11.5	−21.8			(96)
$[Cu(H_2O)_6]^{2+}$ [b]	4.4×10^9	17.2	−44.0	+2.0	I_d	(97,99)
$[Cu(MeOH)_6]^{2+}$ [b]	3.1×10^7	24.3	+8.1	+8.3	I_d	(98)
$[Cu(DMF)_6]^{2+}$	9.1×10^8	9.2	−77	+8.4	I_d	(99)
$[Cu(en)_3]^{2+}$	1.4×10^7				I_d	(100)
$[Zn(H_2O)_6]^{2+}$	$> 5 \times 10^7$					(46)
$[Zn(H_2O)_6]^{2+}$	$< 10^{10}$ (QENS)					(101)

$t_{2g}^6 e_g^3$

[a] Rate constant for the exchange of a particular coordinated solvent molecule (6).
[b] Tetragonal distortion.
[c] 0.02 M HClO₄ in d₂-dichloromethane diluent.
[d] 0.0 M HClO₄ in d₂-dichloromethane diluent.
[e] Average of +9.9, +6.7, and +7.7 cm³ mol⁻¹.

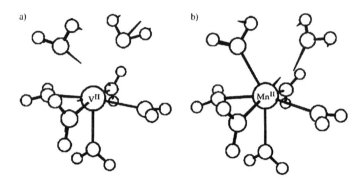

FIG. 6. Structure and imaginary modes of transition states: (a) $[V(OH)_2)_5 \cdots (OH_2)_2^{2+}]^{\ddagger}$ imaginary mode describes concerted entry and leaving of H_2O; (b) $[Mn(OH)_2)_6 \cdots (OH_2)^{2+}]^{\ddagger}$ imaginary mode describes only the motion of the entering H_2O.

on Zn^{2+} was also found by DFT calculations performed by Hartmann et al. (105).

The effect of bulkiness of solvent molecules on exchange mechanisms was studied on Mn^{2+} (77) and Ni^{2+} (92) using a series of nitriles with only small differences in donor numbers. The six nitriles, in the order of increasing molar volume MeCN ($V^0 = 52.9\,cm^3\,mol^{-1}$), EtCN ($V^0 = 70.9\,cm^3\,mol^{-1}$), PrCn ($V^0 = 87.9\,cm^3\,mol^{-1}$), iPrCN ($V^0 = 90.3\,cm^3\,mol^{-1}$), PhCN ($V^0 = 103.1\,cm^3\,mol^{-1}$), and BuCN ($V^0 = 104.6\,cm^3\,mol^{-1}$), form octahedral complexes with both metal ions. Exchange rate constants for solvent exchange are relatively constant within both series (Table VI). The activation volumes measured on Ni^{2+} are all positive and do not vary along the series. It was concluded that the mechanism of the nitrile exchange at the nickel(II) ion is of similar dissociative character for nitriles of different bulkiness (92). The observed increase in $\Delta S^{\ddagger}$ was attributed to changes in solvent–solvent interactions due to the decrease in dipole moment. On Mn^{2+} the exchange of the smallest solvent molecule (MeCN) proceeds via an I_a pathway as concluded from negative $\Delta V^{\ddagger}$ and $\Delta S^{\ddagger}$ values (77). The $\Delta H^{\ddagger}$ and $\Delta S^{\ddagger}$ both increase gradually in the order MeCN < EtCN < PrCN < BuCN < PhCN < iPrCN and the reaction mechanism is concluded to be less associative with increasing bulkiness, but it never becomes as positive as with the sterically hindered solvent DMF.

Solvation of the divalent copper ion requires some special remarks. In solid state hexa-coordinate Cu^{2+} shows Jahn-Teller distortion: the metal–ligand bonds of four ligands in the equatorial plane are shorter than those of the two axial ligands (*31,106,107*). In solution this effect is expected to be dynamic. The distortion axis jumps very rapidly and on average all ligands become in time, axial and equatorial. The axial ligands are only loosely bound to the metal ion which partially explains the very high lability of all copper solvates (Table VI). The positive $\Delta V^{\ddagger}$ values obtained for MeOH (*98*) and DMF (*99*) indicate that these solvent molecules exchange via a **d**-activation mode: molecules at the axial position can easily dissociate. ^{17}O NMR studies on the aqua ion of Cu^{2+} allowed the determination of the water exchange rate and a characteristic time for the inversion of the tetragonal distortion (~ 5 ps at 298 K) (*97*). These results were based on the commonly accepted assumption that Cu^{2+}, in analogy to the solid state, is also six-coordinate in aqueous solution (*108*). The activation volume measured by variable pressure ^{17}O NMR is relatively small ($+2.0 \pm 1.5\,cm^{3}\,mol^{-1}$) (*99*) and, in agreement with results on MeOH and DMF, a **d**-activation mode was proposed for water exchange. However, a complementary experimental and computational study, performed by double difference neutron diffraction and by first principles molecular dynamics, respectively, was published in 2001 and showed evidence that the coordination number of copper(II) is 5 in solution (*96*). The ab-initio molecular dynamics (Car-Parrinello) results show two structures: a square pyramid and a trigonal bipyramid in rapid exchange (Fig. 7). The ^{17}O NMR kinetic data could successfully be reinterpreted in terms of a five-coordinated copper ion with

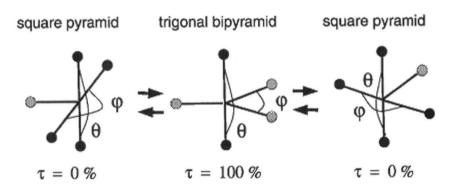

FIG. 7. Structures of five-coordinate Cu^{2+} from first principles molecular dynamics. A Berry twist mechanism for the interconversion of the two structures is shown (from left to right): the reorientation of the main axis of a square pyramidal configuration by pseudo-rotations via a trigonal bipyramidal configuration. The grey atoms in the plane of the trigonal bipyramid are all candidates for becoming apical atoms in a square pyramid.

interconversion between the two limiting structures. More recently two groups tried to use XANES and EXAFS to verify the coordination number in solution. Unfortunately this technique is not conclusive: one group obtained 5 (*109*) and the other 6 coordinated water to Cu^{2+} (*110*).

C. TRIVALENT FIRST-ROW TRANSITION METALS

Solvent exchange on the trivalent transition metal ions shows a similar variation in lability and mechanism with d-electronic configuration as their divalent analogues (Table VII (*111–121*)). The most inert solvent complexes of 1st row transition metal ions are formed by Cr^{3+}. This is a consequence of the large LFAE leading to large $\Delta H^{\ddagger}$ values of more than $100\,kJ\,mol^{-1}$ (Table VII). The interaction of Cr^{3+} with surrounding water molecules is so strong that even second shell molecules are relatively tightly bound: a lifetime of over $100\,ps$ has been measured experimentally by ^{17}O NMR and was confirmed by molecular dynamics simulations (*116*). A ^{1}H NMRD study showed that the exchange time of second shell water protons is longer than both the diffusional correlation time and the rotational correlation time of the complex, both being around $70\,ps$ at room temperature (*122*). Molecular dynamics simulations on trivalent lanthanides in aqueous solution gave second shell lifetimes about 10 times shorter (*123*) which illustrates the exceptionally strong second shell binding in the case of Cr^{3+}. The molecular dynamics simulations on $[Cr(H_2O)_6]^{3+}$ in solution allowed the proposal of a model for the exchange between the second coordination shell and the bulk. In a first step a water molecule enters the second coordination sphere and increases the coordination number temporarily. In a second step a first sphere water molecule, close to the one entered, rotates around its oxygen–chromium bond, one of the hydrogen bonds formed to second sphere waters breaks up, and a new hydrogen bond to the water molecule that has entered, is formed. During the rotation the second hydrogen bond is maintained. The third step is the leaving of the water molecule which lost its hydrogen bond from the second sphere. The activation mode which can be attributed to this reaction from the MD simulation (*116*) is associative.

The important influence of the LFAE is also nicely demonstrated on the labilities of V^{2+} and V^{3+}. The smaller LFAE of t_{2g}^2 $[V(H_2O)_6]^{3+}$ makes it about six times more labile than t_{2g}^3 $[V(H_2O)_6]^{2+}$ despite the smaller charge of the 2+ complex (Fig. 5). Both exchange processes are **a**-activated with negative volumes of activation of -8.9 (V^{3+}) (*113*) and $-4.1\,cm^3\,mol^{-1}$ (V^{2+}) (*68*).

TABLE VII

RATE CONSTANTS AND ACTIVATION PARAMETERS FOR SOLVENT EXCHANGE ON FIRST-ROW TRIVALENT TRANSITION-METAL IONS

$[M(solvent)_6]^{3+}$	d-Electronic configuration	k_1 (298 K)[a] (s^{-1})	$\Delta H^\ddagger$ ($kJ\,mol^{-1}$)	$\Delta S^\ddagger$ ($J\,K^{-1}\,mol^{-1}$)	$\Delta V^\ddagger$ ($cm^3\,mol^{-1}$)	Mechanism	Reference
$[Sc(TMP)_6]^{3+}$	t_{2g}^0	1.21×10^3	37.4	−60.5	−20.1	I_a	(53)
$[Ti(H_2O)_6]^{3+}$	t_{2g}^1	1.8×10^5	43.4	+1.2	−12.1	A, I_a	(111)
$[Ti(DMF)_6]^{3+}$		6.6×10^4	23.6	−73.6	−5.7	I_a	(112)
$[V(H_2O)_6]^{3+}$	t_{2g}^2	5.0×10^2	49.4	−27.8	−8.9	I_a	(113)
$[V(DMSO)_6]^{3+}$		1.31×10^1	38.5	−94.5	−10.1	I_a	(114)
$[Cr(H_2O)_6]^{3+}$	t_{2g}^3	2.4×10^{-6}	108.6	+11.6	−9.6	I_a	(115)
$[Cr(H_2O)_6]^{3+}$	(2nd sphere)	7.8×10^9	21.3	+16.2			(116)
	(2nd sphere)	$6.9\times10^{9\ b}$					(116)
$[Cr(H_2O)_5OH]^{2+}$		1.8×10^{-4}	111.0	+55.6	+2.7	I	(115)
$[Cr(DMSO)_6]^{3+}$		3.1×10^{-8}	96.7	−64.5	−11.3	I_a	(117)
$[Cr(DMF)_6]^{3+}$		3.3×10^{-7}	97.1	−43.5	−6.3	I_a	(118)
$[Fe(H_2O)_6]^{3+}$	$t_{2g}^3 e_g^2$	1.6×10^2	64.0	+12.1	−5.4	I_a	(119,120)
$[Fe(H_2O)_5OH]^{2+}$		1.2×10^5	42.4	+5.3	+7.0	I_d	(119,120)
$[Fe(DMSO)_6]^{3+}$		9.3	62.5	−16.7	−3.1	I_a	(121)
$[Fe(DMF)_6]^{3+}$		6.3×10^1	42.3	−69.0	−0.9	I	(121)

[a] Rate constant for the exchange of a particular coordinated solvent molecule (6).
[b] By molecular dynamics method.

Inspection of the $\Delta V^{\ddagger}$ values for $[MS_6]^{3+}$ shows that they become less negative upon proceeding along the row from Ti^{3+} to Fe^{3+}. If Ga^{3+}, which is not a transition metal but which is in the same row in the periodic table is added, the activation volumes become even positive (Table IV). As in the case of the divalent metal ions, there is a crossover in solvent exchange mechanism from associative to dissociative activation. The $\Delta V^{\ddagger}$ value for water exchange on Ti^{3+} is markedly more negative than the following ones and it is close to the limiting value estimated by Swaddle $(-13.5\,cm^3\,mol^{-1})$ (58,60) for a limiting **A** mechanism (111). The smaller activation volumes measured for V^{3+}, Cr^{3+} and especially Fe^{3+} support the attribution of an $\mathbf{I_a}$ mechanism for these cations. A limiting **A** mechanism for $[Ti(H_2O)_6]^{3+}$ was also concluded from computational studies with the Hartree-Fock method (103) and with DFT (124).

D. SECOND AND THIRD ROW OCTAHEDRAL COMPLEXES

Experimental data for solvent exchange on octahedral *second*- and *third*-row transition metal ions are limited to the $Ru^{2+/3+}$, Rh^{3+} and Ir^{3+} and to water and acetonitrile solvents (Table VIII (3,125–129)).

A comparison of the exchange rates of H_2O and MeCN measured on the low spin t_{2g}^6 Ru^{2+} metal center (Table VIII) with the analogous high-spin $t_{2g}^4 e_g^2$ Fe^{2+} (Table VI) shows that the low-spin complexes are many orders of magnitude less labile, largely as a consequence of their much larger $\Delta H^{\ddagger}$ values. To a certain extent these larger activation enthalpies arise from a smaller ionic radius ($r_M = 73$ pm for low-spin Ru^{2+}, $r_M = 78$ pm for high-spin Fe^{2+}) but mainly from the larger ligand fields generating larger LFAE values.

On the same ion, the lability of $[Ru(MeCN)_6]^{2+}$ is eight orders of magnitude less than that of $[Ru(H_2O)_6]^{2+}$ (125). This is a consequence of a more than $50\,kJ\,mol^{-1}$ increase in $\Delta H^{\ddagger}$, arising probably from very strong back-bonding from the electron-rich t_{2g} Ru^{2+} into the MeCN π^* orbital. For the iron(II) center the change in lability is less than one order of magnitude confirming that back-bonding is less important on $[Fe(MeCN)_6]^{2+}$ (76).

For the low-spin t_{2g}^6 aqua ions $[Ru(H_2O)_6]^{2+}$, $[Rh(H_2O)_6]^{3+}$, and $[Ir(H_2O)_6]^{3+}$ a **d**-activation mode would a priori be predicted. The approach of a seventh water molecule towards a face or edge of the coordination octahedron is electrostatically disfavored by the filled t_{2g} orbitals which are spread out between the ligands. Rate constants for anation reactions of Cl^-, Br^-, and I^- on $[Ru(H_2O)_6]^{2+}$ are very similar, indicating identical steps to reach the transition state, namely the dissociation of a water molecule (130). An extension of this study to a large variety of ligands demonstrated clearly that the rate determining

TABLE VIII

RATE CONSTANTS AND ACTIVATION PARAMETERS FOR SOLVENT EXCHANGE ON SECOND AND THIRD ROW TRIVALENT
TRANSITION-METAL IONS

$[M(solvent)_6]^{2+}$	d-Electronic configuration	k_1 (298 K)^a (s^{-1})	$\Delta H^{\ddagger}$ ($kJ\,mol^{-1}$)	$\Delta S^{\ddagger}$ ($J\,K^{-1}\,mol^{-1}$)	$\Delta V^{\ddagger}$ ($cm^3\,mol^{-1}$)	Mechanism	Reference
$[Ru(H_2O)_6]^{2+}$	t_{2g}^6	1.8×10^{-2}	87.8	+16.1	−0.4	I	(125)
$[Ru(MeCN)_6]^{2+}$		8.9×10^{-11}	140.3	+33.3	+0.4	I	(3,125)
$[Rh_2(MeCN)_{10}]^{4+}$		3.1×10^{-5} (eq)	66	−111	−4.9	I_a, A	(126)
$[Re_2(MeCN)_{10}]^{4+}$		3.1×10^{-7} (eq)	109	+6		I	(127)
$[Ru(H_2O)_6]^{3+}$	t_{2g}^5	3.5×10^{-6}	89.8	−48.3	−8.3	I_a	(125)
$[Ru(H_2O)_5OH]^{2+}$		5.9×10^{-4}	95.8	+14.9	+0.9	I	(125)
$[Rh(H_2O)_6]^{3+}$	t_{2g}^6	2.2×10^{-9}	131	+29	−4.2	I_a	(128)
$[Rh(H_2O)_5OH]^{2+}$		4.2×10^{-5}	103		+1.5	I	(128)
$[Ir(H_2O)_6]^{3+}$	t_{2g}^6	1.1×10^{-10}	130.5	+2.1	−5.7	I_a	(129)
$[Ir(H_2O)_5OH]^{2+}$		5.6×10^{-7}			−0.2	I	(129)

^{a} Rate constant for the exchange of a particular coordinated solvent molecule (6).

step of the monocomplex-formation reaction is independent of the nature of the entering ligand and an $\mathbf{I_d}$ mechanism was proposed (131). A near-zero activation volume was measured for the water exchange on $[Ru(H_2O)_6]^{2+}$ (Table VIII) and an $\mathbf{I}$ mechanism without $\mathbf{a}$ or $\mathbf{d}$ character was attributed (125). However, for water exchange on the isoelectronic $[Rh(H_2O)_6]^{3+}$ and its third row analogue $[Ir(H_2O)_6]^{3+}$, $\mathbf{a}$-activation mechanisms have been assigned from the negative $\Delta V^{\ddagger}$ values (1,128).

Quantum chemical calculations on $[Ru(H_2O)_6]^{2+}$ embedded into a polarizable continuum failed to compute a transition state for an interchange mechanism (132). The calculated energy difference between the transition state and the ground state, $\Delta E^{\ddagger}$, for a $\mathbf{D}$-mechanism $(71.9\,\mathrm{kJ\,mol}^{-1})$ is close to both experimental $\Delta G_{298}^{\ddagger}$ $(83.0\,\mathrm{kJ\,mol}^{-1})$ and $\Delta H^{\ddagger}$ values $(87.8\,\mathrm{kJ\,mol}^{-1})$. This suggests that water exchange on octa-aqua ion of Ru^{2+} proceeds via $\mathbf{d}$-activation, either by $\mathbf{I_d}$ or $\mathbf{D}$ mechanism. Calculations on $[Rh(H_2O)_6]^{3+}$ yielded activation energies of $136.6\,\mathrm{kJ\,mol}^{-1}$ and $114.8\,\mathrm{kJ\,mol}^{-1}$ for $\mathbf{D}$ and $\mathbf{I_a}$ mechanisms, respectively (Fig. 8). This energy difference of $21.8\,\mathrm{kJ\,mol}^{-1}$ together with the negative $\Delta V^{\ddagger}$ measured on this ion are both in favor of an $\mathbf{I_a}$ mechanism. So, why do Rh^{3+} and Ir^{3+} on one hand and Ru^{2+} on the other, undergo water

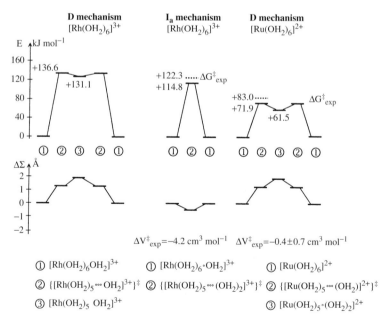

FIG. 8. Energies calculated with a polarizable continuum model, differences of the sums of all metal–oxygen bond lengths, $\Delta\Sigma(M\text{–}O)$, and energy profiles for water exchange on rhodium(III) and ruthenium(II) hexaaqua ions.

exchange via disparate mechanisms although they are isoelectronic? The different ionic charge affects particularly the M–O bond strength. The calculated activation energies for exchange via the **D** mechanism reflects this differences: $71.9\,\text{kJ}\,\text{mol}^{-1}$ for Ru^{2+} and $136.6\,\text{kJ}\,\text{mol}^{-1}$ for Rh^{3+} (*132*) (Fig. 8). The strong Rh–O bonds constrains water exchange on $[Rh(H_2O)_6]^{3+}$ to proceed via an I_a mechanism, whereas the same reaction on $[Ru(H_2O)_6]^{2+}$, which has considerably weaker Ru–O bonds, follows the I_d or **D** mechanism.

The low-spin t_{2g}^5 $[Ru(H_2O)_6]^{3+}$ is four orders of magnitude more labile than the t_{2g}^6 $[Ru(H_2O)_6]^{2+}$ and exchanges water by an I_a mechanism (*125*). The slow water exchange on both complexes allowed the direct measurement of the electron exchange of the $[Ru(H_2O)_6]^{3+/2+}$ couple in acidic solution (Eq. (10)) (*133*).

$$M^*(OH_2)_6^{3+} + M(OH_2)_6^{2+} \rightleftharpoons M^*(OH_2)_6^{2+} + M(OH_2)_6^{3+} \qquad (10)$$

At 298.15 K the rate constant k is $20\,M^{-1}\,s^{-1}$ and the enthalpy and entropy of activation are $\Delta H^{\ddagger} = 46.0\,\text{kJ}\,\text{mol}^{-1}$ and $\Delta S^{\ddagger} = -65.7\,\text{J}\,\text{K}^{-1}\,\text{mol}^{-1}$, respectively, for a solution 2.5 M in ruthenium concentration and 5.0 M in ionic strength, the counter-ion being trifluoro-methanesulfonate. The self-exchange of the $[Ru(H_2O)_6]^{3+/2+}$ couple can be considered as a prototype of an outer-sphere electron transfer reaction (*133*).

E. Effect of Non-Leaving Ligands

Exchange of various solvents has been studied on $[VO(S_{eq})_4(S_{ax})]^{2+}$ (Table IX (*134–143*)). Exchange rates of equatorial solvent molecules increase in the order $H_2O \sim MeOH \sim DMF < DMSO < DMA < MeCN$. For H_2O and DMF rate constants could also be measured for axial ligands. In both cases axial solvent molecules exchange substantially faster than equatorial ones. A similar observation was made on $[TiO(DMSO)]^{2+}$, the only titanyl complex where solvent exchange could be studied. The exchange of the oxo ligand in aqueous solution is much slower on VO^{2+} ($k_O(VO) = 2.4 \times 10^{-5}\,s^{-1}$) (*144*) than on TiO^{2+} ($k_O(TiO) = 1.6 \times 10^4\,s^{-1}$) (*145*) a difference that cannot be assigned to electronic effects alone. Whereas a [H^+] dependence of oxygen exchange on titanium is observed, the exchange of the vanadyl oxygen is base catalysed (*145*). The oxo ligand of TiO^{2+} is easily protonated and thereby labilized.

Replacing several solvent molecules on di- and trivalent transition metal ions by non-leaving ligands can have dramatic effects on the solvent exchange rates of the remaining solvent molecule(s) (Tables X, XI, XII and XIII (*70,71,83,86–88,92,95,96,146–166*) (*115,119–121,167–181*) (*125, 128,129,167,168,181–193*)). For example replacing three MeCN solvent

F.A. DUNAND et al.

TABLE IX

RATE CONSTANTS AND ACTIVATION PARAMETERS FOR AXIAL AND EQUATORIAL SOLVENT EXCHANGE ON FIRST-ROW TRANSITION OXOMETAL IONS MO[(eq-solvent)$_4$(ax-solvent)]$^{2+}$

[MO(solvent)$_5$]$^{2+}$	k_1 (298 K)[a] (s^{-1})	$\Delta H^{\ddagger}$ (kJ mol^{-1})	$\Delta S^{\ddagger}$ (J K^{-1} mol^{-1})	$\Delta V^{\ddagger}$ (cm^3 mol^{-1})	Reference
[TiO(DMSO)$_5$]$^{2+}$ [b]	161 (eq)	57.5	−9.9	+4.8	(134)
	6100 (ax)	59.8	+28.3	+1.6	(134)
[VO(H$_2$O)$_5$]$^{2+}$	500 (eq)	87.8	+16.1	−0.4	(135,136)
	≈10^9 (ax)			+1.9	(135,137)
[VO(MeOH)$_5$]$^{2+}$	565 (eq)	39.6	−59.4		(138)
[VO(MeCN)$_5$]$^{2+}$	2850 (eq)	29.5	−83.7	−1.0	(137,139)
[VO(DMSO)$_5$]$^{2+}$	1760 (eq)	60.1	+18.5	−5.3	(140,141)
[VO(DMF)$_5$]$^{2+}$	575 (eq)	30.3	−87.7		(142)
	200 (eq)	54.8	−17.2	−6.8	(137,143)
	46,000 (ax)	64	+58.0		(143)
[VO(DMA)$_5$]$^{2+}$	4700 (eq)	42.3	−33.1	−9.7	(137,143)

[a] Rate constant for the exchange of a particular coordinated solvent molecule (6).
[b] Studied in d$_3$-nitromethane diluent.

TABLE X

EFFECT OF NON-LEAVING LIGANDS ON THE SOLVENT EXCHANGE RATE CONSTANTS ON FIRST-ROW DIVALENT TRANSITION-METAL IONS

Complex	k_1 (298 K)^a (s^{-1})	Reference
$[Mn(H_2O)_6]^{2+}$	2.1×10^7	(70)
$[Mn(EDTA\text{-}BOM)(H_2O)]^{2+}$	9.3×10^7	(146)
$[Mn(EDTA\text{-}BOM_2)(H_2O)]^{2+}$	1.3×10^8	(146)
$[Co(MeCN)_6]^{2+}$	3.4×10^5	(84,86,87)
$[Co(dmgH)_2(MeCN)_2]^{2+}$	2.3×10^5	(147)
$[Co(dmg)_2(BF_2)_2(MeCN)_2]^{2+}$	1.4×10^5	(147)
$[Co(MeOH)_6]^{2+}$	1.8×10^4	(83,84)
$[Co(dmgH)_2(MeOH)_2]^{2+}$	2.8×10^4	(147)
$[Co(dpgH)_2(MeOH)_2]^{2+}$	7.2×10^5	(147)
$[Ni(H_2O)_6]^{2+}$	3.15×10^4	(70)
$[Ni(NH_3)(H_2O)_5]^{2+}$	25×10^4	(148)
$[Ni(NH_3)_2(H_2O)_4]^{2+}$	61×10^4	(148)
$[Ni(NH_3)_3(H_2O)_3]^{2+}$	250×10^4	(148)
$[Ni(en)(H_2O)_4]^{2+}$	44×10^4	(149)
$[Ni(en)_2(H_2O)_2]^{2+}$	540×10^4	(149)
$[Ni(dien)(H_2O)_3]^{2+}$	120×10^4	(150)
$[Ni(trien)(H_2O)_2]^{2+}$	290×10^4	(150)
$[Ni(tren)(H_2O)_2]^{2+\ b}$	$82/900 \times 10^{4\ b}$	(151)
$[Ni(cyclen)(H_2O)_2]^{2+}$	2100×10^4	(152)
$[Ni(Me_4cyclam)(H_2O)_2]^{2+\ c}$	1600×10^4	(153)
$[Ni(Me_4cyclam)(H_2O)]^{2+\ d}$	160×10^4	(153)
$[Ni(bpy)(H_2O)_4]^{2+}$	4.9×10^4	(154)
$[Ni(bpy)_2(H_2O)_2]^{2+}$	6.6×10^4	(154)
$[Ni(tpy)(H_2O)_3]^{2+}$	5.2×10^4	(155)
$[Ni(H_2O)_5Cl]^+$	14×10^4	(156)
$[Ni(H_2O)_2(NCS)_4]^{2-}$	110×10^4	(157)
$[Ni(H_2O)(EDTA)]^{2-}$	70×10^4	(158)
$[Ni(H_2O)(HEDTA)]^-$	20×10^4	(158)
$[Ni(DMF)_6]^{2+}$	3.8×10^3	(83,88,95)
$[Ni(DMF)_5Cl]^{2+}$	5.3×10^5	(95)
$[Ni(pyMa)(DMF)]^{2+}$	3.3×10^3	(159)
$[Ni(MeCN)_6]^{2+}$	2.8×10^3	(92)
$[Ni(pyMa)(MeCN)]^{2+}$	3.8×10^2	(159)
$[Cu(H_2O)_5]^{2+}$	5.7×10^9	(96)
$[Cu(tmpa)(H_2O)]^{2+}$	8.6×10^6	(160)
$[Cu(tren)(H_2O)]^{2+}$	2.5×10^5	(161)
$[Cu(tpy)(H_2O)_2]^{2+}$	6.6×10^8	(162)

(Continued)

TABLE X

(*Continued*)

Complex	k_1 (298 K)^a (s^{-1})	Reference
$[Cu(fz)_2(H_2O)]^{2+}$	3.5×10^5	(*160*)
$[Cu(tren)(MeCN)]^{2+}$	1.7×10^6	(*163*)
$[Cu(DMF)_6]^{2+}$	9.1×10^8	(*164*)
$[Cu(Me_6tren)(DMF)]^{2+}$	9.1×10^8	(*165,166*)

^{a} Rate constant for the exchange of a particular coordinated solvent molecule (*6*).
^{b} The two inequivalent H_2O exchange at different rates.
^{c} *RRSS* form of the macrocycle.
^{d} Five coordinate *RSRS* form of macrocycle.

molecules on $[Ru(MeCN)_6]^{2+}$ by a (η^6-C_6H_6) or by a (η^5-C_5H_5) accelerates the exchange of the remaining three MeCN by 6 or 11 orders of magnitude, respectively. Replacing three water molecules however on the corresponding hexa-aqua complex by (η^6-C_6H_6) increases the exchange rate of the remaining H_2O molecules only by a factor of 640. Substituting one or two water molecules by small monodentate ligands allowed a study of the influence of axial and equatorial water molecules on the lability. Water exchange proceeds on the substituted complexes by a **d**-activated mechanism. This has been shown on $[Ru(H_2O)_5(H_2C=CH_2)]^{2+}$ where positive $\Delta V^{\ddagger}$ have been measured: +6.5 and +6.1 cm^3 mol^{-1} for the exchange of axial and equatorial waters, respectively (*185*). An increasing *cis*-effect order was established for the lability of the equatorial water molecules in the corresponding mono-substituted Ru(II) complexes: $F_2C=CH_2 \approx CO < Me_2SO < N_2 < H_2C=CH_2 < MeCN < H_2O$. The following increasing *trans*-effect order, established from the lability of the axial water molecule, has been found: N_2, MeCN $< H_2O < CO < Me_2SO < H_2C=CH_2 < F_2C=CH_2$.

F. SQUARE-PLANAR COMPLEXES

Square-planar stereochemistry is mostly confined to the d^8 transition metal ions. The most investigated solvent exchange reactions are those on Pd^{2+} and Pt^{2+} metal centers and the mechanistic picture is well established (Table XIV (*194–203*)). The vast majority of solvent exchange reactions on square-planar complexes undergo an **a**-activated mechanism. This is most probably a consequence of the coordinatively unsaturated four-coordinate 16 outer-shell electron complex achieving noble gas

TABLE XI

EFFECT OF NON-LEAVING LIGANDS ON THE SOLVENT EXCHANGE RATE CONSTANTS ON FIRST-ROW TRIVALENT TRANSITION-METAL IONS

Complex	k_1 (298 K)[a] (s^{-1})	Reference	Complex	k_1 (298 K)[a] (s^{-1})	Reference
$[Cr(H_2O)_6]^{3+}$	2.4×10^{-6}	(115)	$[Co(NH_3)_5(H_2O)]^{3+}$	5.7×10^{-6}	(177)
$[Cr(NH_3)_5(H_2O)]^{3+}$	5.2×10^{-5}	(167)	$[Co(CH_3NH_2)_5(H_2O)]^{3+}$	7.0×10^{-4}	(168)
$[Cr(CH_3NH_2)_5(H_2O)]^{3+}$	4.1×10^{-6}	(168)	cis-$[Co(en)_2(NH_3)(H_2O)]^{3+}$	1.1×10^{-6}	(178)
$[(H_2O)_4Cr(\mu\text{-}OH)_2Cr(H_2O)_4]^{4+}$	3.58×10^{-4} (trans)	(169)	p-$[Co(tren)(NH_3)(H_2O)]^{3+}$	1.1×10^{-5}	(176)
	0.66×10^{-4} (cis)	(169)	t-$[Co(tren)(NH_3)(H_2O)]^{3+}$	1.2×10^{-5}	(176)
	1.14×10^{-5} (bridging)	(169)	$[Co(cyclen)(NH_3)(H_2O)]^{3+}$	$<1.4\times10^{-4}$	(179)
$[(H_2O)_3(OH)Cr(\mu\text{-}OH)_2Cr(OH)(H_2O)_3]^{4+}$	1.24×10^{-2} (trans)	(169)			
	0.475×10^{-2} (cis)	(169)	$[Co(cyclen)(H_2O)_2]^{3+}$	2×10^{-4}	(176)
	1.47×10^{-3} (bridging)	(169)	$[Co(N\text{-}Mecyclen)(H_2O)_2]^{3+}$	2×10^{-4}	(176)
$[Fe(H_2O)_6]^{3+}$	1.6×10^2	(119,120)	$[Co(NH_3)_5(OH)]^{3+}$	$<1.4\times10^{-7}$	(177)
$[Fe(CDTA)(H_2O)]^-$	1.3×10^7	(170)	p-$[Co(tren)(NH_3)(OH)]^{3+}$	1.7×10^{-4}	(176)
$[Fe(EDTA)(H_2O)]^-$	7.2×10^7	(170)	t-$[Co(tren)(NH_3)(OH)]^{3+}$	$<1\times10^{-6}$	(176)
$[Fe(HEDTA)(H_2O)]$	7.8×10^7	(170)	$[Co(cyclen)(NH_3)(OH)]^{3+}$	8	(179)
$[Fe(EDDS)(H_2O)]^-$	4.3×10^5	(170)			
$[Fe(PhDTA)(H_2O)]^-$	1.2×10^7	(171)	cis-$[Co(en)_2(OH)_2]^{3+}$	3.0×10^{-5}	(175)

Complex			Complex		
$[Fe(\alpha\text{-EDDADP})(H_2O)]^-$	$2.5\times10^8 / 2.7\times10^{6\,b}$	(170)	$trans\text{-}[Co(en)_2(OH)_2]^{3+}$	2.3×10^{-6}	(175)
$[Fe(TPPS)(H_2O_2)]^{3-}$	2.0×10^6	(172)	$[Co(tren)(OH)_2]^{3+}$	$9.7\times10^{-5}\,(p)$	(176)
				$2.2\times10^{-7}\,(t)$	(176)
$[Fe(TMPyP)(H_2O_2)]^{5+}$	1.4×10^7	(173)			
	4.5×10^5	(172)			
$[Fe(TMPS)(H_2O_2)]^{3-}$	7.8×10^5	(173)	$cis\text{-}[Co(en)_2(H_2O)OH]^{3+}$	9.3×10^{-4}	(175)
	2.1×10^7	(172)	$trans\text{-}[Co(en)_2(H_2O)OH]^{3+}$	1.2×10^{-3}	(175)
			$[Co(tren)(p\text{-}OH_2)t\text{-}OH]^{3+}$	$0.05(p)$	(176)
$[Fe(DMF)_6]^{3+}$	6.3×10^1	(121)	$[Co(tren)(t\text{-}OH_2)p\text{-}OH]^{3+}$	$0.03(p), 0.01(t)$	(176)
$[Fe(PhDTA)(DMF)]^-$	3.4×10^7	(171,174)	$[Co(cyclen)(H_2O)OH]^{3+}$	12	(176)
			$[Co(N\text{-}Mecyclen)(H_2O)OH]^{3+}$	15	(176)
$cis\text{-}[Co(en)_2(H_2O)_2]^{3+}$	7.5×10^{-6}	(175)			
$trans\text{-}[Co(en)_2(H_2O)_2]^{3+}$	1.1×10^{-5}	(175)	$[Co(CN)_5(H_2O)]^{2-}$	5.9×10^{-4}	(180)
$[Co(tren)(H_2O)_2]^{3+}$	$3.7\times10^{-5}\,(p)$	(176)	$[Co(\eta^5\text{-}C_5Me_5)(H_2O)_3]^{2+}$	60	(181)
	$8.7\times10^{-6}\,(t)$	(176)	$[Co(\eta^5\text{-}C_5Me_5)(bpy)(H_2O)]^{2+}$	0.6	(181)

[a] Rate constant for the exchange of a particular coordinated solvent molecule (6).
[b] Two exchange processes observed.

TABLE XII

EFFECT OF NON-LEAVING LIGANDS ON THE SOLVENT EXCHANGE RATE CONSTANTS ON
SECOND AND THIRD ROW DIVALENT TRANSITION-METAL IONS

Complex	$k_1 (298 \text{ K})^a (\text{s}^{-1})$	Reference
$[Ru(H_2O)_6]^{2+}$	1.8×10^{-2}	(125)
$[Ru(\eta^6\text{-}C_6H_6)(H_2O)_3]^{2+ b}$	11.5	(182)
$[Ru(\eta^6\text{-}C_6H_6)(bpy)(H_2O)]^{2+}$	6.8×10^{-2}	(181)
$[Ru(\eta^6\text{-}C_6Me_6)(bpy)(H_2O)]^{2+}$	10.2×10^{-2}	(181)
$[Ru(\eta^6\text{-cymene})(bpy)(H_2O)]^{2+}$	8.5×10^{-2}	(181)
$[Ru(DPMet)_2(tpmm)(H_2O)]^{2+}$	7.0×10^{-5}	(183)
$[Ru(DPPro)_2(tpmm)(H_2O)]^{2+}$	66	(183)
$[Ru(CO)(H_2O)_5]^{2+}$	2.54×10^{-6} (eq)	(184)
	3.54×10^{-2} (ax)	(184)
$[Ru(CO)_2(H_2O)_4]^{2+}$	1.58×10^{-7} (eq)	(184)
	4.53×10^{-4} (ax)	(184)
$[Ru(MeCN)(H_2O)_5]^{2+}$	1.5×10^{-3} (eq)	(185)
$[Ru(DMSO)(H_2O)_5]^{2+}$	1.9×10^{-5} (eq)	(185)
$[Ru(H_2C=CH_2)(H_2O)_5]^{2+}$	2.8×10^{-4} (eq)	(185)
$[Ru(CO)(H_2O)_5]^{2+}$	2.5×10^{-6} (eq)	(184)
$[Ru(F_2C=CF_2)(H_2O)_5]^{2+}$	9.3×10^{-6} (eq)^c	(185)
$[Ru(MeCN)(H_2O)_5]^{2+}$	3.9×10^{-4} (ax)^d	(185)
$[Ru(DMSO)(H_2O)_5]^{2+}$	6.8×10^{-2} (ax)^d	(185)
$[Ru(H_2C=CH_2)(H_2O)_5]^{2+}$	2.6×10^{-1} (ax)^d	(185)
$[Ru(CO)(H_2O)_5]^{2+}$	3.8×10^{-3} (ax)^d	(184)
$[Ru(F_2C=CF_2)(H_2O)_5]^{2+}$	5.6×10^{-1} (ax)^d	(185)
$[Os(\eta^6\text{-}C_6H_6)(H_2O)_3]^{2+ b}$	11.8	(182)
$trans\text{-}[Os(en)_2(\eta^2\text{-}H_2)(H_2O)]^{2+}$	1.59	(186)
$trans\text{-}[Os(en)_2(\eta^2\text{-}H_2)(MeCN)]^{2+}$	2.7×10^{-4}	(186)
$[Ru(MeCN)_6]^{2+}$	8.9×10^{-11}	(3)
$[Ru(\eta^6\text{-}C_6H_6)(MeCN)_3]^{2+ b}$	4.07×10^{-5}	(187)
$[Ru(\eta^5\text{-}C_5H_5)(MeCN)_3]^{+ b}$	5.6	(187)
$[Ru(Tp)(MeCN)_3]^{+ b}$	1.2×10^{-8}	(188)

^{a} Rate constant for the exchange of a particular coordinated solvent molecule (6).

^{b} All three H_2O or MeCN are equivalent.

^{c} At 308.4 K.

^{d} At 279 K.

configuration in the five-coordinate transition states or reactive intermediate (204).

Solvent exchange on Pd^{2+} and Pt^{2+} complexes shows a variation in
lability of about 16 orders of magnitude and is generally characterized
by either negative or near zero $\Delta V^{\ddagger}$ values. The exchange of non-aqueous
solvents has been studied in inert diluents and was found to have a

F.A. DUNAND et al.

TABLE XIII

EFFECT OF NON-LEAVING LIGANDS ON THE SOLVENT EXCHANGE RATE CONSTANTS ON
SECOND AND THIRD ROW TRIVALENT TRANSITION-METAL IONS

Complex	k_1 (298 K)^a (s^{-1})	Reference
$[Ru(H_2O)_6]^{3+}$	3.5×10^{-6}	(125)
$[Ru(NH_3)_5(H_2O)]^{3+}$	2.3×10^{-4}	(189)
$[Rh(H_2O)_6]^{3+}$	2.2×10^{-9}	(128)
$[Rh(NH_3)_5(H_2O)]^{3+}$	8.4×10^{-6}	(167)
$[Rh(CH_3NH_2)_5(H_2O)]^{3+}$	1.06×10^{-5}	(168)
$[Rh(\eta^5\text{-}C_5Me_5)(H_2O)_3]^{2+}$	1.6×10^5	(181)
$[Rh(\eta^5\text{-}C_5Me_5)(bpy)(H_2O)]^{2+}$	1.6×10^3	(181)
$[(H_2O)_4Rh(\mu\text{-}OH)_2Rh(H_2O)_4]^{4+}$	1.26×10^{-6} (trans)	(190)
	0.85×10^{-6} (trans)	(191)
	0.49×10^{-6} (cis)	(190)
	0.54×10^{-6} (cis)	(191)
	not obs. (bridging)	(190)
$[(H_2O)_3(OH)Rh(\mu\text{-}OH)_2Rh(OH)(H_2O)_3]^{4+}$	3.44×10^{-6} (trans)	(190)
	2.68×10^{-6} (cis)	(190)
	not obs. (bridging)	(190)
$[Rh(\eta^5\text{-}C_5Me_5)(MeCN)_3]^{2+}$	3.7×10^1	(192)
$[Rh(\eta^5\text{-}C_5Me_5)(DMSO)_3]^{2+}$	3.6×10^3	(192)
$[Ir(H_2O)_6]^{3+}$	1.1×10^{-10}	(129)
$[Ir(NH_3)_5(H_2O)]^{3+}$	6.1×10^{-8}	(193)
$[Ir(\eta^5\text{-}C_5Me_5)(H_2O)_3]^{2+}$	2.5×10^4	(181)
$[Ir(\eta^5\text{-}C_5Me_5)(bpy)(H_2O)]^{2+}$	2.2×10^2	(181)
$[Ir(\eta^5\text{-}C_5Me_5)(MeCN)_3]^{2+}$	8.8×10^{-2}	(192)
$[Ir(\eta^5\text{-}C_5Me_5)(DMSO)_3]^{2+}$	2.5×10^3	(192)

^{a} Rate constant for the exchange of a particular coordinated solvent molecule (6).

first-order rate dependence on free solvent concentration. The size of the solvent molecules has an important effect on the lability. Replacing the less sterically hindered $[Pd(Me_2S)_4]^{2+}$ by $[Pd(Et_2S)_4]^{2+}$ increases the exchange rate constant by more than 400-fold. Even more, solvent exchange on $[Pd(DMA)_4]^{2+}$ is slower than on $[Pd(DMF)_4]^{2+}$, despite the stronger nucleophilicity of DMA, as shown by the Gutmann donor numbers of 27.8 and 27.0 for DMA and DMF, respectively. Such steric influence on reaction rates is consistent with the operation of an a-activated mechanism.

Solvent molecules with hard donor atoms, like H_2O, MeCN, DMF, and DMA, have exchange rates on Pd^{2+} which differ less than a factor of 15.

TABLE XIV

Rate Constants and Activation Parameters for Solvent Exchange on Square-planar Transition-metal Ions

Complex	k_2 (298 K)^a (kg mol⁻¹ s⁻¹)	$\Delta H^\ddagger$ (kJ mol⁻¹)	$\Delta S^\ddagger$ (J K⁻¹ mol⁻¹)	$\Delta V^\ddagger$ (cm³ mol⁻¹)	Reference
[Pd(NH₃)₄]²⁺ [b]	0.016	67.3	−54.1		(194)
[Pd(Et₂S)₄]²⁺ [c]	5.0	50.4	−62.8	−11.6	(195)
[Pd(H₂O)₄]²⁺	10.2 [d]	49.5	−60	−2.2	(196)
	560 [e]	49.5	−26	−2.2	(196)
[Pd(DMA)₄]²⁺ [c]	34.8	43.2	−76.2	−2.8	(195)
[Pd(MeCN)₄]²⁺ [c]	48.8	45.4	−60.1	−0.1	(195)
[Pd(CN)₄]²⁻ [f]	82	23.5	−129	−22	(197)
[Pd(HCN)(CN)₃]⁻ [f]	4.5×10³				(197)
[Pd(DMF)₄]²⁺ [c]	153	41.9	−62.3	−0.2	(195)
[Pd(Me₂S)₄]²⁺ [c]	2140	31.9	−74.3 [g]	−9.4	(195)
[Pd(1,4−dithiane)₂]²⁺ [c]	9780	22.9	−91.6	−9.8	(198)
[Pd(MeNC)₄]²⁺ [h]	1.06×10⁶	16.4	−74.5	−3.1	(195)
[Pt(NH₃)₄]²⁺ [b]	9.5×10⁻¹⁰ [d]	125	+4		(199)
[Pt(H₂O)₄]²⁺	7.1×10⁻⁶ [d]	89.7	−43	−4.6	(200)
	3.9×10⁻⁴ [e]	89.7	−9	−4.6	(200)
[Pt(H₂O)₂(C₂O₄)₂] Pt(IV)	7.0×10⁻⁶ [d]	115	+42		(201)
[Pt(Me₂S)₄]²⁺ [c]	1.54	42.1	−100.2	−22.0	(198)
cis[Pt(Me₂SO)₂(Me₂SO)₂]²⁺ [c,i]	2	47	−74	−5	(202)
cis[Pt(Me₂SO)₂(Me₂SO)₂]²⁺ [c,j]	3200	32.8	−62.0	−2.5	(202)
[Pt(CN)₄]²⁻ [f]	11	25.1	−142	−27	(197)

F.A. DUNAND et al.

[Pt(1,4-dithiane)$_2$)]$^{2+\,c}$	28.8	32.9	-106	-12.6	(198)
[Pt(MeNC)$_4$]$^{2+\,h}$	6.2×10^5	13.8	-87.9	-3.7	(195)
[Au(CN)$_4$]$^{-\,f}$	6240	40.0	-38	+2	(203)
[Au(Cl)$_4$]$^{-\,f}$	0.56	65.1	-31	-14	(203)
[Ni(CN)$_4$]$^{2-\,f}$	2.3×10^6	21.6	-51	-19	(197)
[Ni(HCN)(CN)$_3$]$^{-\,f}$	$<10\times10^6$				(197)
[Ni(HCN)$_2$(CN)$_2$]^f	63×10^6	47.3	+63	-6	(197)

[a] Second-order rate constant for the exchange of a particular coordinated solvent molecule (6).
[b] In aqueous solution.
[c] Studied in d$_3$-nitromethane diluent.
[d] Units are dm^3 mol^{-1} s^{-1}.
[e] First-order rate constant for the exchange of a particular water molecule (6).
[f] In water.
[g] Misprint in Ref. (195) corrected in Ref. (198).
[h] Studied in d$_3$-acetonitrile diluent.
[i] Sulfur bonded.
[j] Oxygen bonded.

Solvents such as MeS_2 and MeNC with much softer binding atoms, have much faster exchange rates. In the case of MeNC this may be due to the formation of a five-coordinate intermediate stabilized by π back bonding from the metal center to MeNC. The dependence on the softness of the binding atom is aptly demonstrated by the DMSO exchange on cis-$[Pd(Me_2SO)_2(Me_2SO)_2]^{2+}$ where two Me_2SO are O-bonded and two are S-bonded (202). The soft donor $Me_2S\boldsymbol{O}$ exerts a strong trans effect on the hard donor $Me_2S\boldsymbol{O}$ which is a good leaving ligand, and vice versa. As a consequence, $Me_2S\boldsymbol{O}$ exchanges much more rapidly than $Me_2S\boldsymbol{O}$. The relative labilities of Pd^{2+} and Pt^{2+} complexes vary with the nature of the solvent molecules (Table XIV). Hard solvents exchange six orders of magnitude slower on Pt^{2+} than on Pd^{2+}, while exchange rates of soft solvents, with the exception of 1,4-dithiane and Me_2S, are very similar for both metal ions.

Mechanistic interpretation of activation volumes on square-planar complexes is complicated by the geometry. The sterically less crowded complexes may have loosely bound solvent molecules occupying the axial sites above and below the plane. Replacing them in the formation of a five-coordinate transition state or intermediate may result by compensation in relatively small volume effects. It is therefore difficult to distinguish between $\mathbf{I_a}$ and $\mathbf{A}$ mechanisms from the value of the activation volume. Nevertheless, the $\Delta V^{\ddagger}$ values are negative and together with the second-order rate laws observed, point to an $\mathbf{a}$-activation for those solvent exchange reactions.

However, strong σ-donor ligands such as phenyl and methyl encourage the operation of $\mathbf{d}$-activated mechanisms, with the formation of a three-coordinate intermediate (205). The exchange rate of Me_2SO on $[PtPh_2(Me_2SO)_2]^{2+}$ is nearly independent of free Me_2SO-concentration in $CDCl_3$ diluent. Activation volumes on complexes of the form $[PtR_2S_2]^{2+}$ (R = Ph, Me; S = Me_2SO, Me_2S) are all positive. The overall mechanistic picture has been taken as the first clear-cut evidence for the operation of a $\mathbf{D}$ mechanism involving a 3-coordinate 14-electron intermediate for simple ligand substitution on square-planar complexes (206) (Table XV (196,198,200,202,206–211)).

Deeth et al. have used density functional theory (DFT) to model water exchange on square-planer $[Pd(H_2O)_4]^{2+}$ and $[Pt(H_2O)_4]^{2+}$ (212). Their calculations strongly support that H_2O exchange on these complexes proceeds through an $\mathbf{a}$-activation mechanism, in full agreement with experimental assignments. The agreement between the experimental and calculated activation enthalpy is better than $10 \, kJ \, mol^{-1}$ for an $\mathbf{I_a}$ mechanism, whereas it differs by more than $100 \, kJ \, mol^{-1}$ for a calculated $\mathbf{I_d}$ mechanism.

TABLE XV

EFFECT OF NON-LEAVING LIGAND ON THE RATE CONSTANTS AND ACTIVATION PARAMETERS FOR SOLVENT EXCHANGE ON SQUARE-PLANAR TRANSITION-METAL IONS

Complex	k_2 (298 K)[a] ($\text{kg mol}^{-1}\,\text{s}^{-1}$)	$\Delta V^{\ddagger}$ ($\text{cm}^3\,\text{mol}^{-1}$)	Reference
$[Pd(H_2O)_4]^{2+}$	10.2[b]	−2.2	(196)
	560[c]	−2.2	(207)
$[Pd(dien)(H_2O)]^{2+}$	93[b]	−2.8	(208)
$[Pd(Me_5dien)(H_2O)]^{2+}$	3.4[b]	−7.2	(208)
$[Pd(Et_5dien)(H_2O)]^{2+}$	0.053[b]	−7.7	(200)
$[Pt(H_2O)_4]^{2+}$	7.1×10^{-6}[b]	−4.6	(209)
$[Pt\{C_6H_3(CH_2NMe_2)_2\text{-}2,6\}(H_2O)]^{2+}$	130[b]	−9.2	(198)
$[Pt(Me_2S)_4]^{2+}$[d]	1.54	−22.0	(206)
$cis\text{-}[PtPh_2(Me_2S)_2]^{2+}$[e]	0.21 (in s^{-1})	+4.7	(206)
$cis\text{-}[PtMe_2(Me_2S)_2]^{2+}$[d]	1.38 (in s^{-1})	+4.9	(206)
$cis\text{-}[PtPh_2(DMSO)_2]^{2+}$[h]	1.24 (in s^{-1})	+5.5	(206)
$cis\text{-}[PtMe_2(DMSO)_2]^{2+}$[d]	1.12 (in s^{-1})	+4.9	(206)
$cis\text{-}[Pt(Me_2SO)_2(Me_2SO)_2]^{2+}$[d,f]	2	−5	(202)
$cis\text{-}[Pt(Me_2SO)_2(Me_2SO)_2]^{2+}$[d,g]	3200	−2.5	

Complex	k_2 (298 K)[a] ($\text{kg mol}^{-1}\,\text{s}^{-1}$)	Reference
$[Pt(en)(CH_3)(DMSO)]^{+}$[i]	9.4×10^{-6}	(211)
$[Pt(Me_4en)(CH_3)(DMSO)]^{+}$[i]	1.15×10^{-6}	(211)
$[Pt(2\text{-ampy})(CH_3)(DMSO)]^{+}$[i]	3.5×10^{-3}	(211)
$[Pt(dps)(CH_3)(DMSO)]^{+}$[i]	1.0×10^{-2}	(211)
$cis\text{-}[Pt(py)_2(CH_3)(DMSO)]^{+}$[i]	1.6×10^{-2}	(211)
$[Pt(Me_4phen)(CH_3)(DMSO)]^{+}$[i]	2.66×10^{-2}	(211)
$[Pt(en)(CH_3)(DMSO)]^{+}$[i]	9.4×10^{-6}	(211)
$[Pt(Me_4phen)(CH_3)(DMSO)]^{+}$[i]	2.66×10^{-2}	(211)
$[Pt(dipy)(CH_3)(DMSO)]^{+}$[i]	5.3×10^{-2}	(211)
$[Pt(Ph_2phen)(CH_3)(DMSO)]^{+}$[i]	1.0×10^{-1}	(211)
$[Pt(bpy)(CH_3)(DMSO)]^{+}$[i]	1.6×10^{-1}	(211)
$[Pt(phen)(CH_3)(DMSO)]^{+}$[i]	1.8×10^{-1}	(211)
$[Pt(NO_2phen)(CH_3)(DMSO)]^{+}$[i]	3.3×10^{-1}	(211)
$[Pt(cy_2dim)(CH_3)(DMSO)]^{+}$[i]	9.2×10^{-1}	(211)
$[Pt(pr_2^i dim)(CH_3)(DMSO)]^{+}$[i]	7.9×10^{-1}	(211)
$[Pt(Me_2phen)(CH_3)(DMSO)]^{+}$[i]	3.81×10^{4}	(211)

[a] Second-order rate constant for the exchange of a particular coordinated solvent molecule.
[b] Units are $\text{dm}^3\,\text{mol}^{-1}\,\text{s}^{-1}$.
[c] First-order rate constant for the exchange of a particular water molecule (6).
[d] Studied in d_3-nitromethane diluent.
[e] Studied in benzene diluent.
[f] Sulfur bonded.
[g] Oxygen bonded.
[h] Studied in d_3-chloroform diluent.
[i] Studied in d_6-acetone diluent.

IV. Solvent Exchange on Lanthanides and Actinides

A. TRIVALENT LANTHANIDES

The trivalent lanthanide ions La^{3+}, Ce^{3+}, Pr^{3+}, Nd^{3+}, Pm^{3+}, Sm^{3+}, Eu^{3+}, Gd^{3+}, Tb^{3+}, Dy^{3+}, Ho^{3+}, Er^{3+}, Tm^{3+}, Yb^{3+}, and Lu^{3+}, represent the most extended series of chemically similar metal ions and are denoted commonly as Ln^{3+}. The progressive filling of the 4f orbitals from La^{3+} ($r_M = 121.6$ pm with $CN = 8$, 116.0 pm with $CN = 9$) to Lu^{3+} ($r_M = 103.3$ pm with $CN = 8$, 97.7 pm with $CN = 9$) is accompanied by a smooth decrease in atomic radius. This so called lanthanide contraction is a consequence of the poor shielding of one 4f electron by another owing to the shapes of the orbitals. The repulsions between electrons being added on crossing the f block fail to compensate for the increasing nuclear charge. The dominating effect of the latter is to draw in all the electrons, and to result in a more compact ion (213). The geometry of the 1st coordination shell is little affected by directional effects from the metal center, but by the electrostatic attraction exerted by the strongly charged ion on polar solvent molecules and by electrostatic and steric repulsion between the solvent molecules.

The coordination numbers of the Ln^{3+} ions in water are now well established from different experimental techniques (214–221). The lighter La^{3+}–Nd^{3+} ions are predominantly nine-coordinate, Pm^{3+}–Eu^{3+} exist in equilibria between nine- and eight-coordinate states and the heavier Gd^{3+}–Lu^{3+} are predominantly eight-coordinate. The change in coordination number is also reflected in the absolute partial molar volumes, V_{abs}^0, of several Ln^{3+} ions determined in aqueous solutions (222,223).

For eight-coordinate heavy lanthanides (Gd^{3+} to Yb^{3+}) rate constants for water exchange can be determined by oxygen-17 NMR relaxation and chemical shift measurements (224–226). Water exchange on these lanthanides is characterized by a systematic decrease in k_{H_2O} as the ionic radius decreases from Gd^{3+} to Yb^{3+} (Table XVI (2,224–228)). Both $\Delta V^{\ddagger}$ and $\Delta S^{\ddagger}$ are negative indicating an **a**-activation mode. The directly determined rate constants, k_{H_2O}, are closely correlated with the rate constants for interchange, k_i^{298}, between inner-sphere water and a SO_4^{2-} ion from outer-sphere coordination (Fig. 9) (229). The k_i^{298} values reach a maximum in the middle of the series where the coordination number changes from nine to eight. Unfortunately, for nine-coordinate early lanthanides ($[Nd(H_2O)_9]^{3+}$ and $[Pr(H_2O)_9]^{3+}$), only lower limits for water exchange rate constants could be determined at the available magnetic field of 14.1 T (230).

F.A. DUNAND et al.

TABLE XVI

RATE CONSTANTS AND ACTIVATION PARAMETERS FOR WATER EXCHANGE ON LANTHANIDE AQUA IONS

Complex	k_1 (298 K)^a ($10^7\,\mathrm{s}^{-1}$)	$\Delta H^\ddagger$ ($\mathrm{kJ\,mol}^{-1}$)	$\Delta S^\ddagger$ ($\mathrm{J\,K}^{-1}\,\mathrm{mol}^{-1}$)	$\Delta V^\ddagger$ ($\mathrm{cm}^3\,\mathrm{mol}^{-1}$)	Mechanism	Reference
$[\mathrm{Eu(H_2O)_7}]^{2+}$	500^b	15.7	−7.0	−11.3	$\mathbf{I_a}$, A	(2,227,228)
$[\mathrm{Gd(H_2O)_8}]^{3+}$	83.0	14.9	−24.1	−3.3	$\mathrm{I_a}$	(226)
$[\mathrm{Tb(H_2O)_8}]^{3+}$	55.8	12.1	−36.9	−5.7	$\mathrm{I_a}$	(224,225)
$[\mathrm{Dy(H_2O)_8}]^{3+}$	43.4	16.6	−24.0	−6.0	$\mathrm{I_a}$	(224,225)
$[\mathrm{Ho(H_2O)_8}]^{3+}$	21.4	16.4	−30.5	−6.6	$\mathrm{I_a}$	(224,225)
$[\mathrm{Er(H_2O)_8}]^{3+}$	13.3	18.4	−27.8	−6.9	$\mathrm{I_a}$	(224,225)
$[\mathrm{Tm(H_2O)_8}]^{3+}$	9.1	22.7	−16.4	−6.9	$\mathrm{I_a}$	(224,225)
$[\mathrm{Yb(H_2O)_8}]^{3+}$	4.7	23.3	−21.0	−6.0	$\mathrm{I_a}$	(224,225)

^{a} First-order rate constant for the exchange of a particular coordinated solvent molecule (6).
^{b} The value originally published for a CN=8 in Ref. (227) was corrected for CN=7 in Ref. (228).

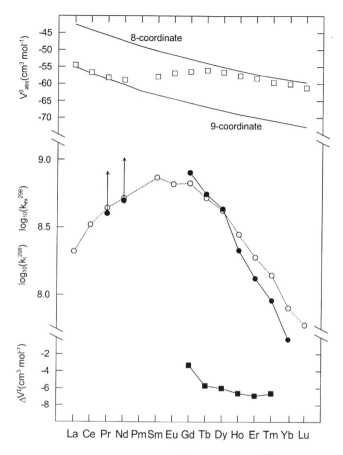

FIG. 9. Absolute partial molar volumes, V^0_{abs} of $[Ln(H_2O)_n]^{3+}$ in aqueous $Ln(ClO_4)_3$ solutions ($\square$), compared with the calculated values for $[Ln(H_2O)_8]^{3+}$ and $[Ln(H_2O)_9]^{3+}$ indicated by the upper and lower solid lines, respectively. Interchange rate constants, k_i, for the substitution of SO_4^{2-} on $[Ln(H_2O)_n]^{3+}$ are shown as $\bigcirc$, and water exchange rate constants, (298 K) for $[Ln(H_2O)_8]^{3+}$ are shown as $\bullet$. Activation volumes, $\Delta V^{\ddagger}$, are shown as $\blacksquare$.

The picture for water exchange on the lanthanide ions, conceived on the basis of structural and kinetic results, is as follows (Fig. 10). Over the whole Ln^{3+} series eight- and nine-coordinate ions are close in energy. Nine coordination is favored for the larger ions and eight coordination for the smaller ones. The slightly less favored coordination state can, however, be similar to the transition state/intermediate in an exchange process. The energy barrier for a water exchange is therefore low, leading to very high reaction rates. Also a change in mechanism is

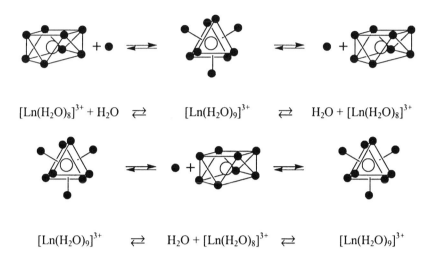

$$[\mathrm{Ln(H_2O)_8}]^{3+} + \mathrm{H_2O} \ \rightleftharpoons \qquad [\mathrm{Ln(H_2O)_9}]^{3+} \qquad \rightleftharpoons \ \mathrm{H_2O} + [\mathrm{Ln(H_2O)_8}]^{3+}$$

$$[\mathrm{Ln(H_2O)_9}]^{3+} \qquad \rightleftharpoons \quad \mathrm{H_2O} + [\mathrm{Ln(H_2O)_8}]^{3+} \ \rightleftharpoons \qquad [\mathrm{Ln(H_2O)_9}]^{3+}$$

FIG. 10. Possible mechanistic paths for water exchange on eight- and nine-coordinate lanthanides.

predicted by this model: the **a**-activated mechanism for the small eight-coordinate ions, determined via the negative activation volumes, should be replaced by a **d**-activated mechanism for the larger nine-coordinate ions. This change in mechanism is supported by a computational study using classical molecular dynamics simulation.

In recent years the subject of water exchange on lanthanide complexes attracted increasing attention owing to the use of $\mathrm{Gd^{3+}}$ as a contrast agent in medical magnetic resonance imaging (MRI). A review of the extensive work performed in this domain exceeds the scope of this article and suitable, comprehensive accounts can be found in recent publications (231–233). Nevertheless, we will present here a brief overview of the results. Nine-coordinate gadolinium(III) complexes with one inner sphere water molecule have water exchange rates which are at least two orders of magnitude slower than that of the aqua-ion (Table XVII (226,228,231,234–249)). The positive $\Delta V^{\ddagger}$ values indicate a **d**-activation mode for water exchange. This could be expected, considering that in a nine-coordinate (240–244,246,248–250) lanthanide complex there is no longer space for a second water molecule to enter before the subsequent departure of the bound water molecule. The eight-coordinate transition state is high in energy resulting in a high energy barrier for the dissociating water molecule to cross. Another important factor is the rigidity of the coordination sphere of the $\mathrm{Gd^{3+}}$: whereas the eight water

TABLE XVII

EFFECT OF NON-LEAVING LIGAND ON THE RATE CONSTANTS AND ACTIVATION PARAMETERS FOR WATER EXCHANGE ON Gd^{3+} AND Eu^{2+} COMPLEXES

Complex	k_1 (298 K)[a] (10^6 s^{-1})	$\Delta H^\ddagger$ (kJ mol^{-1})	$\Delta S^\ddagger$ (J K^{-1} mol^{-1})	$\Delta V^\ddagger$ (cm^3 mol^{-1})	Reference
[Gd(H$_2$O)$_8$]$^{3+}$	830	14.9	−24.1	−3.3	(226)
[Gd(PDTA)(H$_2$O)$_2$]$^-$	102	11.0	−54.6	−1.5	(226)
[Gd(tris(2-hydroxymethyl)-TREN-Me-2,3-HOPO)(H$_2$O)$_2$]$^-$	62.5	2.6			(234)
[Gd(DO2A)(H$_2$O)$_3$]$^-$	1.0	21.3	−39		(235)
[Gd(DO3A)(H$_2$O)$_2$]$^-$	1.1	33.6	+2.4		(236)
[Gd(DTPA)(H$_2$O)]$^{2-}$	4.1^b	52.0	+56.2	+12.5	(237)
	4.0^c	52			(239)
[Gd(DTPA-BMA)(H$_2$O)]	0.43	46.6	+18.9	+7.3	(240)
[Gd(DOTA)(H$_2$O)]$^-$	4.8^b	48.8	+46.6	+10.5	(237)
	5.4 (M)/500 (m)^c	55/50			(239)
[Gd(DOTMA)(H$_2$O)]$^-$	150				(241)
[Gd(DOTA-2DMA)(H$_2$O)]$^-$	0.74 (M)/70 (m)				(242)
[Gd(DOTAM)(H$_2$O)]$^{3+}$	0.5				(243)
[Eu(DOTAM)(H$_2$O)]$^{3+}$ ^{d}	0.008 (M)/0.33 (m)	53.1/44.2	+8.4/+8.8	+4.9/−	(244)
[Eu(H$_2$O)$_7$]$^{2+}$	5000^e	15.7	−7.0	−11.3	(2,227,228)
[Eu(ODDA)(H$_2$O)]	430	22.5	−4.0	−3.9	(246)
[Eu(DTPA)(H$_2$O)]$^{3-}$	1300	26.3	+18.3	+4.5	(247)
[Eu(DOTA)(H$_2$O)]$^{2-}$	2460	21.4	+6.9	+0.1	(248)
[Eu(2.2.2)(H$_2$O)$_2$]$^{3+}$	310	30.6	+20.5	+1	(249)

^{a} First-order rate constant for the exchange of a particular coordinated solvent molecule (6).

^{b} Analysis of ^{17}O NMR data only.

^{c} Combined analysis of EPR, ^{1}H and ^{17}O NMR data.

^{d} Independent analysis of the bound water NMR signal of each species.

^{e} The value originally published for a CN=8 in Ref. (227) was corrected for CN=7 in Ref. (228).

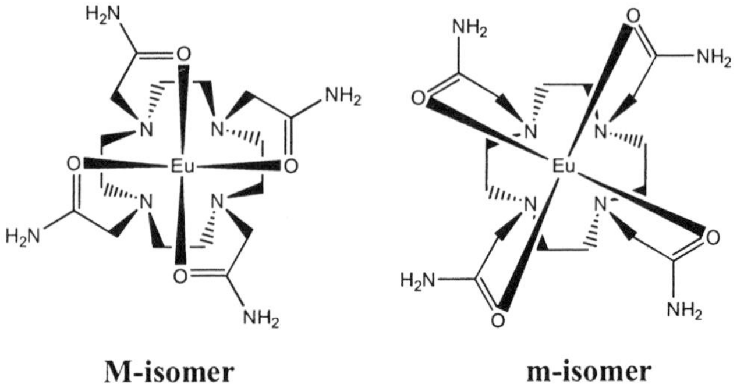

M-isomer   **m-isomer**

Fig. 11. Schematic structure of the two diastereoisomers of the complex [Eu(DOTAM)(H$_2$O)]$^{3+}$. The coordinated water molecule, located above the plain of the four oxygens, has been omitted for clarity.

molecules in the aqua complex can easily rearrange, the poly(amino carboxylate) ligand is very rigid and its rearrangement requires high energy. In summary, the difference in the inner-shell structure and hence the difference in the mechanism, explains why water exchange on nine-coordinate Ln(III) poly(amino carboxylate) complexes is generally much slower when compared to the eight-coordinate [Gd(H$_2$O)$_8$]$^{3+}$ (250).

Cyclic ligands like DOTA or DOTAM are known to exist in two isomeric forms in solution, usually termed M for the major and m for the minor isomer (Fig. 11). They interconvert slowly on the NMR timescale (251–253). It was demonstrated by NMR that the m-isomer exchanges its 1st shell water much faster than the M-isomer, most probably due to steric effects (238,244).

Gadolinium(III) complexes with two inner sphere water molecules have faster water exchange rates than monohydrated chelates. The eight-coordinate complex [Gd(PDTA)(H$_2$O)$_2$]$^-$ (226) exchanges water much faster than the nine-coordinate mono-aqua complexes (250). The mechanism for the former is a-activated with much less associative character than for the latter aqua-ion. This effect is most probably due to the rigidity imposed by the PDTA ligand. A temperature-dependent UV–visible spectrophotometric study on [Eu(DO3A)(H$_2$O)$_n$] indicated the presence of a hydration equilibrium (n=1, 2), strongly shifted towards the bisaqua species ($K_{1-2}^{298} = 7.7$) (236). The limited gain in the water exchange rate compared to that for the DOTA complex (Table XVII) can be explained in terms of a rigid inner sphere structure introduced by the macrocyclic ligand which makes the transition from the reactant to the transition state difficult, and consequently, results in a slower exchange as compared to that for [Gd(PDTA)(H$_2$O)$_2$]$^-$. If two carboxylate arms on the DOTA are replaced by H there is space for two to three H_2O molecules in the first coordination shell of lanthanide complexes (235). For the corresponding Eu^{3+} complex a stability constant $K_{2-3}^{298} = 4.0$ was measured. The small increase in the water exchange rate of $[Gd(DO2A)(H_2O)_{2-3}]^+$ (Table XVII) relative to that of $[Gd(DOTA)(H_2O)]^-$ is a consequence of an unfavorable interplay of charge and hydration equilibria (235).

A promising new class of stable Gd^{3+} complexes based on TREN-Me-3,2-HOPO shows relatively fast water exchange rates. The water soluble tris(2-hydroxymethyl)-TREN-Me-3,2-HOPO complex is eight-coordinate with two water molecules in the first coordination sphere (234). The water exchange rate is more than one order of magnitude faster than on DTPA and DOTA complexes and the mechanism is proposed to be **a**-activated.

The exchange of DMF on lanthanide(III) ions could be observed over the whole series of paramagnetic lanthanide ions in neat solvent and in the diluent CD_3NO_2 (254). For the light lanthanides Ce^{3+}, Pr^{3+}, and Nd^{3+}, an equilibrium between eight- and nine-coordination is observed at low temperature. The reaction volume, ΔV^0, for the addition of a DMF molecule to $[Nd(DMF)_8]^{3+}$ was determined spectrophotometrically to be $-9.8\,\text{cm}^3\,\text{mol}^{-1}$. For the heavier lanthanides, Tb^{3+} to Yb^{3+}, kinetic parameters for DMF exchange were measured by variable temperature and pressure ^1H NMR. The exchange process is characterized by positive $\Delta V^{\ddagger}$ values, a systematic increase in $\Delta H^{\ddagger}$ and a change from negative to positive $\Delta S^{\ddagger}$ values as the ionic radius decreases (Table XVIII (254,255)). **d**-Activation occurs for these exchange processes with a mechanistic crossover from $\mathbf{I_d}$ to limiting $\mathbf{D}$ at erbium (256). Kinetic rate law determinations using CD_3NO_2 as diluent indicate that an interchange mechanism $\mathbf{I_d}$ operates for Tb^{3+} whereas a $\mathbf{D}$ mechanism is operative for Yb^{3+}. For Er^{3+} a mixed rate law with a first- and a second-order term is observed showing that the $\mathbf{I_d}$ and the $\mathbf{D}$ transition states are being similarly favored by this ion.

A decrease in coordination number of Ln^{3+} complexes results, in general, in a decrease in lability. Solvent exchange rates measured on $[Ln(TMU)_6]^{3+}$ are much slower than corresponding values measured on $[Ln(DMF)_8]^{3+}$ (56,257–259) (Table XIX (56,257–259)). The exchange rates, measured in an inert diluent, were found to be independent of free ligand concentration, consistent with a **d**-activation mode.

B. DIVALENT Eu(II)

The only easily accessible divalent lanthanide is Eu^{2+} which is isoelectronic with Gd^{3+} and very similar in size to Sr^{2+} (260). An

TABLE XVIII

RATE CONSTANTS AND ACTIVATION PARAMETERS FOR DMF EXCHANGE ON
$[Ln(DMF)_8]^{3+}$ COMPLEXES

M^{3+}	$r_M{}^a$ (pm)	k_1 (298 K)^b ($10^5\,s^{-1}$)	$\Delta H^{\ddagger}$ (kJ mol^{-1})	$\Delta S^{\ddagger}$ (J K^{-1} mol^{-1})	$\Delta V^{\ddagger}$ (cm^3 mol^{-1})	Mechanism	Ref.
Tb^{3+}	104.0	190	14.1	−58	+5.2	I_d	(254)
Dy^{3+}	102.7	63	13.8	−69	+6.1	I_d	(254)
Ho^{3+}	101.5	36	15.3	−68	+5.2	I_d	(255)
Er^{3+}	100.4	130	23.6	−30	+5.4	D and I_d	(254)
Tm^{3+}	99.4	310	33.2	+10	+7.4	D	(254)
Yb^{3+}	98.5	990	39.3	+40	+11.8	D	(254)

^{a} Six-coordinate ionic radii from Ref. (255).
^{b} First-order rate constant for the exchange of a particular coordinated solvent molecule (6).

TABLE XIX

RATE CONSTANTS AND ACTIVATION PARAMETERS FOR TMU EXCHANGE ON
$[Ln(TMU)_6]^{3+}$ COMPLEXES IN CD$_3$CN DILUENT

M^{3+}	$r_M{}^a$ (pm)	k_1 (298 K)^b (s^{-1})	$\Delta H^{\ddagger}$ (kJ mol^{-1})	$\Delta S^{\ddagger}$ (J K^{-1} mol^{-1})	Mechanism
Tb^{3+}	92.3	1380	38.2	−56.7	−
Dy^{3+}	91.2	1290	38.6	−56.0	−
Ho^{3+}	90.1	510	40.9	−55.9	−
Y^{3+}	90.0	253	27.1	−108	D, I_d
Er^{3+}	89.0	214	35.5	−81.3	D, I_d
Tm^{3+}	88.0	145	29.3	−105	D, I_d
Yb^{3+}	86.8	65.5	38.3	−81.8	D, I_d
Lu^{3+}	86.1	41.9	41.7	−74	D, I_d
Sc^{3+}	74.5	0.90	68.6	−15.7	D, I_d

Adapted from Refs. (56,257–259).
^{a} Six-coordinate ionic radii from Ref. (255).
^{b} First-order rate constant for the exchange of a particular coordinated solvent molecule (6).

EXAFS study on Eu^{2+} and Sr^{2+} in both solid state and aqueous solution gave coordination numbers of 8.0 for strontium(II) and 7.2 for europium(II) (228). The water exchange rate measured on the divalent europium aqua ion is the fastest ever measured by ^{17}O NMR (Table XVI) (2). The activation volume is much more negative (−11.7 cm^3 mol^{-1}) than those determined on trivalent lanthanide aqua ions clearly indicating an **a**-activation mechanism which is most probably a limiting

A mechanism (from comparison with $-12.1\,cm^3\,mol^{-1}$ for $[Ti(H_2O)_6]^{3+}$, Section III.C). The EXAFS study showed that an equilibrium between seven- and eight-coordinate states is possible which would explain the very low energy barrier between the transition state (eight-coordinate for an associative mechanism) and the ground state $[Eu(H_2O)_7]^{2+}$ (228).

It has been shown above that water exchange on poly(aminocarboxy-late) complexes of Gd^{3+} is about three orders of magnitude slower than on the eight-coordinate aqua ion. Such a tremendous loss in lability is not observed for europium(II) complexes (Table XVII). The inner sphere water molecule on $[Eu(DTPA)(H_2O)]^{3-}$ exchanges only ~ 4 times slower than a water molecule on $[Eu(H_2O)_7]^{2+}$ (247). As a consequence of the lower charge and the larger ionic radius, the charge density is significantly smaller on the Eu^{2+} ion as compared to Gd^{3+}. Furthermore, the water exchange has less dissociative character for $[Eu(DTPA)(H_2O)]^{3-}$ ($\Delta V^{\ddagger} = +4.5\,cm^3\,mol^{-1}$, I_d mechanism) as compared to that for $[Gd(DTPA)(H_2O)]^{2-}$ ($\Delta V^{\ddagger} = +12.5\,cm^3\,mol^{-1}$, D mechanism) indicating less steric crowding around the divalent europium ion. On $[Eu(ODDA)(H_2O)]$ however, the negative $\Delta V^{\ddagger}$ ($-3.9\,cm^3\,mol^{-1}$) and the slightly negative $\Delta S^{\ddagger}$ ($-4.0\,J\,K^{-1}\,mol^{-1}$) provide evidence for an a-activated exchange process, most probably via an I_a mechanism (246). The solid state structure of the corresponding Sr^{2+} complex, which is iso-structural with that of europium(II), and solution EXAFS data (261) show that in the case of ODDA the inner shell water molecule coordinates close to the macrocyclic plane and not close to the carboxylate arms, as is the case for the poly(aminocarboxylate) ligands (Fig. 12). The water molecule coordination site is far less crowded and less electro-statically constrained in the $ODDA^{2-}$ than, for example, in the $DOTA^{4-}$ or $DTPA^{5-}$ complexes. The change in mechanism from I_d (for $[Eu(DTPA)(H_2O)]^{3-}$) to I_a (for $[Eu(ODDA)(H_2O)]$) is plausible because a d-activation mode is disfavored in the case of the second complex due to the short $Eu-O_W$ distance of $2.54\,\text{Å}$ (as compared to $2.62\,\text{Å}$ for DTPA), and because of the much more open coordination site on $[Eu(ODDA)(H_2O)]$ allowing a second water molecule to approach the inner sphere.

C. ACTINIDES

Reports on studies of solvent exchange reactions on actinide compounds are very scarce. The most studied cation is UO_2^{2+}. The commonly observed solvated species have five (H_2O, DMSO, DMF, TMP) (262,263)

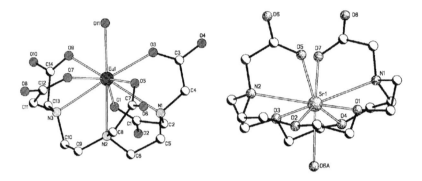

FIG. 12. X-ray structures of $[Eu(DTPA)H_2O]^{3-}$ (left) and $[Sr(ODDA)]$ (right).

or four (HMPA) (264) solvent molecules in the first coordination shell. For complexes of the form $[UO_2S_5]^{2+}$ (S = solvent) the exchange rates measured in inert diluents are nearly independent of concentration of S, and for S = TMP a positive $\Delta V^{\ddagger}$ was measured (6). This is indicative of a **d**-activation mode for solvent exchange on these compounds. The rate law for HMPA exchange on $[UO_2(HMPA)_4]^{2+}$ however, is second-order. This together with the very negative $\Delta V^{\ddagger}$, led to the attribution of an **A** mechanism for the exchange of HMPA. An experimental study of water exchange on $[UO_2(H_2O)_5]^{2+}$ was performed only recently and the experimental data alone do not provide enough information to decide on the mechanism for water exchange (Table XX (265–269)). An ab-initio quantum chemical study together with the experimental data gave strong evidence that the water exchange on $[UO_2(H_2O)_5]^{2+}$ takes place through an **A**- or **I**-mechanism (270). In $[UO_2(C_2O_4)_2(H_2O)]^{2-}$ the water exchange seems to follow an **A**-mechanism with an activation energy similar to that of the penta-aqua complex (270). Water exchange on $[NpO_2(H_2O)_5]^{2+}$ and $[PuO_2(H_2O)_5]^{2+}$ is very rapid and has therefore been studied in an acetone/H_2O mixture at low temperature (269) (Table XX). As the $\Delta S^{\ddagger}$ values become more negative with increasing atomic number, a mechanism with more pronounced associative character was proposed for the heavier species.

The first experimental information on the kinetic parameters for water exchange on a tetravalent metal ion was published in 2000 for U^{4+} and Th^{4+} (265,268,271). The coordination numbers for these two complexes were determined by EXAFS to be 10 ± 1. Based on the high coordination number (there are no complexes known with unidentate ligands and coordination numbers larger than 10) a limiting associative mechanism (**A**) is unlikely and a **d**-activated mechanism is probable. Surprisingly,

TABLE XX

RATE CONSTANTS AND ACTIVATION PARAMETERS FOR WATER EXCHANGE ON ACTINIDES OR ACTINYL AQUA IONS

Complex	k_1 (298 K)[a] (s^{-1})	$\Delta H^{\ddagger}$ (kJ mol^{-1})	$\Delta S^{\ddagger}$ (J K^{-1} mol^{-1})	Mechanism	Reference
$[Th(H_2O)_{10}]^{4+}$	$> 5 \times 10^7$				(265)
$[Th(H_2O)_9]^{4+}$	2.4×10^9 (2nd sphere)				(266)
$[U(H_2O)_{10}]^{4+}$	$\sim 5 \times 10^6$	~ 35	~ 0		(265)
$[U(F)(H_2O)_9]^{4+}$	5×10^6	36	+2.5		(265)
$[UO_2(H_2O)_5]^{2+}$	1.4×10^6	38	-12		(267)
	1.3×10^6	26.1	-40	I or A[f]	(268,270)
$[UO_2(H_2O)_5]^{2+}$	$11.8 \times 10^{3\,b}$	32	-60		(269)
	$460 \times 10^{3\,c}$	31.7	-30		(269)
$[UO_2(C_2O_4)_2(H_2O)]^{2-}$				A[f]	(270)
$[UO_2(C_2O_4)(F)(H_2O)_2]^{-}$	1.8×10^4	45.4	-11.3		(268)
$[NpO_2(H_2O)_5]^{2+}$	$5.3 \times 10^{6\,d}$	20.2	-72		(269)
$[PuO_2(H_2O)_n]^{2+}$	$5.7 \times 10^{4\,e}$	12	-115		(269)

[a] First-order rate constant for the exchange of a particular coordinated solvent molecule (6).
[b] Solution composition (molar ratio): $MO_2^{2+}/H^+/H_2O$/acetone: 1/0.57/69.1/177.9.
[c] Solution composition (molar ratio): $MO_2^{2+}/H^+/H_2O$/acetone: 1/4.1/86.9/161.6.
[d] Solution composition (molar ratio): $MO_2^{2+}/H^+/H_2O$/acetone: 1/3.8/78.7/153.4.
[e] Solution composition (molar ratio): $MO_2^{2+}/H^+/H_2O$/acetone: 1/2.2/67.3/174.7.
[f] By quantum chemical methods.

the coordination of one fluoride or one hydroxide to U(IV) had no detectable effect on the water exchange rate.

ACKNOWLEDGEMENTS

The authors gratefully acknowledge financial support from the Swiss National Science Foundation and the Swiss Office for Education and Science (COST Program). Furthermore, we wish to thank the large number of people who have contributed over the years to the work performed in Lausanne.

V. Appendix: Ligand Abbreviations, Formulae, and Structures

- (2.2.2) = 4,7,13,16,21,24-hexaoxa-1,10-diazabicyclo-[8.8.8]hexacosane

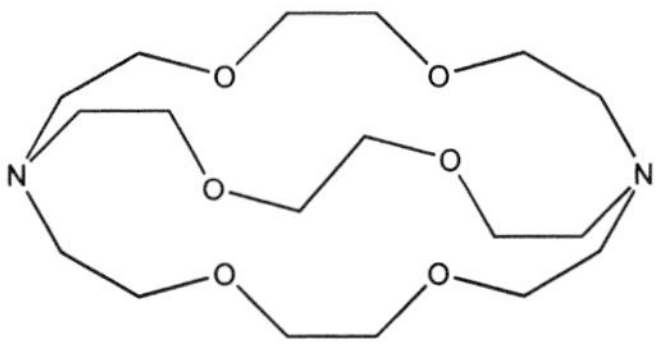

- 2-ampy = 2-(aminomethyl)pyridine

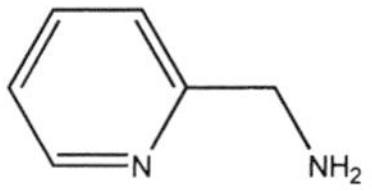

- bpy = 2,2′-bipyridine

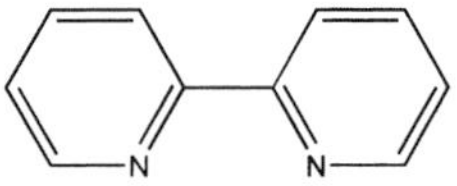

- BuCN = valeronitrile = $NCCH_2CH_2CH_2CH_3$
- $C_2O_4^{2-}$ = oxalate = $^-OOCCOO^-$
- η^5-$C_5Me_5^-$ = pentamethylcyclopentadienyl
- η^6-C_6H_6 = benzene
- η^6-C_6Me_6 = hexamethylbenzene
- $CDTA^{4-}$ = *trans*-1,2-diaminocyclohexanetetraacetate

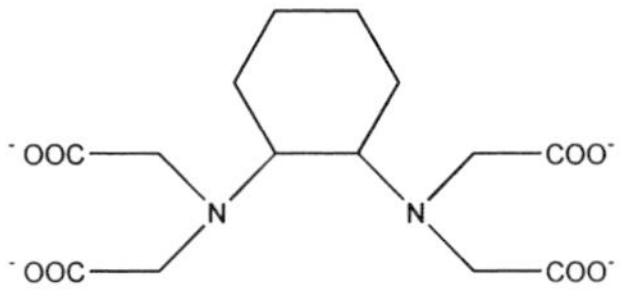

- cyclen = 1,4,7,10-tetraazacyclododecane

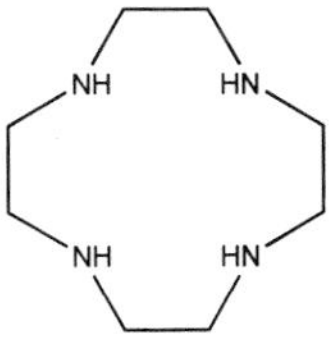

- cy$_2$dim = *N,N'*-dicyclohexylethylenediimine

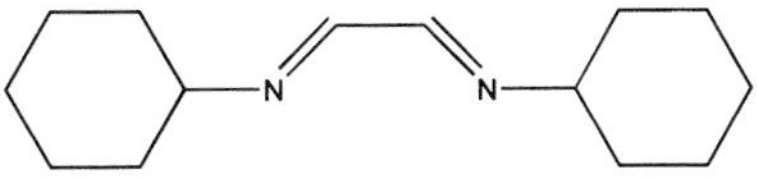

- η^6-cymene = *p*-isopropyltoluene
- DEA = *N,N*-diethylacetamide = OC(Me)N(Et)$_2$
- dien = diethylenetriamine

- dipy = 2,2'-dipyridylamine

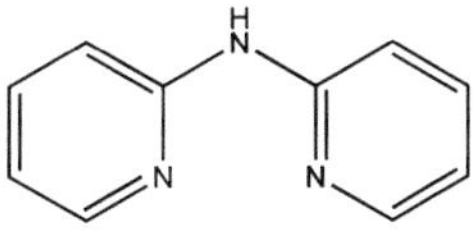

- 1,4-dithiane

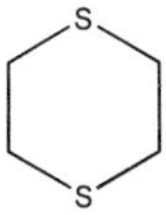

- DMA = *N,N*-dimethylacetamide = OC(Me)N(Me)$_2$
- DMADMP = *O,O'*-dimethyl-*N,N*-dimethylphosphoramidate = OP (Me)$_2$(NMe$_2$)
- DMF = *N,N*-dimethylformamide = OC(H)N(Me)$_2$
- dmgH$^-$ = dimethylglyoxime

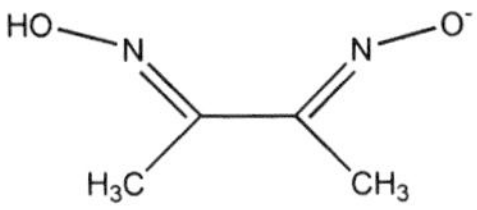

- dmg^{2-}

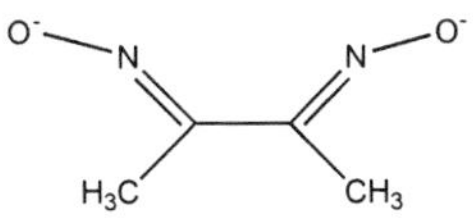

- DMSO = dimethylsulfoxide = OSMe$_2$

- DMMP = dimethyl methylphosphonate = OP(OMe)$_2$(Me)
- DMPU = dimethylpropyleneurea

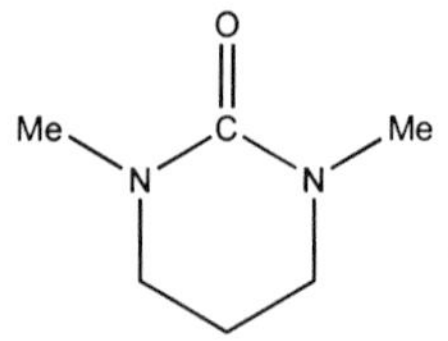

- DO2A^{2-} = 1,4,7,10-tetraaza-1,7-bis-(carboxymethyl)cyclododecane

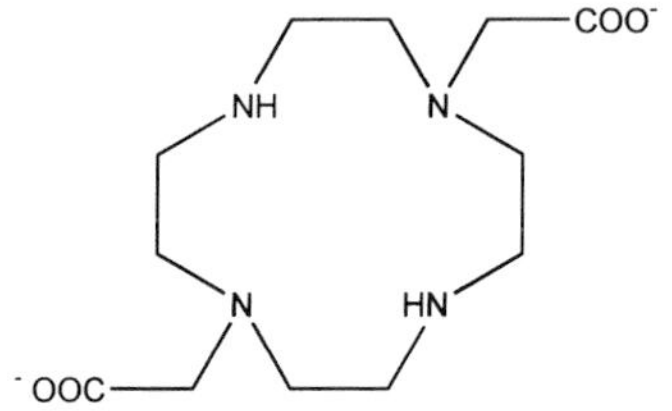

- DO3A^{3-} = 1,4,7,10-tetraaza-1,4,7-tris-(carboxymethyl)cyclododecane

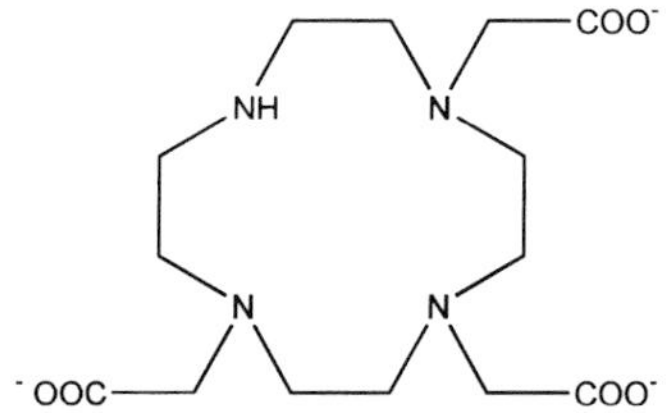

- DOTA^{4-} = 1,4,7,10-tetraaza-1,4,7,10-tetrakis-(carboxymethyl) cyclododecane

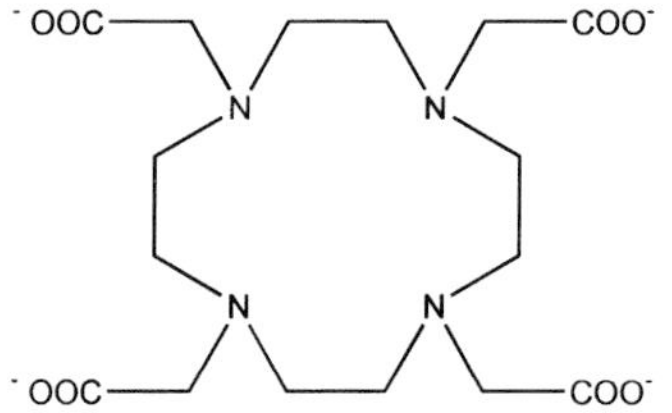

- DOTAM = 1,4,7,10-tetraaza-1,4,7,10-tetrakis-(carbamoylmethyl) cyclododecane

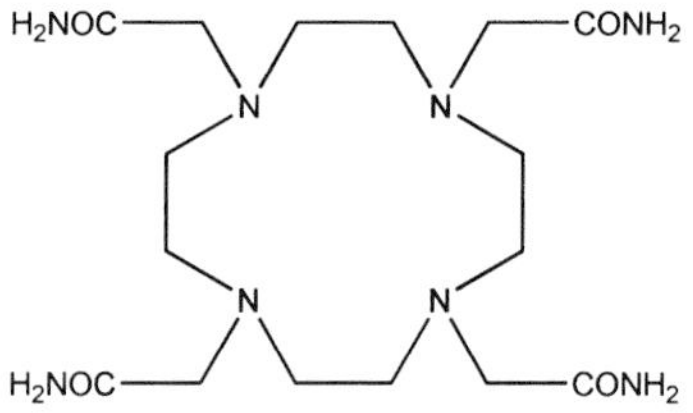

- DOTMA^{4-} = 1,4,7,10-tetraaza-1,4,7,10-tetrakis-(carboxymethyl) cyclododecane

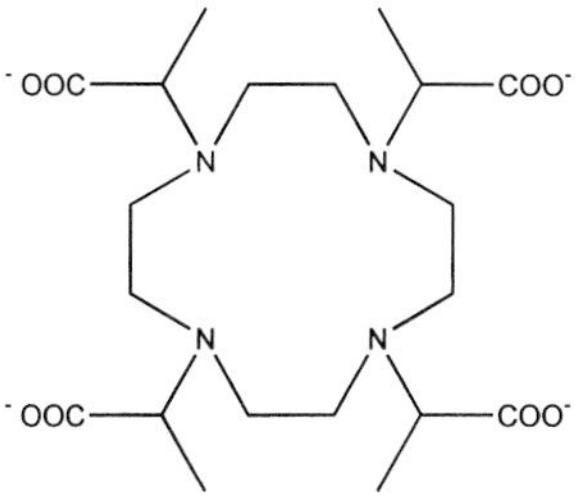

- dpgH$^-$ = diphenylglyoximate

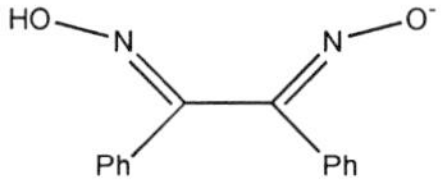

- DPMet = di(1-pyrazolyl)methane

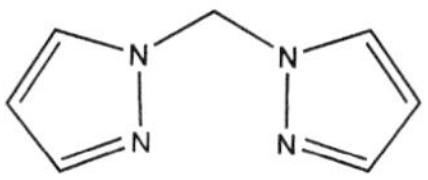

- DPPro = di(1-pyrazolyl)propane

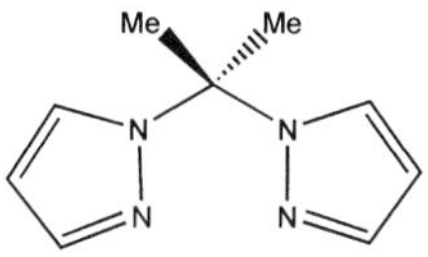

- dps = 2,2′-dipyridylsulfide

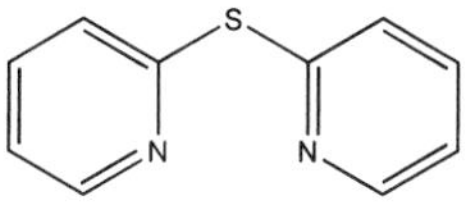

- DTPA^{5-} = 1,1,4,7,7-pentakis(carboxymethyl)-1,4,7-triazaheptane

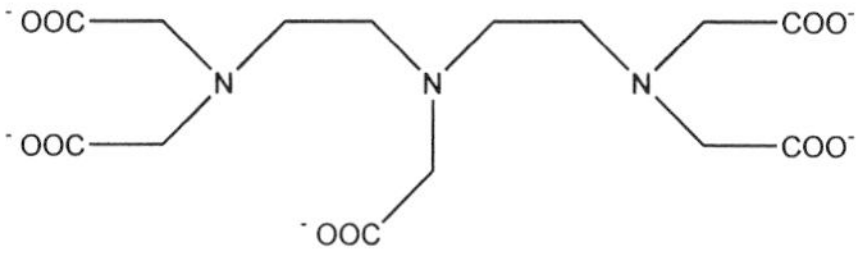

- DTPA-BMA^{3-} = 1,4,7-tri(carboxymethyl)-1,7-bis[(N-methylcarba-moyl)methyl]-1,4,7-triazaheptane

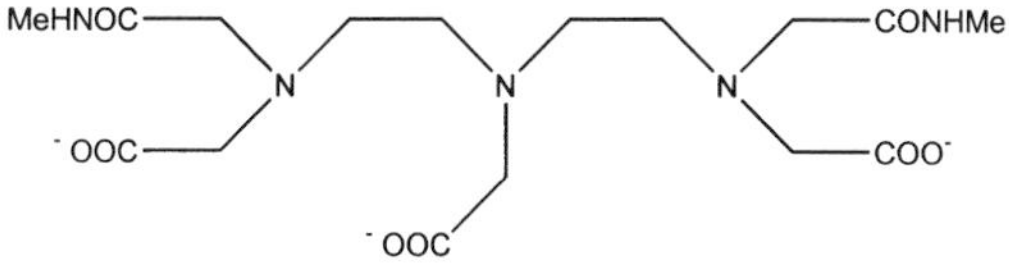

- α-EDDADP^{4-} = α-ethylenediaminediacetatedipropionate

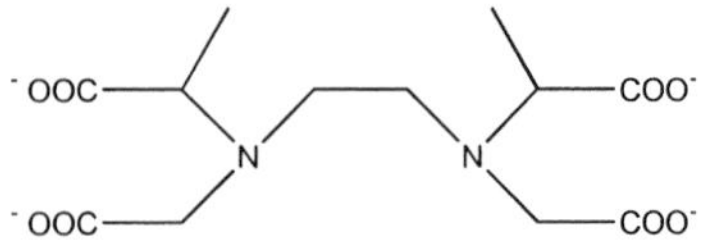

- EDDS^{4-} = s,s-ethylenediaminesisuccinate

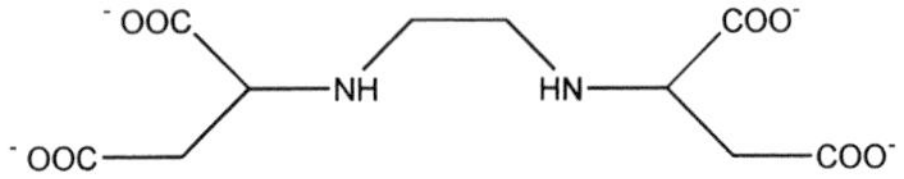

- EDTA^{4-} = ethylenediaminetetraacetate

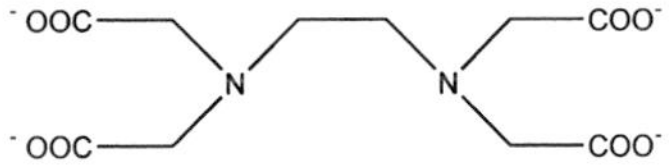

- EDTA-BOM^{4-}: R = H, R' = BOM

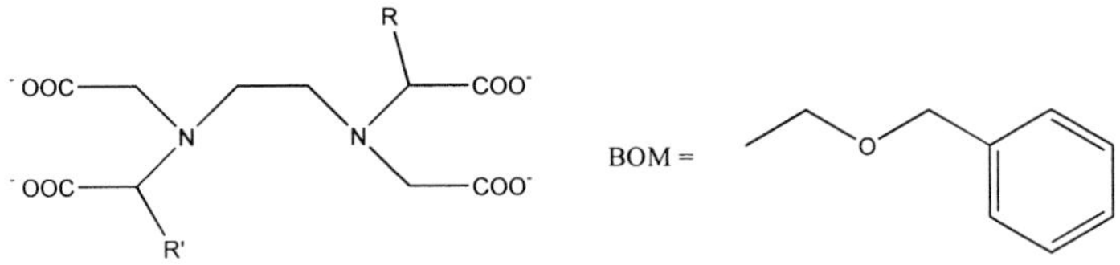

BOM =

- EDTA-BOM$_2^{4-}$: R = R' = BOM
- en = 1,2-diaminoethane

- Et$_5$dien = N,N',N''-pentaethyl-diethylenetriamine

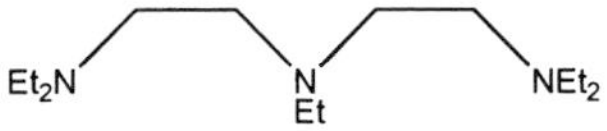

- EtCN = propionitrile = NCCH$_2$CH$_3$
- fz = ferrozine = 3-(2-pyridyl)-5,6-bis(4-phenylsulfonic acid)-1,2,4-triazine

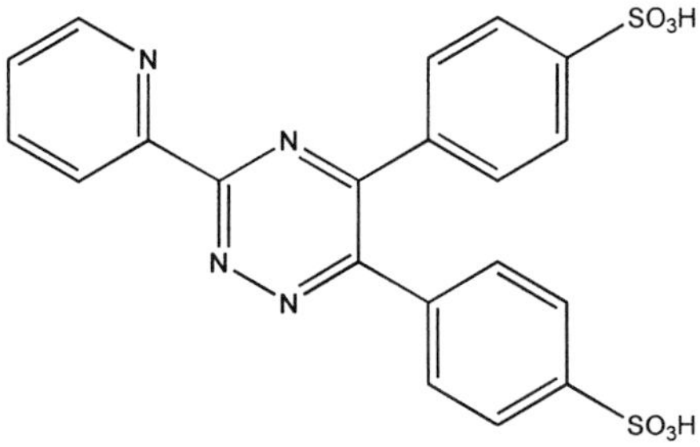

- HMPA = hexamethylphosphoramide = OP(NMe$_2$)$_3$

- HOAc = acetic acid
- ma$^-$ = maltolate = 3-oxy-2-methyl-4-pyrone

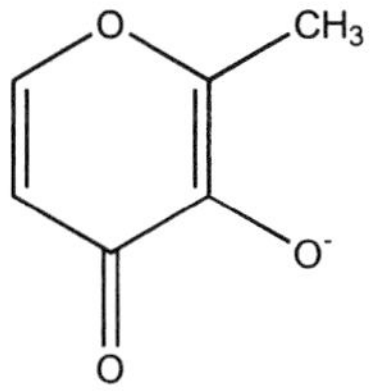

- MeCN = acetonitrile = $NCCH_3$
- Me_4cyclam = N,N',N'',N'''-tetramethyl-1,4,8,11-tetraazacyclotetradecane

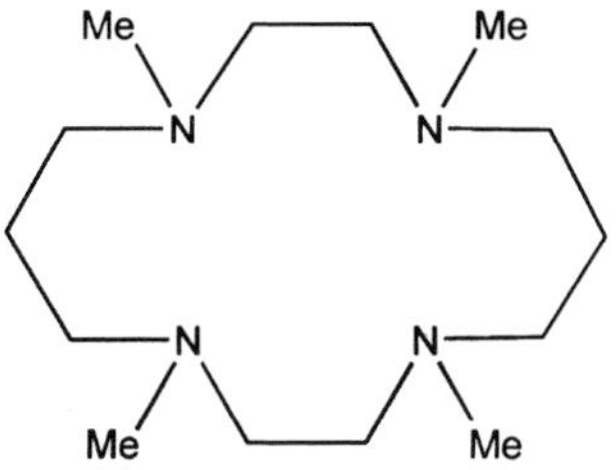

- Me_5dien = N,N,N',N'',N''-pentamethyl-diethylenetriamine

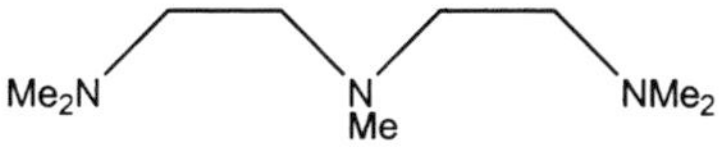

- Me_4en = N,N,N',N'-tetramethyl-1,2-diaminoethane

- Me_2phen = 2,9-dimethyl-1,10-phenanthroline

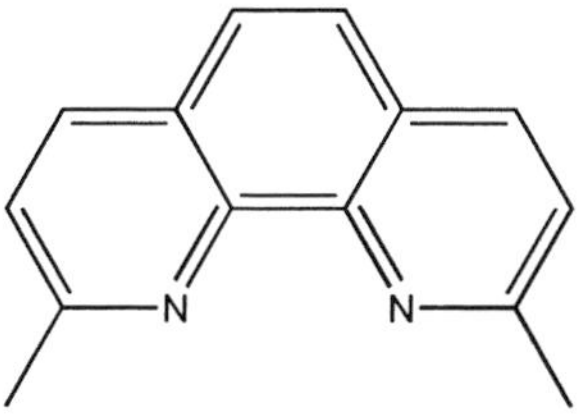

- Me_4phen = 3,4,7,8-tetramethyl-1,10-phenanthroline

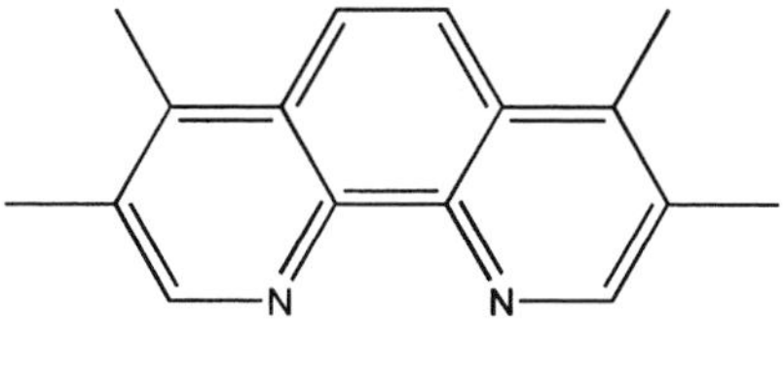

- Me$_6$tren = 2,2′,2″-tri(N,N-dimethylamino)triethylamine

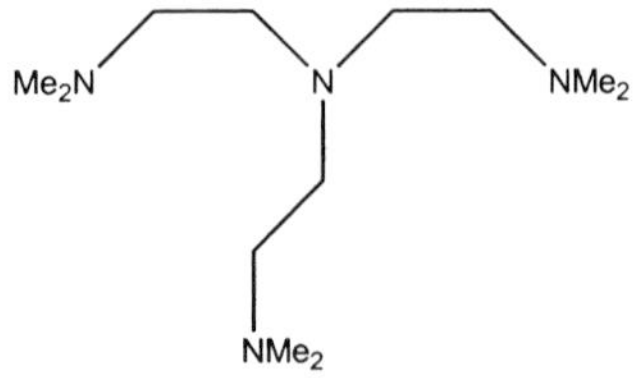

- mMal^{2-} = methylmalonate = $CH_3HC(COO)_2^{2-}$
- MMPP = methyl methylphenylphosphinate = OP(OMe)(Me)(Ph)
- NMA = N-methylacetamide = OC(Me)N(H)(Me)
- N-Mecyclen = N-methyl-1,4,7,10-tetraazacyclododecane
- NO$_2$phen = 5-nitro-1,10-phenanthroline

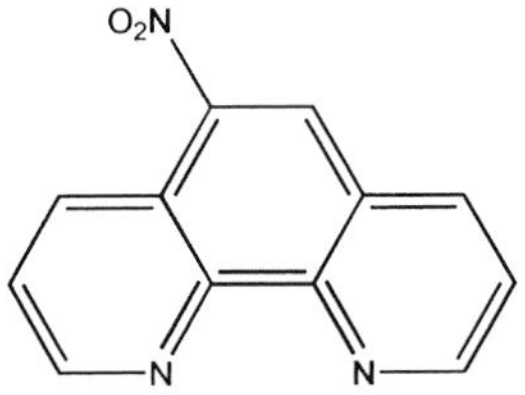

- ODDA^{2-} = 1,4,10,13-tetraoxa-7,16-diazacyclooctadecane-7,16-diacetate

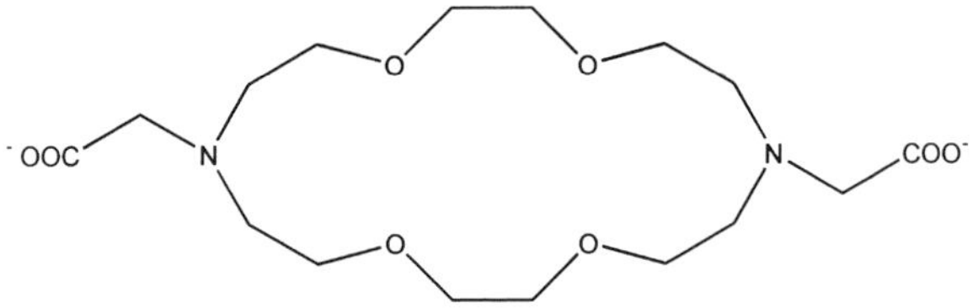

- pa = n-propylamine
- PDTA^{4-} = 1,3-propylenediaminetetraacetate

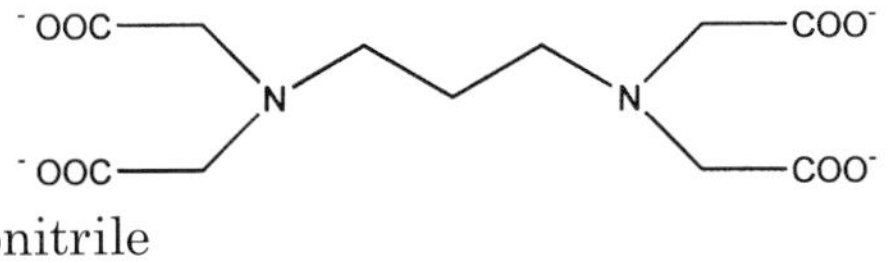

- PhCN = benzonitrile
- PhDTA^{4-} = o-phenylenediamine-N,N,N',N'-tetraacetate

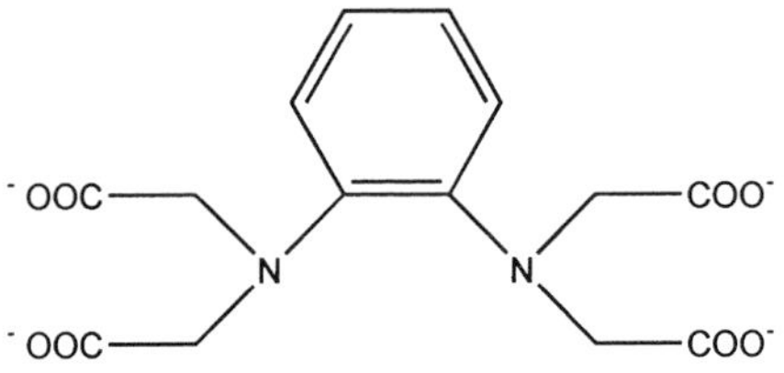

- phen = 1,10-phenanthroline

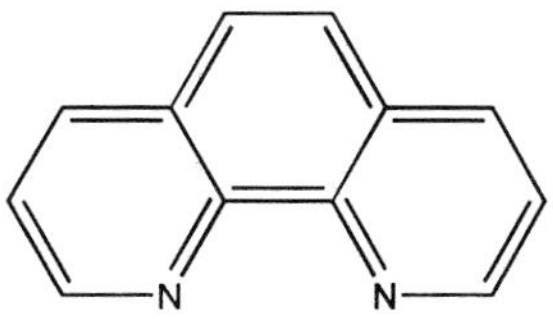

- Ph$_2$phen = 4,7-diphenyl-1,10-phenanthroline

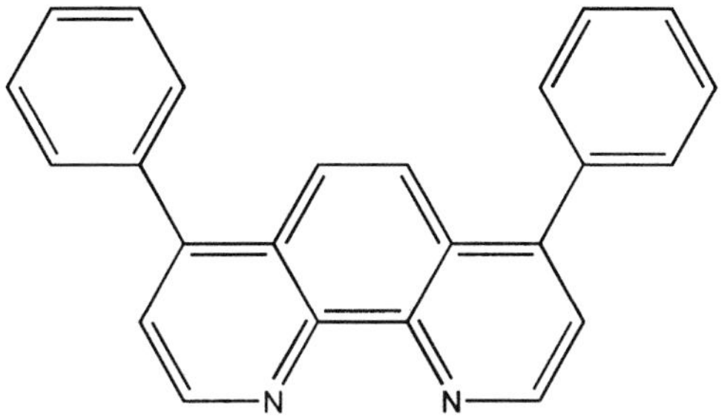

- Pr$_2^i$dim = N,N'-diisopropylethylenediimine

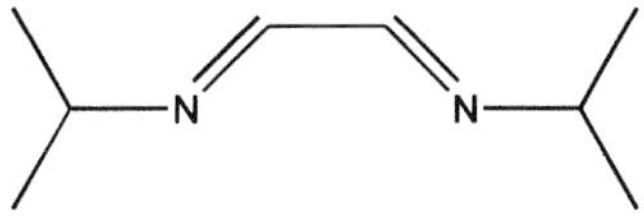

- PrCN = butyronitrile = NCCH$_2$CH$_2$CH$_3$
- Pr^{i}CN = isobutyronitrile

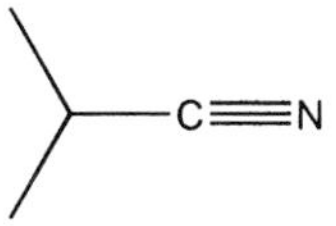

- py = pyridine
- pyMa = 3,7,11-tribenzyl-3,7,11,17-tetraazabicyclo[11.3.1] heptadeca-1(17),13,15-triene

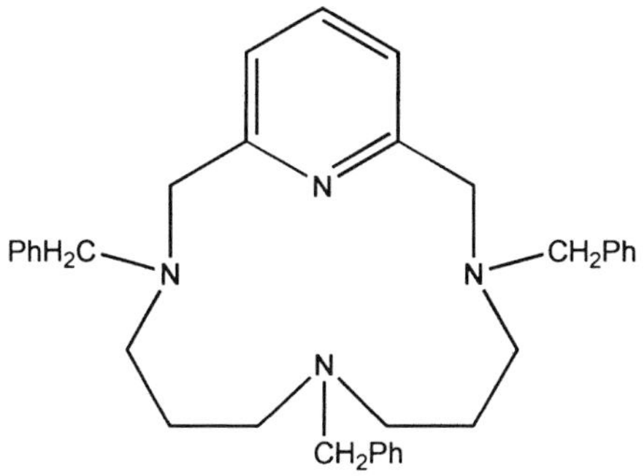

- Sal^{2-} = salicylate

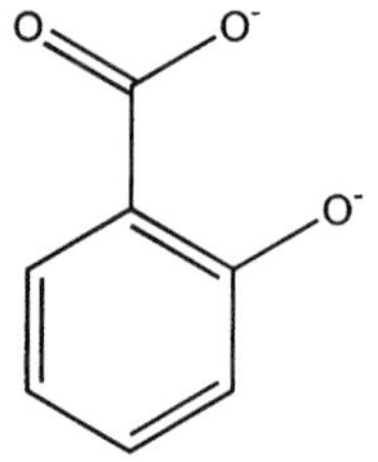

- $sSal^{2-}$ = sulfosalicylate

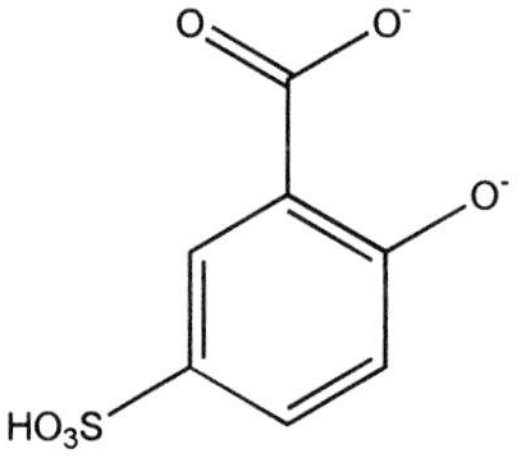

- TMP = trimethylphosphate = $OP(OMe)_3$
- tmpa = tris(2-pyridylmethyl)amine

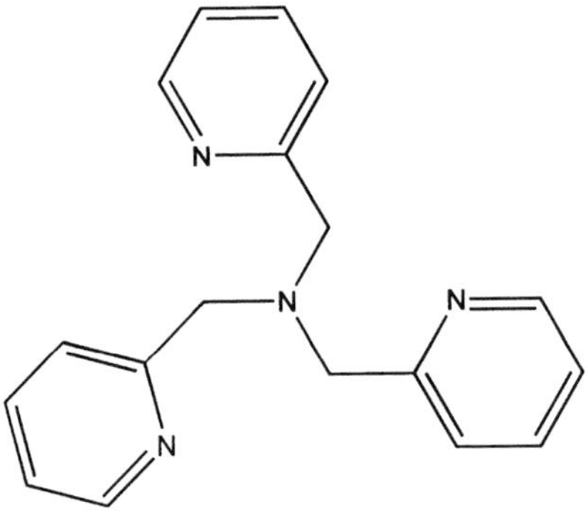

- tpmm = tri(2-pyridyl)methoxymethane

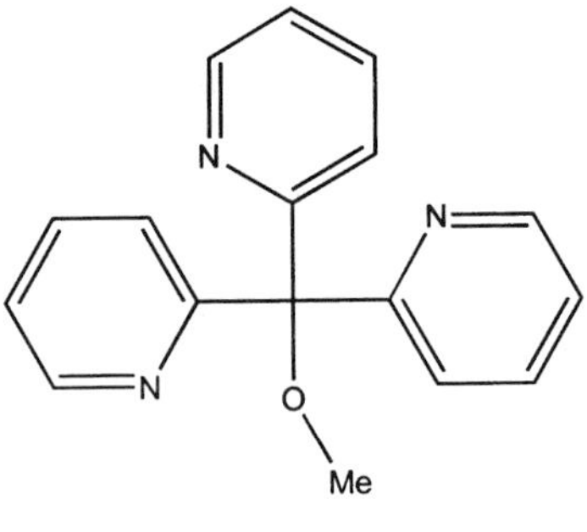

- Tp⁻ = hydridotris(pyrazolyl)borate

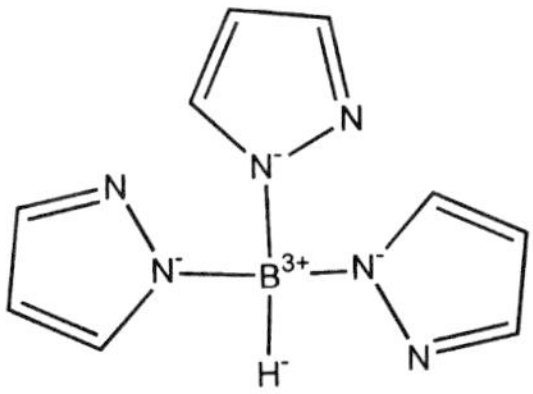

- TPPS = *meso*-tetrakis(*p*-sulfonatophenyl)porphine
- TMPyP = *meso*-tetrakis(*N*-methyl-4pyridyl)porphine
- TMPS = *meso*-tetrakis(sulfonatomesityl)porphine

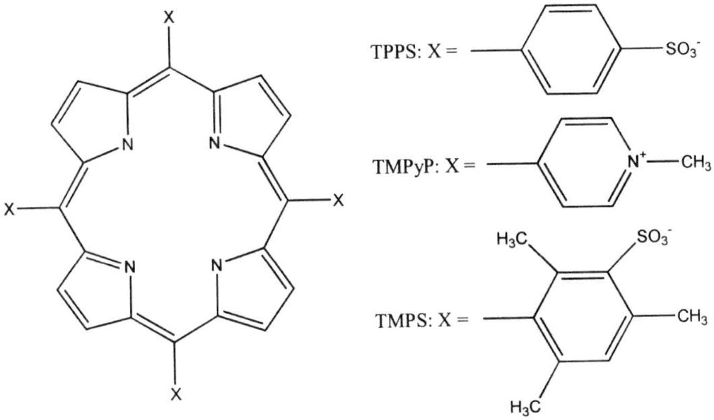

TPPS: X = —⟨ ⟩—SO_3^-

TMPyP: X = —⟨ ⟩—N^+—CH_3

TMPS: X =

- TMU = *N,N,N′,N′*-tetramethylurea = $OC(NMe_2)_2$
- tn = 1,3-propanediamine
- tpy = 2,2′,2″-terpyridine

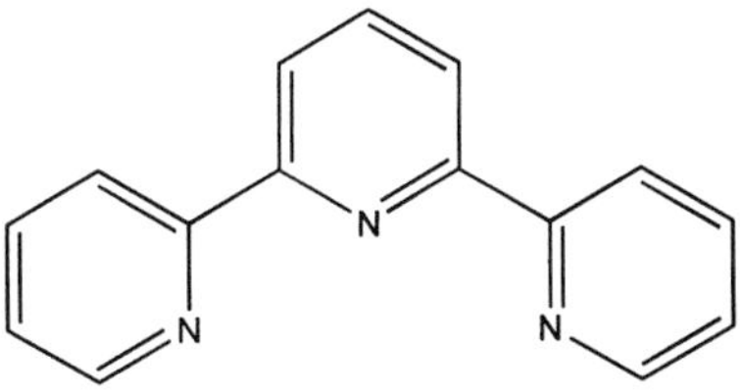

- tren = 2,2′,2″-triaminotriethylamine

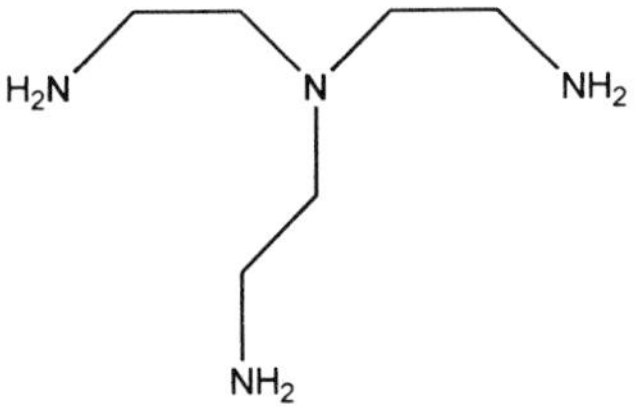

- trien = triethylentetramine

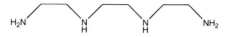

- tris(2-hydroxymethyl)-TREN-Me-2,3-HOPO = tris(2-hydroxymethyl)-tris[(3-hydroxy-1-methyl-2-oxo-1,2-didehydropyridine-4-carbo-xamido)ethyl]amine

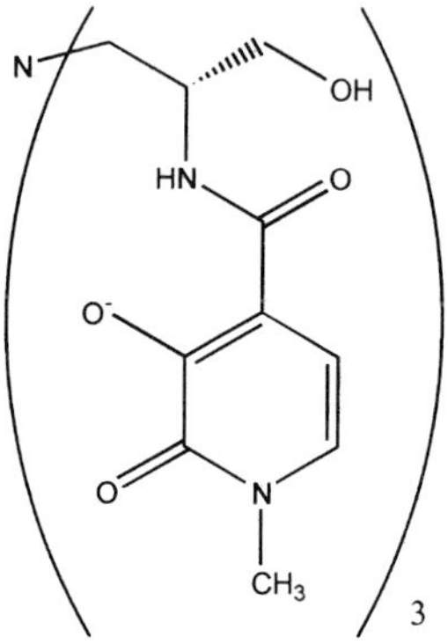

REFERENCES

1. Cusanelli, A.; Frey, U.; Richens, D. T.; Merbach, A. E. *J. Am. Chem. Soc.* **1996**, *118*, 5265–5271.
2. Caravan, P.; Tóth, É.; Rockenbauer, A.; Merbach, A. E. *J. Am. Chem. Soc.* **1999**, *121*, 10403–10409.
3. Ojo, J. F.; Olubuyide, O.; Oyetunji, O. *J. Chem. Soc., Dalton Trans.* **1987**, 957–959.
4. Eigen, M. *Pure Appl. Chem.* **1963**, *6*, 97–115.
5. Diebler, H.; Eigen, M.; Ilgenfritz, G.; Maas, G.; Winkler, R. *Pure Appl. Chem.* **1969**, *20*, 93–115.
6. Lincoln, S. F.; Merbach, A. E. *"Advances in Inorganic Chemistry"*, vol. 42; Ed. Sykes, A. G.; Academic Press: San Diego, 1995, pp. 1–88.
7. Drljaca, A.; Hubbard, C. D.; van Eldik, R.; Asano, T.; Basilevsky, M. V.; le Noble, W. *Chem. Rev.* **1998**, *98*, 2167–2289.
8. Helm, L.; Merbach, A. E. *J. Chem. Soc., Dalton Trans.* **2002**, 633–641.
9. Helm, L.; Merbach, A. E. *"High Pressure Chemistry"*; Eds. van Eldik, R.; Klärner, F.-G.; Wiley-VCH: Weinheim, 2002, pp. 131–160.
10. Helm, L.; Merbach, A. E. *Coord. Chem. Rev.* **1999**, *187*, 151–181.
11. Sandström, J. *"Dynamic NMR Spectroscopy"*; Academic Press: London, 1982.
12. Forsén, S.; Hoffman, R. A. *J. Chem. Phys.* **1963**, *39*, 2892–2901.
13. Led, J. J.; Gesmar, H. *J. Magn. Reson.* **1982**, *49*, 444–463.
14. Salmon, P. S.; Howells, W. S.; Mills, R. *J. Phys. C Solid State Phys.* **1987**, *20*, 5727–5747.
15. Herdman, G. J.; Salmon, P. S. *J. Am. Chem. Soc.* **1991**, *113*, 2930–2939.
16. Merbach, A. E. *Pure Appl. Chem.* **1982**, *54*, 1479–1493.

17. Merbach, A. E. *Pure Appl. Chem.* **1987**, *59*, 161–172.
18. Helm, L.; Merbach, A. E.; Powell, D. H. "*High-pressure Techniques in Chemistry and Physics – A Practical Approach*", vol. 2; Eds. Holzapfel, W. B.; Isaacs, N. S.; Oxford University Press: Oxford, 1997, pp. 187–216.
19. Langford, C. H.; Gray, H. B. "*Ligand Substitution Processes*"; W.A. Benjamin, Inc: New York, 1965.
20. van Eldik, R. "*Inorganic High Pressure Chemistry: Kinetics and Mechanisms*"; Elsevier: Amsterdam, 1986.
21. van Eldik, R. *Pure Appl. Chem.* **1992**, *64*, 1439–1448.
22. van Eldik, R.; Merbach, A. E. *Comments Inorg. Chem.* **1992**, *12*, 341–378.
23. Eckert, C. A. *Ann. Rev. Phys. Chem.* **1972**, *23*, 239–264.
24. van Eldik, R. "*Inorganic High Pressure Chemistry*"; Ed. van Eldik, R.; Elsevier: Amsterdam, 1986, pp. 1–68.
25. Salmon, P. S.; Lond, P. B. *Physica B* **1992**, *182*, 421–430.
26. Salmon, P. S. *Physica B* **1989**, *156&157*, 129–131.
27. Eigen, M.; Wilkins, R. G. *Adv. Chem. Ser.* **1965**, *49*, 55–67.
28. Shannon, R. D.; Prewitt, C. T. *Acta Crystallogr., Sect. B: Struct. Sci.* **1969**, *B25*, 925–946.
29. Shannon, R. D.; Prewitt, C. T. *Acta Crystallogr., Sect. B: Struct. Sci.* **1970**, *B26*, 1046–1048.
30. Pittet, P.-A.; Elbaze, G.; Helm, L.; Merbach, A. E. *Inorg. Chem.* **1990**, *29*, 1936–1942.
31. Ohtaki, H.; Radnai, T. *Chem. Rev.* **1993**, *93*, 1157–1204.
32. Lincoln, S. F.; Tkaczuk, M. *Ber. Bunsenges. Phys. Chem.* **1981**, *85*, 433–437.
33. Delpuech, J. J.; Peguy, A.; Rubini, P.; Steinmetz, J. *Nouv. J. Chim.* **1977**, *1*, 133–139.
34. Tkaczuk, M. N.; Lincoln, S. F. *Ber. Bunsenges. Phys. Chem.* **1982**, *86*, 147–153.
35. Lincoln, S. F.; Tkaczuk, M. N. *Ber. Bunsenges. Phys. Chem.* **1982**, *86*, 221–225.
36. Tkaczuk, M. N.; Lincoln, S. F. *Aust. J. Chem.* **1982**, *35*, 1555–1560.
37. Bleuzen, A.; Pittet, P.-A.; Helm, L.; Merbach, A. E. *Magn. Reson. Chem.* **1997**, *35*, 765–773.
38. Neely, J.; Connick, R. E. *J. Am. Chem. Soc.* **1970**, *92*, 3476–3478.
39. Nakamura, S.; Meiboom, S. *J. Am. Chem. Soc.* **1967**, *89*, 1765–1772.
40. Alger, T. D. *J. Am. Chem. Soc.* **1969**, *91*, 2220–2224.
41. Furrer, P.; Frey, U.; Helm, L.; Merbach, A. E. *High Pressure Research* **1991**, *7*, 144–146.
42. Pisaniello, D. L.; Lincoln, S. F. *Aust. J. Chem.* **1979**, *32*, 715–718.
43. Pisaniello, D. L.; Lincoln, S. F.; Williams, E. H. *Inorg. Chim. Acta* **1978**, *31*, 237–240.
44. Hugi Cleary, D.; Helm, L.; Merbach, A. E. *Helv. Chim. Acta* **1985**, *68*, 545–554.
45. Phillips, B. L.; Casey, W. H.; Crawford, S. N. *Geochim. Cosmochim. Acta* **1997**, *61*, 3041–3049.
46. Kowall, T.; Caravan, P.; Bourgeois, H.; Helm, L.; Rotzinger, F. P.; Merbach, A. E. *J. Am. Chem. Soc.* **1998**, *120*, 6569–6577.
47. Nordin, J. P.; Sullivan, D. J.; Phillips, B. L.; Casey, W. H. *Inorg. Chem.* **1998**, *37*, 4760–4763.
48. Ammann, C.; Moore, P.; Merbach, A. E.; Mc Ateer, C. H. *Helv. Chim. Acta* **1980**, *63*, 268–276.
49. Merbach, A. E.; Moore, P.; Howath, O. W.; McAteer, C. H. *Inorg. Chim. Acta* **1980**, *39*, 129–136.
50. Delpuech, J.-J.; Khaddar, M. R.; Peguy, A. A.; Rubini, P. R. *J. Am. Chem. Soc.* **1975**, *97*, 3373–3379.
51. Hugi Cleary, D.; Helm, L.; Merbach, A. E. *J. Am. Chem. Soc.* **1987**, *109*, 4444–4450.
52. Roderhüser, L.; Rubini, P.; Delpuech, J. J. *Inorg. Chem.* **1977**, *16*, 2837–2840.
53. Helm, L.; Ammann, C.; Merbach, A. E. *Z. Phys. Chem. NF* **1987**, *155*, 145–155.
54. Pisaniello, D. L.; Lincoln, S. F.; Williams, E. H. *J. Chem. Soc., Dalton Trans.* **1979**, 1473–1476.

55. Pisaniello, D. L.; Lincoln, S. F. *Inorg. Chim. Acta* **1979**, *36*, 85–88.
56. Pisaniello, D. L.; Lincoln, S. F. *J. Chem. Soc., Dalton Trans.* **1980**, 699–704.
57. Glass, G. E.; Schwabacher, W. B.; Tobias, S. R. *Inorg. Chem.* **1968**, *7*, 2471–2478.
58. Swaddle, T. W. *Adv. Inorg. Bioinorg. Mech.* **1983**, *2*, 97–138.
59. Swaddle, T. W. *Inorg. Chem.* **1983**, *22*, 2663–2665.
60. Swaddle, T. W.; Mak, M. K. S. *Can. J. Chem.* **1983**, *61*, 472–480.
61. Swaddle, T. W. *Comments Inorg. Chem.* **1991**, *12*, 237–258.
62. Yu, P.; Phillips, B. L.; Casey, W. H. *Inorg. Chem.* **2001**, *40*, 4750–4754.
63. Phillips, B. L.; Crawford, S. N.; Casey, W. H. *Geochim. Cosmochim. Acta* **1997**, *61*, 4965–4973.
64. Casey, W. H.; Phillips, B. L.; Nordin, J. P.; Sullivan, D. J. *Geochim. Cosmochim. Acta* **1998**, *62*, 2789–2797.
65. Yu, P.; Phillips, B. L.; Olmstead, M. M.; Casey, W. H. *J. Chem. Soc., Dalton Trans.* **2002**, 2119–2125.
66. Sullivan, D. J.; Nordin, J. P.; Phillips, B. L.; Casey, W. H. *Geochim. Cosmochim. Acta* **1999**, *63*, 1471–1480.
67. Spees, S. T.; Perumareddi, J. R.; Adamson, A. W. *J. Phys. Chem.* **1968**, *72*, 1822–1825.
68. Ducommun, Y.; Zbinden, D.; Merbach, A. E. *Helv. Chim. Acta* **1982**, *65*, 1385–1390.
69. Li, C.-D.; Jordan, R. B. *Inorg. Chem.* **1987**, *26*, 3855–3857.
70. Ducommun, Y.; Newman, K. E.; Merbach, A. E. *Inorg. Chem.* **1980**, *19*, 3696–3703.
71. Meyer, F. K.; Newman, K. E.; Merbach, A. E. *J. Am. Chem. Soc.* **1979**, *101*, 5588–5592.
72. Breivogel, F. *J. Chem. Phys.* **1969**, *51*, 445–448.
73. Cossy, C.; Helm, L.; Merbach, A. E. *Helv. Chim. Acta* **1987**, *70*, 1516–1525.
74. Ishii, M.; Funahashi, S.; Tanaka, M. *Chem. Lett.* **1987**, *1987*, 871–874.
75. Fielding, L.; Moore, P. *J. Chem. Soc., Chem. Commun.* **1988**, 49–52.
76. Sisley, M.; Yano, Y.; Swaddle, T. W. *Inorg. Chem.* **1982**, *21*, 1141–1145.
77. Inada, Y.; Sugata, T.; Ozutsumi, K.; Funahashi, S. *Inorg. Chem.* **1998**, *37*, 1886–1891.
78. Purcell, W. L.; Marianelli, R. S. *Inorg. Chem.* **1970**, *9*, 1724–1728.
79. Vigee, G.; Watkins, C.; Harris, M. *J. Inorg. Nucl. Chem.* **1980**, *42*, 1441–1445.
80. Ishii, M.; Funahashi, S.; Tanaka, M. *Inorg. Chem.* **1988**, *27*, 3192–3195.
81. Aizawa, S.-I.; Matsuda, K.; Tajima, T.; Maeda, M.; Sugata, T.; Funahashi, S. *Inorg. Chem.* **1995**, *34*, 2042–2047.
82. Aizawa, S.-I.; Iida, S.; Matsuda, K.; Funahashi, S. *Bull. Chem. Soc. Jpn.* **1997**, *70*, 1593–1597.
83. Luz, Z.; Meiboom, S. *J. Chem. Phys.* **1964**, *40*, 2686–2693.
84. Meyer, F. K.; Newman, K. E.; Merbach, A. E. *Inorg. Chem.* **1979**, *18*, 2142–2148.
85. Hioki, A.; Funahashi, M.; Ishii, M.; Tanaka, M. *Inorg. Chem.* **1986**, *25*, 1360–1364.
86. Monnerat, A.; Moore, P.; Newman, K. E.; Merbach, A. E. *Inorg. Chim. Acta* **1981**, *47*, 139–145.
87. Yano, Y.; Fairhurst, M.; Swaddle, T. W. *Inorg. Chem.* **1980**, *19*, 3267–3270.
88. Matwiyoff, N. A. *Inorg. Chem.* **1966**, *5*, 788–795.
89. McAteer, C. H.; Moore, P. *J. Chem. Soc., Dalton Trans.* **1983**, 353–357.
90. Inada, Y.; Itoh, Y.; Kuwabara, H.; Funahashi, S. *Inorg. React. Mech.* **2000**, *2*, 161–168.
91. Aizawa, S.-I.; Iida, S.; Matsuda, K.; Funahashi, S. *Inorg. Chem.* **1996**, *35*, 1338–1342.
92. Ishii, M.; Funahashi, S.; Ishihara, K.; Tanaka, M. *Bull. Chem. Soc. Jpn.* **1989**, *62*, 1852–1858.
93. Frankel, L. S. *Inorg. Chem.* **1971**, *10*, 2360–2361.
94. Babiec, J. S.; Langford, C. H.; Stengle, T. R. *Inorg. Chem.* **1966**, *5*, 1362–1364.
95. Elias, H.; Schumacher, R.; Schwamberger, J.; Wittekopf, T.; Helm, L.; Merbach, A. E.; Ulrich, S. *Inorg. Chem.* **2000**, *39*, 1721–1727.

96. Pasquarello, A.; Petri, I.; Salmon, P. S.; Parisel, O.; Car, R.; Tóth, É.; Powell, D. H.; Fisher, H. E.; Helm, L.; Merbach, A. E. *Science* **2001**, *291*, 856.

97. Powell, D. H.; Helm, L.; Merbach, A. E. *J. Chem. Phys.* **1991**, *95*, 9258–9265.

98. Helm, L.; Lincoln, S. F.; Merbach, A. E.; Zbinden, D. *Inorg. Chem.* **1986**, *25*, 2550–2552.

99. Powell, D. H.; Furrer, P.; Pittet, P.; Merbach, A. E. *J. Phys. Chem.* **1995**, *99*, 16 622–16 629.

100. Inada, Y.; Ozutsumi, K.; Funahashi, S.; Soyama, S.; Kawashima, T.; Tanaka, M. *Inorg. Chem.* **1993**, *32*, 3010–3014.

101. Ohtaki, H. *Yoeki no Kagaku (Chem. of Sol.) In Japanese*; Dai-Nippon Tosho: Tokyo, 1987.

102. Rotzinger, F. P. *J. Am. Chem. Soc.* **1996**, *118*, 6760–6766.

103. Rotzinger, F. P. *J. Am. Chem. Soc.* **1997**, *119*, 5230–5238.

104. Rotzinger, F. P. *Helv. Chim. Acta.* **2000**, *83*, 3006–3020.

105. Hartmann, M.; Clark, T.; van Eldik, R. *J. Am. Chem. Soc.* **1997**, *119*, 7843–7850.

106. Magini, M.; Licheri, G.; Paschina, G.; Piccaluga, G.; Pinna, G. *"X-Ray Diffraction of Ions in Aqueous Solutions: Hydration and Complex Formation"*; CRC Press: Boca Raton, 1988.

107. Magini, M. *Inorg. Chem.* **1982**, *21*, 1535–1538.

108. Salmon, P. S.; Neilson, G. W.; Enderby, J. E. *J. Phys. C Solid State Phys.* **1988**, *21*, 1335–1349.

109. Benfatto, M.; D'Angelo, P.; Della Longa, S.; Pavel, N. V. *Phys. Rev. B* **2002**, *65*, 174205/174201–174205/174205.

110. Persson, I.; Persson, P.; Sandström, M.; Ullström, A.-S. *J. Chem. Soc., Dalton Trans.* **2002**, 1256–1265.

111. Hugi, A. D.; Helm, L.; Merbach, A. E. *Inorg. Chem.* **1987**, *26*, 1763–1768.

112. Dellavia, I.; Helm, L.; Merbach, A. E. *Inorg. Chem.* **1992**, *31*, 2230–2233.

113. Hugi, A. D.; Helm, L.; Merbach, A. E. *Helv. Chim. Acta* **1985**, *68*, 508–521.

114. Dellavia, I.; Sauvageat, P.; Helm, L.; Ducommun, Y.; Merbach, A. E. *Inorg. Chem.* **1992**, *31*, 792–797.

115. Xu, F. C.; Krouse, H. R.; Swaddle, T. W. *Inorg. Chem.* **1985**, *24*, 267–270.

116. Bleuzen, A.; Foglia, F.; Furet, E.; Helm, L.; Merbach, A. E.; Weber, J. *J. Am. Chem. Soc.* **1996**, *118*, 12 777–12 787.

117. Carle, D. L.; Swaddle, T. W. *Can. J. Chem.* **1973**, *51*, 3795–3798.

118. Lo, S.; Swaddle, T. W. *Inorg. Chem.* **1975**, *14*, 1878–1881.

119. Swaddle, T. W.; Merbach, A. E. *Inorg. Chem.* **1981**, *20*, 4212–4216.

120. Grant, M. W.; Jordan, R. B. *Inorg. Chem.* **1981**, *20*, 55–60.

121. Meyer, F. K.; Monnerat, A.; Newman, K. E.; Merbach, A. E. *Inorg. Chem.* **1982**, *21*, 774–778.

122. Bertini, I.; Fragai, M.; Luchinat, C.; Parigi, G. *Inorg. Chem.* **2001**, *40*, 4030–4035.

123. Kowall, T.; Foglia, F.; Helm, L.; Merbach, A. E. *Chem. Eur. J.* **1996**, *2*, 285–294.

124. Hartmann, M.; Clark, T.; van Eldik, R. *J. Phys. Chem. A* **1999**, *103*, 9899–9905.

125. Rapaport, I.; Helm, L.; Merbach, A. E.; Bernhard, P.; Ludi, A. *Inorg. Chem.* **1988**, *27*, 873–879.

126. Pittet, P.-A.; Dadci, L.; Zbinden, P.; Abou-Hamdan, A.; Merbach, A. E. *Inorg. Chim. Acta* **1993**, *206*, 135–140.

127. Døssing, A.; van Lelieveld, A. *Inorg. Chim. Acta* **2001**, *322*, 130–132.

128. Laurenczy, G.; Rapaport, I.; Zbinden, D.; Merbach, A. E. *Magn. Reson. Chem.* **1991**, *29*, S45–S51.

129. Cusanelli, A.; Frey, U.; Richens, D. T.; Merbach, A. E. *J. Am. Chem. Soc.* **1996**, *118*, 5265–5271.

130. Kallen, T. W.; Earley, F. *Inorg. Chem.* **1971**, *10*, 1149–1151.

131. Aebischer, N.; Laurenczy, G.; Ludi, A.; Merbach, A. E. *Inorg. Chem.* **1993**, *32*, 2810–2814.

132. De Vito, D.; Sidorenkova, E.; Rotzinger, F. P.; Weber, J.; Merbach, A. E. *Inorg. Chem.* **2000**, *39*, 5547–5552.

133. Bernhard, P.; Helm, L.; Ludi, A.; Merbach, A. E. *J. Am. Chem. Soc.* **1985**, *107*, 312–317.

134. Dellavia, I.; Helm, L.; Merbach, A. E. *Inorg. Chem.* **1992**, *31*, 4151–4154.

135. Wüthrich, K.; Connick, R. E. *Inorg. Chem.* **1967**, *6*, 583–590.

136. Kuroiwa, Y.; Harada, M.; Tomiyasu, H.; Fukutomi, H. *Inorg. Chim. Acta* **1988**, *146*, 7–8.

137. Kim, J.-S.; Matsuda, T.; Harada, M.; Kim, S.-Y.; Park, Y.-Y.; Tomiyasu, H.; Ikeda, Y. *Inorg. Chim. Acta* **1999**, *294*, 119–122.

138. Angerman, N. S.; Jordan, R. B. *Inorg. Chem.* **1969**, *8*, 1824–1833.

139. Angerman, N. S.; Jordan, R. B. *Inorg. Chem.* **1969**, *8*, 65–69.

140. Kaya, K.; Ikeda, Y.; Fukutomi, H. *Bull. Chem. Soc. Jap.* **1985**, 58.

141. Harada, M.; Ikeda, Y.; Tomiyasu, H.; Fukutomi, H. *Chem. Lett.* **1984**, 1195–1196.

142. Jordan, R. B.; Angerman, N. S. *J. Chem. Phys.* **1968**, *48*, 3983–3988.

143. Miller, G. A.; McClung, R. E. D. *J. Chem. Phys.* **1973**, *58*, 4358–4367.

144. Murmann, R. K. *Inorg. Chim. Acta* **1977**, *25*, L43–L44.

145. Comba, P.; Merbach, A. E. *Inorg. Chem.* **1987**, *26*, 1315–1323.

146. Aime, S.; Anelli, P. L.; Botta, M.; Brocchetta, M.; Canton, S.; Fedeli, F.; Gianolio, E.; Terreno, E. *J. Biol. Inorg. Chem.* **2002**, *7*, 58–67.

147. Wang, K.; Jordan, R. B. *Inorg. Chem.* **1995**, *34*, 5672–5679.

148. Desai, A.; Dodgen, H. W.; Hunt, J. P. *J. Am. Chem. Soc.* **1970**, *92*, 798–801.

149. Desai, A.; Dodgen, H. W.; Hunt, J. P. *J. Am. Chem. Soc.* **1969**, *91*, 5001–5004.

150. Hunt, J. P. *Coord. Chem. Rev.* **1971**, *7*, 1–10.

151. Rablen, D. P.; Dodgen, H. W.; Hunt, J. P. *J. Am. Chem. Soc.* **1972**, *94*, 1771–1772.

152. Coates, J.; Hadi, D.; Lincoln, S. F.; Dodgen, H. W.; Hunt, J. P. *Inorg. Chem.* **1981**, *20*, 707–711.

153. Moore, P. *Pure Appl. Chem.* **1985**, *57*, 347–354.

154. Grant, M.; Dodgen, H. W.; Hunt, J. P. *J. Am. Chem. Soc.* **1970**, *92*, 2321–2323.

155. Rablen, D. P.; Gordon, G. *Inorg. Chem.* **1969**, *8*, 395–397.

156. Lincoln, S. F.; Aprile, F.; Dodgen, H. W.; Hunt, J. P. *Inorg. Chem.* **1968**, *7*, 929–932.

157. Jordan, R. B.; Dodgen, H. W.; Hunt, J. P. *Inorg. Chem.* **1966**, *5*, 1906–1909.

158. Grant, M.; Dodgen, H. W.; Hunt, J. P. *J. Chem. Soc., Chem. Commun.* **1970**, 1446–1447.

159. Fielding, L.; Moore, P. *J. Chem. Soc., Dalton Trans.* **1989**, 873–877.

160. Neubrand, A.; Thaler, F.; Körner, M.; Zahl, A.; Hubbard, C. D.; van Eldik, R. *J. Chem. Soc., Dalton Trans.* **2002**, 957–961.

161. Powell, D. H.; Merbach, A. E.; Fábián, I.; Schindler, S.; van Eldik, R. *Inorg. Chem.* **1994**, *33*, 4468–4473.

162. Priimov, G. U.; Moore, P.; Helm, L.; Merbach, A. E. *Inorg. React. Mech.* **2001**, *3*, 1–23.

163. West, R.; Lincoln, S. F. *J. Chem. Soc., Dalton Trans.* **1974**, 281–284.

164. Powell, D. H.; Furrer, P.; Pittet, P.-A.; Merbach, A. E. *J. Phys. Chem.* **1995**, *45*, 16 622–16 629.

165. Lincoln, S. F.; Coates, J. H.; Doddridge, B. G.; Hounslow, A. M.; Pisaniello, D. L. *Inorg. Chem.* **1983**, *22*, 2869–2872.

166. Lincoln, S. F.; Hounslow, A. M.; Pisaniello, D. L.; Doddridge, B. G.; Coates, J.; Merbach, A. E.; Zbinden, D. *Inorg. Chem.* **1984**, *23*, 1090–1093.

167. Swaddle, T. W.; Stranks, D. R. *J. Am. Chem. Soc.* **1972**, *94*, 8357–8360.

168. González, G.; Moullet, B.; Martinez, M.; Merbach, A. E. *Inorg. Chem.* **1994**, *33*, 2330–2333.

169. Crimp, S. J.; Spiccia, L.; Krouse, H. R.; Swaddle, T. W. *Inorg. Chem.* **1994**, *33*, 465–470.
170. Schneppensieper, T.; Seibig, S.; Zahl, A.; Tregloan, P.; van Eldik, R. *Inorg. Chem.* **2001**, *40*, 3670–3676.
171. Mizuno, M.; Funahashi, S.; Nakasuka, N.; Tanaka, M. *J. Am. Chem. Soc.* **1991**, *30*, 1550–1553.
172. Schneppensieper, T.; Zahl, A.; van Eldik, R. *Angew. Chem. Int. Ed.* **2001**, *40*, 1678–1680.
173. Ostrich, I.; Liu, G.; Dodgen, H. W.; Hunt, J. P. *Inorg. Chem.* **1980**, *19*, 619–621.
174. Mizuno, M.; Funahashi, M.; Nakasuka, N.; Tanaka, M. *Bull. Chem. Soc. Jap.* **1991**, *30*, 1988–1990.
175. Kruse, W.; Taube, H. *J. Am. Chem. Soc.* **1961**, *83*, 1280–1284.
176. Brasch, N. E.; Buckingham, D. A.; Clark, C. R.; Rogers, A. J. *Inorg. Chem.* **1998**, *37*, 4865–4871.
177. Hunt, H. R.; Taube, H. *J. Am. Chem. Soc.* **1958**, *80*, 2642–2646.
178. Martin, D. F.; Tobe, M. L. *J. Chem. Soc.* **1962**, 1388–1396.
179. Buckingham, D. A.; Clark, C. R.; Rogers, A. J.; Simpson, J. *Inorg. Chem.* **1998**, *37*, 3497–3504.
180. Bradley, S. M.; Doine, H.; Krouse, H. R.; Sisley, H. J.; Swaddle, T. W. *Aust. J. Chem.* **1988**, *41*, 1323–1329.
181. Dadci, L.; Elias, H.; Frey, U.; Hörnig, A.; Koelle, U.; Merbach, A. E.; Paulus, H.; Schneider, J. S. *Inorg. Chem.* **1995**, *34*, 306–315.
182. Röthlisberger-Stebler, M.; Hummel, W.; Pittet, P.-A.; Bürgi, H. B.; Ludi, A.; Merbach, A. E. *Inorg. Chem.* **1988**, *27*, 1358–1363.
183. Huynh, M. H. V.; Lasker, J. M.; Wetzler, M.; Mort, B.; Szczepura, L. F.; Witham, L. M.; Cintron, J. M.; Marschilok, A. C.; Ackerman, L. J.; Castellano, R. K.; Jameson, D. L.; Churchill, M. R.; Jircitano, A. J.; Takeuchi, K. J. *J. Am. Chem. Soc.* **2001**, *123*, 8780–8784.
184. Meier, U. C.; Scopelliti, R.; Solari, E.; Merbach, A. E. *Inorg. Chem.* **2000**, *39*, 3816–3822.
185. Aebischer, N.; Sidorenkova, H.; Ravera, M.; Laurenczy, G.; Osella, D.; Weber, J.; Merbach, A. E. *Inorg. Chem.* **1997**, *36*, 6009–6020.
186. Frey, U.; Li, Z.-W.; Matras, A. *Inorg. Chem.* **1996**, *35*, 981–984.
187. Luginbühl-Brand, W.; Zbinden, P.; Pittet, P.-A.; Armbruster, T.; Bürgi, H. B.; Merbach, A. E.; Ludi, A. *Inorg. Chem.* **1991**, *30*, 2350–2355.
188. Rüba, E.; Simanko, W.; Mereiter, K.; Schmid, R.; Kirchner, K. *Inorg. Chem.* **2000**, *39*, 382–384.
189. Doine, H.; Ishihara, K.; Krouse, H. R.; Swaddle, T. W. *Inorg. Chem.* **1987**, *26*, 3240–3242.
190. Drljaca, A.; Spiccia, L.; Krouse, H. R.; Swaddle, T. W. *Inorg. Chem.* **1996**, *35*, 985–990.
191. Drljaca, A.; Zahl, A.; van Eldik, R. *Inorg. Chem.* **1998**, *37*, 3948–3953.
192. Cusanelli, A.; Nicula-Dadci, L.; Frey, U.; Merbach, A. E. *Inorg. Chem.* **1997**, *36*, 2211–2217.
193. Tong, S. B.; Swaddle, T. W. *Inorg. Chem.* **1974**, *13*, 1538–1539.
194. Brønnum, B.; Johansen, H. S.; Skibsted, L. H. *Acta Chem. Scand.* **1989**, *43*, 975–980.
195. Hallinan, N.; Besançon, V.; Forster, M.; Elbaze, G.; Ducommun, Y.; Merbach, A. E. *Inorg. Chem.* **1991**, *30*, 1112–1114.
196. Helm, L.; Elding, L. I.; Merbach, A. E. *Helv. Chim. Acta* **1984**, *67*, 1453–1460.
197. Monlien, F. J.; Helm, L.; Abou-Hamdan, A.; Merbach, A. E. *Inorg. Chem.* **2002**, *41*, 1717–1727.
198. Frey, U.; Elmroth, S.; Moullet, B.; Elding, L. I.; Merbach, A. E. *Inorg. Chem.* **1991**, *30*, 5033–5037.
199. Brønnum, B.; Johansen, H. S.; Skibsted, L. H. *Inorg. Chem.* **1992**, *31*, 3023–3025.
200. Helm, L.; Elding, L. I.; Merbach, A. E. *Inorg. Chem.* **1985**, *24*, 1719–1721.

201. Dunham, S. U.; Abbott, E. H. *Inorg. Chim. Acta* **2000**, *297*, 72–78.
202. Ducommun, Y.; Helm, L.; Merbach, A. E.; Hellquist, B.; Elding, L. I. *Inorg. Chem.* **1989**, *28*, 377–379.
203. Monlien, F. J.; Helm, L.; Abou-Hamdan, A.; Merbach, A. E. *Inorg. Chim. Acta* **2002**, *331*, 257–269.
204. Cross, R. J. *Adv. Inorg. Chem.* **1989**, *34*, 219–292.
205. Romeo, R.; Grassi, A.; Monsu Scolaro, L. *Inorg. Chem.* **1992**, *31*, 4383–4390.
206. Frey, U.; Helm, L.; Merbach, A. E.; Romeo, R. *J. Am. Chem. Soc.* **1989**, *111*, 8161–8165.
207. Berger, J.; Kotowski, M.; van Eldik, R.; Helm, L.; Merbach, A. E.; Frey, U. *Inorg. Chem.* **1989**, *28*, 3759–3765.
208. Helm, L.; Merbach, A. E.; Kotowski, M.; van Eldik, R. *High Pressure Research* **1989**, *2*, 49–55.
209. Frey, U.; Grove, D. M.; van Koten, G. *Inorg. Chim. Acta* **1998**, *269*, 322–325.
210. Aebischer, N.; Churlaud, R.; Dolci, L.; Frey, U.; Merbach, A. E. *Inorg. Chem.* **1998**, *37*, 5915–5924.
211. Romeo, R.; Scolaro, L. M.; Nastasi, N.; Arena, G. *Inorg. Chem.* **1996**, *35*, 5087–5096.
212. Deeth, R. J.; Elding, L. I. *Inorg. Chem.* **1996**, *35*, 5019–5026.
213. Shriver, D. F.; Atkins, P. W.; Langford, C. H. *"Inorganic Chemistry"*; Oxford University Press: Oxford, 1990.
214. Habenschuss, A.; Spedding, F. H. *J. Chem. Phys.* **1979**, *70*, 2797–2806.
215. Habenschuss, A.; Spedding, F. H. *J. Chem. Phys.* **1979**, *70*, 3758–3763.
216. Habenschuss, A.; Spedding, F. H. *J. Chem. Phys.* **1980**, *73*, 442–450.
217. Yamaguchi, T.; Nomura, M.; Wakita, H.; Ohtaki, H. *J. Chem. Phys.* **1988**, *89*, 5153–5159.
218. Cossy, C.; Merbach, A. E. *Pure Appl. Chem.* **1988**, *60*, 1785–1796.
219. Cossy, C.; Barnes, A. C.; Enderby, J. E.; Merbach, A. E. *J. Chem. Phys.* **1989**, *90*, 3254–3260.
220. Helm, L.; Merbach, A. E. *Eur. J. Solid State Inorg. Chem.* **1991**, *28*, 245–250.
221. Cossy, C.; Helm, L.; Powell, D. H.; Merbach, A. E. *New J. Chem.* **1995**, *19*, 27–35.
222. Spedding, F. H.; Shiers, L. E.; Brown, M. A.; Derer, J. L.; Swanson, D. L.; Habenschuss, A. *J. Chem. Eng. Data* **1975**, *20*, 81–88.
223. Spedding, F. H.; Cullen, P. F.; Habenschuss, A. *J. Phys. Chem.* **1974**, *78*, 1106–1110.
224. Cossy, C.; Helm, L.; Merbach, A. E. *Inorg. Chem.* **1988**, *27*, 1973–1979.
225. Cossy, C.; Helm, L.; Merbach, A. E. *Inorg. Chem.* **1989**, *28*, 2699–2703.
226. Micskei, K.; Powell, D. H.; Helm, L.; Brücher, E.; Merbach, A. E. *Magn. Reson. Chem.* **1993**, *31*, 1011–1020.
227. Caravan, P.; Merbach, A. E. *J. Chem. Soc., Chem. Commun.* **1997**, 2147–2148.
228. Moreau, G.; Helm, L.; Purans, J.; Merbach, A. E. *J. Phys. Chem. A* **2002**, *106*, 3034–3043.
229. Fay, D. P.; Litchinsky, D.; Purdie, N. *J. Phys. Chem.* **1969**, *73*, 544–552.
230. Powell, D. H.; Merbach, A. E. *Magn. Reson. Chem.* **1994**, *32*, 739–745.
231. Caravan, P.; Ellison, J. J.; McMurry, T. J.; Lauffer, R. B. *Chem. Rev.* **1999**, *99*, 2293–2352.
232. *"The Chemistry of Contrast Agents in Medical Magnetic Resonance Imaging"*, 1st edn.; Ed. Merbach, A. E.; Tóth, É.; John Wiley & Sons: Chichester, 2001.
233. Tóth, É.; Helm, L.; Merbach, A. E. *"Magnetic Resonance Contrast Agents"*; Ed. Krause, W.; Springer: Heidelberg, 2001.
234. Hajela, S.; Botta, M.; Giraudo, S.; Xu, J.; Raymond, K. N.; Aime, S. *J. Am. Chem. Soc.* **2000**, *122*, 11 228–11 229.
235. Yerly, F.; Dunand, F. A.; Tóth, É.; Figueirinha, A.; Kovacs, Z.; Sherry, A. D.; Geraldes, C. F. G. C.; Merbach, A. E. *Eur. J. Inorg. Chem.* **2000**, *5*, 1001–1006.
236. Tóth, É.; Ni Dhubhghaill, O. M.; Besson, G.; Helm, L.; Merbach, A. E. *Magn. Reson. Chem.* **1999**, *37*, 701–708.
237. Micskei, K.; Helm, L.; Brücher, E.; Merbach, A. E. *Inorg. Chem.* **1993**, *32*, 3844–3850.

238. Dunand, F. A.; Borel, A.; Merbach, A. E. *J. Am. Chem. Soc.* **2002**, *124*, 710–716.
239. Dunand, F. A.; Borel, A.; Helm, L. *Inorg. Chem. Commun.* **2002**, *5*, 811–815.
240. Gonzalez, G.; Powell, D. H.; Tissieres, V.; Merbach, A. E. *J. Phys. Chem.* **1994**, *98*, 53–59.
241. Woods, M.; Aime, S.; Botta, M.; Howard, J. A. K.; Moloney, J. M.; Navet, M.; Parker, D.; Port, M.; Rousseaux, O. *J. Am. Chem. Soc.* **2000**, *122*, 9781–9792.
242. Zhang, S.; Kovacs, Z.; Burgess, S.; Aime, S.; Terreno, E.; Sherry, A. D. *Chem. Eur. J.* **2001**, *7*, 288–296.
243. Aime, S.; Barge, A.; Bruce, J. I.; Botta, M.; Howard, J. A. K.; Moloney, J. M.; Parker, D.; De Sousa, A. S.; Woods, M. *J. Am. Chem. Soc.* **1999**, *121*, 5762–5771.
244. Dunand, F. A.; Aime, S.; Merbach, A. E. *J. Am. Chem. Soc.* **2000**, *122*, 1506–1512.
245. Caravan, P.; Merbach, A. E. *J. Chem. Soc., Chem. Commun.* **1997**, 2147–2148.
246. Burai, L.; Tóth, É.; Seibig, S.; Scopelliti, R.; Merbach, A. E. *Chem. Eur. J.* **2000**, *6*, 3761–3770.
247. Seibig, S.; Tóth, E.; Merbach, A. E. *J. Am. Chem. Soc.* **2002**, *122*, 5822–5830.
248. Burai, L. Private Communication.
249. Burai, L.; Scopelliti, R.; Tóth, É. *J. Chem. Commun.*, **2002**, 2366–2367.
250. Tóth, É.; Helm, L.; Merbach, A. E. "*The Chemistry of Contrast Agents in Medical Magnetic Resonance Imaging*", 1st edn.; Ed. Merbach, A. E.; Tóth, É.; John Wiley & Sons: Chichester, 2001, pp. 45–119.
251. Aime, S.; Botta, M.; Ermondi, G. *Inorg. Chem.* **1992**, *31*, 4291–4299.
252. Spirlet, M.; Rebizant, J.; Desreux, J. F.; Loncin, M. *Inorg. Chem.* **1995**, *23*, 359–363.
253. Marques, M. P. M.; Geraldes, C.; Sherry, A. D.; Merbach, A. E.; Powell, D. H.; Pubanz, D.; Aime, S.; Botta, M. J. *Alloys and Compounds* **1995**, *225*, 303–307.
254. Pisaniello, D. L.; Helm, L.; Meier, P. F.; Merbach, A. E. *J. Am. Chem. Soc.* **1983**, *105*, 4528–4536.
255. Shannon, R. D. *Acta Crystallogr., Sect. A: Found. Crystallogr.* **1976**, *32*, 751–767.
256. Pisaniello, D. L.; Helm, L.; Zbinden, D.; Merbach, A. E. *Helv. Chim. Acta* **1983**, *66*, 1872–1875.
257. Lincoln, S. F.; White, A. H. *Inorg. Chim. Acta* **1990**, *168*, 265–270.
258. Pisaniello, D. L.; Lincoln, S. F.; Williams, E. H.; Jones, A. J. *Aust. J. Chem.* **1981**, *34*, 495–500.
259. Lincoln, S. F.; Hounslow, A. M.; Jones, A. J. *Aust. J. Chem.* **1982**, *35*, 2393–2398.
260. Tóth, É.; Burai, L.; Merbach, A. E. *Coord. Chem. Rev.* **2001**, *216–217*, 363–382.
261. Moreau, G.; Burai, L.; Helm, L.; Purans, J.; Merbach, A. E. *J. Phys. Chem. A* **2003**, *107*, 758–769.
262. Harrowfield, J. M.; Kepert, D. L.; Patrick, J. M.; White, A. H.; Lincoln, S. F. *J. Chem. Soc., Dalton Trans.* **1983**, 393–396.
263. Alcock, N. W.; Esperas, S. *J. Chem. Soc., Dalton Trans.* **1977**, 893–896.
264. Nassimbeni, L. R.; Rodgers, A. L. *Cryst. Struct. Commun.* **1976**, *5*, 301–308.
265. Farkas, I.; Grenthe, I.; Bányai, I. *J. Phys. Chem. A* **2000**, *104*, 1201–1206.
266. Yang, T.; Tsushima, S.; Suzuki, A. *J. Phys. Chem. A* **2001**, *105*, 10 439–10 445.
267. Szabó, Z.; Glaser, J.; Grenthe, I. *Inorg. Chem.* **1996**, *35*, 2036–2044.
268. Farkas, I.; Bányai, I.; Szabó, Z.; Wahlgren, U.; Grenthe, I. *Inorg. Chem.* **2000**, *39*, 799–805.
269. Bardin, N.; Rubini, P.; Madic, C. *Radiochim. Acta* **1998**, *83*, 189–194.
270. Vallet, V.; Wahlgren, U.; Schimmelpfennig, B.; Szabó, Z.; Grenthe, I. *J. Am. Chem. Soc.* **2001**, *123*, 11 999–12 008.
271. Moll, H.; Denecke, M. A.; Jalilehvand, F.; Sandström, M.; Grenthe, I. *Inorg. Chem.* **1999**, *38*, 1795–1799.

LIGAND SUBSTITUTION REACTIONS

JOHN BURGESS[a] and COLIN D. HUBBARD[b]

[a]Department of Chemistry, University of Leicester, Leicester LE1 7RH, UK
[b]Institute for Inorganic Chemistry, University of Erlangen-Nürnberg,
91058 Erlangen, Germany

ADVANCES IN INORGANIC CHEMISTRY
VOLUME 54 ISSN 0898-8838

I. Introduction

In this review we attempt to give an overview of recent progress in the area of kinetics and mechanisms of ligand substitution. We have of necessity had to be very selective (and have, of course, been subjective in our choices for inclusion – they reflect the authors' interests, perforce) – the Royal Society of Chemistry's Annual Reports list several hundred relevant references each year, and thus provide a splendidly comprehensive view of progress (1). We have concentrated on publications which appeared in 2000 and, especially, 2001 and the early part of 2002, but have included a sprinkling of earlier papers, back to the much regretted demise of the series of volumes containing regular surveys devoted to inorganic kinetics and mechanisms (2). Although our emphasis is on classical complexes[1] we have included a selection of bioinorganic reactions, reflecting the growing importance of this bridging area, a few references of relevance to inorganic pharmacology (1999 saw the publication of a complete issue of *Chemical Reviews* devoted to this subject (3)), and a very few organometallic systems where these seem particularly appropriate to our main discussion.

Mechanisms of ligand substitution have been reviewed in several articles, ranging from comprehensive accounts (4) to a succinct retrospective view of substitution in labile octahedral complexes (5). An overall view has been provided in the appropriate Chapters of a book on inorganic reaction mechanisms (6). Mechanisms of ligand, and solvent, exchange for s-, p-, d-, and f-block cations have been discussed in depth, with kinetic parameters fully documented, alongside those for solvent exchange (7). In recent years there has been much use of activation volumes in the establishment of reaction mechanisms. This area forms a major part of a book on high pressure chemistry (8), and has attracted several reviews by some of its main practitioners. These include a comprehensive documentation of activation volumes in inorganic, and in organic, chemistry reported in the period 1987–1997 (9), and an overview of apparatus and techniques, combined with detailed discussion, emphasising volume profiles, of a range of substitution and electron transfer reactions and of small molecule activation (10). The material in this latter review is dealt with rather more briefly elsewhere (11), whereas volume profile analysis is the main concern of a fourth

---

[1]Abbreviations for ligands which make only one appearance are generally defined at the appropriate point in the text, whereas ligand abbreviations which appear in more than one place are listed and defined at the end of this Chapter.

review (12). Further contributions in this area discuss effects of high pressure alongside those of high temperature, and mechanism in parallel with synthetic applications (13), the role of activation volumes in establishing mechanisms in inorganic and bioinorganic chemistry (14), and the use of high pressure in probing photochemical reactions of inorganic complexes and organometallic compounds (15). A review dedicated to reaction mechanisms of coordination complexes, in which activation volumes feature alongside other aspects of the subject (16), is closest in coverage to the present Chapter; it covers the period up to 1998. The significant, sometimes dominant, role played by solvation changes consequent on transition state formation has long been recognized (17). Recently it was claimed that solvent viscosity should not be neglected in the interpretation of $\Delta V^{\ddagger}$ based on an apparent solvent viscosity acceleration of typical Diels-Alder reactions (18). Thus actual $\Delta V^{\ddagger}$ values may be several $cm^3 mol^{-1}$ less negative than indicated by the experimental rate/pressure data. However, it has been shown that the apparent viscosity dependence of such reactions is significantly smaller than claimed, such that the consequence of this effect on reported activation volumes is minor (19). Molecular Dynamics and ab initio calculations have provided support for the use of activation volumes and volume profiles in assignment of mechanism (20). Appropriate computations have supported the assignment of I_d and I_a mechanisms for spontaneous aquation of classical cobalt(III) and chromium(III) complexes, respectively, helping to clarify conclusions from activation volume data (21).

Aquation of a selection of penta- and tetra-ammine or amine complexes of Co(III) and of Cr(III) involving the uncharged leaving groups DMSO, DMF, or DMAC has been studied in order to probe how electronic and steric effects can be used to fine-tune mechanism within the I_a, I, I_d range, using activation parameters $\Delta H^{\ddagger}$, $\Delta S^{\ddagger}$ and $\Delta V^{\ddagger}$ for diagnosis of mechanism (22). Initial steps have been taken towards establishing a general interchange mechanism scale, i.e. scale of "the extent of interchange" in the A, I_a, I, I_d, D continuum for substitution at inert transition metal complexes (23). A general view – a "computational perspective" – on kinetics and mechanisms of reactions of transition metal compounds and complexes has also been offered (24). The assignment of mechanisms to their position in the A, I_a, I, I_d, D sequence remains one of the main aims of kinetic studies of substitution, both for straightforward ligand-by-ligand replacement and for more complicated processes, such as chelate ring closure. Kinetic studies of reactions involving chelate ring formation or opening remain popular, for Co(III), Cr(III) and, especially, Pt(II), with recent examples often involving multidentate ligands of biochemical relevance. Isokinetic plots continue to appear, recent examples involving Rh(III), Pd(II), and several organometallic reactions (25) – most of the plots relating to the organometallic reactions are convincing, those to Rh(III) and Pd(II) less so. An unusual application of isokinetic relationships has been to spin-crossover systems, especially of Fe(II) complexes (26).

Activity in the area of medium effects (27) has declined greatly in recent years, though there has been some interest in kinetics and mechanisms in supercritical fluids (28). Indeed activation volumes for ring closure reactions of diimine-carbonyls $M(CO)_n$(diimine) show some of the most dramatic medium effects. Thus $\Delta V^{\ddagger}$ values range from +66 to +4 $cm^3 mol^{-1}$ on going from 7% benzene in supercritical CO_2 (at 35 °C) to 100% benzene (at 25 °C) (29).

Inorganic photochemistry continues to be a very active field, though recent work has concentrated on redox processes and tended to avoid photosubstitution in classical coordination complexes. However, there have been useful and interesting surveys of such photochemistry in the issue of *Coordination Chemistry Reviews* marking A.W. Adamson's 80th birthday (see Cr(III), Ru(II), and Pt(II) below). A theoretical approach to specific site labilization in photosubstitution in octahedral transition metal complexes deals with differences in behavior, especially those associated with σ and π metal–ligand bond formation, according to d^n configuration (30). A copiously referenced (350 citations) review of linkage isomerization reactions concentrates on pentacyanoferrates and ruthenium complexes (31). The kinetic *trans*-effect in octahedral complexes, including those of Co(III), Rh(III), Ir(III), Fe(II), and Ru(II), is documented descriptively in one section of an extensive review of the *trans*-effect and *trans*-influence (32).

There is increasing interest in so-called hemilabile ligands (a subclass of heterodi-functional ligands (33)), bi- or poly-dentate ligands containing both substitution-inert and -labile donor groups (34,35). The former modify reactivity of the latter, whose loss permits access by, then activation of, small molecules. Reversible loss of the labile moiety may be involved in fluxional or intramolecular rearrangement of complexes containing such ligands. The "Hard and Soft Acids and Bases" principle may be involved, e.g., $MeOCH_2CH_2P$(cyclohexyl)$_2$ (L) binds to soft ruthenium(II) considerably more strongly through P than through O – the bond to the latter donor atom can break fairly easily, generating an active site at the metal in e.g., RuL_2RCl (36). Pyridylphosphines (soft P, hard N) (37,38) provide similar examples, as in designing ligands for Au(I) and Ag(I) to generate potential anti-tumor compounds of high selectivity (39). Hemilabile ligands are perhaps of more interest in organometallic chemistry and homogeneous catalysis than in classical coordination chemistry, but there are a number of recent examples in the latter area (as, e.g., ring closure kinetics of bidentate hemi-labile *P,N* and *P,S* ligands in a Pt(II) complex (*40*)). In a sense complexes of hemilabile ligands are intraligand analogues of ternary complexes such as pentacyanometallates, $[M(CN)_5L]^{n-}$, various hemes and porphyrin complexes, [FeLXY] with L a tetraazamacrocycle, and even ternary cobalt(III)–ammine or amine–anion complexes, where rate constants for the loss of the anionic ligand, or for nitrito to nitro linkage isomerization (*41*), depend considerably on the nature of the non-reacting ligands.

A rapidly developing area where mechanistic information is currently minimal but could usefully be greatly increased is that of self-assembly and supramolecular chemistry. Towards the end of this review we deal briefly with catenanes, rotaxanes, helicates and knots, to give a qualitative view of the development of mechanistic investigation and understanding in this field. Substitution at copper(I) is an important feature of supramolecular chemistry, especially in relation to catenane formation; it is also currently being recognized as of relevance to certain bioinorganic systems. Mechanistic developments should be imminent in relation to the ligand replacement reactions which are an integral part of the entry of copper into cells and of its subsequent central role in metalloprotein mediation of such processes as iron uptake and energy generation through electron transfer.

II. Inert Octahedral M^{II} and M^{III} Complexes

A. COBALT(III)

1. Aquation and Related Reactions (42)

Using the criteria of maximum information and minimum complication, the best set of substrates is provided by the $[Co(NH_3)_5X]^{n+}$ series, for which relevant data, assembled over many years (*4,43*), are collected in Table I.[2] The most labile members included in the list, i.e. $X = CF_3SO_3^-$ or ClO_4^-, are determined by the time taken to get the reaction started. Indeed the ease and rapidity of loss of trifluorosulfonate

---

[2]Throughout this Chapter rate constants and activation volumes are at 298 K, in aqueous solution, unless otherwise indicated (temperatures are given in K or °C as quoted by authors).

TABLE I

RATE CONSTANTS FOR AQUATION AND FOR BASE HYDROLYSIS OF SELECTED
PENTA-AMMINE-COBALT(III) COMPLEXES[a]

Leaving group	k_{aq} (s^{-1})	k_{OH} $(dm^3 mol^{-1} s^{-1})$
ClO_4^-	8.1×10^{-2}	
$CF_3SO_3^-$	2.7×10^{-2}	$> 10^4$
$4\text{-}NO_2C_6H_4SO_3^-$	6.3×10^{-4}	2.7×10^2
$CH_3SO_3^-$	2.0×10^{-6}	5.5×10^1
NO_3^-	2.4×10^{-7}	5.5
Me_2SO	1.8×10^{-7}	5.4
Cl^-	1.8×10^{-6}	2.3×10^{-1}
SO_4^{2-}	8.9×10^{-7}	4.9×10^{-2}
$CH_3CO_2^-$	2.7×10^{-8}	9.6×10^{-4}
N_3^-	2.1×10^{-9}	3.0×10^{-4}
NH_3	5.8×10^{-12}	7.1×10^{-7}

[a] In aqueous solution at 298 K.

ligands is of great value in preparative coordination chemistry, especially when dealing with very inert centers such as Rh(III) or Os(II).
The rate constants in Table I cover an enormous range, of $\sim 10^{10}$.
Nonetheless there is a good correlation – a very good correlation if one
sticks to mononegative ligands – between aquation rate constants and
stability constants for formation of the respective complexes (44,45).
This is hardly surprising since the reverse (anation) process takes place
at rates that are essentially insensitive to the nature of the entering
ligand, other than its charge.

Rate constants and activation volumes for aquation of *trans*-
$[Co(MeNH_2)(NH_3)_4X]^{n+}$, with $X = Cl^-$, Br^-, NO_3^-, SO_4^{2-}, have been
compared with values for their penta-ammine analogues. Both series
illustrate the importance of solvation and electrostriction effects in
these reactions. Thus, for example, the activation volumes for sulfate
loss are 10 or more $cm^3 mol^{-1}$ more negative than for the complexes of
uninegative leaving groups (46). Activation volumes for aquation of a
selection of penta- and tetra-ammine or -amine complexes of Co(III)
and of Cr(III) involving the uncharged leaving groups DMSO, DMF, or
DMAC lie within the range +2 to +17 $cm^3 mol^{-1}$. The observed range
suggests that steric and electronic effects of the various ligands may
have a small but significant effect on the degree of dissociativeness of
the interchange mechanism here (22).

The azide and thiocyanate complexes [CoLX], where L is a dinegative
N_2S_3 donor ligand making the five-coordinate $[CoL]^+$ moiety resemble

the active site in nitrile hydratase, dissociate very rapidly by normal cobalt(III) standards ($k = 2.1 \times 10^{-2}\,\mathrm{s}^{-1}$ for loss of azide, $7.2 \times 10^{-1}\,\mathrm{s}^{-1}$ for thiocyanate (but cf. ClO_4^-, $CF_3SO_3^-$ in Table I). Such rates are much more appropriate for biological activity than those characterizing standard cobalt(III) complexes (47).

The kinetics and mechanisms of formation, acid-catalyzed aquation, reversible anation and photochemistry of the *trans*-(aqua)-(sulfito-S)[*N,N*-ethylenebis(salicylidiniminato)]-cobaltate(III) anion have been investigated in aqueous media (48). The *trans*-effect of the sulfite ligand is unusually small in these reactions. On the other hand, the *trans*-effect of carbon-bonded ligands is generally particularly high, as may be illustrated by a pulse radiolysis study of kinetics and reaction mechanisms of complexes with Co–carbon sigma bonds of the type $[(NH_3)_5Co-R]^{n+}$ in aqueous solution (49) and by the high lability and large positive $\Delta V^{\ddagger}$ for substitution at $[Co(NH_3)_5Me]^{2+}$, which clearly has a limiting dissociative (D) mechanism (see also Section IV.B.4 below). Replacement of ammine ligands in $[Co(NH_3)_5Me]^{2+}$ by ethane-1,2-diamine is characterized by $\Delta V^{\ddagger} = +14$ and $+24\,\mathrm{cm}^3\,\mathrm{mol}^{-1}$ for the two stages. The subsequent slow *cis→trans* isomerization of the $[Co(en)_2(NH_3)Me]^{2+}$ produced has $\Delta S^{\ddagger} = +86\,\mathrm{J\,K}^{-1}\,\mathrm{mol}^{-1}$ and $\Delta V^{\ddagger} = +14\,\mathrm{cm}^3\,\mathrm{mol}^{-1}$. Dissociative activation is thus indicated for all three stages (11,50).

The mechanism for acid-catalyzed aquation of $[Co(en)L_2(O_2CO)]^+$ (L = (methyl)-imidazole) (51) and of $[Co(imid)_4(O_2CO)]^+$ (52) is of rapid reversible ligand protonation followed by rate-determining carbonate chelate ring opening, in turn followed by rapid loss of monodentate $-OCO_2H^-$. An extensive study of non-leaving ligands effects on acid-catalyzed aquation of $[CoL_4(O_2CO)]^+$, L_4 = a tetradentate ligand such as trien, 2,3,2-tet, cyclam, cyclen, or edda, revealed a rate constant range of over 10^4-fold. Reactivity depends on chelate ring size and the relation of this to intramolecular hydrogen-bonding. In turn this determines the ease and extent of weakening Co–ligand bonding and labilization of the leaving protonated carbonate. Activation parameters range from 55 to $83\,\mathrm{kJ\,mol}^{-1}$ for $\Delta H^{\ddagger}$, and from -52 to $+7\,\mathrm{J\,K}^{-1}\,\mathrm{mol}^{-1}$ for $\Delta S^{\ddagger}$ (53). Ligand denticity effects have also been documented, for $[CoL_4(O_2CO)]^+$ with $L_4 = (NH_3)_4$, $(en)_2$, trien, tren, nta (54); this study also compared $[Co(NH_3)_5(OCO_2)]^+$ with its Rh and Ir analogues. Relevant to acid-catalyzed aquation of carbonato–cobalt(III) complexes is the demonstration of both mono- and bi-dentate bonding modes of bicarbonate in Cu(II) complexes (55). The recent demonstration that carbonic acid is, at least under certain conditions, considerably less unstable than heretofore generally believed (56) may also be relevant to the overall reaction in such systems.

Another rather unusual case of rapid aquation is provided by those oxoanion ligands MO_n^{x-} in which the M–O bond is sufficiently weak for it to break rather than the Co–O bond. This time we have an intramolecular reaction of a coordinated ligand. A recent example is provided by aquation of the $[Co(NH_3)_5(OMoO_3)]^+$ cation (57). This aquation pathway is actually more commonly encountered for chromium(III) complexes.

The rate constant for aquation of $[Co(CN)_5(ONOO)]^{3-}$, the first stable transition metal complex of peroxonitrite, is $4.9 \times 10^{-6} \, s^{-1}$ (58). Acid-catalyzed hydrolysis in daylight gives $[Co(CN)_5(H_2O)]^{2-}$, plus decomposition products of the peroxonitrite; in the dark there is isomerization of coordinated $ONOO^-$ to coordinated nitrate (59). Peroxonitrite may be of importance in vivo, where it could be formed from nitric oxide plus superoxide. Peroxonitrite forms an adduct with carbon dioxide, $ONOOCO_2^-$, whose dissociation rate constant is about $400 \, s^{-1}$. It is unreactive towards flavonoids and towards many organic carbonyl compounds, but reacts with such substrates as catechol or hydroquinone to give the respective quinones. The rate law for these reactions is zeroth-order in organic substrate, but reaction with acetone (in dilute aqueous solution) is first-order in acetone (60). There is a similar dichotomy in relation to inorganic substrates, for reaction with $[Fe(CN)_6]^{4-}$ or $[Mo(CN)_8]^{4-}$ is zeroth-order in complex, with $[Ni(cyclam)]^{2+}$ is a bimolecular reaction. The activation volume for this last reaction is $-7 \, cm^3 \, mol^{-1}$, but for peroxonitrite decomposition in the presence of varying amounts of nitrite, and for the zeroth-order-in-complex oxidations $\Delta V^{\ddagger}$ is between $+6$ and $+14 \, cm^3 \, mol^{-1}$, indicating rate-controlling dissociative decomposition (61).

2. Biochemical Models and Systems

Ligand substitution at trans-$[Co(en)_2(Me)(H_2O)]^{2+}$, a coenzyme B_{12} model, and trans-$[Co(en)_2(Me)(NH_3)]^{2+}$ with cyanide, thiocyanate, azide, or imidazole (X) gives trans-$[Co(en)_2(Me)(X)]^{n+}$, reactions of the ammine complex proceeding via the aqua complex. For trans-$[Co(en)_2(Me)(H_2O)]^{2+} \rightarrow$ trans-$[Co(en)_2(Me)(NH_3)]^{2+}$ $\Delta S^{\ddagger} = +3 \, J \, K^{-1} \, mol^{-1}$ and $\Delta V^{\ddagger} = +5.7 \, cm^3 \, mol^{-1}$; for the reverse reaction $\Delta S^{\ddagger} = +40 \, J \, K^{-1} \, mol^{-1}$ and $\Delta V^{\ddagger} = +9.0 \, cm^3 \, mol^{-1}$. These reactions take place by dissociative interchange, as for water replacement by pyridine in vitamin B_{12} (aquacobalamin(III)) (62), with some participation by the entering group in the transition state (63). An equilibrium study of five coordination and adduct formation in cobalt(III) corrinoids is relevant, indeed important, to kinetics and mechanism in this area, in facilitating the dissection of ligand substitution into its component steps (*64*). It established the volume change on removing a water molecule from cobalt(III) in vinylcobinamide to be $+12 \, cm^3 \, mol^{-1}$, a value which is reassuringly close to the calculated activation volume of $+13.1 \, cm^3 \, mol^{-1}$ for water loss from a metal(III) center by a limiting dissociative (*D*) mechanism.

5′-Deoxyadenosylcobinamide reacts with cyanide to form dicyanocobinamide plus adenine plus 1-cyano-D-*erythro*-2,3-dihydroxy-4-pentenol. Saturation kinetics were observed at high cyanide concentrations. This kinetic behavior, coupled with solvent effects on reactivity, suggests a two-step mechanism in which rapid equilibrium formation of a (β-adenosyl)(α-cyano)cobinamide intermediate is followed by solvent-assisted rate-determining cleavage of a cobalt–adenosyl bond. The use of a DMF-rich medium, in which the intermediate is relatively long-lived, permitted answering the long-standing question as to whether cyanide attacks at the α or at the β site of the corrinoid moiety of coenzyme B_{12} (5′-deoxyadenosylcobalamin) (*65*). Thermodynamic and kinetic studies on the reaction between the vitamin B_{12} derivative β-(*N*-methylimidazoyl)cobalamin and *N*-methylimidazole indicate ligand displacement at the α-axial site of cobalamins. The inverse dependence of rate constants on incoming ligand concentration suggests the intermediacy of a solvento-intermediate; activation entropies and volumes indicate a dissociative mechanism (*66*). There is kinetic and direct ^1H spectroscopic evidence for a (β-5-deoxyadenosyl)(α-cyano)cobalamin intermediate in the reaction between coenzyme B_{12} and cyanide when this is conducted in 92% DMF. Rapid reversible addition of cyanide is followed by slow ($k = 9 \times 10^{-5} \, s^{-1}$) Co–C bond cleavage (in water the first step is rate-limiting) (*67*).

Equilibrium and kinetic studies on the reaction of alkylcobalamins with cyanide reveal that the cyanide first substitutes the 5,6-dimethylbenzimidazole moiety in the α-position, then the alkyl group in the β-position. The first step proved too fast to monitor for standard alkyl groups, but for their CF_3 and CH_2CF_3 analogues saturation kinetics and activation parameters suggest *D* and I_d mechanisms, respectively (*68*). Exceptionally, the carbon-bonded ligand in coenzyme B_{12} does not guarantee dissociative activation – the markedly negative activation volume ($-10.0 \, cm^3 \, mol^{-1}$) for replacement of adenosyl by cyanide points to considerable associative character to substitution at cobalt(III) here (*69*).

The activation parameters $\Delta S^{\ddagger}$ and $\Delta V^{\ddagger}$ for formation and dissociation of the nitric oxide adduct of cobalamin

$$\text{cobalamin(II)} \cdot H_2O + NO \rightleftarrows \text{cobalamin(III)} \cdot NO^- + H_2O$$

are positive, consistent with substitution by dissociative interchange in both directions (70). The reverse aquation is of Co(III) if one uses the formalism of NO⁻ as ligand in the adduct and thus couples substitution with internal electron transfer.

Aquacobalamin (vitamin B_{12}) does not bind NO in aqueous solution – nitrite impurities account for the reaction observed (71).

Steric factors play an important role in reactions of bromomethylaquacobaloxime with substituted imidazoles – 1-methylimidazole reacts at approximately the same rate as imidazole itself, but the presence of an alkyl substituent in the 2-position of the incoming imidazole reduces the rate dramatically (72).

3. Base Hydrolysis

Base hydrolysis is much faster, at any significant hydroxide ion concentration, than aquation but, as is apparent from Table I the two reactions exhibit comparable ranges of rate constants. Indeed the two sets of rate constants correlate very well, over more than nine orders of magnitude, with a slope close to unity for a correlation plot.

The long-running dispute over the mechanism of base hydrolysis of cobalt(III)-ammine and -amine complexes, S_N2 versus S_N1CB (better termed D_{cb}), was several years ago resolved in favor of the latter (73). Recent activity on reactions of this type has concentrated on attempting to locate the precise site of deprotonation of the complex, an exercise successfully accomplished for the complexes syn,anti-[Co(cyclen)(NH$_3$)$_2$]$^{3+}$ and syn,anti-[Co(cyclen)(diamine)]$^{3+}$ (diamine=H$_2$N(CH$_2$)$_2$NH$_2$,H$_2$N(CH$_2$)$_3$NH$_2$) (74).

The kinetics of base hydrolysis of several complexes of the type [Co(NH$_3$)$_3$L$_3$] have been examined in order to see whether the mechanism for these uncharged complexes is the same as that operating for base hydrolysis of the standard cationic complexes (75). A comparison of kinetic parameters – a small selection is given in Table II (76,77) – suggests that all cobalt(III)-nitro-amine complexes, charged and uncharged, undergo base hydrolysis by the S_N1CB (D_{cb}) mechanism.

Rate constants and activation parameters ($\Delta H^{\ddagger}$, $\Delta S^{\ddagger}$, and $\Delta V^{\ddagger}$) for base hydrolysis of trans-[Co(MeNH$_2$)(NH$_3$)$_4$X]$^{n+}$, with X = Cl⁻, Br⁻, NO$_3^-$, SO$_4^{2-}$, have been compared with values for their penta-ammine analogues. The trans-MeNH$_2$ ligand appears to increase the dissociativeness of these S_N1CB (D_{cb}) reactions. As in the case of aquation (see above), both series show important solvation and electrostriction effects (46,78).

Non-leaving ligand effects, for a series of 13-, 14- and 15-membered tetraaza-macrocycles with amine or carboxylate pendant groups, have

TABLE II

Kinetic Parameters for Base Hydrolysis of Selected Cobalt(III)-Nitro-Amine Complexes [a]

	k (dm^3 mol^{-1} s^{-1})	$\Delta H^{\ddagger}$ (kJ mol^{-1})	$\Delta S^{\ddagger}$ (J K^{-1} mol^{-1})	Ref.
$[\text{Co(NH}_3)_5(\text{NO}_2)]^{2+}$	7.1×10^{-7}	153	+149	(76)
cis-$[\text{Co(en)}_2(\text{NH}_3)(\text{NO}_2)]^{2+}$	5.0×10^{-5}	141	+145	(77)
trans-$[\text{Co(en)}_2(\text{NH}_3)(\text{NO}_2)]^{2+}$	7.5×10^{-5}	145	+164	(77)
cis-$[\text{Co(NH}_3)_4(\text{NO}_2)_2]^{+}$	4.4×10^{-7}	147	+126	(76)
trans-$[\text{Co(NH}_3)_4(\text{NO}_2)_2]^{+}$	1.5×10^{-6}	143	+125	(76)
cis-$[\text{Co(en)}_2(\text{NO}_2)_2]^{+}$	3.8×10^{-5}	134	+119	(77)
trans-$[\text{Co(en)}_2(\text{NO}_2)_2]^{+}$	1.5×10^{-4}	134	+130	(77)
mer-$[\text{Co(NH}_3)_3(\text{NO}_2)_3]$	1.2×10^{-6}	135	+96	(76)
mer-$[\text{Co(dien)}(\text{NO}_2)_3]$	1.6×10^{-1}	105	+93	(75)
mer-$[\text{Co(en)}(\text{NH}_3)(\text{NO}_2)_3]$	1.2×10^{-4}	119	+77	(75)

[a] In aqueous solution at 298 K.

been documented for the stereoretentive base hydrolysis of their respective chlorocobalt(III) complexes (79). Chlorocobalt(III) complexes containing tetrapodal pentaamine ligands have provided new insight into details of base hydrolysis mechanisms, particularly for relatively rigid complexes. These tetrapodal ligands are comprised of two –C(CH$_2$NH$_2$)$_2$CH$_3$ groups attached either to –CH$_2$NHCH$_2$– or to 2- and 6-CH$_2$ substituents on pyridine. The former ligand gives a chlorocobalt (III) complex with unusually high base hydrolysis reactivity (80); activation parameters, especially $\Delta S^{\ddagger}$ and $\Delta V^{\ddagger}$, for base hydrolysis of the chlorocobalt(III) complex of the latter strongly indicate a D_{cb} mechanism (81).

B. Chromium(III)

A review of recent advances in chromium chemistry (82) supplements earlier comprehensive reviews of kinetics and mechanisms of substitution in chromium(III) complexes (83). This recent review tabulates kinetic parameters for base hydrolysis of some Cr(III) complexes, mentions mechanisms of formation of polynuclear Cr(III) species, and discusses current views on the question of the mechanism(s) of such reactions. It seems that both CB (conjugate base) and S_N2 mechanisms operate, depending on the situation. The important role played by ion-pairing in base hydrolysis of macrocyclic complexes of chromium(III) has been stressed. This is evidenced by the observed order, greater than 1, with respect to hydroxide in the rate law for base hydrolysis of $[Cr(cyca)(H_2O)X]^{2+}$ (X = NCS, N_3, Cl) and of its cyc$b$ analogue in the presence of high concentrations of inert electrolytes. The base hydrolysis mechanism involves a reactive conjugate base formed by proton transfer within the ion-pair (*84*).

Kinetic parameters for aquation at corresponding Cr(III) and Co(III) complexes have been compared for a series of complexes cis-$[ML_4XY]^{n+}$, where $L_4 = (NH_3)_4$ or (en)$_2$, X = Cl$^-$ or H_2O, and Y=an uncharged leaving group (DMSO, DMF, or DMAC). The uniformly negative activation volumes ($\Delta V^{\ddagger}$ between -2 and $-11\,cm^3\,mol^{-1}$) for the chromium complexes contrast with uniformly positive activation volumes ($\Delta V^{\ddagger}$ between $+3$ and $+12\,cm^3\,mol^{-1}$) for the cobalt complexes – $\Delta V^{\ddagger}$ values provide a more clear-cut contrast than $\Delta S^{\ddagger}$ values here (*22*).

Rate constants for formation of Cr(III) complexes with such ligands as pyridine carboxylates, anthranilate, salicylate, and amino acids range from $k_f = 6$ to $660 \times 10^{-5}\,M^{-1}\,s^{-1}$. Both the range and the fact that the rates for all these formation reactions are considerably faster than that corresponding to water exchange indicate that a dissociative (interchange) mechanism is unlikely to operate (*85*) – i.e. provide further evidence for associative activation in formation reactions of Cr^{3+}_{aq}. The very slow rate of ligand substitution at $[Cr(tren)(H_2O)(OH)]^{2+}$ ($\tau_{1/2}$ for water exchange is nearly 2 weeks) makes this complex an attractive model probe for interaction of Fe^{2+}_{aq} with ferritin. The chromium complex binds to ferritin at surface $-CO_2^-$ sites (*86*). cis-$[Cr(bpy)_2(H_2O)_2]^{3+}$ reacts with $[M(CN)_8]^{4-}$ (M = Mo, W) in two kinetically distinct steps, respectively first-order and zeroth-order in $[M(CN)_8]^{4-}$. The rate law and activation entropy suggest associative activation in the second, ring closure, step (*87*).

A mechanistic study of acid and metal ion (Ni^{2+}, Cu^{2+}, Zn^{2+}) promoted hydrolysis of [N-(2-carboxyphenyl)iminodiacetate](picolinato)chromate (III) indicated parallel H$^+$- or M^{2+}-dependent and -independent pathways. Solvent isotope effects indicate that the H$^+$-dependent path involves rapid pre-equilibrium protonation followed by rate-limiting ring opening. Similarly, the M^{2+}-dependent path involves rate-determining Cr–O bond breaking in a rapidly formed binuclear intermediate. The relative catalytic efficiencies of the three metal ions reflect the Irving-Williams stability order (*88*).

Mercury(II) reacts with organochromium complexes by electrophilic substitution. Rate constants have been reported for Hg^{2+} attack at a series of alkylchromium complexes with the macrocyclic ligand 1,4,8,11-tetrazacyclotetradecane, CrR(H_2O)([14]aneN$_4$). The Hammett relationship established for a series of meta and para substituted benzyl analogues is consistent with attack of the Hg^{2+} at α-carbon (*89*).

Three reviews in the A. W. Adamson 80th birthday commemoration volume deal with mechanisms of photochemical reactions of chromium complexes (*90*); rate constants for photochemical and thermal solvolyses have been compared for $[Cr(NCS)_6]^{3-}$ and for $[Cr(NCS)_5(H_2O)]^{2-}$ in DMF and in MeCN (*91*).

A few comments on substitutional reactivity of $Cr[CH(SiMe_3)_2]_3$ are buried within a 150 page review of three-coordinate complexes (*92*).

C. RHODIUM(III) AND IRIDIUM(III)

Values of $63\ kJ\ mol^{-1}$ for $\Delta H^{\ddagger}$ and $-129\ J\ K^{-1}\ mol^{-1}$ for $\Delta S^{\ddagger}$ for reaction with cytidine extend yet further the activation parameter ranges $(72 < \Delta H^{\ddagger} < 93\ kJ\ mol^{-1};\ -91 < \Delta S^{\ddagger} < -49\ J\ K^{-1}\ mol^{-1})$ for complex formation from $Rh(OH)_{aq}^{2+}$. These activation parameters (*93*), and those for reaction with adenosine (*94*), were claimed to support associative activation as the mechanism for substitution at this hydroxo-aqua-cation. Alkyl group variation has a big effect on rate constants for water replacement at *trans*-$[Rh(dmgH)_2(R)(H_2O)]$ (Table III), whereas the nature of the incoming group has only a small effect on kinetic parameters. Replacement of $-CH_3$ by $-CH_2CF_3$ (cf. cyanide reaction with alkylcobalamins above (*68*)) results in a changeover from a dissociative to an associative mechanism of substitution (*95*).

The imidazole complex *trans*-$[Ir(imid)_2Cl_4]^-$ is stable for days in neutral aqueous solution, and for hours in the presence of added thiocyanate. Addition of silver nitrate precipitates the silver salt of the complex, with no indication of Ag^+-catalyzed removal of coordinated chloride. Thus this iridium(III) complex is substitutionally much more inert than its much-studied (because (potentially) anti-tumor) ruthenium(III) analogue (*96*).

TABLE III

KINETIC PARAMETERS FOR WATER REPLACEMENT IN ALKYL-AQUA-DIMETHYLGLYOXIMATO-RHODIUM(III)

Alkyl group	Entering ligand(s)[a]	$\Delta S^{\ddagger}$ $(J\ K^{-1}\ mol^{-1})$	$\Delta V^{\ddagger}$ $(cm^3\ mol^{-1})$
CH_3	I^-, py, tu, tmtu	-22 to -61	$+2.1$ to $+7.2$
CH_2Cl	tu	-51	-0.6
CH_2CF_3	tu	-16	-4.0

[a] py, pyridine; tu, thiourea; tmtu, tetramethylthiourea.

The mechanism of ligand substitution in a cationic hydridotris(pyrazolyl)borate-iridium(III) complex has been established as dissociative (D) on the basis of a kinetic study (rate law; reactivity comparisons) in dichloromethane (97). The mechanism of water substitution at half-sandwich iridium(III) aqua cations $[(\eta^5\text{-}C_5Me_5)Ir(L)(H_2O)]^{n+}$ (L is an N,N or N,O ligand) is, based on small positive $\Delta V^{\ddagger}$ values, an I_d mechanism. However, because of arguments presented a full D mechanism cannot be ruled out (98). The $\eta^5\text{-}C_5Me_5$ group causes a great increase in lability of the coordinated water, as in a number of complexes $[(\eta^5\text{-}C_5Me_5)M(H_2O)_3]^{n+}$. Other recent kinetic and mechanistic studies on substitution in organometallic aqua-cations have involved ruthenium(II) (see Section II.E.1 below), the technetium and rhenium species fac-$[M(CO)_3(H_2O)_3]^+$ reacting with Schiff bases (e.g., pyridine 2-CH=NNH₂) (99), aromatic amines and with thioethers, and fac-$[Tc(CO)_3(H_2O)_3]^+$ reacting with carbon monoxide (100). The technetium and rhenium studies represent the beginnings of systematic investigations of the kinetics and mechanisms of this type of reaction, with the eventual objective of the efficient design and synthesis of biopharmaceutical agents incorporating ^{99m}Tc (for diagnosis) (101) or ^{186}Re or ^{188}Re (for radiotherapy) (102).

D. LOW-SPIN IRON(II) COMPLEXES

1. Spin Cross-over

Spin cross-over kinetics for Fe(II) complexes (26,103) have been reviewed. The l.s. ⇄ h.s. ($^1A_1$ ⇄ $^5T_2$) transition in solution is characterized by values between 2 and 29 kJ mol^{-1} for $\Delta H^{\ddagger}$, −8 and −75 J K^{-1} mol^{-1} for $\Delta S^{\ddagger}$, for Fe^{2+} in various environments (i.e. ligands; solvents). These activation parameters may be primarily determined by the ease with which the cooordinated ligands can twist; the ease of twisting may be determined by solvation effects as much as by intrinsic rigidity of the complex itself (104). Activation volumes are, apart from one designedly exceptional system, between 0 and +10 cm^3 mol^{-1} for l.s. → h.s. and close to 6 cm^3 mol^{-1} for the reverse, for a range of tris-diimine (e.g., 2-pyridylimidazole), bis-terimine, and hexadentate linked-bis-terimine complexes. The derived volume profiles thus indicate that transition states are approximately midway between initial and final states (105). XANES gives useful information on structural changes associated with pressure-induced spin-crossover, sometimes showing that the change in spin state is not a simple process. Thus increasing the pressure on a high-spin sample of the hydrate or ethanol solvate of the 2-(amino-methyl)pyridine, amp, complex [Fe(amp)$_3$]Cl$_2$ has a major effect on the hydrogen-bonding network in the solid as well as on the iron–ligand bond distances, while increasing the pressure applied to high-spin [Fe(btr)$_2$(NCS)$_2$], where btr = 4,4′-bis-1,2,4-triazole, produces a second high-spin modification before the change to the low-spin state (106).

2. Diimine Complexes

Activation volumes for aquation of Schiff base complexes [Fe(C$_5$H$_4$NCH=NHR)$_3$]$^{2+}$ (R = Me, Et, ^{n}Pr, ^{n}Bu) are between +11 and +14 cm^3 mol^{-1} (107), and thus within the range established earlier (108) for (substituted) tris-1,10-phenanthroline-iron(II) complexes, viz. +11 to +22 cm^3 mol^{-1}. These positive values are consistent with dissociative activation. Kinetic studies of the reaction of a CH$_2$S(CH$_2$)$_3$SCH$_2$-linked bis(terpy) ligand (L^6) with [Fe(terpy)$_2$]$^{2+}$ showed a very slow two-step process. The suggested mechanism consisted of slow loss of one terpy, rapid formation of [Fe(terpy)(L^6)], and finally slow displacement of the second terpy as the partially-bonded L^6 becomes hexadentate (109).

The mechanism of replacement of benzylideneacetone (PhCH= CHCOMe, bza) in [Fe(CO)$_3$(bza)] by diimines is determined by the nature of the incoming ligand, with bpy or diacetyldianil reacting by parallel associative and dissociative pathways, but 2-acetylpyridine anil reacting solely by a limiting dissociative mechanism (110). Replacement of the 2,4,6-tri(2′-pyridyl)-s-triazine, tptz, in [Fe(tptz)$_2$]$^{2+}$ by phen or bpy is alleged to occur by an associative mechanism (111). Cyanide attack at [Fe(ttpz)$_2$]$^{2+}$, where ttpz is the terdentate ligand 2,3,5,6-tetrakis(2-pyridyl)pyrazine, follows a simple second-order rate law; activation parameters are comparable with those for other iron(II)-diimine plus cyanide reactions (112). Although the mechanism of hydroxide or cyanide attack at such complexes, i.e. nucleophilic attack at iron versus attack at the coordinated ligand, is still not unequivocally established there is no doubt of the bimolecular nature of the major or sole second-order term in the rate law. Such second-order rate constants provide a useful probe for medium effects, as can be seen in the gathering together of ligand effects on reactivity for base hydrolysis of low-spin iron(II)-diimine complexes in methanol–water (113) and in DMSO–water (114), and in the effects of methyl-D-glycopyranosides on reactant hydration and thus on kinetics of base hydrolysis of [Fe(bpy)$_3$]$^{2+}$ (115). Attenuation of substituent effects on reactivities of low-spin iron(II) complexes of Schiff base ligands has been assessed through base hydrolysis kinetics (116), as have

salt effects on reactivities of low-spin iron(II) complexes of diazabuta-
diene and Schiff base ligands (*117*).

The cylindrical helical binuclear complex $[Fe_2L_3]^{4+}$, where L is the bis-
diimine Schiff base **1**, exists as two enantiomers, whose interactions
with DNA differ markedly. The more strongly interacting enantiomer
targets the major groove (cf. ruthenium(II) complexes, Section II.E.1
below) and induces dramatic tightening of the DNA coil (*118*).

1

3. Diphosphine Complexes

Ligand concentration dependences, activation parameters, and solvent
effects for reaction of *trans*-$[FeH(H_2)(dppe)_2]^+$ with MeCN in MeCN,
THF or acetone present a not altogether consistent mechanistic picture.
In particular the large negative activation volumes are unexpected.
These reactions cannot be simple dissociative, in contrast to earlier-studied similar reactions of iron(II)-phosphine complexes (*119*). The suggestion is that the rate-limiting step is some sort of associative ring closure
after easy initial opening of the dppe chelate ring (*120*).

4. Complexes of Macrocyclic Ligands

Detailed kinetic studies of the reaction of Fe(II) in cyclophane hemes
with O_2 and with CO probed polarity and steric effects; the effects of
deformation of the porphyrin skeleton from planarity were assessed for
one compound (*121*). Volume profiles have been established for reactions
of a lacunar Fe(II) complex with CO (*122*), for myoglobin with O_2 and
with CO and for hemerythrin with O_2 (*123*).

Investigations continue of ligand replacement reactions of the two
coordinated solvent molecules of iron(II) phthalocyanine in DMSO
(*124*). It has been confirmed that both steps can be monitored for the
incoming ligands pyridine, 4-aminopyridine, or imidazole. Rate and equilibrium constants are presented for both stages for the reactions with
all three incoming groups (*125*). The relatively weak electron-donating
properties of alkylthio-substituents are reflected in the rate constants
for addition of a second cyanide (addition of the first is very fast) to
octa-n-butylthiophthalocyanine-iron(II) (*126*).

5. Pentacyanoferrates(II)

Kinetics of formation and dissociation of several pentacyanoferrate(II) complexes $[Fe(CN)_5L]^{n-}$, and of their formation from $[Fe(CN)_5(H_2O)]^{n-}$, have been established. Ligands L include several S- and N-donor hetero-cycles (127), 3- and 4-hydroxy- and methoxy-pyridines (128), and 3-Me and 3-Ph sydnones (129), cyanopyridine complexes of pentaammine-ruthenium-(II) and -(III) (130), and some bis-pyridine ligands (131). A selection of rate constants is given in Table IV; comparative values for many other ligands L are tabulated elsewhere (132). Both formation and dissociation rate constants for the sydnones are very much lower than for most other uncharged ligands. This exceptional behavior may arise from the mesoionic nature of these ligands (formation), and from consid-erably more π-bonding than in, e.g., pyridine–pentacyanoferrate(II) complexes (dissociation). Formation of the 4-hydroxypyridine complex is also exceptionally slow, because 4-hydroxypyridine exists almost completely in the *keto* form ($K_{(keto/enol)} = 1310$). The bis-pyridine species were investigated in relation to the generation of $[Fe(CN)_5L]$-capped

TABLE IV

KINETIC PARAMETERS FOR FORMATION AND DISSOCIATION OF PENTACYANOFERRATES(II) $[Fe(CN)_5L]^{n-}$

L	k_f (dm^3 mol^{-1} s^{-1})	k_d (s^{-1})	Ref.
Ammonia	4.6×10^2	3.2×10^{-2}	(133)
py [a]	3.7×10^2	1.1×10^{-3}	[e]
3OH–py	1.7×10^2	2.3×10^{-3}	(202)
4OH–py	2.7×10^{-1} [b]	1.1×10^{-3}	(202)
3OMe–py	4.2×10^2	2.4×10^{-3}	(202)
4OMe–py	3.8×10^2	2.1×10^{-3}	(202)
py–X–py [c]	5.7×10^2 to 7.6×10^2	5.9×10^{-4} to 8.0×10^{-4}	(131)
py–X–py·cyclodextrin	3.0×10 to 2.2×10^2		(131)
$Fe^{II}(CN)_5(py–X–py)^{3-}$ [d]	1.6×10^2		(131)
$Ru^{II}(NH_3)_5(py3CN)^{2+}$	1.9×10^3	3.9×10^{-3}	(130)
$Ru^{II}(NH_3)_5(py4CN)^{2+}$	2.0×10^3	2.4×10^{-3}	(130)
$Ru^{III}(NH_3)_5(py3CN)^{3+}$	3.0×10^3	10×10^{-3}	(130)
$Ru^{III}(NH_3)_5(py4CN)^{3+}$	3.0×10^3	8×10^{-3}	(130)
3Me-sydnone	5.4×10^{-2}	1.4×10^{-5}	(129)
3Ph-sydnone	8.8×10^{-2}	4.7×10^{-5}	(129)

[a] py, pyridine.
[b] See text concerning this exceptionally slow reaction.
[c] X=one of four bridging groups (–CH=CH–; –CH=N–N=CH–; –N=N–; –CMe=N–N=CMe–) link-ing the 4-positions of two pyridine rings.
[d] X here is –CH=N–N=CH–.
[e] See pp. 186, 308 of Ref. (6).

rotaxanes based on cyclodextrins (see Section VII.E.1 below). Kinetics and mechanisms of the formation, dissociation, and interconversion of linkage isomers of pentacyanoferrate(II) complexes of ambidentate ligands have been discussed in detail (31).

The volume profile has been established for the reaction:

$$[Fe(CN)_5(H_2O)]^{3-} + NH_3 \rightleftarrows [Fe(CN)_5(NH_3)]^{3-} + H_2O$$

$\Delta V^{\ddagger} = +14 \, cm^3 \, mol^{-1}$ for both the forward and the reverse reaction. That this $\Delta V^{\ddagger}$ value is markedly less than the partial molar volumes of water and of ammonia (25 and $18 \, cm^3 \, mol^{-1}$, respectively) indicates limiting dissociative (D) activation (133), as do the $\Delta S^{\ddagger}$ values of close to $+70 \, J \, K^{-1} mol^{-1}$ in both directions. Overall, the current situation with regard to thermal substitution at pentacyanoferrates(II) appears to be that an I_d mechanism can also operate for reactions of $[Fe(CN)_5(H_2O)]^{3-}$, whereas the D mechanism operates for all other $[Fe(CN)_5L]^{n-}$ complexes (134).

The hexadecyltrimethylammonium cation causes a modest increase in rate constant for the anion–anion reaction $[Fe(CN)_5(4\text{-CNpy})]^{3-} + CN^-$. This can equally well be interpreted according to the pseudophase model developed from the Olson-Simonson treatment of kinetics in micellar systems or by the classical Brønsted equation (135).

Examination of the photolability of nitroprusside as a function of wavelength revealed significant release of cyanide only below 480 nm; above 480 nm there is almost exclusively loss of nitric oxide (136). Synthesis of ^{57}Fe-enriched $[Fe(CN)_5(NO)]^{2-}$ may facilitate mechanistic studies by permitting readier Mössbauer examination of metastable states and intermediates (137). EPR and relativistic density functional calculations suggest that although $[Fe(CN)_5(NO)]^{2-}$ is, as usually assumed, best regarded as an NO complex of iron(II), there is substantial electron transfer from metal to ligand in the ground state, i.e. Fe^{III}– NO^- (138) (cf. the Fe–edta–NO system mentioned in Section V.A.2 above).

It has been known since the 19th century that hexacyanoferrates(II) react with concentrated sulphuric acid to give carbon monoxide; the preparation of $[Fe(CN)_5(CO)]^{3-}$ from these reagents was reported in 1913. It now seems likely that the conversion of coordinated –CN into –CO proceeds through $-CONH_2$, here as in certain iron-containing hydrogenases (139). The stereospecific formation of fac-$[Fe(CN)_3(CO)_3]^-$ then cis-$[Fe(CN)_4(CO)_2]^{2-}$ from $[Fe(CO)_4I_2]$, as well as the stereospecific production of trans-$[Fe(CN)_4(CO)_2]^{2-}$ from iron(II) chloride plus cyanide in an atmosphere of CO, appears to be under kinetic control. cis-$[Fe(CN)_4(CO)_2]^{2-}$ decomposes in minutes in aqueous solution, whereas the *trans* isomer, first reported in 2001 (*140*), decomposes considerably more slowly (*141*).

E. RUTHENIUM(II) AND RUTHENIUM(III)

Linkage isomerism has been discussed for O,S and N,O ligands both at Ru(II) and at Ru(III), with particular mention of relevance to redox reactions of DMSO complexes (*31*).

1. *Ruthenium(II)*

Dissection of observed second-order rate constants, k_f, into pre-association equilibrium and interchange rate constants, K_{os} and k_i, for a range of uncharged, 1−, and 2− incoming ligands reacting with Ru^{2+}_{aq} demonstrates, through the near-equality of all the k_i values, dissociative activation (*142*). The volume profile for dmf substitution at Ru^{2+}_{aq} also supports an I_d mechanism (*143*) for complex formation in this solvent. These approaches to the diagnosis of an I_d mechanism for water replacement at Ru^{2+} are considerably more convincing than arguments based on $\Delta S^{\ddagger}$ values, which are often surprisingly negative for a dissociative process (*144,145*). However, it must be said that activation entropies ($\Delta S^{\ddagger}_f$) for reaction of Ni^{2+}_{aq} (*146*) and Cu^{2+}_{aq} (*147*) with simple ligands are between −45 and +70 J K^{-1} mol^{-1} and between −110 and −10 J K^{-1} mol^{-1}, respectively, depending on the charge on the incoming ligand; values of $\Delta S^{\ddagger}_i$ for Ni^{2+}_{aq} are typically around zero (*148*).

The plot of rate constant versus incoming ligand concentration for reaction of *cis*-[Ru(bpy)$_2$(H$_2$O)$_2$]$^{2+}$ with dimethylglyoxime (perforce in 10% ethanol rather than in water) shows marked curvature. This was interpreted in terms of slow dissociation of one water ligand followed by more rapid, bpy-assisted, loss of the second and relatively rapid chelate ring closure of the incoming dmg. This system is very similar to reactions of *cis*-[Ru(bpy)$_2$(H$_2$O)$_2$]$^{2+}$ with acetylacetone, 8-hydroxyquinoline, cysteine, and pyridine 2-aldoxime; activation parameters for the five incoming ligands span relatively small ranges ($59 < \Delta H^{\ddagger} < 67$ kJ mol^{-1} and $-101 < \Delta S^{\ddagger} < -124$ J K^{-1} mol^{-1}). This pattern is consistent with dissociative activation (*149*), even if the $\Delta S^{\ddagger}$ values are again (cf. preceding paragraph) remarkably negative for dissociative activation.

Enormous acceleration of substitution at ruthenium(II) can be obtained by appropriate choice of ligands. Following the demonstration of remarkably rapid water exchange and complex formation (I_d or D in mechanism) at the organometallic aqua-ion [Ru(η^5-C$_5$Me$_5$)(H$_2$O)$_3$]$^{2+}$ (*150*), comparably rapid substitution has been demonstrated at

[Ru(LLL)(LL)(H$_2$O)]$^{2+}$, where LLL = mer-terpy or fac-tpmm {tris(2-pyri-dyl)-methoxy-methane} and LL = a bidentate N,N-donor. Thus there is a 10^7-fold increase in rate constant on going from the complex with LLL = tppm and LL = phen, where the slow rate is characteristic of this normally substitution-inert center, to that with LLL = dppm and LL = the flexible aromatic ligand 2,2′-(dipyrazol-1-yl)propane (dpzPr, **2**). The explanation offered invokes "super-scorpionate" (*151*) behavior aris-ing from the relative geometric dispositions and consequent synergy between the tppm and dpzPr ligands (*152*). Rate constants for replace-ment of the coordinated water in two 2,2′-(dipyrazol-1-yl)alkane com-plexes [RuL(tpmm)(H$_2$O)]$^{2+}$, where L = dpzMe or dpzPr (**2**), differ by a factor of almost 10^6, with correspondingly large differences in activa-tion parameters ($\Delta H^{\ddagger} = 85$ kJ mol^{-1}, $\Delta S^{\ddagger} = -38$ J K^{-1} mol^{-1} and $\Delta H^{\ddagger} = 68$ kJ mol^{-1}, $\Delta S^{\ddagger} = 17$ J K^{-1} mol^{-1} for methane and propane derivatives, respectively). The differences are attributed to the propane-based ligand exerting a large steric effect facilitating departure of the coordi-nated water (*153*).

dpzMe : R = H
dpzPr : R = Me

2

Activation volumes for the forward and the reverse reactions:

$$[Ru(NH_3)_4\{P(OEt)_3\}(H_2O)]^{2+} + L \rightleftharpoons [Ru(NH_3)_4\{P(OEt)_3\}L]^{2+} + H_2O$$

are $\Delta V^{\ddagger} = +4.2, +1.9$ and $+2.0$ cm^3 mol^{-1} for reaction of the aqua complex with imidazole, isonicotinamide and pyrazine and $\Delta V^{\ddagger} = +7.5$ and $+10.4$ cm^3 mol^{-1} for hydrolysis of the isonicotinamide and pyrazine com-plexes. These values indicate dissociative (I_d) activation in every case (*154*). This is as expected, both in the light of the above discussion and of the large labilizing effect of P(OEt)$_3$. A rate constant of 3.2×10^{-5} s^{-1} is assigned to loss of the second chloride from $trans$-[Ru{P(OEt)$_3$}$_4$Cl$_2$] in 67% and in 100% ethanol. In neutral solution the product is $trans$-[Ru{P(OEt)$_3$}$_4$(solvent)$_2$]; in the presence of 10^{-2} M CF$_3$CO$_2$H solvolysis is nearly 10 times faster and the final product is cis-[Ru{P(OEt)$_3$}$_2$(solvent)$_4$]$^{2+}$ (*155*).

There are both geometric and kinetic differences between the interac-tions of Δ- and Λ-[Ru(bpy)$_2$(hpimp)]$^{2+}$ {hpimp=2-(2-hydroxyphenyl)imi-dazo[4,5-f]-1,10-phenanthroline} and the hexanucleotide d(GTCGAC)$_2$. The Λ-isomer penetrates more deeply into the major groove of the poly-nucleotide; it also binds more rapidly than the Δ-isomer to DNA (*156*).

This behavior may be compared with that of $[Ru(bpy)_3]^{2+}$ (157) and $[Pd(dien)Cl]^+$ (Section IV.C), which associate with anionic regions of DNA, of Δ-$[Ru(phen)_3]^{2+}$ which is believed to bind in the minor groove (158), and the helical binuclear iron(II) complex mentioned in Section II.D.2, which is accommodated in the major groove (118).

The rate constant for aquation of the 4,4-dithiodipyridine complex $[Ru(NH_3)_5(dtdp)]^{2+}$, $k = 4.5 \times 10^{-5}\,s^{-1}$, is almost exactly the same as that for $[Ru(NH_3)_5(py)]^{2+}$, and only slightly slower than that for dissociation of $[Ru(CN)_5(dtdp)]^{2+}$ in aqueous DMSO. Dissociation of $[Ru(CN)_5(dtdp)]^{2+}$ is, unusually, only 10 times slower than that of its iron(II) analogue $[Fe(CN)_5(dtdp)]^{2+}$ (159). Rate constants for formation and dissociation of $[(H_3N)_5Ru(NCpy)Fe(CN)_5]$ were given and referenced in Table IV (Section II.D.5); a useful summary of rate constants for formation and dissociation of pentacyanoruthenates (D mechanism in all cases) forms part of a review of pentacyanometallates(II) $[M(CN)_5L]^{n-}$, M = Fe, Ru, and Os (134).

Three different isomers are formed consecutively on reacting $[RuCl_2(PPh_3)_3]$ with 4,6-bis(pyrazol-1-yl)pyrimidine, bpzpm; the final product is $cis,trans$-$[RuCl_2(PPh_3)_2(bpzpm)]$ (160). Nuclear magnetic resonance spectroscopy (NMR) shows facile $cis \rightleftarrows trans$ interconversion for $[H(dtbp)Ru(\mu\text{-}Cl)_2Ru(dtbp)H]$, where dtbp = bis[di(t-butyl)phosphano-methane or -ethane. Crossover experiments implicate mononuclear intermediates in these isomerizations (161).

The photochemistry of ammine-ruthenium(II) complexes, mainly photosubstitution at $[Ru(NH_3)_5L]^{n+}$ and $[Ru(NH_3)_4LL']^{n+}$ (162), and of polypyridyl-ruthenium(II) complexes (163), has been reviewed.

2. Ruthenium(III)

Kinetic studies of aquation, at pH 4.6, of cis-$[RuCl_2(cyclen)]^+$ showed Cl^- to be released much more readily than from cis-$[RuCl_2(cyclam)]^+$; $k = 3.6 \times 10^{-3}\,s^{-1}$, $\Delta H^{\ddagger} = 45\,kJ\,mol^{-1}$, $\Delta S^{\ddagger} = +142\,J\,K^{-1}\,mol^{-1}$ for the cyclen complex (164).

Kinetic parameters for forward and reverse complexation and dissociation reactions:

$$[Ru(edta)(H_2O)]^- + LH \underset{k_{-2}}{\overset{k_2}{\rightleftharpoons}} [Ru(edta)L]^{2-} + H_3O^+ \qquad (1)$$

for LH = the potential thio-ligands cysteine $\{HSCH_2CH(NH_3^+)CO_2^-\}$, N-acetylcysteine $\{HSCH_2CH(NH_3^+COMe)CO_2^-\}$, 2-mercaptoethanol $\{HSCH_2CH_2OH\}$, glutathione $\{HO_2CCH_2NHCOCH(CH_2SH)NHCOCH_2CH_2CH(NH_3^+)CO_2^-\}$, and penicillamine $\{HSCMe_2CH(NH_3^+)CO_2^-\}$ are

TABLE V

$\textsc{Kinetic Parameters for the Forward and Reverse Reactions of Eq. (1)}^{a}$

Ligand	k_2 (dm^3 mol^{-1} s^{-1})	$\Delta H_2^{\ddagger}$ (kJ mol^{-1})	$\Delta S_2^{\ddagger}$ (J K^{-1} mol^{-1})	$\Delta V_2^{\ddagger}$ (cm^3 mol^{-1})	k_{-2} (s^{-1})	Ref.
Nitric oxide	$\sim 10^{8\ b}$					(171)
Pyrazine	2×10^4	24	-84		2×10^{-6}	(166)
2-Mercaptopyridine	1.1×10^4	24	-84		3.4×10^{-3}	(167)
Adenosine	8.9×10^3					(165)
Adenine	8.8×10^3					(165)
Pyridine	6.3×10^3				6.1×10^{-2}	(166)
Thiourea	3.0×10^3	22	-105	-6.8		(168)
adp		37	-55			(165)
amp	2.9×10^3	33	-67			(165)
azide	2.1×10^3	6		-9.5		(168)
4-Sulfanylpyridine	1.5×10^3	37	-60	-6.6		(170)
Me$_2$thiourea	1.4×10^3	25	-107	-8.8		(168)
atp	1.1×10^3	38	-59			(165)
2-Mercaptoethanol	5.2×10^2	37	-67		2.3×10^{-3}	(169)
Thiocyanate	2.7×10^2	37	-75	-9.6	5×10^{-7}	(168)
Glutathione	2.6×10^2	38	-71		5×10^{-4}	(169)
N-Acetylcysteine	2.4×10^2	38	-75		3×10^{-4}	(169)
Cysteine	1.7×10^2	40	-71		5.7×10^{-4}	(169)
Penicillamine	1.6×10	46	-59		3×10^{-4}	(169)

^{a} In aqueous solution at 298 K.

^{b} Buffers may have a marked effect by converting the [Ru^{III}(edta)(H$_2$O)]$^-$ into less reactive species such as [Ru^{III}(edta)(OAc)]$^{2-}$.

compared in Table V with values for some analogous reactions, including reactions with DNA bases (165), and with related potential ligands such as pyrazine (166), 2-mercaptopyridine (167), and thiourea (168). Rate constants span a moderate range, dissociation constants a much larger range; activation parameters for the forward reaction are rather similar for all the ligands mentioned. The negative $\Delta S_2^{\ddagger}$ values, and the reasonably marked dependence of forward rate constants (k_2) on the nature of the entering ligand, are consistent with an associative mechanism for the formation reactions. There are also dissections of observed rate constants into their components for variously (de)protonated forms of aqua-complex and incoming group (169). A comparison of the effects of S- and N-donor atoms was conducted through a kinetic study of reaction of [Ru(edta)(H$_2$O)]$^-$ with 4-sulfanylpyridine (4-HS–C$_5$H$_4$N). Rate constants for formation of the S- and N-bonded products are 1460 and 4950 M^{-1} s^{-1}, respectively; the N-bonded product subsequently isomerizes to the S-bonded isomer. The N-bonded form is favored kinetically, but the thermodynamically favored isomer is the S-bonded form. Isomerization of $[\text{Ru(edta)(fspy)}]^{2-}$ with N-bonded 4-sulfanylpyridine (fspy) to the thermodynamically-favored isomer with the fspy S-bonded has a rate constant of $0.07\,\text{s}^{-1}$ (170).

[Ru(edta)(H$_2$O)]$^-$ reacts very rapidly with nitric oxide (171). Reaction is much more rapid at pH$\sim$5 than at low and high pHs. The pH/rate profile for this reaction is very similar to those established earlier for reaction of this ruthenium(III) complex with azide and with dimethylthiourea. Such behavior may be interpreted in terms of the protonation equilibria between [Ru(edtaH)(H$_2$O)], [Ru(edta)(H$_2$O)]$^-$, and [Ru(edta)(OH)]$^{2-}$; the [Ru(edta)(H$_2$O)]$^-$ species is always the most reactive. The apparent relative slowness of the reaction of [Ru(edta)(H$_2$O)]$^-$ with nitric oxide in acetate buffer is attributable to rapid formation of less reactive [Ru(edta)(OAc)]$^-$; [Ru(edta)(H$_2$O)]$^-$ also reacts relatively slowly with nitrite. Laser flash photolysis studies of [Ru(edta)(NO)]$^-$ show a complicated kinetic pattern, from which it is possible to extract activation parameters both for dissociation of this complex and for its formation from [Ru(edta)(H$_2$O)]$^-$. Values of $\Delta S^{\ddagger} = -76\,\text{J\,K}^{-1}\text{mol}^{-1}$ and $\Delta V^{\ddagger} = -12.8\,\text{cm}^3\,\text{mol}^{-1}$ for the latter are compatible with $\Delta S^{\ddagger}$ values between -76 and $-107\,\text{J\,K}^{-1}\text{mol}^{-1}$ and $\Delta V^{\ddagger}$ values between -7 and $-12\,\text{cm}^3\,\text{mol}^{-1}$ for other complex-formation reactions of [Ru(edta)(H$_2$O)]$^-$ (168) and with an associative mechanism. In contrast, activation parameters for dissociation of [Ru(edta)(NO)]$^-$ ($\Delta S^{\ddagger} = -4\,\text{J\,K}^{-1}\text{mol}^{-1}$; $\Delta V^{\ddagger} = +10\,\text{cm}^3\,\text{mol}^{-1}$) suggest a dissociative interchange mechanism (172).

The reactions of [Ru(edta)(H$_2$O)]$^-$ with adenine, and with adenosine and its phosphate derivatives amp, adp, and atp, involve ring closure in a reversibly-formed intermediate containing a unidentate incoming ligand. Both formation (cf. above) and aquation of the intermediates are, on the evidence of the $\Delta S^{\ddagger}$ values for the amp, adp and atp systems, associative. Rate constants for ring closure are between 0.6 and $4.4\,\text{s}^{-1}$ (165).

The mechanism of pyridine exchange at a series of μ-oxo-bis(μ-acetato)diruthenium(III) complexes $[\text{Ru}_2(\mu\text{-O})(\mu\text{-OAc})_2\text{L}_6]^{2+}$ appears to depend strongly on the supporting ligands. Activation parameters for the complexes with $\text{L}_6 = (\text{py})_6$ or $(4\text{-acetylpy})_4(\text{py})_2$ are $\Delta H^{\ddagger} = 116$, $119\,\text{kJ\,mol}^{-1}$ and $\Delta S^{\ddagger} = +46, +45\,\text{J\,K}^{-1}\text{mol}^{-1}$, indicating dissociative activation (I_d or D). But when $\text{L}_6 = (\text{bpy})_2(\text{py})_2$ or $(4,4'\text{-Me}_2\text{bpy})_2(\text{py})_2$ the lower $\Delta H^{\ddagger}$ values of 100, 95 kJ mol^{-1} and the negative $\Delta S^{\ddagger}$ values of -10, $-25\,\text{J\,K}^{-1}\text{mol}^{-1}$ indicate significant associative character (173).

Acid-catalyzed breakdown of the trinuclear cation $[(\text{H}_3\text{N})\text{Ru}(\mu\text{-X})\text{Ru}(\text{NH}_3)_4(\mu\text{-X})\text{Ru}(\text{NH}_3)_4(\text{SCN})]^{6+}$ (X = Cl or Br) takes place by associative attack of water at the central Ru(III) (174).

F. OSMIUM

Kinetics of ligand interchange in pentacyano-osmate(II) complexes $[Os(CN)_5L]^{3-}$ with $L = H_2O$, NH_3, or N-heterocyclic ligands are consistent with the limiting dissociative (D) mechanism characteristic of pentacyanometallates (134,175). For $[Os(CN)_5(pyrazine)]^{3-}$ k_{diss} is $10^{-8}\,s^{-1}$, about 10^4 times lower than the rate constant for dissociation of its Ru(II) analogue. The rate constant for dissociation of $[Os(CN)_5(NH_3)]^{3-}$, $1.1 \times 10^{-7}\,s^{-1}$, is dramatically less than that of $[Fe(CN)_5(NH_3)]^{3-}$, which gives the aqua-complex immediately on dissolution in water.

III. High Oxidation State Complexes

A. PLATINUM(IV)

Platinum(IV) provides a link between this and previous sections, its low-spin d^6 (t_{2g}^6) electronic configuration relating the substitution behavior of its complexes to that of e.g., Co(III), Rh(III), and low-spin Fe(II) complexes. The kinetics of reactions of $[PtMe_2X(LL)(SR_2)]$, where LL is a C,N-donor imine with a series of phosphines, in particular the dependences of rate constants on concentrations of phosphine and of leaving SR_2, suggest the operation of a D mechanism. However, $\Delta S^{\ddagger}$ ranges from a value of $+48\,J\,K^{-1}\,mol^{-1}$ typical of D substitution to unexpectedly large negative values; hydrogen-bonding and steric effects are invoked to explain the latter. For the special case of $[PtMe_2F(C_5CF_4CH =NCH_2Ph)(SMe_2)]$ it is claimed that reaction with P(cyclohexylamine)$_3$ is dissociative, but with PPh$_3$ or PPh$_2$Me associative (176).

An NMR investigation of water exchange at $[Pt(H_2O)_2(oxalate)_2]$ is relevant to the mechanism of formation of one-dimensional mixed valence oxalatoplatinum polymers. In fact the rate constant for this presumably dissociative ($\Delta S^{\ddagger} = +42\,J\,K^{-1}\,mol^{-1}$) reaction is considerably too low for water loss to be, as recently proposed, the first step in formation of these polymers. The mechanism of *trans* to *cis* isomerization for this oxalate complex, and for its (2-methyl)malonate analogues, is intramolecular (Bailar or Rây-Dutt twist), since there is no concurrent incorporation of labeled solvent (177).

Kinetics and mechanisms of substitution at Pt(IV) are occasionally mentioned in relation to those complexes which may have anti-tumor properties. An article on molecular modeling of interactions between platinum complexes and nucleotides or DNA includes a brief mention of Pt(IV) (178).

B. Vanadium(IV) and Vanadium(V)

Oxygen exchange with the axial O in $V^{IV}O^{2+}$ complexes is usually slow, as for example in the case of a series of Schiff base complexes of tetradentate salen-type ligands. Here rate constants are between 6×10^{-7} and $6 \times 10^{-5} \, s^{-1}$ for exchange with a dilute solution of $H_2^{18}O$ in DMSO. Rate constants vary with electronic effects and ligand hydrophobicity; the ^{18}O-exchanging water molecule approaches predominantly from the same side of the vanadium as the vanadyl-oxygen, in contrast to vanadium(V) analogues where approach is *trans* to vanadyl-oxygen (*179*).

Interest is developing in high oxidation state vanadium complexes in relation to their role in biological systems, especially their apparent potential in the treatment of diabetes (*180*). Both substitution and electron transfer may be important in the mechanisms of action of 3-hydroxypyridinone complexes (*181*). Citrate complexes are obvious candidates for investigation in a biochemical or pharmacological context. A first step in such investigation is represented by a study of isomerization in $V_2O_2(citrate)_n$ complexes, whose interconversion, involving various citrate coordination modes, is facile and rapid on the NMR timescale (*182*).

C. Binary and Ternary Cyanide Complexes

The photochemistry of octacyanometallates, and of mixed cyano-diimine complexes of the type $[W(CN)_6(diimine)]^{2-}$ and $[MO(CN)_3(bpy)]^-$ $\{M = Mo, W\}$, has been reviewed (*183*). The authors pay particular attention to the role of the counterion in this type of reaction; they also call attention to questions which were, at the time of writing, unresolved. A mainly structural and redox review of octacyano-, nitridotetracyano-, and oxotetracyano-metallates (Nb, Ta; Mo, W; Tc, Re) contains some kinetic and mechanistic information on thermal and photochemical substitution in these complexes, with the main conclusion being that much more needs to be done on such reactions (*184*).

Following on from these two reviews, both of which indicated the need for further study of key unanswered questions, a few relevant contributions have appeared. These have included an investigation of oxygen exchange, cyanide exchange, and inversion at $[MO(L)(CN)_4]^{n-}$ $(M = Mo^{IV}, W^{IV}; Tc^V, Re^V; Os^{VI}; L = H_2O, OH^-, O^{2-})$ and several other anions (*185*), and proton exchange at protonated forms of $[MO_2(CN)_4]^{n-}$ $(M = Mo^{IV}, W^{IV}; Tc^V, Re^V)$, examined by ^{13}C and ^{17}O NMR to obtain rate laws and rate constants (*186*). Photochemical reaction of $[M(CN)_8]^{4-}$, $M = Mo$ or W, in the presence of bpy or phen, is now believed to give

[MO(CN)$_3$(diimine)]$^-$, rather than [M(CN)$_6$(diimine)]$^{2-}$ as proposed earlier. The [MO(CN)$_3$(diimine)]$^-$ produced may react further, e.g., to give [MO(CN)$_2$(diimine)$_2$] by subsequent thermally activated substitution. A general reaction scheme is presented (187). The primary photoproduct from irradiation of [Mo(CN)$_8$]$^{4-}$ within the range 400–480 nm has been confirmed, partly through its reaction with trans-[Os(en)$_2$O$_2$]$^{2+}$ to give the heterotrinuclear anion [(NC)$_7$Mo–O–Os(en)$_2$–O–Mo(CN)$_7$]$^{4-}$, as [Mo(CN)$_7$(H$_2$O)]$^{3-}$ (188).

The [WO(CN)$_5$]$^{3-}$ anion reacts with molecular oxygen, either in the solid state or in ethanol–acetone solution, to give the peroxo complex [WO(O$_2$)(CN)$_4$]$^{2-}$, as confirmed by X-ray structural characterization of the product. In solution this reaction is a second-order process, with activation parameters $\Delta H^{\ddagger} = 55\,\text{kJ mol}^{-1}$ and $\Delta S^{\ddagger} = -46\,\text{J K}^{-1}\text{mol}^{-1}$. The mechanism must be more complicated than implied by the simple second-order rate law, though the first step is believed to involve initial nucleophilic attack by O$_2$, displacing cyanide. Electron transfer and rate-limiting further reaction with O$_2$ give the intermediate [W^{VI}O(O$_2$)$_2$(CN)$_4$]$^{2-}$ (189). The first step in the reaction of the pyrazine complexes [MO(CN)$_4$(pz)]$^{2-}$ with O$_2$ is, similarly, attack by O$_2$ at the metal (Mo or W) with displacement of pz (190). The rate-limiting step in the reaction of trans-[MO$_2$(CN)$_4$]$^{4-}$ with salicaldehyde is chelate ring closure subsequent to rapid displacement of water, probably dissociatively activated (I_d or D mechanism) from the doubly protonated form of the substrate, viz. trans-[MO(H$_2$O)(CN)$_4$]$^{2-}$ which is in equilibrium with trans-[MO(OH)(CN)$_4$]$^{3-}$ in acidic solution (191). There is kinetic evidence for a ring closure step in the reaction of cis-[Cr(bpy)$_2$(H$_2$O)$_2$]$^{3+}$ and [M(CN)$_8$]$^{4-}$ (M = Mo, W) (87).

IV. Square-Planar Complexes

A. GENERAL

Kinetic parameters for substitution at four square-planar metal centers, viz. Pt(II), Pd(II), and Ni(II) (192), and also Au(III) (193), can be compared in terms of cyanide exchange at the respective [M(CN)$_4$]$^{2-}$ complexes (Table VI).

Kinetics and mechanisms of substitution at Pt(II) and Pd(II) have been reviewed and compared with respect to reactions of nitrogen bases such as imidazole, pyrazole, inosine, adenosine, and guanosine-5′-monophosphate with ammine, amine, pyridine carboxylate, and

TABLE VI

KINETIC PARAMETERS FOR CYANIDE EXCHANGE ^{a} WITH $[M(CN)_4]^{n-}$

	k_2 (dm^3 mol^{-1} s^{-1})	$\Delta H_2^{\ddagger}$ (kJ mol^{-1})	$\Delta S_2^{\ddagger}$ (J K^{-1} mol^{-1})	$\Delta V_2^{\ddagger}$ (cm^3 mol^{-1})
$[Ni(CN)_4]^{2-\,b}$	2.3×10^6	21.6	-51	-19
$[Pd(CN)_4]^{2-}$	8.2×10	23.5	-129	-22
$[Pt(CN)_4]^{2-}$	1.1×10	25.1	-142	-27
$[Au(CN)_4]^-$	6.2×10^3	40.0	-38	$+2$

^{a}These values are for the second-order pathway for reaction with CN$^-$ – see Refs. (*192,193*) for other pathways.
^{b}Kinetic parameters are for formation of $[Ni(CN)_5]^{3-}$.

pyridoxine complexes (*194*) and to substitution and isomerization of aqua-ammine, -en, or -dien complexes with histidine and histidyl-containing peptides (*195*).

B. PLATINUM(II)

1. Substitution

It has long been known that substitution at the anion of Zeise's salt, $[Pt(CH=CH_2)Cl_3]^-$, is, thanks to the high *trans* effect of the coordinated ethene, very fast. Recent developments in low-temperature stopped-flow apparatus have now permitted the study of the kinetics of substitution at Zeise's and other $[Pt(alkene)Cl_3]^-$ anions in methanol solution. These substitutions obey the customary two-term rate law (i.e. with $k_{obs} = k_1 + k_2[nucleophile]$), with large negative $\Delta S^{\ddagger}$ values for the k_2 term as expected for S_N2 processes (*196*).

The nucleophilicities of undissociated thiols towards square-planar platinum(II) complexes have been compared with those of thiourea for displacement of chloride from $[Pt(bpy)(NO_2)Cl]$ and $[Pt(terpy)Cl]^+$. The observed trends can be understood in terms of the inductive and steric effects of the various groups on the sulfur; thiols containing such groups as $-OH$, $-CO_2H$, or $-NH_3^+$ are more reactive thanks to favorable hydrogen-bonding with the leaving chloride (*197*). Reactions of $[PtL(H_2O)]^{2+}$, where L = one of a range of five terdentate nitrogen donors (from fully aromatic terpy through e.g., 2,6-bis-(aminomethyl)pyridine to fully aliphatic dien) with thiourea and its dimethyl and tetramethyl derivatives, follow essentially second-order rate laws, as expected from the high affinity of these S-donor ligands for Pt(II). The kinetic results permitted a separation of spectator ligand *cis* and *trans* effects. Perhaps unexpectedly, the *cis*-π-acceptor effect was found to be larger than the *trans* – at least in this system. Even more unexpected was the discovery that it is electronic interaction or *communication* between π-acceptor ligands in the five-coordinate intermediate or transition state, rather than simple primary π-acceptor effects, that leads to enhanced reactivity (*198*). Base hydrolysis of $[Pt(terpy)Cl]^+$ follows the usual two-term rate law,

$$\text{rate} = \{k_1 + k_2[OH^-]\}[Pt(terpy)Cl^+]$$

k_1 here has a value over 200 times larger than for $[Pt(dien)Cl]^+$, quantifying the differing effects of the dien and terpy ligands (*199*). The strongly labilizing effects of coordinated diimine and terimine ligands are well illustrated by the k_2 value ($> 10^5 \, M^{-1} s^{-1}$) for reaction of $[Pt(terpy)Cl]^+$ with PPh_3 in dichloromethane (*200*).

Substitution (hydrolysis $\rightleftarrows$ anation by X^-) reactions $[Pt(LL)X_2] \rightleftarrows [Pt(LL)(H_2O)X]^+ \rightleftarrows [Pt(LL)(H_2O)_2]^{2+}$, where LL = *cis*- or *trans*-$(NH_3)_2$, $H_2NCH_2CH_2NH_2$, or $H_2N(CH_2)_3NH_2$, and X=Cl or Br, have activation volumes in the range -5 to $-11 \, cm^3 mol^{-1}$, entirely consistent with the established associative activation (*201*). Activation volumes between -4 and $-10 \, cm^3 mol^{-1}$ for the forward and reverse reactions comprising the hydrolysis equilibria of $[Pt(R_1\text{-en})Cl_2]$, where R_1-en $= (1R,2R,4S)$-*exo*-2-(aminomethyl)-2-amino-7-oxabicyclo[2.2.1]heptane, a ligand chosen in the light of the anti-tumor properties of several Pt-substituted en complexes, indicate the expected associative interchange mechanism for both aquation and both chloride anation steps (*202*). The kinetics of the reversible hydrolysis of $[PtCl(dien)]^+$ and of $[PtCl(NH_3)_3]^+$ have been re-examined, in order to resolve a conflict of opinion as to the mechanisms of these reactions (*203*). It now appears that a combination of relatively fast anation and of equilibrium compositions very close to 100% chloro-complex prevent the detection of the hydrolytic contribution under certain experimental conditions. The hydrolysis rate constant determined here for $[PtCl(dien)]^+$ agrees well with that estimated earlier from kinetics of reaction of this complex with guanosine-5′-monophosphate; the rate constant for reaction of $[Pt(dien)(H_2O)]^{2+}$ with chloride is consistent with previously published values.

Kinetic studies of aquation of dinuclear $[\{trans\text{-}PtCl(NH_3)_2\}_2 (\mu\text{-}NH_2(CH_2)_6NH_2)]^{2+}$ established rate constants for the loss of the first and second chloride ligands (7.9×10^{-5} and $10.6 \times 10^{-4} s^{-1}$), and for the reverse anations (1.2 and $1.5 \, M^{-1} s^{-1}$). Reactivities here are very similar to those in analogous mononuclear systems $[Pt(amine)_3Cl]^+$ (*204*). A kinetic and equilibrium study of axial ligand substitution reactions of the head-to-tail α-pyridonate-bridged *cis*-diammine-platinum(III) dinuclear complex has shown transmission of the *trans* effect of one axial ligand through the Pt–Pt bond to the opposite axial ligand (*205*). Rate constants for loss of the first and second chlorides from the bifunctional trinuclear complex $Cl(NH_3)_2Pt–NH_2(CH_2)_6NH_2–Pt(NH_3)_2–NH_2(CH_2)_6NH_2–Pt(NH_3)_2Cl$, a potential chemotherapeutic agent currently undergoing clinical trials, are essentially identical. At $7 \times 10^{-5} s^{-1}$ they are only slightly higher than the rate constant for aquation (loss of first chloride) of cisplatin, which is $2.4 \times 10^{-5} s^{-1}$, at 25 °C (*206*). Likewise the two anation rate constants for the reverse reaction are almost identical. Clearly the two terminal platinum atoms are very well insulated from each other by the long alkyl chains of the bridging ligands (*207*).

Irradiation of $[Pt(hfac)_2]$ with an excess of ethene at 350 nm produced an X-ray-characterized five-coordinated intermediate containing both hfac ligands still bidentate (i.e. limiting *A* mechanism) (*208*). Quenching of photoexcited $[Pt(terpy)Cl]^+$, and 4′-substituted terpy and NCS^- analogues variously, by a range of Lewis bases, such as MeCN, DMSO, py, acetone, has been documented (*209*).

Relative reactivities of dichlorobis(sulfide)platinum(II) complexes and of their dimethyl sulfoxide analogues with the oximes of acetophenone and of benzaldehyde illustrate that sulfoxides are the better leaving groups from platinum(II). Orthoplatination of acetophenone oxime by *cis*-$[Pt(dmso)_2Cl_2]$ exhibits a kinetic pattern of consecutive but coupled reactions. Essentially instantaneous replacement of one dimethyl sulfoxide ligand is followed by chloride loss with concurrent ring closure, in turn followed by proton loss from the oxime ligand, postulated to occur through a five-coordinated intermediate containing a hydride ligand (*210*).

2. Ring Closing and Ring Opening

$[Pt(en)(H_2O)_2]^{2+}$ reacts with indole-3-acetamide, ita, to give $[Pt(en)(ita)]^{2+}$ in which the ita is *C,O*-bonded. The rate law is zeroth-order in platinum complex, first-order in ita. The interpretation of this unexpected rate law is of rapid replacement of one water by unidentate ita, with subsequent ring closure and loss of the second water ligand rate-determining ($k = 0.002 s^{-1}$) (*211*). Reaction of $[Pt(en)(H_2O)_2]^{2+}$ with L-glutamine at pH 4 involves predominantly the diaqua form of the complex and the zwitterionic form of the entering ligand. The observed kinetic pattern corresponds to ligand-assisted anation ($\Delta H^{\ddagger} = 44 \text{ kJ mol}^{-1}$; $\Delta S^{\ddagger} = -123 \text{ J K}^{-1} \text{mol}^{-1}$) followed by chelate ring closure

($\Delta H^{\ddagger} = 40\,\text{kJ}\,\text{mol}^{-1}$; $\Delta S^{\ddagger} = -204\,\text{J}\,\text{K}^{-1}\,\text{mol}^{-1}$). The activation parameters indicate associative activation for both steps (*212*). Associative chelate ring closure here is consistent with the mechanistic pattern established in organometallic systems (*213*), where this process is associative for [M(CO)$_5$(diimine)] where M = Mo or W (though dissociative for M = first-row Cr (*214*)) and the diimine does not carry bulky substituents.

In reactions of the platinum(II) complex *cis*-[PtPh$_2$(CO)(SEt$_2$)] with diphosphines Ph$_2$P(CH$_2$)$_n$PPh$_2$ (pp), when $n = 2$ or 3 the first observable product is [PtPh$_2$(pp)] – ring closure must be rapid compared with the initial attack of the ligand at the platinum. But for $n = 1$ or 4 the kinetic pattern is two-stage, with initial formation of [PtPh$_2$(CO)(pp)] followed by slow ring closure to give [PtPh$_2$(pp)] (Table VII). There is a remarkable difference between the activation parameters for formation of the four- and seven-membered chelate rings, reflecting both the highly strained nature of the four-membered ring and the much lower probability of seven-membered ring closure (*215*). In an extension of this study to complexes of bidentate hemi-labile *P,N* and *P,S* ligands {Ph$_2$P(CH$_2$)$_n$X with X = e.g., SPh, SEt, NH$_2$, NMe$_2$} it was found that ring closure kinetics were much affected by the nature of the *N*- or *S*-donor group, rather little by the size of the chelate ring formed ($n = 2$ or 3). The reverse ring-opening reactions are mechanistically noteworthy for their two-term rate law which indicates parallel dissociative (*D*) and nucleophilic attack by dppe pathways (*40*).

Reactions of [Pt(dipic)Cl]$^-$, dipic = dipicolinate, with 1-methylimidazole or with 1,2-diaminoethane, monitored in DMF solution, involve replacement of chloride followed by opening of the dipic chelate ring (*216*). Kinetic data for acid-catalyzed ring opening in hydrolysis of [Pt(dipic)Cl]$^-$ and of [Pt(glygly)Cl]$^-$ (*217*) are compared with those for carboplatin (*218*) in Table VIII.

Ring opening reactions are the main feature of a brief review (though with 69 references) of kinetics and mechanisms of hydrolysis and substitution reactions of platinum(II) complexes (*219*).

TABLE VII

Kinetic Parameters for Chelate Ring Closure of [PtPh$_2$(CO)–(Ph$_2$P(CH$_2$)$_n$PPh$_2$)], $n = 1$ (dppm) or 4 (dppb), in CDCl$_3$ solution (298 K)

	k_{rc} (s^{-1})	$\Delta H^{\ddagger}$ (kJ mol^{-1})	$\Delta S^{\ddagger}$ (J K^{-1} mol^{-1})
[PtPh$_2$(CO)(dppm)]	2.0×10^{-2}	88	+17
[PtPh$_2$(CO)(dppb)]	2.7×10^{-3}	38	−169

TABLE VIII

KINETIC PARAMETERS FOR ACID-CATALYZED RING OPENING OF
PLATINUM(II) CHELATES

	k^a $(dm^3\,mol^{-1}\,s^{-1})$	$\Delta H^{\ddagger}$ $(kJ\,mol^{-1})$	$\Delta S^{\ddagger}$ $(J\,K^{-1}\,mol^{-1})$	Ref.
[Pt(glygly)Cl]$^-$	2.3×10^{-2}	36 ± 5	-158	(217)
[Pt(dipic)Cl]$^-$	1.6×10^{-3}	67 ± 7	-78	(217)
Carboplatin^b	5.9×10^{-4}	88	-15	(218)

^{a} At 303 K.
^{b} Carboplatin, cis-diammine(cyclobutane-1,1-dicarboxylato)platinum(II).

3. Systems of Biochemical and Pharmacological Relevance

Two examples of aquation/anation studies of chloro-platinum(II) complexes of possible medical relevance appeared in subsection 1 above (202,207). Aquation of cisplatin is slower in the presence of DNA but not in the presence of phosphate (220). DNA also inhibits substitution in [Pt(terpy)(py)]$^{2+}$ and related complexes. For reaction of these charged complexes with iodide ion inhibition is attributable to electrostatic interactions – the complex is concentrated on the double helix and thus separated from the iodide, which distances itself from the helix. Intercalation of these complexes within the helix also serves to make nucleophilic approach by neutral reagents such as thiourea more difficult (221).

Rate constants for reaction of cis-[Pt(NH$_3$)$_2$(H$_2$O)Cl]$^+$ with phosphate and with 3'- and 5'-nucleotide bases are 4.6×10^{-3}, 0.48, and 0.16 M^{-1} s^{-1}, respectively, with ring closure rate constants of 0.17×10^{-5} and 2.55×10^{-5} s^{-1} for subsequent reaction in the latter two cases (220). Kinetic aspects of interactions between DNA and platinum(II) complexes such as [Pt(NH$_3$)$_3$(H$_2$O)]$^{2+}$, cis-[Pt(NH$_3$)$_2$(H$_2$O)$_2$]$^{2+}$, and cis-[Pt(NH$_3$)$_2$(H$_2$O)Cl]$^+$, of loss of chloride from Pt–DNA–Cl$^-$ adducts, and of chelate ring formation of cis-[Pt(NH$_3$)$_2$(H$_2$O)(oligonucleotide)]$^{n+}$ intermediates implicate cis-[Pt(NH$_3$)$_2$(H$_2$O)$_2$]$^{2+}$ rather than cis-[Pt(NH$_3$)$_2$(H$_2$O)Cl]$^+$, as usually proposed, as the most important Pt-binder (222). The role of aquation in the overall scheme of platinum(II)/DNA interactions has been reviewed (223), and platinum(II)–nucleotide–DNA interactions have been the subject of molecular modeling investigations (178).

Glutathione readily replaces the GSMe on platinum in the reaction with [Pt(dien)(GSMe)]$^{2+}$ (GSMe = S-methylglutathione) – this system is claimed to be an effective model for cisplatin–protein interaction (224). Rate constants and activation parameters have been determined for reaction of cis-$[Pt(NH_3)_2L^1L^2]^{2+}$, where L^1, L^2 = adenosine or guanosine, N1- or N7-bonded to the platinum, with thiourea. Large negative $\Delta S^{\ddagger}$ values in all cases indicate associative activation (*225*). N7 to N1 linkage isomerization is slow, both for cis-$[Pt(NH_3)_2L^1L^2]^{2+}$ with L^1, L^2 = adenosine or guanosine and, especially ($\tau_{1/2}$ of several days at 85 °C), for cis-$[Pt(NH_3)_2(9\text{-methyladenine})_2]^{2+}$ (*226*). The formation of N7 linkage isomers from diaqua precursors is kinetically favored, but the N1 isomers are thermodynamically favored. The same applies to $[Pt(dien)L]^{2+}$, where L = N7-adenosine or N7-guanosine, reacting with thiourea or with iodide – both to the k_1 and the k_2 pathways (*227*). The most interesting aspect of this system is that after dissociation of $[Pt(dien)L]^{2+}$ in acid solution in the presence of thiourea, the $[Pt(N,N\text{-}dienH)(tu)L]^{2+}$ produced reforms $[Pt(dien)L]^{2+}$ on raising the pH – the *trans* effect of the thiourea ligand is overshadowed by the much larger nucleophilic power of the incompletely chelating amine. Semi-quantitative kinetic information on the conversion of carboplatin into cisplatin in aqueous chloride media (> 18% Cl^-) can be gleaned from the Figures in Ref. (*228*).

In relation to chemotherapy involving cisplatin, a number of sulfur-donor ligands are being tested as "rescue agents" – complexes which protect healthy tissues from toxic side effects of cisplatin and other platinum anti-tumor agents. In this connection relevant kinetic information on reactions of model complexes and ligands, involving variously cisplatin, $[Pt(dien)Cl]^+$, $[Pt(dien)(GSMe)]^{2+}$ {GSMe = S-methylglutathione}, or cis-$[Pt(NH_3)_2(GSMe)_2]^{2+}$ reacting with glutathione, thiourea, thiosulfate, or diethyldithiocarbamate (ddtc), has been published. All reactions obey a simple second-order rate law, as expected for ligands with such high affinities for platinum(II). The second-order rate constants span a rather small range, from 0.2 to 8 $M^{-1}s^{-1}$, with ddtc being both the fastest and the most effective rescue agent. Despite the small range of rate constants the activation parameters are very ligand-dependent, with $\Delta H^{\ddagger}$ values from 47 to 73 kJ mol^{-1}, $\Delta S^{\ddagger}$ values from -25 to -92 J K^{-1} mol^{-1} for the reactions of $[Pt(dien)Cl]^+$. Negative activation entropies are consistent with the S_N2 mechanism indicated by the rate law (*229*).

Rotamer distribution and interconversion rates give biochemically-relevant information on hydrogen-bonding and steric interligand interactions in $[Pt(Me_3dien)(guanine)]^{2+}$ (*230*).

4. Dissociative Substitution

Dissociative substitution at square-planar centers (*231*), particularly platinum(II), was a matter of controversy for several years, with the initial report (232) allegedly being held up by disbelieving referees (194,233). Early attempts used the strong trans-labilizing effects of certain sulfur-donor ligands in efforts to promote ligand dissociation to dominate over nucleophilic attack,[3] and steric barriers, in the form of polyalkyl substitution in the dien ligand of [Pd(dien)Cl]$^+$, to the approach of the incoming nucleophile. The latter approach certainly led to greatly reduced reactivity, but the mechanism remained associative (234). A recent variant on this steric hindrance approach utilized phosphine exchange at trans-Pt(SnPh$_3$)$_2$(PMe$_2$Ph)$_2$. This follows a second-order rate law and has $\Delta S^\ddagger = -131\,\mathrm{J\,K^{-1}\,mol^{-1}}$. The mechanism is again associative despite the presence of the bulky –SnPh$_3$ groups – the π-acidity of –SnPh$_3$ may favor interaction with the incoming group (235). The failure of the S-donor and R$_n$-dien routes to impel dissociative substitution led to attention being switched to C-donor ligands as powerful strong trans-labilizers. Kinetics of ligand exchange and of water replacement at [Pt(N∩C∩N)(H$_2$O)]$^+$, where N∩C∩N is the ligand 2,6-(Me$_2$NCH$_2$)$_2$C$_6$H$_3$ (see Ref. (236) for a review of this and related ligands), were studied in the hope that the Pt–C bond might promote dissociation of trans-water and thus a dissociative mechanism. Substitution rates were dramatically increased, but activation volumes were decidedly negative, indicating associative activation yet again (237).

Further efforts along these lines to promote dissociative substitution at platinum(II) have combined this use of trans-labilization by carbon, silicon, and antimony σ-donors with the established (238) labilization by π-acceptor ligands (e.g., [Pt(phen)(H$_2$O)$_2$]$^{2+}$ reacts 100 times faster than [Pt(en)(H$_2$O)$_2$]$^{2+}$ with thiourea). Although activation volumes are now small, they remain negative (239), suggesting that the changeover to dissociative activation has been approached but not yet quite achieved. However, the isomerizations cis- to trans-[PtR(PEt$_3$)$_2$(MeOH)]$^+$ have activation volumes between +16 and +20 cm^3 mol^{-1}, in marked contrast to the negative values for the formation of cis-[PtR(PEt$_3$)$_2$(MeOH)]$^+$ from cis-[PtRR'(PEt$_3$)$_2$] (R, R' are alkyl groups). The isomerization is therefore believed to proceed by dissociation of MeOH to give a three-coordinated intermediate (240). A kinetic and computational study has been carried out of substitution and phosphine exchange at tetrahedrally distorted – allegedly by electronic rather than steric factors – cis-Pt(SiMePh$_2$)$_2$(PMe$_2$Ph)$_2$ (241). The evidence for a dissociative mechanism is the observed dependence of rate constants

---

[3]See Section IV.B.2 above for a recent example of the balance between dissociative (D) and associative mechanisms, in the case of ring opening of platinum(II) chelates of hemi-labile ligands.

on nucleophile concentration for reaction with dppe, the fact that PMe_2Ph exchange in toluene has rate $\infty [PMe_2Ph]^0$, and the activation parameters, notably the high $\Delta H^{\ddagger}$ ($118\,kJ\,mol^{-1}$) and very positive $\Delta S^{\ddagger}$ ($+120\,J\,K^{-1}\,mol^{-1}$). The three-coordinate species may be a transition state or an intermediate.

Returning to the strongly labilizing $N \cap C \cap N$ ligand, sulfur dioxide exchange with the five-coordinate adduct $[Pt(N \cap C \cap N)I(SO_2)]$ is zeroth-order in SO_2, and is therefore believed to procede by a D mechanism – a conclusion supported by the negative activation entropy ($\Delta S^{\ddagger} = +33\,J\,K^{-1}\,mol^{-1}$) for this exchange (and indeed by some ab initio studies of this system) (242).

Whereas C-bonded ligands tend to promote dissociative substitution, suitable diimine and terpyridyl ligands encourage associative substitution (200,237). When these two opposing effects are combined in one ligand, as in 6-substituted-C-deprotonated alkyl, aryl, or benzyl 2,2′-bipyridyl ligands ($N \cap N \cap C$, e.g., 3 and 4), it has been established that the diimine moiety dominates, as the reaction of $[Pt(N \cap N \cap C)Cl]$ with PPh_3 gives $[Pt(N \cap N \cap C)(PPh_3)]^+$ and Cl^- in a simple second-order process. This behavior contrasts with the analogous reactions of $[Pt(bpy)RCl]$ (R = Me or Ph), which involve a four stage reaction sequence of chloride replacement, a monodentate bpy intermediate, $cis \rightarrow trans$ isomerization, and finally loss of bpy. These associative processes also contrast with dissociative ligand exchange at $[Pt(bphy)(SR_2)_2]$, where bphy is the chelating two-carbon donor ligand doubly deprotonated 2,2′-biphenyl (200). A combined kinetic and theoretical approach (243) shows that cis-$PtR_2(SR')_2+CO$ is not dissociative, thanks to the strong π-acceptor properties of CO.

3

4

The reaction of cis-$[PtMe_2(dmso)_2]$ with pyridine in toluene exhibits two-stage kinetics. The first stage has parallel A (established through saturation kinetics; $\Delta V^{\ddagger} = -11.4\,cm^3\,mol^{-1}$) and D (limiting dissociative established by effects of added dmso, i.e. by dependences of rate constants on leaving and incoming ligand concentrations) pathways. The second is predominantly dissociative ($\Delta S^{\ddagger} = +86\,J\,K^{-1}\,mol^{-1}$), though there is also a minor associative path at high pyridine concentrations (*244*).

5. *Intramolecular Processes*

Intramolecular coordination site exchange reactions in square-planar platinum(II) complexes $[Pt(tbte)Cl]^+$ of the potentially four-donor-atom tripodal ligand tris[2-(t-butyl-thio)ethyl]amine (tbte) *N,S,S*-bonded occur through a trigonal bipyramidal five-coordinate intermediate (*245*). Inversion at sulfur (also selenium and, occasionally, tellurium) coordinated to Pt(II), and indeed to a number of other transition metal centers, e.g., W(0), Re(I), Rh(III), and especially Pt(IV), has been studied extensively over several decades (*246*). Observation of the kinetics of such processes is often complicated by concurrent changes elsewhere in the complex, for example by hindered rotation about C–S in the square-planar complexes $[(XS)_2Pt(\mu\text{-}SX)_2Pt(SX)_2]^{2-}$ with $X = C_6F_4H$, $C_6F_5, C_6F_4(4\text{-}CF_3)$. Resolution of the observed kinetics here indicated barriers ($\Delta G^{\ddagger}$) of between 54 and 59 kJ mol^{-1} for inversion at coordinated sulfur, between 40 and 60 kJ mol^{-1} for the hindered rotation (*247*).

C. PALLADIUM(II)

Pd^{2+}_{aq} reacts reversibly with (bi)sulfate; the forward reaction follows a simple second-order rate law, with $k_2(HSO_4^-) = 119 \, M^{-1}s^{-1}$ (*248*). For reactions of Pd^{2+}_{aq} with 13 carboxylic acids, where stability constants in the region of unity permit the ready determination of forward and reverse rate constants, $\Delta S^{\ddagger}$ values for the formation reactions are in the range -7 to $-78 \, J \, K^{-1} mol^{-1}$, for the dissociations -12 to $-42 \, J \, K^{-1}$ mol^{-1}. $\Delta V^{\ddagger}$ values, for formation and for dissociation of five of these complexes, are within the range -1 to $-9 \, cm^3 \, mol^{-1}$. These results, $\Delta S^{\ddagger}$ and $\Delta V^{\ddagger}$ values reported earlier (and conveniently collected here) for formation reactions of Pd^{2+}_{aq}, and volume profiles are all consistent with the expected associative activation. The slope of a correlation plot of $\Delta V^{\ddagger}$ with ΔV° indicates that bond-making dominates the transition state for complex formation. There is a good isokinetic plot for these Pd^{2+}_{aq} plus carboxylate reactions (*249*).

Activation parameters $\Delta H^{\ddagger}$, $\Delta S^{\ddagger}$ and $\Delta V^{\ddagger}$ have been determined for all nine combinations of $[Pd(LLL)(H_2O)]^{2+}$, LLL = dien, terpy, or bis(2-pyridylmethyl)amine reacting with L-cysteine (cys), DL-penicillamine (pen), or glutathione (glu) (*250*). All $\Delta S^{\ddagger}$ and $\Delta V^{\ddagger}$ values are negative, consistent with associative activation, but there is no $\Delta S^{\ddagger}/\Delta V^{\ddagger}$ correlation, except for the three reactions with pen. Indeed all the reactions with cys and glu have essentially the same $\Delta V^{\ddagger}$, although $\Delta S^{\ddagger}$ ranges from -54 to $-102 \, \mathrm{J \, K^{-1} mol^{-1}}$. $[\mathrm{Pd(glycyl\text{-}L\text{-}methionine)Cl}]^+$ reacts more slowly than $[\mathrm{Pd(dien)Cl}]^+$ with such nucleophiles as glu, cys, or pen (reactivities glu > cyst > pen); the range of $\Delta H^{\ddagger}$ values (38–$60 \, \mathrm{kJ \, mol^{-1}}$) and negative $\Delta S^{\ddagger}$ values (-20 to $-76 \, \mathrm{J \, K^{-1} mol^{-1}}$) indicate associative activation (251).

Electrostatically-controlled pre-association interactions have an important effect on rates for $[\mathrm{Pd(dien)Cl}]^+$ reacting with thione-containing nucleosides, nucleotides and oligonucleotides, as is often the case for reactions between metal complexes and this type of biological ligand. Interaction between the charged complex and the polyanionic oligonucleotide surface leads to an increase in both enthalpy and entropy of activation in the DNA or model environment (252).

Products of substitution of inosine and guanosine $5'$-monophosphate for chloride or for water on ternary aminocarboxylate complexes such as $[\mathrm{Pd(mida)(D_2O)}]$, where mida = N-methyliminodiacetate, or $[\mathrm{Pd_2(hdta)Cl_2}]^{2-}$, where hdta = 1,6-hexanediamine-N,N,N',N'-tetraacetate, is subject to mechanistic controls in terms of number of coordinated donor atoms and pendant groups and of the length of the chain joining the functional groups in the bis-iminodiacetate ligands. These factors determine the nature and stereochemistry of intermediates and the relative amounts of mono- and bi-nuclear products (253).

Diimines such as bpy and phen replace 2-(arylazo)pyridine ligands (aap) in $[\mathrm{Pd(aap)Cl_2}]$ by a simple second-order process, whose detailed mechanism may depend on the nature of the incoming ligand (254). Three phthalocyanine units, each containing Zn^{2+}, can be bonded to tetrahedral phosphorus, to give $[\mathrm{PPh(pc\text{-}Zn)_3}]^+$. Mechanistic proposals are advanced for this novel exchange reaction in which palladium-bound phthalocyanine replaces phenyl on phosphorus (255).

Brief mentions of kinetics and mechanisms of reactions of nitrogen bases with a selection of palladium(II) complexes with ammine, amines, pyridine-2-carboxylate, pyridoxine, and related ligands are included in a review of analogous platinum(II) reactions (194).

Reaction of $[\mathrm{Pd(pica)(H_2O)_2}]^{2+}$ (pica = 2-picolylamine) with cbdc, cyclobutane-1,1-dicarboxylate, to give $[\mathrm{Pd(pica)(cbdc)(H_2O)}]$, containing monodentate cbdc, is characterized by an activation volume close to zero, indicating a balance between a negative contribution from associative activation and a positive contribution from solvational changes associated with transition state formation (256).

Reaction of the binuclear complex $[(\mathrm{bpy})\mathrm{Pd}(\mu\text{-}\mathrm{OH})_2\mathrm{Pd}(\mathrm{bpy})]^{2+}$ with DL-methionine (met) obeys a simple second-order rate law ($\Delta H^{\ddagger} = 46 \, \mathrm{kJ \, mol^{-1}}$; $\Delta S^{\ddagger} = -101 \, \mathrm{J \, K^{-1} mol^{-1}}$). The mechanism suggested is rate-determining associative attack of the methionine-sulfur to give [(bpy)Pd(met)(μ-OH)Pd(OH)(bpy)]$^{2+}$ followed by rapid attachment of the coordinated sulfur to the second Pd to give [(bpy)Pd(μ-met)(μ-OH)Pd(bpy)]$^{3+}$ (257).

For intramolecular coordination site exchange reactions in [Pd(tbte)Cl]$^{+}$ (tbte is the four-donor tripodal ligand tris[2-(t-butyl-thio)ethyl]amine) it is possible, in acetone-d$_6$ solution, to see the four $\rightleftharpoons$ five coordinate equilibrium involved (245). Barriers to isomer interconversion in [Pd(pzpym)XCl], pzpym = pyrazolylpyrimidine, X = Me or COMe, have been determined from NMR coalescence temperatures (160). Activation parameters ($\Delta H^{\ddagger}$ and $\Delta S^{\ddagger}$) for fluxional behavior of pyramidal (apical Br, I) tetranuclear complexes [Pd$_4$(2,4,6-tris(trifluoromethyl)phenyl)$_4$X$_5$]$^{-}$ have been discussed in terms of three possible dynamic processes (258). Binuclear complexes X–Pd–terpy–spacer–terpy–Pd–X, where X is pyridine with an anthracene-bearing carboxylate substituent, may act as molecular receptors. Such a function would require intramolecular stereochemical rearrangement, of one X group with respect to the other; NMR shows this to be a rapid process with the barrier to interconversion $\sim 40\,\mathrm{kJ\,mol^{-1}}$ (259).

D. NICKEL(II)

Large negative $\Delta S^{\ddagger}$ values for redistribution and replacement reactions of planar Ni(II) complexes of N-alkyl-α-aminoxime ligands:

$$\mathrm{NiL_2^+ + NiL_2'^+ \rightarrow 2NiLL'^+} \quad \text{and} \quad \mathrm{NiL_2^+ + L' \rightarrow NiLL'^+ + L}$$

indicate a rate-determining step of dissociative Ni–L bond breaking in each case (260).

The replacement of both N-alkylsalicylaldiminate ligands in a bis-ligand complex NiL$_2$ by a tetradentate ligand of the salen type follows a simple second-order rate law, interpreted in terms of rapid reversible addition of the incoming ligand to the nickel, to give a six-coordinate intermediate. This intermediate loses the two bidentate ligands in two relatively slower steps which may be kinetically distinguishable. Such reactions involving optically active complexes show modest chiral discrimination (261).

Formation and dissociation kinetics of nine Ni(II)-macrocyclic tetrathiaether complexes (eight macrocyclic, one linear; in acetonitrile) have been compared with those for Cu(II) analogues and for Ni(II) complexes with macrocylic tetramines (262). Whereas for the tetramine complexes conformational changes may be apparent in the kinetics this is not the case for the tetrathiaether complexes, where there is no kinetic evidence for slow conformational changes after initial bonding of the ligand to the nickel. Ni(II) and Cu(II) complexes exhibit similar patterns in their formation kinetics, but not in dissociation.

Kinetic parameters for formation of $[Ni(CN)_5]^{3-}$, which strictly is addition to $[Ni(CN)_4]^{2-}$ rather than substitution, may be found in Table VI in Section IV.A above.

E. GOLD(III)

A review of the photochemistry of gold(III), and of gold(I), complexes deals both with substitution and redox reactions (263).

Studies of substitution at gold(III) are severely restricted by the oxidizing properties of this metal center. However, chloride and cyanide both resist oxidation by gold(III), and it is therefore possible to establish kinetic parameters for chloride and for cyanide exchange. Rate constants are 10^2 to 10^3 times larger than for platinum(II). Chloride exchange obeys a second-order rate law, with $\Delta S^{\ddagger}$ and $\Delta V^{\ddagger}$ values of $-31\,J\,K^{-1}\,mol^{-1}$ and $-14\,cm^3\,mol^{-1}$ indicating an associative mechanism. The situation in respect of cyanide is more complicated, with a strong pH dependence – the overall rate law indicates three parallel pathways. Values of $-38\,J\,K^{-1}\,mol^{-1}$ and $+2\,cm^3\,mol^{-1}$ for $\Delta S^{\ddagger}$ and $\Delta V^{\ddagger}$ suggest, but do not unequivocally indicate, an interchange with modest associative character for cyanide exchange (193). Kinetic parameters for cyanide exchange with $[Au(CN)_4]^-$ are compared with those for cyanide exchange at other square-planar centers in Table VI (see Section IV.A above).

Kinetic and equilibrium studies have established a rate constant of $0.018\,s^{-1}$ for hydrolysis of $[Au(terpy)Cl]^{2+}$ and provided an estimated value of $\geq 16\,000\,M^{-1}\,s^{-1}$ for the reverse reaction, formation of $[Au(terpy)Cl]^{2+}$ from $[Au(terpy)(H_2O)]^{3+}$ (264).

F. RHODIUM(I) AND IRIDIUM(I)

The mechanism of substitution at these centers can conveniently be probed through CO exchange at the cis-$[M(CO)_2X_2]^-$ anions (M = Rh, Ir; X = Cl, Br or I). The rate law for these exchanges, and the activation parameters shown in Table IX, suggest the operation of a limiting A mechanism (265).

Replacements of ferrocene-substituted β-diketone ligands, β-dik, in cyclooctadiene-rhodium(I) complexes [Rh(β-dik)(cod)] by 1,10-phenanthroline are characterized by large negative activation entropies, indicating the operation of the expected associative mechanism, although the

TABLE IX

ACTIVATION ENTROPIES AND VOLUMES FOR CARBON MONOXIDE EXCHANGE
WITH cis-$[M(CO)_2X_2]^-$

X	$\Delta S^{\ddagger}$ (J K^{-1} mol^{-1})		$\Delta V^{\ddagger}$ (cm^3 mol^{-1})	
	$[Rh(CO)_2X_2]^-$	$[Ir(CO)_2X_2]^-$	$[Rh(CO)_2X_2]^-$	$[Ir(CO)_2X_2]^-$
Cl	−125	−135	−17.2	−20.9
Br	−123	−122		
I	−98	−107		

authors feel that Rh–N bond making actually contributes rather little to transition state formation. Rate constants are largely controlled/determined by substituent electronegativities and ligand pK values (*266*).

V. Reactions at Labile Transition Metal Centers

A. METAL(II) CATIONS

1. General

Ni^{2+} was very popular in the early days of the investigation of mechanisms of complex formation, since the time-scale for its reactions with simple ligands was so convenient for the then recently developed stopped-flow technique. However, interest has now moved on to other first-row cations, especially to Cu^{2+}. A review of the kinetics and mechanisms of formation of tetraazamacrocyclic complexes concentrates on Ni^{2+} and Cu^{2+}, and their reactions with cyclam and similar ligands (*267*). The tetra(4-sulfonatophenyl)porphyrin complexes of Ni^{2+} and of Cu^{2+} react immeasurably slowly with cyanide, but their N-methyl derivatives do react, albeit extremely slowly. The relevant time scales are hours for removal of Ni^{2+}, months for the removal of Cu^{2+}, by 10^{-4} M cyanide at pH 7.4 (*268*).

Rate constants for complex formation between Mn^{2+}, Co^{2+}, Ni^{2+}, Cu^{2+} and Zn^{2+} and 5,10,15,20-tetraphenylporphyrin in acetonitrile correlate with rate constants for acetonitrile exchange at these cations, though of course the reactions with the porphyrins are much (10^4 to 10^6 times) slower than the corresponding rates of solvent exchange (*269*).

The kinetics of reaction of Co^{2+}, Ni^{2+}, Cu^{2+}, or Zn^{2+} with terpy in DMSO–water mixtures are complicated, possibly both by the terdentate nature of this flexible ligand and by the fact that both solvent components are potential ligands. Extensive study and discussion lead to a proposed mechanism centered on rate-determining ring closure in a bidentate-terpy intermediate (270). Kinetic studies of the transfer of a quadridentate N_2S_2 ligand from M^{2+} (M=Cr, Mn, Fe, Co or Ni) to Cu^{2+} (271) and of reaction of Cr^{2+}_{aq} with bipym (272) provide rare examples of ligand substitution studies involving the very labile, air-sensitive, and strongly reducing Cr^{2+}. The N_2S_2 ligand-transfer study revealed unexpected reactivities of bimetallic intermediates. The rate constant for reaction of Cr^{2+}_{aq} with bipym ($k_f = 1.6 \times 10^8 \, M^{-1} s^{-1}$) is similar to those reported for the few previously-studied complex-formation reactions of Cr^{2+}_{aq}.

It is by now well established that $\Delta V^{\ddagger}$ trends for complex formation and for solvent exchange show a change of mechanism from associative to dissociative as one goes from left to right across the first row of the transition metal 2+ cations. A recent illustration is afforded by amine and diamine reactions at Mn^{II}, Fe^{II}, Ni^{II}; exchange of 1,3-propanediamine also shows how steric bulk may affect mechanism (273).

2. Iron(II)

Activation volumes for formation ($\Delta V^{\ddagger}_f = +6.1 \, cm^3 \, mol^{-1}$) and for dissociation ($\Delta V^{\ddagger}_d = +1.3 \, cm^3 \, mol^{-1}$) of $Fe(NO)^{2+}_{aq}$ indicate dissociative interchange mechanisms in both directions, though the very small negative values for the activation entropies ($\Delta S^{\ddagger}_f = -3 \, J \, K^{-1} \, mol^{-1}$, $\Delta S^{\ddagger}_d = -15 \, J \, K^{-1} \, mol^{-1}$) suggest that the character of the process is close to the pure interchange boundary – not altogether unexpectedly for Fe^{2+}. Interestingly, Mössbauer and electron paramagnetic resonance (EPR) spectra of the product suggest that a formulation based on Fe^{III}–NO^- (cf. cobalamin(III)–NO^- in Section II.A.2 above (71)) is to be preferred to the Fe^I–NO^+ form usually given in text books (274). Aminocarboxylatoaqua-complexes of iron(II) react somewhat more quickly with nitric oxide than does Fe^{2+}_{aq} (275); there is good agreement between equilibrium constants for formation of these NO complexes determined from kinetic measurements and values determined spectrophotometrically (276). Activation volumes for reactions of most aminocarboxylatoaqua-complexes of iron(II) with NO indicate dissociative interchange, though the negative value of $\Delta V^{\ddagger}$ for reaction of $[Fe(nta)(H_2O)_2]^-$ with NO suggests a mechanism of associative interchange in this case (277).

Reactions of NO with water-soluble Fe(II), and Co(II), porphyrin complexes are very fast ($k \sim 10^9 \, M^{-1} \, s^{-1}$) and characterized by small positive values of $\Delta S^{\ddagger}$ and $\Delta V^{\ddagger}$, suggesting diffusion-dominated dissociative interchange (278). Reversible binding of NO to iron(II) may be tuned by using aminocarboxylate and related complexes in aqueous solution (276). Formation of $Fe(tpp)(NO_2)$ occurs with traces of NO_2; coordinated $-NO_2$ labilizes the coordinated $-NO$ in the reversibly-formed intermediate $[Fe(tpp)(NO)(NO_2)]$ (279).

3. Cobalt(II)

Kinetic studies on complex formation reactions of the tripodal tetramine complex $[Co(Me_6tren)(H_2O)]^{2+}$ with pyridine, 4-methylpyridine, and imidazole yielded activation parameters $\Delta H^{\ddagger}$ and $\Delta S^{\ddagger}$. Activation parameters and dependences of rate constants on incoming ligand concentration indicated that the formation mechanism ranged from dissociative for the weaker and bulkier incoming ligands (py, 4-Mepy) to associative for the more basic and less bulky imidazole; 2-methylimidazole occupies an intermediate position (280).

Rate constants for aquation of the trigonal bipyramidal $[M(Me_3tren)(H_2O)]^{2+}$ complexes of Co^{2+} and Cu^{2+} are remarkably similar, at 34 and $106\,s^{-1}$, respectively (281). Co_{aq}^{2+} reacts with a typical zinc finger peptide with $k_f = 7.5 \times 10^4\,M^{-1}s^{-1}$; the product reacts with Zn_{aq}^{2+} rather slowly ($k = 5.3 \times 10^{-2}\,s^{-1}$) by a dissociative mechanism (282).

Redistribution of 6-borneolterpy (brnterpy) ligands on cobalt(II):

$$[Co(S\text{-brnterpy})_2]^{2+} + [Co(R\text{-brnterpy})_2]^{2+}$$
$$= 2[Co(S\text{-brnterpy})(R\text{-brnterpy})]^{2+}$$

is remarkably slow for a d^7 metal center. The system requires about a day to reach equilibrium ($k_2 \sim 6 \times 10^{-5}\,M^{-1}s^{-1}$). Analogous reactions for complexes of 4'-substituted terpyridyls are much faster ($k_2 \sim 10^3\,M^{-1}s^{-1}$), as reported earlier for 5,5''-disubstituted analogues ($k_2 \sim 10^3\,M^{-1}s^{-1}$ (283)), except for the 4'-ferrocenylterpy system, where redistribution is again slow ($k_2 \sim 6 \times 10^{-5}\,M^{-1}s^{-1}$) (284).

A plot of k_{obs} against DNA concentration for reaction of tetraphenylporphyrin-1-tryptophan-cobalt(II) with calf thymus DNA is linear; $k_f \sim 2 \times 10^4\,M^{-1}s^{-1}$ at room temperature. The suggested mechanism is of rapid reversible addition of water to 5-coordinate $[Co(trp)(tpp)]$ followed by rate-limiting reaction of $[Co(trp)(tpp)(H_2O)]$ with the DNA (285).

4. Nickel(II)

Reaction of Ni^{2+} with 1,4,7-triazacyclononane-N,N',N''-triacetate ($tcta^{3-}$) is relatively slow (k_{288} between 20 and $90\,s^{-1}$; pH dependent);

the rate-limiting step is a first-order rearrangement, coupled to proton release, of rapidly and reversibly formed [Ni(tctaH)] (*286*). Ni(dmf)$_6^{2+}$ reacts with tetra-*N*-alkylated cyclam ligands, as with cyclam, in two kinetically-distinct stages. Rapid complex formation, with formation of the second Ni–N bond probably rate-limiting, is followed by slow isomerization of the kinetically-favored to the thermodynamically-favored form of the product. These reactions are markedly accelerated by added chloride, due to rapid formation of [Ni(dmf)$_5$Cl]$^+$, which is more labile than [Ni(dmf)$_6$]$^{2+}$ (*287*).

Kinetics of H$^+$-promoted dissociation of the Ni^{2+} complex of a tetra-dentate aza-oxa-cryptate derived from tren, conducted in acidic aqueous acetonitrile, indicate that its dissociation rate is smaller than that of [Ni(tren)(H$_2$O)$_2$]$^{2+}$, despite the much higher thermodynamic stability of the tren complex – a "kinetic cryptate effect" is invoked to rationalize this (*288*).

Kinetic evidence obtained for intramolecular proton transfer between nickel and coordinated thiolate, in a tetrahedral complex containing the bulky triphos ligand (Ph$_2$PCH$_2$CH$_2$)$_2$PPh to prevent interference from binuclear μ-thiolate species, is important with respect to the mechanisms of action of a number of metalloenzymes, of nickel (cf. urease, Section VII.B.4) and of other metals (*289*).

5. Copper(II)

a. Formation Reactions of Cu$_{aq}^{2+}$ Cu$_{aq}^{2+}$ reacts with the polyamine Me$_2$octaen (L) to give 1:1 and 2:1 complexes, in varying degrees of proto-nation. Reaction of Cu$_{aq}^{2+}$ with H$_6$L^{6+} seems to proceed by the ICB mechanism, with the rather small ICB effect here (very much smaller than for Ni$_{aq}^{2+}$+Me$_2$octaen) attributed to the rapidity of the reaction at Jahn-Teller-distorted Cu^{2+} (*290*). A kinetic study of complex formation with 12 tripodal aminopolythiaether and aminopolypyridyl ligands in aqueous solution indicated formation of the first Cu–N or Cu–S bond to be rate-limiting, except when all three arms contain S-donor atoms. In this case it is formation of the second Cu–S bond which is rate-limiting (*291*).

Formation kinetics for eight tetraaza macrocycles of the cyclam type reacting with copper(II) have been analyzed in terms of rate constants for reaction with [Cu(OH)$_3$]$^-$ and with [Cu(OH)$_4$]$^{2-}$. There is a detailed discussion of mechanism and of specific steric effects (*292*). Complex formation from cyclam derivatives containing –NH$_2$ groups on the ring –CH$_2$CH$_2$CH$_2$– units proceeds by formation followed by kinetically-dis-tinct isomerization. The dramatic reactivity decreases consequent on protonating these $-NH_2$ substituents suggests that they play a key role in the initial formation step (*293*). Formation kinetics for several pendant arm macrocycles reacting in strongly alkaline solution with copper(II), i.e. with $[Cu(OH)_3]^-$ and $[Cu(OH)_4]^{2-}$ indicate a two-stage process, of formation and isomerization (*294*). Equilibrium studies on the protonation and Cu(II) complexation by an hexaaza macrocycle containing *p*-xylyl spacers between dien units are complemented by the crystal structure of the hexaprotonated ligand and the kinetics of decomposition of its mono- and bi-nuclear Cu(II) complexes. Differences between the complexes with *p*-xylyl and *m*-xylyl spacers are remarkably small (*295*). Comparison of rate constants for formation of mono- and bi-nuclear complexes of a flexible octa-azamacrocycle, at high pHs, demonstrates that the kinetics of formation of the binuclear complex are statistically controlled (*296*).

In a study of the mechanism of Cu^{2+} incorporation into a series of four bis-tetraazamacrocycles (L) with various combinations of 12- and 14-membered rings it proved necessary to isolate intermediates, and to characterize them spectroscopically and kinetically, in order to analyze the kinetic data fully. The mechanism is particularly interesting for bis-macrocycles which contain a 12-membered ring, for here a Cu^{2+} ion may be partially coordinated by both rings at an intermediate stage, with a kinetically-significant intramolecular rearrangement prior to reaction with the second Cu^{2+} to form the final CuL_2 product (*297*).

Formation kinetics have been established for Cu_{aq}^{2+} reacting with the aminoglycoside neamine (**5**) and with 2-deoxystreptamine (**6**). Despite the complicated nature of neamine, its reaction with Cu^{2+} in water at pH∼7 is a simple two-step process, in methanol a single-step reaction (*298*). These reactions are remarkably slow for complex formation from Cu_{aq}^{2+}.

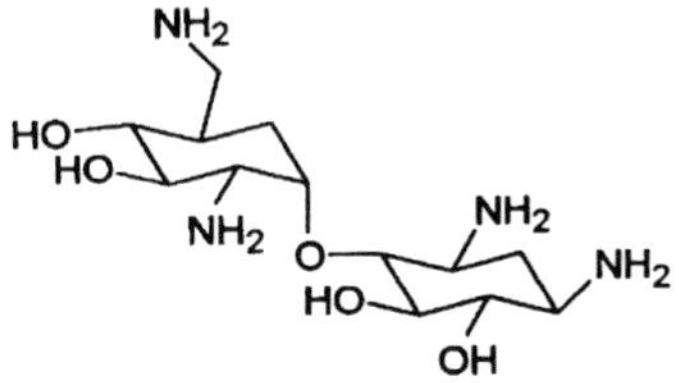

5

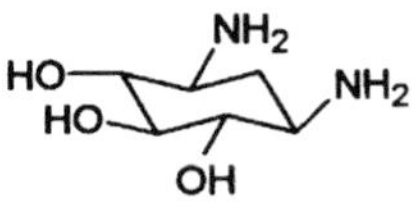

6

Chelate ring opening is suggested to be the rate limiting step in the reaction of Cu^{2+} with tris-((4-methoxy)-2-diphenyl-dithiolenato) tungsten(VI), and with its 4-dimethylamino analogue. These reactions provide interesting examples of a coordination complex as incoming ligand in complex formation kinetics (*299*).

The transfer of a quadridentate N_2S_2-donor ligand from M^{2+} (M = Cr, Mn, Fe, Co or Ni) to Cu^{2+} (*271*), already mentioned in Section V.A.1, has a formal connection with an investigation of the mechanism of copper delivery to metalloproteins, such as copper zinc superoxide dismutase. Both are ligand exchange reactions of the type $ML + CuL' \rightarrow ML' + CuL$ (*300*).

The kinetics for Cu^{2+} reacting with four porphyrins, to form Sitting-Atop (SAT) complexes, in aqueous acetonitrile illustrate both the difference in reactivity between $[Cu(MeCN)_6]^{2+}$ and $[Cu(MeCN)_5(H_2O)]^{2+}$ and the effects of peripheral substituents. There is a range of at least 10^3 in rate constants as a consequence of substituent effects on porphyrin flexibility (*301*). Rate constants for deprotonation of the copper(II)-5,10,15,20-tetraphenylporphyrin SAT complex by bases such as 3-picoline, DMF, or DMSO, in aqueous acetonitrile, correlate linearly with the σ-donor properties of the respective bases. The kinetic results indicate a mechanism of nucleophilic attack by the base at a pyrrole proton, and also provide evidence germane to mechanisms of porphyrin metallation (*302*). Intercalation of tetrakis-(*N*-methylpyridinium-4-yl) porphinecopper(II) into DNA is very rapid, despite this complex being overlarge and thus forcing structural modification of the DNA. However, aggregation of this complex on the surface of DNA is much slower, typically with a half-life of a few hundred seconds, and is autocatalytic (*303*).

b. Dissociation and Ligand Replacement The copper(II) complex of the *N*-methylene-(phenylphosphinate) derivative of cyclam dissociates much more rapidly than its cyclen analogue, which in turn dissociates much more rapidly than its parent $[Cu(cyclen)]^{2+}$ cation. The dissociation mechanism for the phosphinate derivatives involves protonation prior to dissociation, with the transfer of a proton from a phosphinic pendant arm to the azamacrocyclic ring being an important feature of the mechanism (*304*).

Acid-promoted aquation of the binuclear complex Cu_2L of the hexaaza macrocycle L = 2,5,8,17,20,23-hexaaza[9.9]paracyclophane, whose half-life is of the order of a second, exhibits simple one-stage first-order kinetics. This is attributed to parallel reactions at each Cu(II) center having identical rate constants (*305*). The kinetics of dissociation of mono- and bi-nuclear Cu(II) complexes of a hexaaza macrocycle containing p-xylyl spacers have already been mentioned (295). Comparison of rate constants for acid-promoted dissociation of mono- and bi-nuclear Cu(II) complexes of an octaaza macrocycle show that kinetics of dissociation of the binuclear complex are, as for its formation (v.s.), statistically controlled. Apparently this macrocycle is both flexible enough and large enough (no Cu $\cdots$ Cu interaction) for substitution at two coordination centers to be independent processes (296). The effect of SCN$^-$ as ancillary ligand on kinetics of decomposition of binuclear copper(II) complexes with a symmetrical hexaaza macrocycle has been analyzed in terms of contributions from the species Cu_2L^{4+}, $Cu_2L(OH)^{3+}$, $Cu_2L(OH)_2^{2+}$, $Cu_2L(NCS)^{3+}$, and $CuL(NCS)_2^{2+}$ (306).

There are parallel ligand-dependent (claimed to be associative) and ligand-independent paths in the reactions of $[Cu(bigH)]^{2+}$, big = biguanide, with aminoacids (aaH); rate-limiting formation of the intermediate $[Cu(bigH)(aa)]^+$, is followed by rapid formation of $[Cu(aa)_2]$ (307).

c. Five-coordinate Copper(II)   The most interesting aspects of ternary aqua-copper(II) complexes are the great changes both in reactivity and in mechanism on going from Cu_{aq}^{2+} to aqua-complexes containing certain polydentate ligands, for instance tren, tris-(aminoethyl)amine, $N(CH_2CH_2NH_2)_3$. Rate constants for substitution at such complexes are very much smaller than for Cu_{aq}^{2+}, there being a 10^3 to 10^4 reduction in rate constants on going from Jahn-Teller distorted octahedral Cu_{aq}^{2+} to five-coordinate ternary aqua-Cu^{2+}. The tren ligand restricts motion within $[Cu(tren)(H_2O)]^{2+}$ and thus diminishes the effects of the Jahn-Teller distortion. Activation volumes of between -7 and $-10\,cm^3\,mol^{-1}$ indicate a changeover from dissociative to associative activation $(I_d \rightarrow I_a)$. However, the incorporation of six methyl substituents into the tren ligand causes sufficient steric crowding for substitution at the Me$_6$tren complex to be around a thousand times smaller and for the mechanisms to revert to dissociative activation (281,308,309) (see also Ref. (158) in Chapter 1, Table X). $[Cu(Me_3tren)(H_2O)]^{2+}$ dissociation in acidic aqueous solution has $\Delta S^\ddagger$ and $\Delta V^\ddagger$ values of $-6\,J\,K^{-1}\,mol^{-1}$ and $+0.3\,cm^3\,mol^{-1}$. It is suggested that these near-zero values represent dissociative interchange with the effects of one ligand arm coming away from the copper compensated by partial bonding of the incoming water molecule to the Cu^{2+} ion (281). Volume profiles for the $[Cu(tren)(H_2O)]^{2+}/$ pyridine and $[Cu(Me_3tren)(H_2O)]^{2+}/$pyridine systems indicate I_a mechanisms. For the Me$_3$tren complexes, there are similar contractions, of -8.7 and $-6.2\,cm^3\,mol^{-1}$, for replacement of water and of pyridine, respectively (309). Activation volumes of -3.0 and $-4.7\,cm^3\,mol^{-1}$ for water exchange at $[Cu(tmpa)(H_2O)]^{2+}$ and at $[Cu(fz)_2(H_2O)]^{2+}$ {tmpa = tris(2-pyridylmethyl)amine; fz = ferrozine} (310) (see also Ref. (157) in Chapter 1, Table X) are relevant here, for they emphasize the contrast between I_d substitution at Jahn-Teller-distorted Cu_{aq}^{2+} and I_a substitution at five-coordinated Cu^{2+}.

The weakly bonded axial water in five-coordinate $[CuL(H_2O)]$, where L = N,N'-bis(2-pyridylmethylene)-1,3-diamino-2,2-dicarboxyethylpropane (7), is rapidly replaced by chloride in aqueous hydrochloric acid, with subsequent slow dissociation in two kinetically-distinct steps. The acid-independent pathway of the first stage is suggested to involve Cu–N bond rupture preceded by rate-limiting displacement of axial chloride by a ligand –NH– group. The acid-dependent pathway has an $[H^+]^2$ dependent term, suggesting that both coordinated –NH– groups have to be protonated prior to ligand loss; again there is axial site involvement (311).

7

Rate constants for the replacement of water by azide or thiocyanate from the five coordinate (2tyr; 2his, $1H_2O$) copper center in *Fusarium* galactose oxidase decrease with increasing pH, due to the greater difficulty of displacing OH^- (312).

6. Rhodium(II)

There is continuing interest in substitution at binuclear rhodium(II) complexes, both for intrinsic interest and in relation to organometallic catalysis kinetics (313). The weakness of the metal–metal bond and relative stability of Rh–C and other Rh–L bonds favor Rh–Rh bond breaking as the first step in such catalysis. However, substitution at the rhodium(II) without Rh–Rh bond fission is possible, and indeed activation parameters $\Delta H^{\ddagger}$ and $\Delta S^{\ddagger}$ have been reported for substitution of equatorial acetonitrile in the $[Rh_2(OAc)_2(MeCN)_4]^{2+}$ cation. For reaction with PMe_3 or P(cyclohexyl$_3$) the substitution mechanism comprises fast reversible replacement of one axial MeCN followed by relatively slow formation of a bis-substituted (one axial, one equatorial) derivative, these ligand replacements being of an interchange character (314).

Activation parameters for Rh–Rh bond fission in a tetra-t-butyl-salicylaldiminatodirhodium(II) complex are $\Delta H^{\ddagger} = 65\,\text{kJ}\,\text{mol}^{-1}$ and $\Delta S^{\ddagger} = +22\,\text{J}\,\text{K}^{-1}\,\text{mol}^{-1}$. $\Delta H^{\ddagger}$ is about $10\,\text{kJ}\,\text{mol}^{-1}$ higher than ΔH° for breaking the Rh–Rh bond, a difference within the normal range. This binuclear complex is remarkably more reactive, e.g., with respect to reaction with ethene, in benzene than in THF (315). The octaethyltetraazaporphyrinatorhodium dimer, [Rh(oetap)]$_2$, reacts more slowly than its octaethylporphyrin parent, [Rh(oep)]$_2$, with ligands such as trimethylphosphite – and indeed fails to react with some ligands that react with [Rh(oep)]$_2$. These observations are interpreted in terms of a stronger rhodium–rhodium bond in [Rh(oetap)]$_2$ (316).

7. Zinc(II); Cadmium(II); Mercury(II)

Ligand release in the reaction of bis(N-alkylsalicylaldiminato)zinc(II) complexes with ammonium ions in acetonitrile follows first-order kinetics, albeit in double exponential form for the t-butyl complex. The first bond to break is Zn–O (317).

The identification of different carbonate binding modes in copper(II) and in zinc(II)/2,2'-bipyridine or tris(2-aminoethyl)amine/(bi)carbonate systems, specifically the characterization by X-ray diffraction techniques of both η^1 and η^2 isomers of [Cu(phen)$_2$(HCO$_3$)]$^+$ in their respective perchlorate salts, supports theories of the mechanism of action of carbonic anhydrase which invoke intramolecular proton transfer and thus participation by η^1 and by η^2 bicarbonate (55,318).

Relative reactivies of the species Zn^{2+}, Zn(OH)$^+$, Zn(OH)$_3^-$, and Zn(OH)$_4^{2-}$ have been established for the reaction of zinc(II) with tetra(N-methyl-4-pyridyl)porphyrins in basic solution (319). The rate constant for reaction of a typical zinc finger peptide with Zn$_{aq}^{2+}$ has been estimated as $2.8 \times 10^7\,\text{M}^{-1}\,\text{s}^{-1}$, for dissociation of this complex $1.6 \times 10^4\,\text{s}^{-1}$ (282).

The demonstration of the formation of a hexanuclear zinc complex with the S-donor ligand 2-aminoethanethiolate, containing Zn$_3$S$_3$ and Zn$_4$S$_4$ cyclic units, contributes to the building up of a pattern of polynuclear complex formation based on coordination preferences of the metal ions involved (320) – reaction of Zn^{2+} with salicylideneamino ligands and pyrazine can give linear tetranuclear complexes (321). Another hexanuclear zinc complex appears in the section on supramolecular chemistry below (Section VII.D).

Reactivities of Zn^{2+} and Cd^{2+} have been compared with those of Ni^{2+} in a study of kinetics of dissociation of their respective 1,7-diaza-4,10,13-trioxacyclopentadecane-N,N'-diacetate and 1,10-diaza-4,7,13,16-tetraoxacyclooctadecane-N,N'-diacetate complexes. Cu^{2+} was used as scavenger, as its complexes are, as usual, more stable than their Ni^{2+}, Zn^{2+} and Cd^{2+} analogues. Measured rate constants are independent of the Cu^{2+} concentration, indicating rate-limiting complex dissociation prior to the released ligand reacting relatively rapidly with (the very labile) Cu^{2+}_{aq} species. Interpretation and comparisons are, however, complicated by differing relative importances of H^+-dependent and H^+-independent terms (322). A study of metallothionein turnover in mussels (*Mytilus edulis*) has given some idea of the timescale for the release of Cd^{2+} from the relevant cadmium–metallothionein complex. This complex is very inert, at least in situ, with the half-life for Cd^{2+} release of the order of 300 days (the half-life for Cd^{2+}+metallothionein is a mere week or so) (323).

A rare example of substitution kinetics at mercury(II) is provided by the adaptation of replacement of coordinated 4-(2-pyridylazo)resorcinol by 1,2-cyclohexanediamine-N,N,N',N'-tetraacetate to analysis for sulfite or thiosulfite (324).

8. sp-Block Elements

Relatively slow incorporation of Mg^{2+} into porphyrins can be attributed to relatively slow water loss from Mg^{2+}_{aq}; incorporation of Mg^{2+} into three isomeric water-soluble near-planar porphyrins is an order of magnitude slower than into bent porphyrins (325).

The Mg^{2+}, Ca^{2+}, Sr^{2+}, and Ba^{2+} ions in their complexes with 1,4,7,10-tetrakis(2-hydroxyethyl)- and 1,4,7,10-tetrakis(2-methoxyethyl)-1,4,7,10-tetraazacyclododecane, thec and tmec, respectively, are eight-coordinate, except for seven-coordinate Mg^{2+} in $[Mg(tmec)]^{2+}$. Rate constants for Δ/Λ enantiomerization for the $[M(thec)]^{2+}$ complexes range from $445 \, s^{-1}$ for M = Ba to $2310 \, s^{-1}$ for M = Mg; activation entropies are negative. The mechanism is certainly intramolecular overall, but there could be some transitory M–N bond-breaking to facilitate the numerous inversions at nitrogen donor atoms required for enantiomerization. The rate constant associated with Mg–N bond-breaking and -making in seven-coordinate $[Mg(tmec)]^{2+}$ is $1.6 \times 10^5 \, s^{-1}$ (very similar to that for water exchange at Mg^{2+}_{aq}); $\Delta S^{\ddagger} = -25 \, J \, K^{-1} \, mol^{-1}$ (326).

The role of Ca^{2+} in inducing refolding of α-lactalbumin is reflected in clean two-stage kinetics, with rate constants 6.0 and $1.3 \, s^{-1}$. The maximum concentration of the intermediate, monitored by stopped-flow fluorescence and time-resolved photo-CIDNP NMR, occurs at about 200 ms (327).

B. METAL(III) CATIONS

1. Iron(III)

Interest continues in kinetics and mechanisms of complex formation from iron(III) species in aqueous solution, with ligands ranging from relatively simple species such as gallic acid (3,4,5-trihydroxybenzoic acid) (328) to complicated species of biochemical or pharmacological interest. It is well established that Fe_{aq}^{3+} generally reacts with potential ligands by an associative mechanism (I_a), but that $Fe(OH)_{aq}^{2+}$ generally reacts by dissociative interchange (I_d) (329). Iron(III) complexation by 5-nitrotropolone (330) follows this pattern, with dinuclear $Fe_2(OH)_2 \, _{aq}^{4+}$ reacting by an I_d mechanism as does $Fe(OH)_{aq}^{2+}$. The half-life of $Fe_2(OH)_2^{4+}$ at room temperature is a few seconds. An improved model for the kinetics of dissociation of this dinuclear cation recognizes significant participation by $Fe_2(OH)_3^{3+}$ at higher pHs, thus clearing up earlier slight anomalies in this area (331). The kinetics of reaction of $Fe(OH)^{2+}$ and of $Fe_2(OH)_2^{4+}$ with variously protonated forms of phosphate, phosphite, hypophosphite, sulfate, and selenite have been investigated, mainly at 283 K. The formation mechanism from the dimer is somewhat complicated, e.g., by formation of mononuclear complexes, probably via μ-hydroxo-μ-oxoanion di-iron intermediates, after the initial I_d complexation step (332). The kinetics and mechanism of complex formation in the iron(III)-phosphate system in the presence of a large excess of iron(III) involve the formation of a tetranuclear complex, proposed to be $Fe_4(PO_4)(OH)_2(H_2O)_{16}^{7+}$ (333).

A new aminocarboxylate chelator of potential therapeutic value, N-(2-hydroxybenzyl)-N'-benzylethylenediamine-N,N'-diacetate, reacts as LH_4^+ and LH_3 with $Fe(OH)_{aq}^{2+}$ by dissociative activation with rate constants of 770 and 13 300 $M^{-1}s^{-1}$, respectively. These rate constants are similar to those for reaction of $Fe(OH)_{aq}^{2+}$ with edta and with nta. These formation reactions are, however, considerably faster than with simple ligands of identical charge thanks to the zwitterionic properties of aminocarboxylates (334).

Formation of L-lysinehydroxamato-iron(III) complexes occurs by an interchange mechanism; formation and dissociation (acid-catalyzed) are significantly affected by charge repulsion {the ligand is $H_3N^+(CH_2)_4CH(NH_3^+)CONHOH$} (335). Rate constants for complex formation between Fe_{aq}^{3+} and two synthetic chelators of the dicatecholspermidine family are, at 450 and 500 $M^{-1}s^{-1}$ (336), similar to that for desferrioxamine.

The acid-catalyzed aquation of iron(III)-(substituted) oxinate complexes involves iron–oxygen bond breaking and concomitant proton transfer in transition state formation. The latter aspect contrasts with the much slower acid-catalyzed aquation of hydroxamates, where proton transfer seems not to take place in the transition state. Reactivities, with and without proton assistance, for various stages in dissociation of a selection of bidentate and hexadentate hydroxamates, oxinates, and salicylates are compared and discussed – the overall theme is of dissociative activation (337). A multiple-path mechanism has been elaborated for dissociation of the mono- and bi-nuclear tris(hydroxamato)-iron(III) complexes with dihydroxamate ligands in aqueous solution (338). Iron removal by edta from mono-, bi-, and tri-nuclear complexes with model desferrioxamine-related siderophores containing one, two, or three tris-hydroxamate units generally follows first-order kinetics. Iron removal from the trinuclear tripodal complex follows a pattern of consecutive first-order reactions, characterized by seemingly remarkably similar rate constants (339). Kinetic patterns for the proton-driven dissociation of iron(III) from mononuclear and binuclear complexes of the tetradentate dihydroxamate siderophores alcaligin (cyclic) and rhodotorulate (linear) have been compared with each other and with the analogous process for the iron complex of desferrioxamine. The high degree of organization of alcaligin (alc) has a marked effect on the dissociation kinetics of $[Fe_2(alc)_3]$, and causes dissociation of mononuclear $[Fe(alc)(H_2O)_2]^+$ to be very slow ($\tau_{1/2}$ is a matter of hours) (340). The kinetics of removal of iron(III) from its complexes with the aminocarboxylate-anthraquinone analytical reagent calcein and with the anti-tumor anthracycline doxorubicin by the potential iron(III)-chelating pharmaceutical agent 1,2-dimethyl-3-hydroxy-4-pyridinone (dppm, known familiarly as CP20, or L1) have been monitored. Rate constants for metal removal are almost independent of the concentration of the replacing ligand, indicating dissociative mechanisms; they are approximately $1 \times 10^{-2}\,s^{-1}$ for displacement from doxorubin and between 1 and $2 \times 10^{-2}\,s^{-1}$ from calcein (341). The removal of iron from ferritin is, as one would expect, considerably slower. Rate constants are between 1.5 and $7.5 \times 10^{-5}\,s^{-1}$ for such removal by a series of hexadentate ligands each consisting of three substituted N-hydroxypyrimidinone or N-hydroxypyrazinone units, the rate decreasing with increasing substituent bulk. The slowest rate approximates to that for removal of iron from ferritin by desferrioxamine. The influence of chirality on the kinetic barrier provides insight into the detailed mechanism of removal in these systems (342).

Rate constants for incorporation of Fe(III) into tripodal hydroxamates containing [Ala-Ala-β-(HO)Ala] and [Ala-Ala-β-(HO)Ala]$_2$ units, and of

Fe(III) displacement of Al(III), Ga(III), or In(III) from their respective complexes with these tripodal ligands, have been determined. The M(III)-by-Fe(III) displacement processes are controlled by the ease of dissociation of Al(III), Ga(III), or In(III); Fe(III) may in turn be displaced from these complexes by edta (removal from the two non-equivalent sites gives rise to an appropriate kinetic pattern) (343). Kinetics and mechanism of a catalytic chloride ion effect on the dissociation of model siderophore-hydroxamate iron(III) complexes – chloride and, to lesser extents, bromide and nitrate, catalyze ligand dissociation through transient coordination of the added anion to the iron (344). A catechol derivative of desferrioxamine has been found to remove iron from transferrin about 100 times faster than desferrioxamine itself; it forms a significantly more stable product with Fe^{3+} (345).

The first step in the reaction of $trans$-[Fe(salpn)(H$_2$O)$_2$]$^+$, salpn=N,N'-propylene-1,2-bis-salicylidiniminate, with sulfur(IV) is the formation of [Fe(SO$_3$)(salpn)(H$_2$O)]$^-$, with the pH-rate profile showing greater $trans$-labilization by hydroxide than by water, in that $trans$-[Fe(salpn)(H$_2$O)$_2$]$^+$, reacts 10 times less rapidly than $trans$-[Fe(salpn)(OH)(H$_2$O)]. A limiting dissociative (D) mechanism is proposed for reaction of the latter; formation of the sulfito complex is followed by a slow intermolecular redox reaction (346). A similar situation prevails for the analogous $trans$-[Fe(salen)(H$_2$O)$_2$]$^+$/sulfur(IV) system (347).

Activation parameters for water exchange at three substituted porphyrin complexes [Fe(porph)(H$_2$O)$_2$]$^{3/5+}$, $57 < \Delta H^{\ddagger} < 71 \, \text{kJ mol}^{-1}$, $+60 < \Delta S^{\ddagger} < +100 \, \text{J K}^{-1} \, \text{mol}^{-1}$, and $+7 < \Delta V^{\ddagger} < +12 \, \text{cm}^3 \, \text{mol}^{-1}$ indicate dissociative activation for water exchange and, by implication, for complex formation reactions of these complexes (348). The Fe(III) tetra-meso-(4-sulfonatophenyl)porphinate complex and its sulfonatomesityl analogue react with NO in aqueous solution with large positive $\Delta S^{\ddagger}$ and $\Delta V^{\ddagger}$ values; values for dissociation of the adducts formed are similar. A D mechanism is thus believed to operate for each of these reactions (see also Section V.A.2). Less rapid reaction, and negative $\Delta S^{\ddagger}$ and $\Delta V^{\ddagger}$, for reaction of these sulfonatoporphyrin complexes with CO indicate associative interchange here (278). Reactions of metalloporphyrins with NO have been reviewed (349). See also Chapter 4.

Kinetics of formation of the dinuclear iron(III) complex [(tpa)Fe(μ-O)(μ-urea)Fe(tpa)]$^{3+}$ {tpa = tris(2-pyridylmethyl)amine} were investigated in relation to the suggestion that urease action in vivo involves an intermediate containing –Ni(μ-OH)(μ-urea)Ni–. The mechanism of formation of the di-iron species from [(tpa)(H$_2$O)Fe(μ-O)Fe(OH)(tpa)]$^{3+}$ is proposed to involve three reversible steps (350). Three kinetically distinct steps are also involved in the deposition of FeO(OH) in bacterioferritin (from *E. coli*), this time with two intermediate μ-oxo-di-iron species. The first is formed in milliseconds from its mononuclear precursor; it is oxidized in a few seconds to the second, which is deposited as FeO(OH) in a matter of minutes. The first step is indubitably substitution – whether or not electron transfer accompanies substitution in the second and third stages depends on the assignment of oxidation states in the second intermediate (*351*). Self-assembly of chiral dinuclear binaphthol-linked iron(III)-porphyrin complexes occurs through inter-molecular formation of μ-oxo dimers (*352*).

   In contrast to the numerous I_d mechanisms mentioned above for sub-stitution at hydroxoiron(III) species, substitution by H_2O_2 at [Fe(Rtpen) (OMe)]$^{2+}$ {Rtpen=**8**), to form the relatively stable [Fe(Rtpen)(η^1-OOH)]$^{2+}$, is, like complex formation from Fe_{aq}^{3+}, I_a in character (*353*).

8

2. *Aluminum(III); Gallium(III); Indium(III)*

   In a rare kinetic study of complex formation from aluminum(III) rate constants have been determined for its reactions with edta and dtpa. The results for the protonated ligands edtaH$_2^{2-}$ and dtpaH$_3^{2-}$ were ana-lyzed in terms of the standard Eigen-Wilkins I_d formation mechanism (*354*). Evidence has been presented for a dissociative interchange mechanism on $Ga(H_2O)_5(OH)^{2+}$ in the reaction of gallium(III) with 4-nitrocatechol (*355*), in conformity with several other I_d complex-forma-tion reactions from $Ga(OH)^{2+}$ (*356*), and indeed with complex formation from $Al(OH)^{2+}$ and $Fe(OH)^{2+}$. In recent years similarities between the kinetic behavior of Fe(III) and its Group 13 relatives Al(III), Ga(III), and In(III) – especially Ga(III), as Ga^{3+} has a very similar ionic radius to Fe^{3+} – have often been emphasized (*1,343*). The formation of multi-connected polynuclear species can give remarkably rigid structures, as has been demonstrated for a number of bis-catecholate complexes of Al^{3+}, Ga^{3+}, In^{3+}, as well as of Fe^{3+}. Thus, for example, k_{racn} for Ga(catecholate)$_3$ is $\sim 10\,s^{-1}$, for a helical dinuclear Ga^{3+} complex $k_{racn} \sim 0.1\,s^{-1}$, but Ga$_4$(bis-catecholate)$_6$ shows no signs of racemizing over a period of many months (*357*).

3. Bismuth(III)

Ligand exchange of nitrilotriacetate and of N-(2-hydroxyethyl)imino-diacetate at their respective bis-ligand bismuth(III) complexes occurs at rates amenable to following by NMR line-broadening techniques. In view of the high coordination number of these complexes in aqueous solution, believed to be eight, a dissociative mechanism (probably involving transient aqua-intermediates) seems probable, despite the large negative activation entropies (358).

4. Nickel(III)

Chloride substitution kinetics of $[Ni^{III}L(H_2O)_2]^{3+}$, and its protonated form $[Ni^{III}L(H_2O)(H_3O)]^{4+}$, where L = 14-oxa-1,4,8,11-tetraazabicy-clo[9.5.3]nonadecane, yield $k_{(H_2O)_2} = 1400\,M^{-1}\,s^{-1}$ and $k_{(H_2O)(H_3O^+)} = 142\,M^{-1}s^{-1}$. The reverse, chloride dissociation, reactions have $k_{(H_2O)Cl} = 2.7\,s^{-1}$; $k_{(H_3O^+)Cl} = 0.22\,s^{-1}$. All four reactions occur through dissociative interchange mechanisms, like earlier-studied substitutions at nickel(III) (359).

5. Lanthanides(III)

a. General   A few fragments of information concerning kinetics and mechanisms of reactions of lanthanide complexes may be found scattered through the nearly 700 pages of 22 articles in the recent thematic issue of *Chemical Reviews* (360) devoted to lanthanide chemistry. The article of most relevance to this present review devotes several pages to complex formation and to ring inversion in macrocyclic complexes (361). Barriers to stereochemical change, in free and in complexed polydentate ligands often have a considerable effect on kinetics of formation and dissociation. Likely transition states for isomer interconversions in tetra-azacarboxylate complexes of La^{3+} and of Y^{3+} have been explored by computational methods – NMR and X-ray diffraction have revealed two accessible geometrical isomers in solution and in the solid state (362).

b. Complex Formation from Ln^{3+}_{aq} Despite the rather large uncertainties in activation parameters, the large positive values for $\Delta S^{\ddagger}$ for both the $La^{3+}+HA$ and the $La^{3+}+A^-$ pathways for complexation of La^{3+} by acethy-droxamate indicate dissociative activation (as for analogous reactions of iron(III), see Section V.B.1 above) (363). Kinetics of formation of lanthanide(III) complexes of *trans*-1,2-diaminocyclohexane-N,N,N′,N′-tetraacetate (364) and of cyclen derivatives with two or three $CH_2CO_2^-$-derivatized pendant arms (365) have been reported. The key rate-limiting step in formation of these cyclen complexes is deprotonation and rearrangement of an intermediate complex; formation rate constants increase from Ce^{3+} along to Yb^{3+}. Both $Gd(dotp)^{5-}$ and $Gd(dotpmb)^{-}$ form somewhat more slowly than their carboxy analogue $Gd(dota)^{-}$ (see 9–11 below); again (de)protonation and rearrangement of intermediates are kinetically important (366). Perhaps the most complicated complex-formation reactions currently under investigation are those between ternary aqua-lanthanide(III) complexes of tetraazamacrocyclic ligands bearing pendant arms with oligonucleotides and other models for DNA. Rate constants for such reactions are controlled by the rate of displacement of the water ligand(s), which in turn may be controlled by steric constraints imposed by bulky pendant arm substituents (367). There appears, both from kinetic (relaxivities and exchange rates for gadolinium complexes) and structural (for a lanthanum complex) observations, to be a particularly low barrier to eight $\rightleftarrows$ nine coordination number change in complexes with hexadentate tripodal ligands containing hydroxypyridinone moieties in two of their three legs (368). This could well be important in relation to substitution mechanisms for this type of complex.

9 dotp : R =PO_3^{2-}
10 dotpmb : R = $RO_2(OBu)^{-}$
11 dota : R = CO_2^{-}

9-11

c. *Complex–Substrate Interactions*   Currently a main interest in this type of complex is examination of the nature of the interactions of approved and potential MRI (Magnetic Resonance Imaging) contrast agents with model and actual substrates, especially oligonucleotides and DNA. Kinetic factors loom large, both in the formation of such complexes and in their chemical and physical properties. The lifetimes of coordinated water ligands have an important effect on relaxivities and therefore on effectiveness as contrast agents. High inertness of the polyazacarboxylate, polyazaphosphonate, or other macrocyclic ligand, both with respect to dissociation and to racemization, is essential for studies of detailed mechanisms of substrate–contrast agent interactions. Configurational stability, at least on the luminescence and NMR timescales, is required for stereochemical studies. Lifetimes of coordinated water on Gd^{3+} in six such complexes are between 30 and 270 ns, with fastest displacement of water on the sole di-aqua-complex examined (369). Water relaxivities for lipophilic gadolinium complexes of the dotp (9) derivative with one $-CH_2PO_3^{2-}$ replaced by $-CHRPO_3^{2-}$ where R = $-(CH_2)_7CH_3$ or $-(CH_2)_{10}CH_3$ increase significantly in the presence of albumin; kinetic and equilibrium studies increase understanding of complex-albumin interactions in these systems (370). It should be noted that luminescence studies of europium analogues suggest it may not be necessary to lose coordinated water in order to interact with oligonucleotides. ^{1}H NMR TOCSY and NOESY studies of interaction of a gadolinium complex interacting with a dodecameric oligonucleotide have located the most favorable intercalation site at the center of the oligonucleotide; the Δ-isomer of the complex binds more strongly than the Λ-isomer (371).

d. *Dissociation of Macrocyclic Complexes*   The established widespread use of complexes of gadolinium in MRI, and actual and potential uses of certain isotopes of yttrium and lanthanides in diagnosis and therapy, have led to a search for very stable and inert complexes (preferably uncharged and with an appropriate hydrophilic/lipophilic balance) for such medicinal uses. What can be achieved may be illustrated by the 4-nitrobenzyldota complex of Y^{3+}, whose dissociation half-life is more than 200 days – dramatically longer than the half-life for water exchange at Y_{aq}^{3+} ($\sim 10^{-9}$ s) and indeed significantly longer than the half-life for radioactive decay of the relevant isotope, ^{90}Y – at 37 °C (372).

Slow solvolytic dissociation of lanthanide(III) complexes with acyclic polyamino-carboxylate ligands in aqueous solution (probed via ligand exchange and metal exchange kinetics) can be attributed to the large negative activation entropies, which can be as large as +250 J K^{-1} mol^{-1} (373). This study complements several earlier studies on kinetics of formation of such complexes, where the slow rates of formation may be attributed to ligand strain. The gadolinium complex of a cyclohexyl derivative of dota has a dissociation half-life twice that of the parent dota complex – the cyclohexyl ring increases ligand rigidity and makes it harder for the inversions required to lose the ligand (374). Dissociation kinetics of $Gd(dotp)^{5-}$ and $Gd(dotpmb)^-$ have been studied, the latter in particular because its much lower charge makes it a viable contrast agent for MRI. These phosphinate complexes undergo acid-catalyzed aquation (at similar rates) about a thousand times faster than $Gd(dota)^-$. Protonation and protonated species and intermediates are important in dissociation as they are in formation (366). Cu^{2+} and Zn^{2+}, as well as H^+, assist dissociation of $[Gd(dtpa)]^{2-}$, with the respective rate constants 0.93, 0.056, 0.58 M^{-1}s^{-1}. The two metal ions are believed to give binuclear intermediates with a dtpa-glycinate bonded to the M^{2+} (375).

Kinetics of dissociation of one yttrium and seven gadolinium complexes of polyazaphosphinic acid ligands, **9** with the four $CH_2PO_3^{2-}$

groups either replaced by four $CH_2PRO_2^-$ or by three $CH_2PRO_2^-$ and one $CH_2CONR'R''$ have been studied. It is easier to tailor these ligands than dota as they have variable R, and the latter also R' and R'', groups which the $CH_2CO_2^-$ pendant arms of dota lack. All these complexes dissociate slowly enough to have in vivo potential, with the uncharged complexes aquating more slowly than their anionic analogues. Indeed the inertness of one of the Y^{3+} complexes approaches that of the complex mentioned in the first paragraph of this section. Both the Gd^{3+} and Y^{3+} complexes of the tetrakis-$CH_2PMeO_2^-$ complex dissociate only slightly less slowly than their dota analogues (*376*). Minimal information on half-lives for dissociation of lanthanide complexes of dota and its phosphinic analogues is vouchsafed in a review on the inertness of lanthanide complexes of macrocyclic ligands (*377*).

e. Polynuclear Cation Formation   Formation of polynuclear (Ln_{12} and Ln_{15}) oxo-hydroxo-lanthanide(III) species at high pHs in the presence of, inter alios, glutamate or tyrosine proceeds through intermediates containing such entities as $[Ln_4(\mu^3\text{-}OH)_4]^{8+}$ or $[Ln_6(\mu^6\text{-}O)(\mu^3\text{-}OH)_8]^{8+}$ (*378*). These intermediates are reminiscent of such well-established species as $[Ni_4(OH)_4]^{4+}$, $[Pb_4(OH)_4]^{4+}$, or $[Ln_6(\mu^6\text{-}O)(\mu^3\text{-}OH)_6]^{4+}$. The formation of $[Gd_4(\mu^3\text{-}OH)_4]^{8+}$ is controlled by L-valine (*379*), of octahedral oxo-centered $[Ln_6(\mu^6\text{-}O)(\mu^3\text{-}OH)_8]^{8+}$ (Ln = Nd or Gd) by L-serine (*380*). $[Ln_{15}(\mu^3\text{-}OH)_{20}(\mu^5\text{-}X)]^{24+}$ (Ln = Eu, Nd, Gd, Pr; X = Cl, Br) cations are made up from five $[Ln_4(\mu^3\text{-}OH)_4]^{8+}$ units, vertex-sharing around a central halide template, assembled under L-tyrosine and Cl^- or Br^- control (*381*). The underlying principles are set out in the L-valine-control paper – "the hydrophilic groups lie within the cluster core to hold the metal ions together and the hydrophobic groups take up positions on the periphery, preventing the core from further aggregation". So there is method and a degree of control, even if not mechanistic understanding, involved in these assembly reactions.

VI. Transition Metal Triangles and Clusters

Flash photolysis of the dianion of Roussin's Red Salt, $[Fe_2S_2(NO)_4]^{2-}$, in particular the initial photoinitiated loss of NO (*382*) and the reverse recombination reaction, en route to the eventual product, the anion of Roussin's Black Salt, $[Fe_4S_3(NO)_7]^-$, has been documented (*383*). A 4-$RC_6H_4S^-$ group (R = H, Me, OMe, Cl, or CF_3) replaces one of the chloride ligands in $[Fe_4S_4Cl_4]^{2-}$ via a five-coordinated intermediate, with the detailed sequence of steps acid-dependent (*384*). Loss of chloride is pH-dependent, with the rate depending on the electron-withdrawing properties of the substituent R (385).

Reactions of $[Fe_4S_4Cl_4]^{2-}$ and of $[Fe_4S_4(SPh)_4]^{2-}$ with diethyldithiocarbamate are dissociatively activated (386). Reaction of $[M_4(SPh)_{10}]^{2-}$, M = Fe or Co, with PhS^- takes place by initial associative attack at one of the tetrahedral M atoms, to give $[M_4(SPh)_{11}]^{3-}$, followed by a sequence of rapid reactions to give the final product $[M(SPh)_4]^{2-}$. In contrast the first step in reaction of these $[M_4(SPh)_{10}]^{2-}$ cluster anions with $[MoS_4]^{2-}$ is, at least at low $[MoS_4]^{2-}$ concentrations, dissociative in character. However, a second-order, presumably associative, term is discernable for $[Co_4(SPh)_{10}]^{2-}$ at high $[MoS_4]^{2-}$ concentrations (387).

$[Cr_3(\mu_3\text{-}O)(\mu\text{-}O_2CCH_3)_6]^{2+}$ is considerably more reactive than $[Cr_2(\mu\text{-}O)(\mu\text{-}O_2CCH_3)_2]^{2+}$ (388). Activation parameters ($\Delta H^{\ddagger}$ and $\Delta S^{\ddagger}$) suggest that the acid-dependent path for aquation of $[Cr_2Rh(\mu_3\text{-}OH)(\mu_2\text{-}OH)_3]^{2+}$ is I_d but that the acid-dependent path involves a more associative dissociation of a protonated form of the complex (389). Both electronic and steric factors are important in determining reactivity for pyridine exchange at a series of nine complexes of the $[Cr_3(\mu_3\text{-}O)(\mu\text{-}O_2CR)_6(py)_3]^{2+}$ type (R = alkyl or chloroalkyl). The activation parameters ($\Delta S^{\ddagger}$ between +35 and +97 $J\,K^{-1}\,mol^{-1}$ and $\Delta V^{\ddagger}$ between +9.6 and +14.3 $cm^3\,mol^{-1}$) suggest a dissociative mechanism (390).

The dependence of rate constants for approach to equilibrium for reaction of the mixed oxide-sulfide complex $[Mo_3(\mu_3\text{-}S)(\mu\text{-}O)_3(H_2O)_9]^{4+}$ with thiocyanate has been analyzed into formation and aquation contributions. These reactions involve positions *trans* to μ-oxo groups, mechanisms are dissociative (391). Kinetic and thermodynamic studies on reaction of $[Mo_3MS_4(H_2O)_{10}]^{4+}$ (M = Ni, Pd) with CO have yielded rate constants for reaction with CO. These were put into context with substitution by halide and thiocyanate for the nickel-containing cluster (392). A review of the chemistry of $[Mo_3S_4(H_2O)_9]^{4+}$ and related clusters contains some information on substitution in mixed metal derivatives $[Mo_3MS_4(H_2O)_n]^{4+}$ (M = Cr, Fe, Ni, Cu, Pd) (393). There are a few asides of mechanistic relevance in a review of synthetic Mo–Fe–S clusters and their relevance to nitrogenase (394).

The mechanism of formation of corner-shared double cubes $[Mo_6HgE_8(H_2O)_{18}]^{8+}$ (E = S, Se) involves reaction of $[Mo_3E_4(H_2O)_9]^{4+}$ with Hg^0 or Hg_2^{2+}. The first-order dependence on $[Mo_3E_4(H_2O)_9]^{4+}$ suggests that the mechanism involves reaction of $[Mo_3E_4]^{4+}$ with Hg, followed by reaction with the second $[Mo_3E_4]^{4+}$ unit, rather than reaction of two $[Mo_3E_4]^{4+}$ units followed by incorporation of the mercury (395).

The clusters $[Re_6Se_8(PEt_3)_5(MeCN)]^{2+}$ and $[Re_6Se_8(PEt_3)_5(dmso)]^{2+}$ are substitutionally inert, with MeCN and dmso exchange or replacement being characterized by rate constants of the order of $10^{-6} s^{-1}$ at laboratory temperatures. Activation entropies of +62 and +28 J K^{-1} mol^{-1} for MeCN and dmso exchange indicate that substitution at these clusters is dissociative in mechanism; activation enthalpies of 130 and 120 kJ mol^{-1} are high, as expected for dissociative substitution (*396*).

Two hexanuclear zinc clusters are mentioned elsewhere in this Chapter, one in Section V.A.7 (on zinc(II) reactions) above, the other in Section VII.D (on supramolecular chemistry) below.

VII. Reactions of Coordinated Ligands

A. NUCLEOPHILIC ATTACK AT COORDINATED LIGANDS

Rate and equilibrium constant data, including substituent and isotope effects, for the reaction of [Pt(bpy)$_2$]$^{2+}$ with hydroxide, are all consistent with, and interpreted in terms of, reversible addition of the hydroxide to the coordinated 2,2'-bipyridyl (*397*). Equilibrium constants for addition of hydroxide to a series of platinum(II)-diimine cations [Pt(diimine)$_2$]$^{2+}$, the diimines being 2,2'-bipyridyl, 2,2'-bipyrazine, 3,3'-bipyridazine, and 2,2'-bipyrimidine, suggest that hydroxide adds at the 6 position of the coordinated ligand (*398*). Support for this covalent hydration mechanism for hydroxide attack at coordinated diimines comes from crystal structure determinations of binuclear mixed valence copper(I)/copper(II) complexes of 2-hydroxylated 1,10-phenanthroline and 2,2'-bipyridyl (*399*).

Nucleophilic attack at the carbonyl-carbon of 4,5-diazafluoren-9-one (dzf, **12**) coordinated to ruthenium has been demonstrated, and a mechanism outlined, for reaction of [Ru(bpy)$_2$(dzf)]$^{2+}$ with 1,8-diazabicyclo[5,4,0] undec-7-ene (dbu)**13**, which gives a tris-2,2'-bipyridyl derivative (*400*).

Whereas heating [Pt(terpy)(N$_3$)]$^+$ in the gas phase gives dinitrogen, heating in solution in acetonitrile or benzonitrile gives the tetrazolato complexes **14** (R = Me, Ph), presumably by nucleophilic attack by RCN at coordinated azide and subsequent closure of the N$_4$C ring (*401*).

12

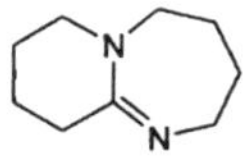

13

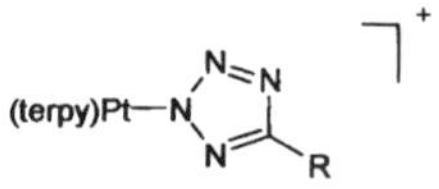

14

New NMR information on the state of coordinated 2-formylglycinate in [Co(en)$_2$(formgly)], showing it to be in hydrate plus enol form rather than in aldehyde form, has led to a new theory for the mechanism of the reaction of coordinated formylglycinate with penicillamine (402).

cis-[Rh(L)(H$_2$O)$_2$]$^+$, with L = (en)$_2$ or cycb, catalyzes the conversion of methylglyoxal or, less efficiently, of 1,3-dihydroxyacetone into coordinated lactate – mechanisms are presented for the former, and for rhodium(III)-catalyzed interconversion between 1,3-dihydroxyacetone and glyceraldehyde (403).

Hydrolysis of coordinated ligands is a special case of nucleophilic attack. Two examples involving inorganic ligands have already been given in Section II.A on aquation of cobalt(III) complexes. Many further examples will be found in the following Section VII.B on catalysis of hydrolysis of organic substrates by metal ions and complexes.

B. METAL-CATALYZED HYDROLYSIS OF ORGANIC COMPOUNDS

1. General

The high levels of activity in recent years – for instance the Royal Society of Chemistry's Annual Reports for 1998 cite some 80 references in this area, with the Reports for 1998 and 1999 each citing more than 30 references on phosphate ester hydrolysis alone (1) – continue unabated. Much of this activity is biologically-related, but the establishment of optimum conditions and of mechanisms involved depends heavily on the particular properties of the metal ions or complexes involved. In the following paragraphs we shall cite a selection of recent references to give some idea of the range of activity, both in respect of substrates and of metals involved. Phosphate esters continue to dominate, but there is considerable activity in relation to carboxy esters, and indeed to a variety of biologically-relevant classes of compound.

Inert centers facilitate the detection and characterization of intermediates. Recently investigated systems involving cobalt(III) complexes include [Co(L)(OH)(H$_2$O)]$^{2+}$-promoted hydrolysis of 2,4-dinitrophenyl diethyl phosphate (*404*) and hydrolysis of 4-nitrophenyl phosphate by dinuclear cobalt(III) species (*405*). Molecular mechanics calculations have been carried out on the mechanism of action of cobalt(III) complexes as catalysts for phosphate ester hydrolysis (*406*).

Labile centers are much more frequently encountered. Now the plausible mechanism of hydrolysis of substrate subsequent to its coordination to the metal center, through nucleophilic attack by water or hydroxide at an atom whose electron density has been reduced by the nearby metal cation, is often assumed but rarely demonstrated unequivocally. The water or hydroxide may also be coordinated, giving a template reaction. Increasingly it is being found that dinuclear species, generally with bridging hydroxide or oxide ligands, are particularly effective catalysts. M–OH–M units have been established at the active sites of a variety of enzymes, including phosphatases, aminopeptidases, and urease. Phosphate ester hydrolysis by a di-iron complex has now been shown to involve nucleophilic attack by bridging hydroxide (as proposed but not conclusively demonstrated for several M–OH–M-containing catalytic species) rather than by hydroxide bonded just to one Fe (*407*). The activity of a di-copper(II) catalyst has been shown to be enhanced in the presence of a third metal ion, Ni^{2+}, Pd^{2+}, or a third Cu^{2+}. Hydrolytic cleavage of the phosphodiester 2-hydroxypropyl-4-nitrophenylphosphate exhibits allosteric regulation in the system **15**, where the activities of the two copper ions bound to the bipyridyl moieties are modulated by the third metal ion, whose radius affects the $Cu \cdots Cu$ distance in the active site. The kinetics of the tris-Cu^{2+} system were examined in detail (*408*).

● = Ni, Pd, or Cu

15

2. Phosphate Esters

In addition to the systems just mentioned, recent kinetic and mechanistic studies have included those involving copper(II) (*409,410*) and zinc(II) (*411*) species, various binuclear metal(II) complexes of first row transition elements (*412–414*), especially iron (*407*), cobalt (*415*), copper (*305,416*), and zinc (*417,418*), yttrium (*419,420*) and lanthanide (*421,422*) species, and thorium(IV) (*423*).

The 10^4-fold enhancement of ribonuclease activity (uridine $2',3'$-cyclic monophosphate hydrolysis) by a di-zinc bis-imidazole complex is attributed to a combination of Lewis acid and general base catalysis (*424*). The unusual third-order dependence of phosphate diester hydrolysis by mono- and di-nuclear lanthanum complexes suggests a mechanism where a metal-coordinated hydroxide attacks the phosphorus of the substrate on the side opposite the negatively charged oxygens (*422*). Phosphodiester hydrolysis is also markedly catalyzed by triazacyclo nonanecopper(II) complexes. There is a half-order dependence on the concentration of copper complex, ascribed to equilibrium between catalytically-active $[Cu(tiptcn)(OH)(H_2O)]^+$, tiptcn = tri-isopropyltriazacyclononane, and a this-time ineffective dimer (*410*).

3. Carboxylate Esters

The use of a lipophilic zinc(II) macrocycle complex, 1-hexadecyl-1,4,7,10-tetraazacyclododecane, to catalyze hydrolysis of lipophilic esters, both phosphate and carboxy (*425*), links this Section to the previous Section. Here, and in studies of the catalysis of hydrolysis of 4-nitrophenyl acetate by the Zn^{2+} and Co^{2+} complexes of tris(4,5-di-n-propyl-2-imidazolyl)phosphine (*426*) and of a phosphate triester, a phosphonate diester, and *O*-isopropyl methylfluorophosphonate (Sarin) by $[Cu(N,N,N'-\text{trimethyl-}N'\text{-tetradecylethylenediamine})]$ (*427*), various micellar effects have been brought into play. Catalysis of carboxylic ester hydrolysis is more effectively catalyzed by *N*-methylimidazole-functionalized gold nanoparticles than by micellar catalysis (*428*). Other reports on mechanisms of metal-assisted carboxy ester hydrolyses deal with copper(II) (*429*), zinc(II) (*430,431*), and palladium(II) (*432*).

4. Other Substrates

Kinetic and mechanistic studies of other metal-catalyzed hydrolysis (or methanolysis) have included the following systems:

- The tripodal Schiff base ligand tris-[2-(salicylideneamino)ethyl] amine, by Cu(II), Zn(II), Sn(II) (*433*). Whereas it is possible to isolate and characterize intermediate complexes containing partially hydrolyzed ligand when Cu(II) or Zn(II) are catalysts, there is no indication of analogous intermediates when Sn(II) is catalyst; Pb(II) has negligible catalytic effect.
- The activated amide acetylimidazole, by Zn^{2+}, Co^{2+}, La^{3+} (*434*).
- Thiamine (vitamin B_1) in the presence of Me_2TlOH (*435*). Here the kinetics are much affected, through coordination of $TlMe_2^+$ to the sulfur of the thiamine in its enethiolate form.

- Peptide hydrolysis by platinum(II) (*436*) and palladium(II) complexes (*437*). In the latter case there is selective hydrolysis of the unactivated peptide bond in *N*-acetylated L-histidylglycine; the hydrolysis rate depends on the steric bulk of the catalyst.
- DNA cleavage by copper(II) complexes (*410,438*).

Hydrolysis rates for lactam rings in nickel(II) cages of the sarcophagine type are similar to those for uncoordinated analogues. A detailed study involving a bis-carboxymethyl-diamino-sarcophagine-nickel(II) complex showed clean two-stage kinetics (though not an exact first-order dependence on $[OH^-]$), despite the complicated sequence of inversion-at-nitrogen and ring opening steps which must be involved (*439*). Catalysis of hydrolysis of β-lactams by dinuclear zinc complexes (*440*), and of nitrocefin, the standard β-lactam (antibiotic) probe, by Zn^{2+} (*441*), have been investigated. For the latter there are two concurrent acyl transfer pathways, involving one and two Zn^{2+} ions, respectively.

Urease is a spectacularly efficient metalloenzyme, converting urea into hydrolysis products $> 10^{14}$ times faster than the uncatalyzed rate. The terminal urea of a bis-urea-dinickel complex with μ-hydroxy-μ-urea-μ-bdptz (bptdz = **16**) triple bridging undergoes hydrolysis in acetonitrile solution in two steps, the second being hydrolysis of cyanate formed in the first step. The characterization of the cyanate-containing dinickel intermediate lends support to theories of urease action which involve cyanate-containing intermediates (*442*). The possibility of the intermediacy of μ-urea species was mentioned in Section V.B.1 above in connection with an iron(III) model system (*350*). Functional and structural influences of tetradentate N_4-donor tripodal ligands on the base-promoted catechol cleaving activity of iron(III)-containing model compounds for catechol 1,2-dioxygenase have been assessed. There is a correlation between rate constants for cleaving the coordinated catechol and reduction potentials of the respective iron(III) complexes, over a range of 300-fold in rate constant (*443*).

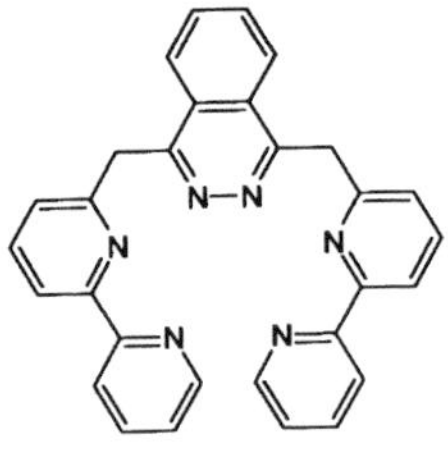

16

Marcus theory, first developed for electron transfer reactions, then extended to atom transfer, is now being applied to catalytic systems. Successful applications to catalysis by labile metal ions include such reactions as decarboxylation of oxaloacetate, ketonization of enolpyruvate, and pyruvate dimerization (444).

C. TEMPLATE REACTIONS

Here a metal ion, complex, or anion is used as a reaction center to facilitate condensation of appropriate precursors. A long-standing example is the preparation of Schiff bases, some of which form much more readily, cleanly, and in higher yield in the presence of an appropriate metal ion. Thus methylamine readily condenses with biacetyl in the presence of Fe^{2+} to give a high yield of $[Fe(MeN=CMeCMe=NMe)_3]^{2+}$. A recent example is afforded by the formation of the rhombohedral cage compounds $[Ag_{14}(C \equiv C^tBu)_{12}X]^+$, which form on anion templates $X^- = Cl^-$ or Br^-, whereas reaction with $AgBF_4$ gives an ill-defined polymeric product (445). A review of azamacrocyclic ligands includes some mechanistic discussion of the formation of such species by template or condensation reactions at metal ion centers (446). The determination of the crystal structure of (1,11-bis-benzyl-5,5,7-trimethyl-1,4,8,11-tetraaza-undeca-4-ene)nickel(II) perchlorate, the product of the reaction of N-benzylethylenediamine with acetone on a Ni^{2+} template permitted the proposal of a mechanism for this reaction which may help in understanding some details of the original Curtis reaction (447) of $[Ni(en)_3]^{2+}$ with acetone (448).

D. SUPRAMOLECULAR CHEMISTRY

Currently the main interest in template reactions lies in their key role in the controlled synthesis or the self-assembly of a variety of supramolecular entities (449). One needs "a combination of intuition, conjecture, and serendipity" (450); a recent example of successfully combining serendipity and rational design is provided by the silver(I)-promoted assembly of one-dimensional stranded chains (451). One also needs an understanding of mechanism in order to optimize the selection and design of building blocks and templates for the generation of yet more sophisticated supramolecular structures – references cited in this present review contain at least some kinetic or mechanistic information or speculation. Template routes to interlocked molecular structures have been reviewed (452), while a discussion of switching by transition metal contains a little about the kinetics and mechanisms of this aspect of template reactions, i.e. about metal ion translocation in two-compartment ligands (453).

Anion templates have been popular in recent years, with reports of their use often including at least a suggested reaction sequence as a minimal mechanistic sketch. Examples of their use include the formation of the silver complex mentioned above (445), of pseudorotaxanes (454), rotaxanes, and knots (455) (see below), and of polynuclear hydroxolanthanide cations (see Section V.B.5.e above). Anion control of self-organization may be illustrated by the case of $[Cu(arginineH)_2]^{2+}$ (456). Benzene-1,3-disulfonate promotes a layer structure, sulfate (through anion hydrogen-bonding to guanidinium groups of coordinated arginine) a single-helical and benzene-1,3-dicarboxylate a double-helical structure. Concentration control of assembly may be illustrated by the $Cd^{2+}/4,4'$-bipyridyl system, where crystallization of solutions having a $Cd^{2+}:4,4'$-bpy ratio >3 gives a linear aggregate, at low $Cd^{2+}:4,4'$-bpy ratios square $Cd_4(4,4'$-bpy$)_4$ aggregates, except at low overall concentrations when binuclear $Cd_2(4,4'$-bpy$)_5$ is obtained (457).

Copper(I) plays a key role in supramolecular chemistry, though currently there are efforts to extend the principles established at this tetrahedral center to octahedral centers – control may then be based on coordination preferences of metal ions (320). Copper(I) may be important in biomimetic templates (458); many metal ions and complexes have effects on the hydrogen-bonding and ring-stacking which are so important in organization and recognition processes in biological supramolecular chemistry (456). The roles of Cr, Fe, Co, Cu, Zn, and Ru in supramolecular chemistry and self-assembly are mentioned at various points in a collection of some 80 articles covering preparative, thermodynamic, and mechanistic aspects of both inorganic and organic systems (459).

Controlled cleavage of the three-dimensional coordination polymer formed by reaction of zinc acetate with di-2-pyridyl ketone to give a hexanuclear cluster complex provides a rare example of designed synthesis, rather than hit-and-miss self-assembly, in supramolecular chemistry. This controlled breakdown approach may well permit the development of rational synthetic methods for clusters of high nuclearity (460). The production of coordination polymers is therefore relevant here, and may be exemplified by grid formation from copper(I) and the bis-1,10-phenanthroline ligand **17** (461) or by the coordination polycatenanes $M(bpe)_2(NCS)_2 \cdot MeOH$, M = Fe or Co, bpe = *trans*-1,2-bis(4-pyridyl)ethene (462). The assembly of "molecular panelling", i.e. the generation of three-dimensional structures, with spaces for the inclusion of guests, from two-dimensional (or even one-dimensional) structures may show

striking specificity, as in the generation of just one isomer (of the hexahe-
dral $M_{15}L_6$ product) from 21 individual components in the reaction of
[Pd(en)(NO$_3$)$_2$] with **18** (*463*). Molecular boxes have been generated from
self-organization of 1,2-bis(2-pyridylethynyl)benzene ligands, containing
bulky substituents suitably tailored for their mutual steric interference
to impose the box shape, on a copper(I) template (*464*). The one-dimensional stranded chains mentioned above, assembled as coordination
polymers on silver(I), enclose nanometric cavities whose size may be
varied through ligand modification and also possibly through mediation
by weakly coordinating anions (*451*).

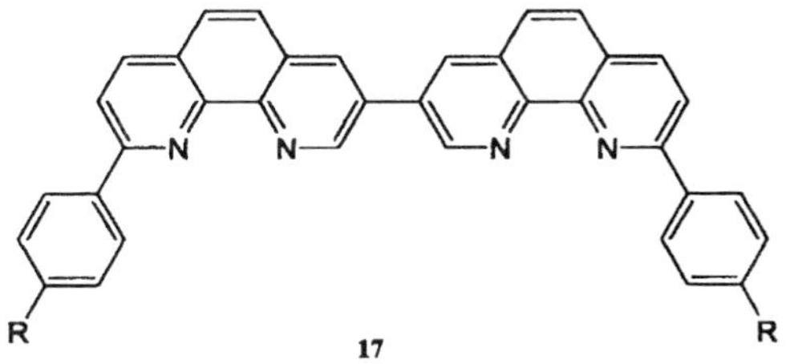

17

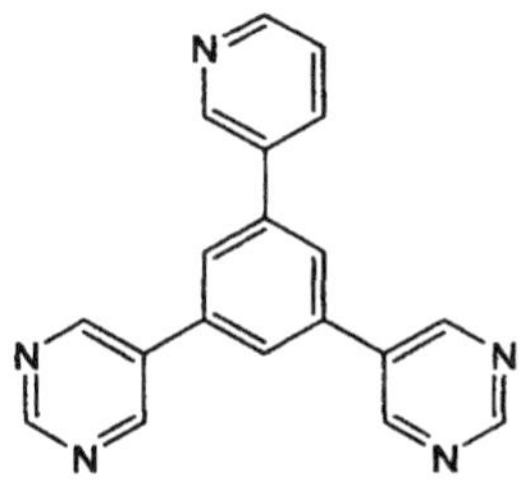

18

There may be difficulties in predicting three-dimensional networks –
copper(I) bromide reacts with 3,3′-dipyridylethyne to form a one-dimensional polymer, but this kinetically favored product slowly ($\tau_{1/2}$ ∼ a
week) rearranges to a three-dimensional network which could be
described as containing interwoven ribbons linked by Cu$\cdots$Br chains
(*465*). Self-assembly of a supramolecular triangle, rather than the
expected square tetramer in a Pt/pyrazine/phosphine system indicates
possible limitations to "rational design" in this area (*466*). Other unexpected products from self-assembly reactions include a hexanuclear
zinc complex containing Zn$_3$S$_3$ and Zn$_4$S$_4$ cyclic units derived from
the *S*-donor ligand 2-aminoethanethiolate (*320*) and a decanuclear
Cu(II) wheel (*467*). The discussion of factors leading to the self-assembly

of the latter product constitute a minimal review of mechanistic possibilities.

E. ROTAXANES, CATENANES, AND KNOTS

A useful summary of the various and numerous types of rotaxanes, catenanes, and knots can be found in a review of template routes to interlocked molecular structures (*468*). Inorganic chemistry is centrally involved in the templating involved in self-assembly and in controlled synthesis of such species.

1. Pseudorotaxanes and Rotaxanes

Pseudorotaxanes are precursors of both rotaxanes and catenanes; they consist of a guest molecule threaded through a macrocyclic host. Stoppering both ends of the threaded molecule gives a rotaxane, cyclization of the thread gives a catenane. Pseudorotaxane formation may occur by spontaneous self-assembly, or may be template-controlled. Anion size can be of paramount importance for such templates – Cl^- is effective, Br^-, I^- less good, and PF_6^- ineffective when the "recognition motif" demands a small template (*454*).

Bis-pyridine compounds containing the conjugated spacers $-CH=CH-$, $-CH=N-N=CH-$, $-N=N-$, or $-CMe=N-N=CMe-$ linking the 4-positions of the two pyridine rings and having a pentacyanoferrate(II) group can be threaded through a cyclodextrin; coordination of a second pentacyanoferrate(II) moiety to the other pyridine then produces a rotaxane. Kinetic and mechanistic features of these systems have been described (see also Section II.D.5 above) (*131*). A 1:1 intermediate has been characterized and monitored in the first example of rotaxane formation using ferrocenyl-β-cyclodextrin stoppers (*469*).

Ligands comprising a tetraazamacrocycle with a pyridine-containing pendant arm coordinate strongly through the former, relatively weakly through the latter. These characteristics result in a monomer ⇌ dimer equilibrium for a zinc-porphyrin complex of this type which has been exploited in the design and generation of a rotaxane, by reaction of the equilibrium mixture with terephthalic acid, in which two porphyrin entities form the stoppers (*470*). Other porphyrin-stopped rotaxanes containing $Ru^{II}CO$, $Rh^{III}I$ moieties complement such zinc-containing compounds to give a very wide range of formation and dissociation rate constants. These various systems exhibit kinetic barriers by remote control, through metal coordination to the porphyrin, of rotaxane formation (*471*).

2. Catenanes

Designed assembly may be illustrated by the copper(I)-templated [2]-catenane containing zinc- and gold(III)-porphyrin rings (472), by the use of the complementary binding properties of ruthenium(II) (to nitrogen donors) and tin(IV) (to oxygen donors) (473), and by synthesis using "pre-determined" μ-oxo-di-iron-dimers (352). The mechanism of formation of catenanes using copper(I) templates has been followed by utilizing ruthenium-catalyzed ring closure (474). The use of octahedral templates rather than the ubiquitous tetrahedral copper(I) for catenand assembly has very rarely been exploited (320) but recently the reaction sequence for the construction of benzylic imine catenates using octahedral metal centers (Mn^{2+}, Fe^{2+}, Co^{2+}, Ni^{2+}, Zn^{2+}, Cd^{2+}, Hg^{2+}) as templates has been established (475).

Redox control of ring gliding in a copper-complexed [2]-catenane, makes use of the common substitution-lability but differing coordination number preferences of copper(I) and copper(II). The synthesis of this species depends on the design of, and controlled assembly from, precursors containing bidentate and terdentate polypyridyl moieties – each ring has to contain one bpy and one terpy unit. The gliding mechanism almost certainly involves transient five-coordinate Cu^{I} and Cu^{II} species; ring-gliding is considerably easier and faster in DMF than in water (476).

3. Helicates and Knots

Detailed knowledge of the mechanisms of formation of helical complexes is still sparse (477). The mixture of mono- and poly-nuclear copper(I) helices obtained in the preparatively straightforward reaction of copper(I) with 2,7-diimino-1,8-naphthyridine (478) illustrates the complications generally involved. A thermodynamic, kinetic, and mechanistic study of self-assembly of a tricopper(I) double helicate of the ligands 19 ($X=CO_2Et$ or $CONEt_2$) resolved the reaction sequence into four distinct steps. Intermediates containing Cu^{+} in stereochemically unfavorable coordination lead to the final product containing three tetrahedrally coordinated Cu^{+} ions. The CO_2Et and $CONEt_2$ derivatives react at markedly different rates, though the respective final products appear to be of comparable stability (479). Selective self-assembly has been demonstrated in the formation of double-stranded binuclear copper(I) helicate (480). Directionality in double-stranded helicates – head-to-tail (C_2 symmetry) versus head-to-head – may be controlled by the electronic configuration of the heterocycles in the ligands involved. This is suggested by the use of a mixed bipyridine–bipyrazine ligand here – bipyrazine has much less aromatic character than most diimines. Previously substituent bulk has been used to control directionality. Stability constants for copper(I)-helicate systems are relevant to formation and assembly mechanisms (481).

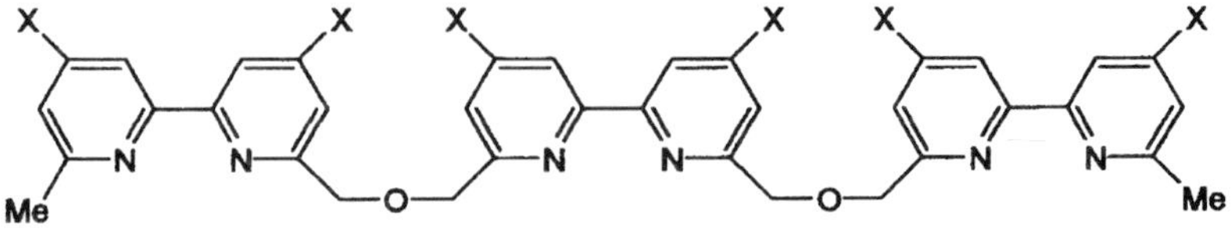

X = CONEt$_2$ or CO$_2$Et

19

There are two levels of self-assembly in the formation of tetra-, penta- and hexa-nuclear products from the poly-bipyridyls (L) **20** and **21** and iron(II) salts FeCl$_2$, FeBr$_2$ or FeSO$_4$ – the products are anion-dependent. The coordination of three bpy units, from different ligand molecules, to the Fe^{2+} centers produces a helical structure; interaction of these helical strands with anions results in further molecular organization to form the final toroidal product. The discussion draws parallels between the helical and toroidal structures here and secondary and tertiary structure in biological systems (482). Thermodynamic and kinetic intermediates have been characterized in the self-assembly of a di-iron triple stranded helicate with bis(2,2′-bipyridyl) ligands (483).

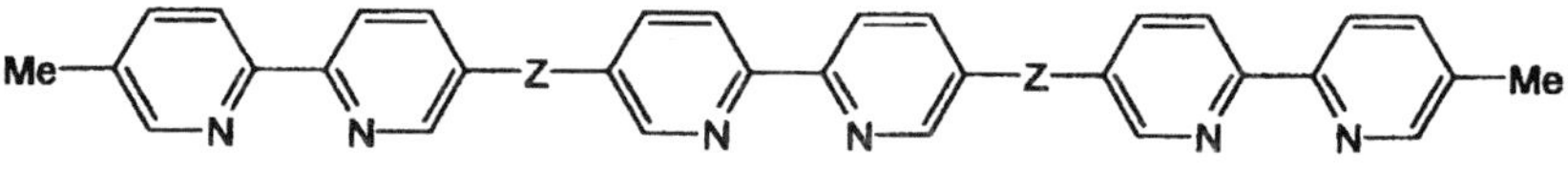

20   —Z—   =   —CH$_2$CH$_2$—

21   —Z—   =   —CH$_2$OCH$_2$—

Most helicates have linear axes, though a few helicates with circular axes are known – indeed the chiral (D_4) molecular squares formed from Zn^{2+} and 2,5-bis(2,2′-bipyridin-6-yl)pyrazine, **22**, may be regarded as circular helicates (450). The formation of circular or linear forms seems to depend on balances between kinetic and thermodynamic control; iron(II)-poly-2,2′-diimine systems with their substitutionally-inert metal centers provide useful systems for disentangling thermodynamic and kinetic contributions. The mechanism of formation of circular helicates of this type is believed to entail a kinetically favored triple helicate intermediate (484). Self-assembly of chiral-twisted iron(III)-porphyrin dimers into extended polynuclear species takes place through the intermediacy of μ-oxo dimers (*352*). The choice of building blocks is the key to success here, with the properties (electron and energy transfer; catalysis) of the product determined by this choice. The use of somewhat flexible building blocks, rather than the rigid ones normally favored, may give opportunities for fine-tuning products, as discussed in relation to manganese-containing helixes (*485*). The assembly of a triple-stranded catechol/8-hydroxyquinoline/gallium(III) helix is highly stereoselective, under stringent steric and electronic control (*486*).

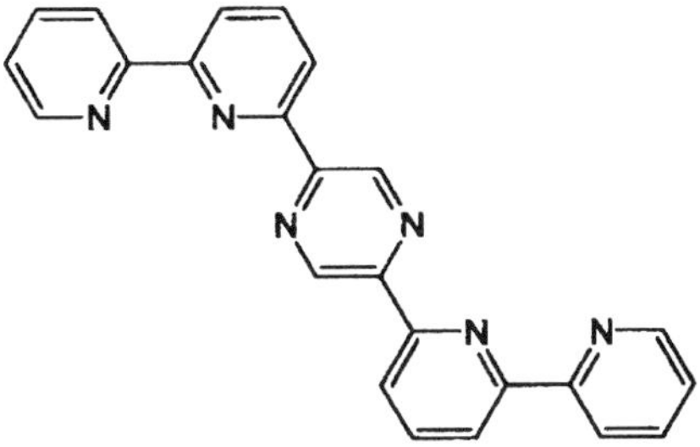

22

The mechanism of formation of molecular knots via catenanes, using copper(I) templates, has been investigated through ruthenium-catalyzed ring closure. This latter is the novel feature, which now provides an easy route to trefoil knot species (*474*). A bascule (coupled pivot) mechanism is proposed for the generation of a trinuclear copper(I) trefoil knot precursor. The synthesis of the starting materials for this species involves palladium-dependent aryl coupling (*487*). The high symmetry of this precursor indicates a very ordered mechanism for its formation – which state of affairs may be contrasted with the "entangled precursors" invoked in an earlier study of knot formation (*488*).

Ligand Abbreviations

Abbreviations for ligands which make only one appearance are generally defined at the appropriate point in the text; ligand abbreviations which appear in more than one place are listed and defined below.

amp   adenosine monophosphate
        (also adp, atp for di- and tri-phosphates)
bpy   2,2′-bipyridyl (4,4′-bipyridyl is abbreviated as 4,4′-bpy}
cod   cyclo-octa-1,5-diene

cyc*a*	*meso*-5,5,7,12,12,14-hexamethyl-1,4,8,11-tetraazacyclotetradecane
cyc*b*	*rac*-5,5,7,12,12,14-hexamethyl-1,4,8,11-tetraazacyclotetradecane
cyclam	1,4,8,11-tetraazacyclotetradecane
cyclen	1,4,7,10-tetraazacyclododecane
dien	diethylenetriamine
dmg	dimethylglyoxime
dppe	1,2-diphenylphosphinoethane
dtpa	diethylenetriaminepentacetate
edta	ethylenediaminetetraacetate
en	1,2-diaminoethane
glygly	glycylglycine
hfac	hexafluoroacetone
imid	imidazole
nta	nitrilotriacetate
oep	octaethylporphyrin
phen	1,10-phenanthroline
py	pyridine
salen	N,N'-ethylene-1,2-salicylidiniminate
terpy	$2,2',6',2''$-terpyridyl
tpp	tetraphenylporphyrin
tren	tris(aminoethyl)amine
ttpz	2,3,5,6-tetrakis(2-pyridyl)pyrazine
tu	thiourea

Solvent Abbreviations

DMAC	N,N'-dimethylacetamide
DMF	N,N'-dimethylformamide
DMSO	dimethylsulfoxide
MeCN	acetonitrile
THF	tetrahydrofuran

REFERENCES

1. Winterton, N. *Ann. Rep., Sect. A* **1994**, *89*, 475; **1995**, *90*, 515; **1996**, *92*, 481; **1997**, *93*, 541; **1998**, *94*, 537; **1999**, *95*, 535; *Ann. Rep. Prog. Chem., Sect. A* **2000**, *96*, 557.
2. "*Mechanisms of Inorganic and Organometallic Reactions*", vol. 8; Ed. Twigg, M. V.; Plenum Press: New York, 1994.
3. Orvig, C.; Abrams, M. J., Eds. *Chem. Rev.* **1999**, *99*, 2201–2842.
4. E.g., Swaddle, T. W. *Adv. Inorg. Bioinorg. Chem.* **1983**, *2*, 95; Swaddle, T. W. *Comments Inorg. Chem.* **1991**, *12*, 237; Lay, P. A. *Coord. Chem. Rev.* **1991**, *110*, 213.
5. Wilkins, R. G. *Transition Met. Chem.* **1998**, *23*, 735.
6. Tobe, M. L.; Burgess, J. "*Inorganic Reaction Mechanisms*"; Addison-Wesley-Longman: Harlow, 1999.

7. Lincoln, S. F.; Merbach, A. E. *Adv. Inorg. Chem.* **1995**, *42*, 1.
8. *"High Pressure Chemistry – Synthetic, Mechanistic, and Supercritical Applications"*, Eds. van Eldik, R.; Klärner, F.-G.; Wiley-VCH: Weinheim, 2002.
9. Drljaca, A.; Hubbard, C. D.; van Eldik, R.; Asano, T.; Basilevsky, M. V.; le Noble, W. J. *Chem. Rev.* **1998**, *98*, 2167.
10. van Eldik, R.; Hubbard, C. D. *S. Afr. J. Chem.* **2000**, *52*, 139.
11. van Eldik, R. *"High Pressure Molecular Science"*, Ed. Winter, R.; Jonas, J.; Kluwer: Dordrecht, 1999, p. 267.
12. Stochel, G.; van Eldik, R. *Coord. Chem. Rev.* **1999**, *187*, 329.
13. van Eldik, R.; Hubbard, C. D. *New J. Chem.* **1997**, *21*, 825.
14. van Eldik, R.; Dücker-Benfer, C.; Thaler, F. *Adv. Inorg. Chem.* **2000**, *49*, 1.
15. Stochel, G.; van Eldik, R. *Coord. Chem. Rev.* **1997**, *159*, 153; van Eldik, R.; Ford, P. C. *Adv. Photochem.* **1998**, *24*, 61.
16. van Eldik, R. *Coord. Chem. Rev.* **1999**, *182*, 373.
17. See, e.g., pp. 343–346 of Ref. (*6*).
18. Swiss, K. A.; Firestone, R. A. *J. Phys. Chem. A* **2000**, *104*, 3057; **2002**, *106*, 6909.
19. Weber, C. F.; van Eldik, R. *J. Phys. Chem. A* **2002**, *106*, 6904.
20. Kowall, T.; Caravan, P.; Bourgeois, H.; Helm, L.; Rotzinger, F. P.; Merbach, A. E. *J. Am. Chem. Soc.* **1998**, *120*, 6569; Helm, L.; Merbach, A. E. *Coord. Chem. Rev.* **1999**, *187*, 151.
21. Rotzinger, F. P. *Inorg. Chem.* **1999**, *38*, 5730.
22. Benzo, F.; Bernhardt, P. V.; González, G.; Martinez, M.; Sienra, B. *J. Chem. Soc., Dalton Trans.* **1999**, 3973.
23. House, D. A. *Comments Inorg. Chem.* **1994**, *16*, 229.
24. Bray, M. R.; Deeth, R. J.; Paget, V. J. *Prog. React. Kinet.* **1996**, *21*, 169.
25. Hao, J.; Poë, A. J. *Transition Met. Chem.* **1998**, *23*, 739.
26. Linert, W.; Kudryavtsev, A. B. *Coord. Chem. Rev.* **1999**, *190–192*, 405.
27. Burgess, J.; Pelizzetti, E. *Gazz. Chim. Ital.* **1988**, *118*, 803; Burgess, J. *Coll. Surfaces* **1990**, *48*, 185; Burgess, J.; Pelizzetti, E. *Prog. React. Kinet.* **1992**, *17*, 1; and see Chap. 8 of Ref. (*6*).
28. Darr, J. A.; Poliakoff, M. *Chem. Rev.* **1999**, *99*, 495 – e.g., p. 525.
29. Ji, Q.; Lloyd, C. R.; Eyring, E. M.; van Eldik, R. *J. Phys. Chem. A* **1997**, *101*, 243.
30. Zhang, T.-L.; Jin, D.-M. *Indian J. Chem.* **2001**, *40A*, 783.
31. Toma, H. E.; Rocha, R. C. *Croat. Chem. Acta* **2001**, *74*, 499.
32. Coe, B. J.; Glenwright, S. J. *Coord. Chem. Rev.* **2000**, *203*, 5.
33. E.g. Katti, K. V.; Cavell, R. G. *Comments Inorg. Chem.* **1990**, *10*, 53.
34. Slone, C. S.; Weinberger, D. A.; Mirkin, C. A. *Prog. Inorg. Chem.* **1999**, *48*, 233.
35. Braunstein, P.; Naud, F. *Angew. Chem. Int. Ed.* **2001**, *40*, 680.
36. Jung, S.; Brandt, C. D.; Werner, H. *New J. Chem.* **2001**, *25*, 1101.
37. Espinet, P.; Soulantica, K. *Coord. Chem. Rev.* **1999**, *193–195*, 499.
38. Clarke, M. L.; Cole-Hamilton, D. J.; Woollins, J. D. *J. Chem. Soc., Dalton Trans.* **2001**, 2721.
39. Berners-Price, S. J.; Bowen, R. J.; Galettis, P.; Healy, P. C.; McKeage, M. J. *Coord. Chem. Rev.* **1999**, *185–186*, 823.
40. Romeo, R.; Monsu' Scolaro, L.; Plutino, M. R.; Romeo, A.; Nicolo', F.; Del Zotto, A. *Eur. J. Inorg. Chem.* **2002**, 629.
41. Bozoglián, F.; González, G.; Martínez, M.; Queirolo, M.; Sienra, B. *Inorg. Chim. Acta* **2001**, *318*, 191.
42. Lawrance, G. A. *Adv. Inorg. Chem.* **1989**, *34*, 145.
43. House, D. A. *Coord. Chem. Rev.* **1977**, *23*, 223; Uzice, J. L.; Lopez de la Vega, R. *Inorg. Chem.* **1990**, *29*, 382; van Eldik, R. *Pure Appl. Chem.* **1992**, *64*, 1439.

44. Basolo, F.; Pearson, R. G. *"Mechanisms of Inorganic Reactions"*, 2nd edn.; Wiley: New York, 1967, p. 69.
45. Rodgers, G. E. *"Introduction to Coordination, Solid State, and Descriptive Inorganic Chemistry"*; McGraw-Hill: New York, 1994, p. 101.
46. Benzo, F.; González, G.; Martínez, M.; Sienra, B. *Inorg. React. Mech.* **2001**, *3*, 25.
47. Shearer, J.; Kung, I. Y.; Lovell, S.; Kaminsky, W.; Kovacs, J. A. *J. Am. Chem. Soc.* **2001**, *123*, 463.
48. Das, A.; Dash, A. C. *J. Chem. Soc., Dalton Trans.* **2000**, 1949.
49. Shaham, N.; Masarwa, A.; Matana, Y.; Cohen, H.; Meyerstein, D. *Eur. J. Inorg. Chem.* **2002**, 87.
50. Dücker-Benfer, C.; Hamza, M. S. A.; Eckhardt, C.; van Eldik, R. *Eur. J. Inorg. Chem.* **2000**, 1563.
51. Chatlas, J.; Danilchuk, E.; Nasiadko, M.; Raszkowska, T. *Transition Met. Chem.* **2002**, *27*, 346.
52. Danilczuk, E.; Chatlas, J. *Transition Met. Chem.* **1996**, *21*, 432.
53. Kitamura, Y.; Hyodoh, K.; Hayashi, M.; Nagawo, Y.; Sasaki, Y.; Shibata, A. *Inorg. React. Mech.* **2001**, *3*, 197.
54. Kitamura, Y.; Yano, L.; Fujimori, K.; Mizuki, R.; Hayashi, M.; Shibata, A. *Bull. Chem. Soc. Jpn.* **2000**, *73*, 3025.
55. Mao, Z.-W.; Liehr, G.; van Eldik, R. *J. Am. Chem. Soc.* **2000**, *122*, 4839.
56. Loerting, T.; Tautermann, C.; Kroemer, R. T.; Kohl, I.; Hallbrucker, A.; Mayer, E.; Leidl, K. R. *Angew. Chem. Int. Ed.* **2000**, *39*, 891; Ludwig, R.; Kornath, A. *Angew. Chem. Int. Ed.* **2000**, *39*, 1421.
57. Holder, A. A.; Dasgupta, T. P. *J. Chem. Soc., Dalton Trans.* **1996**, 2637.
58. Wick, P. K.; Kissner, R.; Koppenol, W. H. *Helv. Chim. Acta* **2000**, *83*, 748.
59. Wick, P. K.; Kissner, R.; Koppenol, W. H. *Helv. Chim. Acta* **2001**, *84*, 3057.
60. Tibi, S.; Koppenol, W. H. *Helv. Chim. Acta* **2000**, *83*, 2412.
61. Goldstein, S.; Meyerstein, D.; van Eldik, R.; Czapski, G. *J. Phys. Chem. A* **1999**, *103*, 6587; Coddington, J. W.; Wherland, S.; Hurst, J. K. *Inorg. Chem.* **2001**, *40*, 528.
62. Meier, M.; van Eldik, R. *Inorg. Chem.* **1993**, *32*, 2635.
63. Alzoubi, B. M.; Hamza, M. S. A.; Dücker-Benfer, C.; van Eldik, R. *Eur. J. Inorg. Chem.* **2002**, 968; Hamza, M. S. A.; Dücker-Benfer, C.; van Eldik, R. *Inorg. Chem.* **2000**, *39*, 3777.
64. Hamza, M. S. A.; van Eldik, R.; Harper, P. L. S.; Pratt, J. M.; Betterton, E. A. *Eur. J. Inorg. Chem.* **2002**, 580.
65. Brasch, N. E.; Cregan, A. G.; Vanselow, M. E. *J. Chem. Soc., Dalton Trans.* **2002**, 1287.
66. Cregan, A. G.; Brasch, N. E.; van Eldik, R. *Inorg. Chem.* **2001**, *40*, 1430.
67. Brasch, N. E.; Haupt, R. J. *Inorg. Chem.* **2000**, *39*, 5469.
68. Hamza, M. S. A.; Zou, X.; Brown, K. L.; van Eldik, R. *Inorg. Chem.* **2001**, *40*, 5440.
69. Brasch, N. E.; Hamza, M. S. A.; van Eldik, R. *Inorg. Chem.* **1997**, *36*, 3216.
70. Wolak, M.; Zahl, A.; Schneppensieper, T.; Stochel, G.; van Eldik, R. *J. Am. Chem. Soc.* **2001**, *123*, 9780.
71. Wolak, M.; Stochel, G.; Hamza, M.; van Eldik, R. *Inorg. Chem.* **2000**, *39*, 2018.
72. Sridhar, V.; Satyanarayana, S. *Indian J. Chem.* **2001**, *40A*, 165.
73. Tobe, M. L. *Adv. Inorg. Bioinorg. Chem.* **1983**, *2*, 1; see also Section 4.6 of Ref. (*6*).
74. Clarkson, A. J.; Buckingham, D. A.; Rogers, A. J.; Blackman, A. G.; Clark, C. R. *Inorg. Chem.* **2000**, *39*, 4769.
75. MacDonald, C. J.; Browning, J.; Steel, P. J.; House, D. A. *Inorg. React. Mech.* **2000**, *2*, 289.
76. Balt, S.; Breman, J.; Dekker, C. *J. Inorg. Nucl. Chem.* **1976**, *38*, 2023.
77. Clark, J. B.; Pilkington, K. A.; Staples, P. J. *J. Chem. Soc. (A)* **1966**, 153.
78. Benzo, F.; Capparelli, A. L.; Mártire, D. O.; Sienra, B. *Inorg. React. Mech.* **2000**, *1*, 319.

79. Baran, Y.; Hambley, T. W.; Lawrance, G. A.; Wilkes, E. N. *Aust. J. Chem.* **1997**, *50*, 883; Baran, Y.; Lawrance, G. A.; Martínez, M.; Wilkes, E. N. *Inorg. React. Mech.* **2000**, *1*, 315.
80. Fabius, B.; Geue, R. J.; Hazell, R. G.; Jackson, W. G.; Larsen, F. K.; Qin, C. J.; Sargeson, A. M. *J. Chem. Soc., Dalton Trans.* **1999**, 3961.
81. Poth, T.; Paulus, H.; Elias, H.; van Eldik, R.; Grohmann, A. *Eur. J. Inorg. Chem.* **1999**, 643.
82. House, D. A. *Adv. Inorg. Chem.* **1997**, *44*, 341.
83. House, D. A. *"Mechanisms of Inorganic and Organometallic Reactions"*; Ed. Twigg, M. V.; Plenum Press: New York, 1994, vol. 8; 1991, vol. 7; 1989, vol. 6.
84. Madej, E.; Mønsted, O.; Kita, P. *J. Chem. Soc., Dalton Trans.* **2002**, 2361.
85. Behera, J.; Brahma, G. S.; Mohanty, P. *J. Indian Chem. Soc.* **2001**, *78*, 123.
86. Barnés, C. M.; Theil, E. C.; Raymond, K. N. *Proc. Natl. Acad. Sci. (USA)* **2002**, *99*, 5195.
87. Kania, R.; Siecklucka, B. *Polyhedron* **2000**, *19*, 2225.
88. Chatterjee, C.; Stephen, M. *Polyhedron* **2001**, *20*, 2917.
89. Butković, V.; Bakač, A.; Espenson, J. H. *Croat. Chem. Acta* **2001**, *74*, 735.
90. Zinato, E.; Riccieri, P. *Coord. Chem. Rev.* **2001**, *211*, 5; Irwin G.; Kirk, A. D. *Coord. Chem. Rev.* **2001**, *211*, 25; Derwahl, A.; Wasgestian, F.; House, D. A.; Robinson, W. T. *Coord. Chem. Rev.* **2001**, *211*, 45.
91. Yang, G.; Leng, W.; Zhang, Y.; Chen, Z.; Van Houten, J. *Polyhedron* **1999**, *18*, 1273.
92. Cummins, C. C. *Prog. Inorg. Chem.* **1998**, *47*, 685.
93. Mukhopadhyay, S. K.; Ghosh, A. K.; De, G. S. *Indian J. Chem., Sect. A* **1999**, *38*, 895.
94. Ghosh, A. K. *Transition Met. Chem.* **1998**, *23*, 269.
95. Dücker-Benfer, C.; Dreos, R.; van Eldik, R. *Angew. Chem. Int. Ed.* **1995**, *34*, 2245.
96. Mura, P.; Casini, A.; Marcon, G.; Messori, L. *Inorg. Chim. Acta* **2001**, *312*, 74.
97. Tellers, D. M.; Bergman, R. G. *Can. J. Chem.* **2001**, *79*, 525.
98. Poth, T.; Paulus, H.; Elias, H.; Dücker-Benfer, C.; van Eldik, R. *Eur. J. Inorg. Chem.* **2001**, 1361.
99. Alberto, R.; Schibli, R.; Schubiger, A. P.; Abram, U.; Pietzsch, H.-J.; Johannsen, B. *J. Am. Chem. Soc.* **1999**, *121*, 6076.
100. Aebischer, N.; Schibli, R.; Alberto, R.; Merbach, A. E. *Angew. Chem. Int. Ed.* **2000**, *39*, 254; Metzler-Nolte, N. *Angew. Chem. Int. Ed.* **2001**, *40*, 1040.
101. See e.g. Wald, J.; Alberto, R.; Ortner, K.; Candreia, L. *Angew. Chem. Int. Ed.* **2001**, *40*, 3062.
102. Alberto, R.; Schibli, R.; Waibel, R.; Abram, U.; Schubiger, A. P. *Coord. Chem. Rev.* **1999**, *190–192*, 901.
103. Hauser, A.; Jeftić, J.; Romstedt, H.; Hinek, R.; Spiering, H. *Coord. Chem. Rev.* **1999**, *190–192*, 471.
104. Toftlund, H. *Monatsh.* **2001**, *132*, 1269.
105. McGarvey, J. J.; Lawthers, I.; Heremans, K.; Toftlund, H. *Inorg. Chem.* **1990**, *29*, 252.
106. Boillot, M.-L.; Zarembowitch, J.; Itié, J.-P.; Polian, A.; Bourdet, E.; Haasnoot, J. G. *New J. Chem.* **2002**, *26*, 313.
107. de Carvalho, I. N.; Tubino, M. *J. Braz. Chem. Soc.* **1991**, *2*, 56.
108. Lucie, J.-M.; Stranks, D. R.; Burgess, J. *J. Chem. Soc., Dalton Trans.* **1975**, 245; Burgess, J.; Galema, S. A.; Hubbard, C. D. *Polyhedron* **1991**, *10*, 703.
109. Priimov, G. U.; Moore, P.; Maritim, P. K.; Butalanyi, P. K.; Alcock, N. W. *J. Chem. Soc., Dalton Trans.* **2000**, 445.
110. Squizani, F.; Stein, E.; Vichi, E. J. S. *J. Braz. Chem. Soc.* **1996**, *7*, 127.
111. Rao, G. V.; Sridhar, Y.; Hela, P. G.; Padhi, T.; Anipindi, N. R. *Transition Met. Chem.* **1999**, *24*, 566.

112. Haines, R. I.; Hutchings, D. R.; Strickland, D. W. *Inorg. React. Mech.* **2000**, *2*, 223.
113. Abu Gharib, E. A.; Gosal, N.; Burgess, J. *Croat. Chem. Acta* **2001**, *74*, 545.
114. Alousy, A.; Burgess, J.; Elvidge, D.; Hubbard, C. D.; Radulović, S. *Inorg. React. Mech.* **2000**, *2*, 249.
115. Milde, S. P.; Blandamer, M. J.; Burgess, J.; Engberts, J. B. F. N.; Galema, S. A.; Hubbard, C. D. *J. Phys. Org. Chem.* **1999**, *12*, 227.
116. Alshehri, S.; Burgess, J.; Shaker, A. M. *Transition Met. Chem.* **1998**, *23*, 689.
117. Shaker, A. M.; Alshehri, S.; Burgess, J. *Transition Met. Chem.* **1998**, *23*, 683.
118. Meistermann, I.; Moreno, V.; Prieto, M. J.; Moldrheim, E.; Sletten, E.; Khalid, S.; Rodger, P. M.; Peberdy, J. C.; Isaac, C. J.; Rodger, A.; Hannon, M. J. *Proc. Natl. Acad. Sci. (USA)* **2002**, *89*, 5069.
119. Henderson, R. A. *J. Chem. Soc., Dalton Trans.* **1988**, 509.
120. Basallote, M. G.; Durán, J.; Fernández-Trujillo, M. J.; González, G.; Máñez, M. A.; Martínez, M. *Inorg. Chem.* **1998**, *37*, 1623.
121. David, S.; James, B. R.; Dolphin, D.; Traylor, T. G.; Lopez, M. A. *J. Am. Chem. Soc.* **1994**, *116*, 6.
122. Buchalova, M.; Warburton, P. R.; van Eldik, R.; Busch, D. H. *J. Am. Chem. Soc.* **1997**, *119*, 5867; Buchalova, M.; Busch, D. H.; van Eldik, R. *Inorg. Chem.* **1998**, *37*, 1116.
123. Projahn, H.-D.; van Eldik, R. *Inorg. Chem.* **1991**, *30*, 3288; Projahn, H.-D.; Schindler, S.; van Eldik, R.; Fortier, D. G.; Andrew, C. R.; Sykes, A. G. *Inorg. Chem.* **1995**, *34*, 5935.
124. Farrington, D. J.; Jones, J. G.; Robinson, N. D.; Twigg, M. V. *Transition Met. Chem.* **1999**, *24*, 697; Baldacchini, T.; Monacelli, F. *Inorg. Chim. Acta* **1999**, *295*, 200.
125. Jones, J. G.; Farrington, D. J.; McDonald, F. M.; Mooney, P. M.; Twigg, M. V. *Transition Met. Chem.* **1998**, *23*, 693.
126. Ozoemena, K.; Nyokong, T. *J. Chem. Soc., Dalton Trans.* **2002**, 1806.
127. Baran, Y.; Ulgen, A. *Int. J. Chem. Kinet.* **1998**, *30*, 415.
128. Chen, C.; Wu, M.; Yeh, A.; Tsai, T. Y. R. *Inorg. Chim. Acta* **1998**, *267*, 81.
129. Shi, C.-H.; Wang, S.-T.; Yang, S.-Y.; Yeh, A.; Tien, H.-J. *J. Chin. Chem. Soc. (Taipei)* **1998**, *45*, 77 (*Chem. Abstr.* **1998**, *128*, 252063r).
130. Almaraz, A. E.; Gentil, L. A.; Baraldo, L. M.; Olabe, J. A. *Inorg. Chem.* **1996**, *35*, 7718.
131. Baer, A. J.; Macartney, D. H. *Inorg. Chem.* **2000**, *39*, 1410.
132. See, e.g., pp. 186 and 308 of Ref. (*6*).
133. Maciejowska, I.; van Eldik, R.; Stochel, G.; Stasicka, Z. *Inorg. Chem.* **1997**, *36*, 5409.
134. Baraldo, L. M.; Forlano, P.; Parise, A. R.; Slep, L. D.; Olabe, J. A. *Coord. Chem. Rev.* **2001**, *219–221*, 881.
135. Muriel-Delgado, F.; Jiménez, R.; Gómez-Herrera, C.; Sánchez, F. *Langmuir* **1999**, *15*, 4344; Muriel-Delgado, F.; Jiménez, R.; Gómez-Herrera, C.; Sánchez, F. *New J. Chem.* **1999**, *23*, 1203.
136. Shishido, S. M.; Ganzarolli de Oliveira, M. *Prog. React. Kinet. Mech.* **2001**, *26*, 239.
137. Janiak, C.; Dorn, T.; Paulsen, H.; Wrackmeyer, B. *Z. Anorg. Allg. Chem.* **2001**, *627*, 1663.
138. Wanner, M.; Scheiring, T.; Kaim, W.; Slep, L. D.; Baroldo, L. M.; Olabe, J. A.; Zális, S.; Baerends, E. J. *Inorg. Chem.* **2001**, *40*, 5704.
139. Jiang, J.; Acunzo, A.; Koch, S. A. *J. Am. Chem. Soc.* **2001**, *123*, 12 109.
140. Jiang, J.; Koch, S. A. *Angew. Chem. Int. Ed.* **2001**, *40*, 2629; Rauchfuss, T. B.; Contakes, S. M.; Hsu, S. C. N.; Reynolds, M. A.; Wilson, S. R. *J. Am. Chem. Soc.* **2001**, *123*, 6933.
141. Jiang, J.; Koch, S. A. *Inorg. Chem.* **2002**, *41*, 158.
142. Grundler, P. V.; Laurenczy, G.; Merbach, A. E. *Helv. Chim. Acta* **2001**, *84*, 2854.

143. Aebischer, N.; Merbach, A. E. *Inorg. React. Mech.* **1999**, *1*, 233.
144. See, e.g., Richens, D. T. "*The Chemistry of Aqua Ions*"; Wiley: Chichester, 1997, p. 390, and Table 7.20 of Ref. (*6*).
145. Aebischer, N.; Churlaud, R.; Dolci, L.; Frey, U.; Merbach, A. E. *Inorg. Chem.* **1998**, *37*, 5915.
146. Johnson, W. A.; Wilkins, R. G. *Inorg. Chem.* **1970**, *9*, 1917; Hewkin, D. J.; Prince, R. H. *Coord. Chem. Rev.* **1970**, *5*, 45.
147. Roche, T. S.; Wilkins, R. G. *J. Am. Chem. Soc.* **1974**, *96*, 5082.
148. See, e.g., Burgess, J. "*Metal Ions in Solution*"; Ellis Horwood: Chichester, 1978, p. 354.
149. Das, T.; De, G. S.; Ghosh, A. K. *J. Indian Chem. Soc.* **2001**, *78*, 451.
150. Cayemittes, S.; Poth, T.; Fernandez, M. J.; Lye, P. G.; Becker, M.; Elias, H.; Merbach, A. E. *Inorg. Chem.* **1999**, *38*, 4309.
151. Trofimenko, S. *Chem. Rev.* **1993**, *93*, 943; Trofimenko, S. "*Scorpionates: The Coordination Chemistry of Polypyrazolylborate Ligands*"; Imperial College Press: London, 1999.
152. Huynh, M. H. V.; Smyth, J.; Wetzler, M.; Mort, B.; Gong, P. K.; Witham, L. M.; Jameson, D. L.; Geiger, D. K.; Lasker, J. M.; Charepoo, M.; Gornikiewicz, M.; Cintron, J. M.; Imahori, G.; Sanchez, R. R.; Marschilok, A. C.; Krajkowski, L. M.; Churchill, D. G.; Churchill, M. R.; Takeuchi, K. J. *Angew. Chem. Int. Ed.* **2001**, *40*, 4469.
153. Huynh, M. H. V.; Lasker, J. M.; Wetzler, M.; Mort, B.; Szczepura, L. F.; Witham, L. M.; Cintron, J. M.; Marschilok, A. C.; Ackerman, L. J.; Castellano, R. K.; Jameson, D. L.; Churchill, M. R.; Jircitano, A. J.; Takeuchi, K. J. *J. Am. Chem. Soc.* **2001**, *123*, 8780.
154. Lima Neto, B. S.; Franco, D. W.; van Eldik, R. *J. Chem. Soc., Dalton Trans.* **1995**, 463.
155. Mazzetto, S. E.; Gambardella, M. T. do P.; Santos, R. H. A.; Lopes, L. G. de F.; Franco, D. W. *Polyhedron*, **1999**, *18*, 979.
156. Liu, J.-G.; Ji, L.-N. *J. Indian Chem. Soc.* **2000**, *77*, 539.
157. Kelly, J. M.; Tossi, A. B.; McConnell, D. J.; Ohlingin, C. *Nucleic Acid Res.* **1985**, *13*, 6017.
158. Coggan, D. Z. M.; Haworth, I. S.; Bates, P. J.; Robinson, A.; Rodger, A. *Inorg. Chem.* **1999**, *38*, 4486.
159. de Sousa Moreira, I.; Franco, D. W. *Inorg. Chem.* **1994**, *33*, 1607.
160. Elguero, J.; Guerrero, A.; Gómez de la Torre, F.; de la Hoz, A.; Jalón, F. A.; Manzano, B. R.; Rodríguez, A. *New J. Chem.* **2001**, *25*, 1050.
161. Volland, M. A. O.; Hofmann, P. *Helv. Chim. Acta* **2001**, *84*, 3456.
162. Tfouni, E. *Coord. Chem. Rev.* **2000**, *196*, 281.
163. Balzani, V.; Juris, A. *Coord. Chem. Rev.* **2001**, *211*, 97; Shan, B.-Z.; Zhao, Q.; Goswami, N.; Eichhorn, D. M.; Rillema, D. P. *Coord. Chem. Rev.* **2001**, *211*, 117.
164. Ferreira, K. Q.; Lucchesi, A. M.; de Rocha, Z. N.; da Silva, R. S. *Inorg. Chim. Acta* **2002**, *328*, 147.
165. Chatterjee, D.; Mitra, A.; Hamza, M. S. A.; van Eldik, R. *J. Chem. Soc., Dalton Trans.* **2002**, 962.
166. Matsubara, T.; Creutz, C. *Inorg. Chem.* **1979**, *18*, 1956.
167. Toma, H. E.; Santos, P. S.; Mattioli, M. P. D.; Oliveira, L. A. A. *Polyhedron* **1987**, *6*, 603.
168. Bajaj, H. C.; van Eldik, R. *Inorg. Chem.* **1988**, *27*, 4052.
169. Povse, V. G.; Olabe, J. A. *Transition Met. Chem.* **1998**, *23*, 657.
170. Bajaj, H.; Das, A.; van Eldik, R. *J. Chem. Soc., Dalton Trans.* **1998**, 1563.
171. Davies, N. A.; Wilson, M. T.; Slade, E.; Fricker, S. P.; Murrer, B. A.; Powell, N. A.; Henderson, G. R. *Chem. Commun.* **1997**, 47.

172. Wanat, A.; Schneppensieper, T.; Karocki, A.; Stochel, G.; van Eldik, R. *J. Chem. Soc., Dalton Trans.* **2002**, 941.
173. Abe, M.; Mitani, A.; Ohsawa, A.; Herai, M.; Tanaka, M.; Sasaki, Y. *Inorg. Chim. Acta* **2002**, *331*, 158.
174. Chakravarty, B.; Bhattacharya, R.; Bisai, S.; Rahman, M. *Indian J. Chem., Sect. A* **1999**, *38*, 686.
175. Slep, L. D.; Alborés, P.; Baraldo, L. M.; Olabe, J. A. *Inorg. Chem.* **2002**, *41*, 114.
176. Bernhardt, P. V.; Gallego, C.; Martínez, M.; Parella, T. *Inorg. Chem.* **2002**, *41*, 1747.
177. Dunham, S. U.; Abbott, E. H. *Inorg. Chim. Acta* **2000**, *297*, 72.
178. Hambley, T. W.; Jones, A. R. *Coord. Chem. Rev.* **2001**, *212*, 35.
179. Tsuchimoto, M.; Hoshina, G.; Uemura, R.; Nakajima, K.; Kojima, M.; Ohba, S. *Bull. Chem. Soc. Jpn.* **2000**, *73*, 2317.
180. Thompson, K. H.; McNeill, J. H.; Orvig, C. *Chem. Rev.* **1999**, *99*, 2561.
181. Rangel, M. *Transition Met. Chem.* **2001**, *26*, 219.
182. Tsaramyrsi, M.; Kaliva, M.; Salifoglou, A.; Raptopoulou, C. P.; Terzis, A.; Tangoulis, V.; Giapintzakis, J. *Inorg. Chem.* **2001**, *40*, 5772.
183. Samotus, A; Szklarzewicz, J. *Coord. Chem. Rev.* **1993**, *125*, 63.
184. Leipoldt, J. G.; Basson, S. S.; Roodt, A. *Adv. Inorg. Chem.* **1993**, *40*, 241.
185. Roodt, A.; Abou-Hamdan, A.; Engelbrecht, H. P.; Merbach, A. *Adv. Inorg. Chem.* **2000**, *49*, 59.
186. Roodt, A.; Leipoldt, J. G.; Helm, L.; Merbach, A. E. *Inorg. Chem.* **1994**, *33*, 140.
187. Szklarzewicz, J.; Samotus, A. *Transition Met. Chem.* **1995**, *20*, 174.
188. Murmann, R. K.; Barnes, C. L. *Inorg. Chem.* **2001**, *40*, 6514.
189. Szklarzewicz, J.; Matoga, D.; Samotus, A.; Burgess, J.; Fawcett, J.; Russell, D. R. *Croat. Chem. Acta* **2001**, *74*, 529.
190. Matoga, D.; Szklarzewicz, J.; Samotus, A.; Burgess, J.; Fawcett, J.; Russell, D. R. *Polyhedron* **2000**, *19*, 1503.
191. Szklarzewicz, J.; Samotus, A. *Transition Met. Chem.* **1998**, *23*, 807.
192. Monlien, F. J.; Helm, L.; Abou-Hamdan, A.; Merbach, A. E. *Inorg. Chem.* **2002**, *41*, 1717.
193. Monlien, F. J.; Helm, L.; Abou-Hamdan, A.; Merbach, A. E. *Inorg. Chim. Acta* **2002**, *331*, 257.
194. Banerjee, P. *Coord. Chem. Rev.* **1999**, *190–192*, 19.
195. Tsiveriotis, P.; Malandrinos, G.; Hadjiliadis, N. *Rev. Inorg. Chem.* **2000**, *20*, 305.
196. Otto, S.; Elding, L. I. *J. Chem. Soc., Dalton Trans.* **2002**, 2354.
197. Annibale, G.; Brandolisio, M.; Bugarcić, Z.; Cattalini, L. *Transition Met. Chem.* **1998**, *23*, 715.
198. Jaganyi, D.; Hofmann, A.; van Eldik, R. *Angew. Chem. Int. Ed.* **2001**, *40*, 1680.
199. Annibale, G.; Cattalini, L.; Cornia, A.; Fabretti, A.; Guidi, F. *Inorg. React. Mech.* **2000**, *2*, 185.
200. Romeo, R.; Plutino, M. R.; Monsù Scolaro, L.; Stoccoro, S.; Minghetti, G. *Inorg. Chem.* **2000**, *39*, 4749.
201. Hindmarsh, K.; House, D. A.; van Eldik, R. *Inorg. React. Mech.* **1999**, *1*, 107.
202. Jestin, J.-L.; Chottard, J.-C.; Frey, U.; Laurenczy, G.; Merbach, A. E. *Inorg. Chem.* **1994**, *33*, 4277; Kondo, Y.; Ishikawa, M.; Ishihara, K. *Inorg. Chim. Acta* **1996**, *241*, 81.
203. Marti, N.; Hoa, G. H. B.; Kozelka, J. *Inorg. Chem. Commun.* **1998**, *1*, 439.
204. Davies, M. S.; Cox, J. W.; Berners-Price, S. J.; Barklage, W.; Qu, Y.; Farrell, N. *Inorg. Chem.* **2000**, *39*, 1710.
205. Saeki, N.; Hirano, Y.; Sasamoto, Y.; Sato, I.; Toshida, T.; Ito, S.; Nakamura, N.; Ishihari, K.; Matsumoto, K. *Eur. J. Inorg. Chem.* **2001**, 2081; Saeki, N.; Hirano, Y.;

Sasamoto, Y.; Sato, I.; Toshida, T.; Ito, S.; Nakamura, N.; Ishihari, K.; Matsumoto, K. *Bull. Chem. Soc. Jpn.* **2001**, 74, 861.
206. Miller, S. E.; House, D. A. *Inorg. Chim. Acta* **1989**, *161*, 131; **1989**, *166*, 189.
207. Davies, M. S.; Thomas, D. S.; Hegmans, A.; Berners-Price, S. J.; Farrell, N. *Inorg. Chem.* **2002**, *41*, 1101.
208. Wang, F.; Wu, X.; Pinkerton, A. A.; Kumaradhas, P.; Neckers, D. C. *Inorg. Chem.* **2001**, *40*, 6000.
209. Crites Tears, D. K.; McMillin, D. R. *Coord. Chem. Rev.* **2001**, *211*, 195.
210. Ryabov, A. D.; Otto, S.; Samuleev, P. V.; Polyakov, V. A.; Alexandrova, L.; Kazankov, G. M.; Shova, S.; Revenco, M.; Lipkowski, J.; Johansson, M. H. *Inorg. Chem.* **2002**, *41*, 4286.
211. Kaminskaia, N. V.; Ullmann, G. M.; Fulton, D. B.; Kostić, N. M. *Inorg. Chem.* **2000**, *39*, 5004.
212. Sengupta, P. S.; Sinha, R.; De, G. S. *Indian J. Chem.* **2001**, *40A*, 509.
213. Zhang, S.; Zang, V.; Dobson, G. R.; van Eldik, R. *Inorg. Chem.* **1991**, *30*, 355; Cao, S.; Bal Reddy, K.; Eyring, E. M.; van Eldik, R. *Organometallics* **1994**, *13*, 91.
214. Bal Reddy, K.; van Eldik, R. *Organometallics* **1990**, *9*, 1418; Dücker-Benfer, C.; Grevels, F.-W.; van Eldik, R. *Organometallics* **1998**, *17*, 1669.
215. Romeo, R.; Monsù Scolaro, L.; Plutino, M. R.; Del Zotto, A. *Transition Met. Chem.* **1998**, *23*, 789.
216. Ray, M.; Dey, S.; Banerjee, P. *Inorg. React. Mech.* **2000**, *1*, 281.
217. Dey, S.; Ray, M.; Banerjee, P. *Inorg. React. Mech.* **2000**, *2*, 267.
218. Hay, R. W.; Miller, S. E. *Polyhedron* **1988**, *17*, 2337.
219. Ray, M.; Banerjee, P. *J. Indian Chem. Soc.* **2000**, *77*, 511.
220. Davies, M. S.; Berners-Price, S. J.; Hambley, T. W. *Inorg. Chem.* **2000**, *39*, 5603.
221. Cusumano, M.; Di Pietro, M. L.; Giannetto, A.; Romano, F. *Inorg. Chem.* **2000**, *39*, 50.
222. Kozelka, J.; Legendre, F.; Reeder, F.; Chottard, J.-C. *Coord. Chem. Rev.* **1999**, *190–192*, 61.
223. Hambley, T. W. *J. Chem. Soc., Dalton Trans.* **2001**, 2711.
224. Teuben, J.-M.; Rodriguez i Zubiri, M.; Reedijk, J. *J. Chem. Soc., Dalton Trans.* **2000**, 369.
225. Mikola, M.; Klika, K. D.; Arpalahti, J. *Chem. Eur. J.* **2000**, *6*, 3404.
226. Arpalahti, J.; Klika, K. D.; Molander, S. *Eur. J. Inorg. Chem.* **2000**, 1007.
227. Mikola, M.; Klika, K. D.; Hakala, A.; Arpalahti, J. *Inorg. Chem.* **1999**, *38*, 571.
228. Curis, E.; Provost, K.; Nicolis, I.; Bouvet, D.; Bénazeth, S.; Crauste-Manciet, S.; Brion, F.; Brossard, D. *New J. Chem.* **2000**, *24*, 1003.
229. Lemma, K.; Elmroth, S. K. C.; Elding, L. I. *J. Chem. Soc., Dalton Trans.* **2002**, 1281.
230. Carlone, M.; Fanizzi, F. P.; Intini, F. P.; Margiotta, N.; Marzilli, L. G.; Natile, G. *Inorg. Chem.* **2000**, *39*, 634.
231. Romeo, R. *Comments Inorg. Chem.* **1990**, *11*, 21;; see also Section 3.3.11 of Ref. (*6*) for a succinct summary.
232. Faraone, G.; Ricevuto, V.; Romeo, R.; Trozzi, M. *J. Chem. Soc. (A)* **1971**, 1877.
233. Basolo, F. *Coord. Chem. Rev.* **1996**, *154*, 151.
234. See p. 153 of Ref. (*10*).
235. Fischer, A.; Wendt, O. L. *J. Chem. Soc., Dalton Trans.* **2001**, 1266.
236. Albrecht, M.; van Koten, G. *Angew. Chem. Int. Ed.* **2001**, *40*, 3750.
237. Schmülling, M.; Ryabov, A. D.; van Eldik, R. *J. Chem. Soc., Dalton Trans.* **1994**, 1257; Schmülling, M.; Grove, D. M.; van Koten, G.; van Eldik, R.; Veldman, N.; Spek, A. L. *Organometallics* **1996**, *15*, 1384; Frey, U.; Grove, D. M.; van Koten, G. *Inorg. Chim. Acta* **1998**, *269*, 322.

238. Fekl, U.; van Eldik, R. *Eur. J. Inorg. Chem.* **1998**, 389.
239. Wendt, O. F.; Elding, L. I. *J. Chem. Soc., Dalton Trans.* **1997**, 4725; Wendt, O. F.; Oskarsson, Å.; Leipoldt, J. G.; Elding, L. I. *Inorg. Chem.* **1997**, *36*, 4514; Wendt, O. F.; Elding, L. I. *Inorg. Chem.* **1997**, *36*, 6028; Wendt, O. F.; Scodinu, A.; Elding, L. I. *Inorg. Chim. Acta* **1998**, *277*, 237.
240. Romeo, R.; Plutino, M. R.; Elding, L. I. *Inorg. Chem.* **1997**, *36*, 5909.
241. Wendt, O. F.; Deeth, R. J.; Elding, L. I. *Inorg. Chem.* **2000**, *39*, 5271.
242. Albrecht, M.; Gossage, R. A.; Frey, U.; Ehlers, A. W.; Baerends, E. J.; Merbach, A. E.; van Koten, G. *Inorg. Chem.* **2001**, *40*, 850.
243. Plutino, M. R.; Monsù Scolaro, L.; Romeo, R.; Grassi, A. *Inorg. Chem.* **2000**, *39*, 2712.
244. Schmülling, M.; van Eldik, R. *Chem. Ber. Receuil* **1997**, *130*, 1791.
245. Nakajima, K.; Kajino, T.; Nonoyama, M.; Kojima, M. *Inorg. Chim. Acta* **2001**, *312*, 67.
246. Abel, E. W.; Bhargava, S. K.; Orrell, K. G. *Prog. Inorg. Chem.* **1984**, *32*, 1; Abel, E. W. *Chem. Brit.*, **1990**, 148.
247. Rivera, G.; Bernès, S.; Rodriguez de Barbarin, C.; Torrens, H. *Inorg. Chem.* **2001**, *40*, 5575.
248. Shi, T.; Elding, L. I. *Acta Chem. Scand.* **1998**, *52*, 897.
249. Shi, T.; Elding, L. I. *Inorg. Chem.* **1997**, *36*, 528.
250. Bugarcić, Z. D.; Liehr, G.; van Eldik, R. *J. Chem. Soc., Dalton Trans.* **2002**, 951.
251. Bugarcić, Z. D.; Ilic, D.; Djuran, M. I. *Aust. J. Chem.* **2001**, *54*, 237.
252. Ericson, A.; Iljina, Y.; McCary, J. L.; Coleman, R. S.; Elmroth, S. K. C. *Inorg. Chim. Acta* **2000**, *297*, 56.
253. Stringfield, T. W.; Shepherd, R. E. *Inorg. Chim. Acta* **2000**, *309*, 28.
254. Roy, R.; Das, D.; Mahapatra, A.; Sinha, C. *Inorg. React. Mech.* **2000**, *2*, 213.
255. de la Torre, G.; Gouloumis, A.; Vázquez, P.; Torres, T. *Angew. Chem. Int. Ed.* **2001**, *40*, 2895.
256. Rau, T.; Shoukry, M.; van Eldik, R. *Inorg. Chem.* **1997**, *36*, 1454.
257. Moi, S. C.; Ghosh, A. K.; De, G. S. *Indian J. Chem.* **2001**, *40A*, 1187.
258. Bartolomé, C.; de Blas, R.; Espinet, P.; Martín-Álvarez, J. M.; Villafañe, F. *Angew. Chem. Int. Ed.* **2001**, *40*, 2521.
259. Goshe, A. J.; Crowley, J. D.; Bosnich, B. *Helv. Chim. Acta* **2001**, *84*, 2971.
260. Brownback, M. A.; Murmann, R. K.; Barnes, C. L. *Polyhedron* **2001**, *20*, 2505.
261. Haus, A.; Raidt, M.; Link, T. A.; Elias, H. *Inorg. Chem.* **2000**, *39*, 5111.
262. Kulatilleke, C. Z.; Goldie, S. N.; Heeg, M. J.; Ochrymowycz, L. A.; Rorabacher, D. B. *Inorg. Chem.* **2000**, *39*, 1444.
263. Vogler, A.; Kunkely, H. *Coord. Chem. Rev.* **2001**, *219–221*, 489.
264. Pitteri, B.; Marangoni, G.; Visentin, F.; Bobbo, T.; Bertolasi, V.; Gilli, P. *J. Chem. Soc., Dalton Trans.* **1999**, 677.
265. Churlaud, R.; Frey, U.; Metz, F.; Merbach, A. E. *Inorg. Chem.* **2000**, *39*, 304.
266. Vosloo, T. G.; du Plessis, W. C.; Swarts, J. C. *Inorg. Chim. Acta* **2002**, *331*, 188.
267. Elias, H. *Coord. Chem. Rev.* **1999**, *187*, 37.
268. Rahimi, R.; Hambright, P. *J. Porphyrins Phthalocyanines* **1998**, *2*, 493.
269. Inada, Y.; Nakano, Y.; Inamo, M.; Nomura, M.; Funahashi, S. *Inorg. Chem.* **2000**, *39*, 4793.
270. Priimov, G. U.; Moore, P.; Helm, L.; Merbach, A. E. *Inorg. React. Mech.* **2001**, *3*, 1.
271. Barclay, J. E.; Davies, S. C.; Evans, D. J.; Fairhurst, S. A.; Fowler, C.; Henderson, R. A.; Hughes, D. L.; Oglieve, K. E. *Transition Met. Chem.* **1998**, *23*, 701.
272. Bonifacić, M.; Lovrić, J.; Orhanović, M. *J. Chem. Soc., Dalton Trans.* **1998**, 2879.
273. Aizawa, S.; Iida, S.; Matsuda, K.; Funahashi, S. *Bull. Chem. Soc. Jpn.* **1997**, *70*, 1593; and references therein.

274. Wanat, A.; Schneppensieper, T.; Stochel, G.; van Eldik, R.; Bill, E.; Wieghardt, K. *Inorg. Chem.* **2002**, *41*, 4.
275. Schneppensieper, T.; Wanat, A.; Stochel, G.; Goldstein, S.; Meyerstein, D.; van Eldik, R. *Eur. J. Inorg. Chem.* **2001**, 2317.
276. Schneppensieper, T.; Finkler, S.; Czap, A.; van Eldik, R.; Heus, M.; Nieuwenhuizen, P.; Wreesmann, C.; Abma, W. *Eur. J. Inorg. Chem.* **2001**, 491.
277. Schneppensieper, T.; Wanat, A.; Stochel, G.; van Eldik, R. *Inorg. Chem.* **2002**, *41*, 2565.
278. Laverman, L. E.; Ford, P. C. *J. Am. Chem. Soc.* **2001**, *123*, 11 614.
279. Lorkavić, I. M.; Ford, P. C. *Inorg. Chem.* **2000**, *39*, 632.
280. Lu, Z.-L.; Hamza, M. S. A.; van Eldik, R. *Eur. J. Inorg. Chem.* **2001**, 503.
281. Thaler, F.; Hubbard, C. D.; van Eldik, R. *Inorg. React. Mech.* **1999**, *1*, 83.
282. Buchsbaum, J. C.; Berg, J. M. *Inorg. Chim. Acta* **2000**, *297*, 217.
283. Goral, V.; Nelen, M. I.; Eliseev, A. V.; Lehn, J.-M. *Proc. Natl. Acad. Sci. (USA)* **2001**, *98*, 1347.
284. Constable, E. C.; Housecroft, C. E.; Kulke, T.; Lazzarini, C.; Schofield, E. R.; Zimmermann, V. *J. Chem. Soc., Dalton Trans.* **2001**, 2864.
285. Tabassum, S.; Athar, F.; Arjmand, F. *Transition Met. Chem.* **2002**, *27*, 256.
286. Größ, S.; Elias, H. *Inorg. Chim. Acta* **1996**, *251*, 347.
287. Elias, H.; Schumacher, R.; Schwamberger, J.; Wittekopf, T.; Helm, L.; Merbach, A. E.; Ulrich, S. *Inorg. Chem.* **2000**, *39*, 1721.
288. Ghosh, T.; Bandyopadhyay, P.; Bharadwaj, P. K.; Banerjee, R. *Polyhedron* **2001**, *20*, 477.
289. Clegg, W.; Henderson, R. A. *Inorg. Chem.* **2002**, *41*, 1128.
290. Biver, T.; Secco, F.; Tinè, M. R.; Venturini, M. *Polyhedron* **2001**, *20*, 1953.
291. Ambundo, E. A.; Deydier, M.-V.; Ochrymowycz, L. A.; Rorabacher, D. B. *Inorg. Chem.* **2000**, *39*, 1171.
292. Hassan, M. M.; Hay, R. W. *Inorg. React. Mech.* **2000**, *2*, 361.
293. Lye, P. G.; Lawrance, G. A.; Maeder, M. *J. Chem. Soc., Dalton Trans.* **2001**, 2376.
294. Lye, P. G.; Lawrance, G. A.; Maeder, M. *Inorg. React. Mech.* **1999**, *1*, 153.
295. Basallote, M. G.; Durán, J.; Fernández-Trujillo, M. J.; Máñez, M. A.; Quirós, M.; Salas, J. M. *Polyhedron* **2001**, *20*, 297.
296. Basallote, M. G.; Durán, J.; Fernández-Trujillo, M. J.; Máñez, M. A. *J. Chem. Soc., Dalton Trans.* **1999**, 3817.
297. Siegfried, L.; Urfer, A.; Kaden, T. A. *Inorg. Chim. Acta* **1996**, *251*, 177.
298. Baran, Y.; Kau, P.M.; Lawrance, G.A.; Sutrisno; von Nagy-Felsobuki, E.I. *Inorg. React. Mech.* **2001**, *3*, 31.
299. Mitsopoulou, C.-A.; Lyris, E.; Veltsos, S.; Katakis, D. *Inorg. React. Mech.* **2001**, *3*, 99.
300. Rosenzweig, A. C. *Acc. Chem. Res.* **2001**, *34*, 119.
301. Inamo, M.; Kamiya, N.; Inada, Y.; Nomura, M.; Funahashi, S. *Inorg. Chem.* **2001**, *40*, 5636.
302. Inada, Y.; Yamaguchi, T.; Satoh, H.; Funahashi, S. *Inorg. React. Mech.* **2000**, *2*, 277.
303. Pasternack, R. F.; Ewen, S.; Rao, A.; Meyer, A. S.; Freedman, M. A.; Collings, P. J.; Frey, S. L.; Ranen, M. C.; de Paula, J. C. *Inorg. Chim. Acta* **2001**, *317*, 59.
304. Lubal, P.; Kývala, M.; Hermann, P.; Holubová, J.; Rohovec, J.; Havel, J.; Lukes, I. *Polyhedron* **2001**, *20*, 47.
305. Clifford, T.; Danby, A. M.; Lightfoot, P.; Richens, D. T.; Hay, R. W. *J. Chem. Soc., Dalton Trans.* **2001**, 240.
306. Basallote, M. G.; Durán, J.; Fernández-Trujillo, M. J.; Máñez, M. A. *Polyhedron* **2001**, *20*, 75.

307. Das, A. K. *Inorg. React. Mech.* **2000**, *1*, 309.

308. Powell, D. H.; Merbach, A. E.; Fábián, I.; Schindler, S.; van Eldik, R. *Inorg. Chem.*
      1994, *33*, 4468; Dittler-Klingemann, A. M.; Orvig, C.; Hahn, F. E.; Thaler, F.;
      Hubbard, C. D.; van Eldik, R.; Schindler, S.; Fábián, I. *Inorg. Chem.* **1996**, *35*, 7798.

309. Thaler, F.; Hubbard, C. D.; Heinemann, F. W.; van Eldik, R.; Schindler, S.; Fábián, I.;
      Dittler-Klingemann, A. M.; Hahn, F. E.; Orvig, C. *Inorg. Chem.* **1998**, *37*, 4022.

310. Neubrand, A.; Thaler, F.; Körner, M.; Zahl, A.; Hubbard, C. D.; van Eldik, R. *J. Chem.
      Soc., Dalton Trans.* **2002**, 957.

311. McConnell, A.; Lightfoot, P.; Richens, D. T. *Inorg. Chim. Acta* **2002**, *331*, 143.

312. Wright, C.; Im, S.-C.; Twitchett, M. B.; Saysall, C. G.; Sokolowski, A.; Sykes, A. G.
      Inorg. Chem. **2001**, *40*, 294.

313. See, e.g., Pirrung, M. C.; Liu, H.; Morehead, A. T. *J. Am. Chem. Soc.* **2002**, *124*, 1014.

314. Chisholm, M. H.; Huffmann, J. C.; Iyer, S. S. *J. Chem. Soc., Dalton Trans.* **2000**, 1483.

315. Bunn, A. G.; Ni, Y.; Wei, M.; Wayland, B. B. *Inorg. Chem.* **2000**, *39*, 5576.

316. Ni, Y.; Fitzgerald, J. P.; Carroll, P.; Wayland, B. B. *Inorg. Chem.* **1994**, *33*, 2018.

317. Carbonaro, L.; Guogang, L.; Isola, M.; Senatore, L. *Inorg. Chim. Acta* **2000**, *303*, 40.

318. Mao, Z.-W.; Liehr, G.; van Eldik, R. *J. Chem. Soc., Dalton Trans.*, **2001**, 1593; Mao,
      Z.-W.; Heinemann, F.W.; Liehr, G.; van Eldik, R. *J. Chem. Soc., Dalton Trans.*, **2001**,
      3652.

319. Islam, M. Q.; Hambright, P. *Transition Met. Chem.* **1998**, *23*, 727.

320. Yamada, Y.; Miyashita, Y.; Fujisawa, K.; Okamoto, K. *Bull. Chem. Soc. Jpn.* **2001**, *74*,
      97.

321. Mikuriya, M.; Ikema, S.; Lim, J.-W. *Bull. Chem. Soc. Jpn.* **2001**, *74*, 99; Mikuriya, M.;
      Ikenuoe, S.; Nukada, R.; Lim, J.-W. *Bull. Chem. Soc. Jpn.* **2001**, *74*, 101.

322. Chang, C. A. *J. Chin. Chem. Soc. (Taipei)* **1996**, *43*, 419.

323. Bebianno, M. J.; Langston, W. J. *BioMetals* **1993**, *6*, 239.

324. Yamada, S.; Umika, F.; Nakamura, M.; Nakamura, S. *Talanta* **1996**, *43*, 1715.

325. Bailey, S. L.; Hambright, P. *Inorg. React. Mech.* **2001**, *3*, 51.

326. Dhillon, R.; Lincoln, S. F.; Madbak, S.; Stephens, A. K. W.; Wainwright, K. P.;
      Whitbread, S. L. *Inorg. Chem.* **2000**, *39*, 1855.

327. Wirmer, J.; Kühn, T.; Schwalbe, H. *Angew. Chem. Int. Ed.* **2001**, *40*, 4248.

328. Qureshi, M. S.; Kazmi, S. A. *J. Chem. Soc. Pak.* **1998**, *20*, 175 (*Chem. Abstr.* **1999**, *130*,
      187657n).

329. See Section 7.5.3 of Ref. (*6*).

330. Secco, F.; Venturini, M. *Polyhedron* **1999**, *18*, 3289.

331. Lente, G.; Fábián, I. *Inorg. Chem.* **1999**, *38*, 603.

332. Lente, G.; Fábián, I. *Inorg. Chem.* **2002**, *41*, 1306.

333. Lente, G.; Magalhães, M. E. A.; Fábián, I. *Inorg. Chem.* **2000**, *39*, 1950.

334. Serratice, G.; Galey, J.-B.; Aman, E. S.; Dumats, J. *Eur. J. Inorg. Chem.* **2001**, 471.

335. Wirgau, J. I.; Spasojević, I.; Boukhalfa, H.; Batinić-Haberle, I.; Crumbliss, A. L.
      Inorg. Chem. **2002**, *41*, 1464.

336. El Hage Chahine, J.-M.; Bauer, A.-M.; Baraldo, K.; Lion, C.; Ramiandrasoa, F.;
      Kunesch, G. *Eur. J. Inorg. Chem.* **2001**, 2287.

337. Boukhalfa, H.; Thomas, F.; Serratrice, G.; Béguin, C. G. *Inorg. React. Mech.* **2001**, *3*,
      153.

338. Boukhalfa, H.; Crumbliss, A. L. *Inorg. Chem.* **2000**, *39*, 4318.

339. Tsubouchi, A.; Shen, L.; Hara, Y.; Akiyama, M. *New J. Chem.* **2001**, *25*, 275.

340. Boukhalfa, H.; Brickman, T. J.; Armstrong, S. K.; Crumbliss, A. L. *Inorg. Chem.* **2000**,
      39, 5591.

341. Barnabé, N.; Zastre, J. A.; Venkataram, S.; Hasinoff, B. B. *Free Radic. Biol. Med.* **2002**, *33*, 266.

342. Ohkanda, J.; Katoh, A. *"Reviews on Heteroatom Chemistry"*, Eds. Oae, S.; Ohno, A.; Okuyama, T. MYU: Tokyo, 1998, p. 87.

343. Hara, Y.; Akiyama, M. *J. Am. Chem. Soc.* **2001**, *123*, 7247.

344. Boukhalfa, H.; Crumbliss, A. L. *Inorg. Chem.* **2001**, *40*, 4183.

345. Hou, Z.; Whisenhunt, D. W.; Xu, J.; Raymond, K. N. *J. Am. Chem. Soc.* **1994**, *116*, 840.

346. Das, A.; Dash, A. C. *Indian J. Chem.* **2001**, *40A*, 65.

347. Das, A.; Dash, A. C. *Inorg. React. Mech.* **2000**, *2*, 101.

348. Schneppensieper, T.; Zahl, A.; van Eldik, R. *Angew. Chem. Int. Ed.* **2001**, *40*, 1678.

349. Hoshino, M.; Laverman, L.; Ford, P. C. *Coord. Chem. Rev.* **1999**, *187*, 75.

350. Kryatov, S. V.; Nazarenko, A. Y.; Robinson, P. D.; Rybak-Akimova, E. V. *Chem. Commun.* **2000**, 921.

351. Yang, X.; Le Brun, N. E.; Thomson, A. J.; Moore, G. R.; Chasteen, N. D. *Biochemistry* **2000**, *39*, 4915.

352. Kimura, M.; Kitamura, T.; Sano, M.; Muto, T.; Hanabusa, K.; Shirai, H.; Kobayashi, N. *New J. Chem.* **2000**, *24*, 113.

353. Hazell, A.; McKenzie, C. J.; Nielsen, L. P.; Schindler, S.; Weitzer, M. *J. Chem. Soc., Dalton Trans.* **2002**, 310.

354. Tomany, C. T.; Hynes, M. J. *Inorg. React. Mech.* **1999**, *1*, 137.

355. Macdonald, C. J. *Inorg. Chim. Acta* **2000**, *311*, 33.

356. See p. 305 of Ref. (*6*).

357. Terpin, A. J.; Ziegler, M.; Johnson, D. W.; Raymond, K. N. *Angew. Chem. Int. Ed.* **2001**, *40*, 157.

358. Asato, E.; Kamamuta, K.; Imade, R.; Yamasaki, M. *Inorg. React. Mech.* **2000**, *2*, 57.

359. Rodopoulos, M.; Rodopoulos, T.; Bridson, J. N.; Elding, L. I.; Rettig, S. J.; McAuley, A. *Inorg. Chem.* **2001**, *40*, 2737.

360. Kagan, H. B., Ed. *Chem. Rev.* **2002**, *102*, 1805–2476.

361. Parker, D.; Dickins, R. S.; Puschmann, H.; Crossland, C.; Howard, J. A. K. *Chem. Rev.* **2002**, *102*, 1977.

362. Di Vaira, M.; Stoppioni, P. *New J. Chem.* **2002**, *26*, 136.

363. Birus, M.; van Eldik, R.; Gabricvić, M.; Zahl, A. *Eur. J. Inorg. Chem.* **2002**, 819.

364. Szilágyi, E.; Brücher, E. *J. Chem. Soc., Dalton Trans.* **2000**, 2229.

365. Szilágyi, E.; Tóth, E.; Kovács, Z.; Platzek, J.; Radüchel, B.; Brücher, E. *Inorg. Chim. Acta* **2000**, *298*, 226.

366. Burai, L.; Király, R.; Lázár, I.; Brücher, E. *Eur. J. Inorg. Chem.* **2001**, 813.

367. Lowe, M. P.; Parker, D.; Reany, O.; Aime, S.; Botta, M.; Castellano, G.; Gianolio, E.; Pagliarin, R. *J. Am. Chem. Soc.* **2001**, *123*, 7601.

368. Cohen, S. M.; Xu, J.; Radkov, E.; Raymond, K. N.; Botta, M.; Barge, A.; Aime, S. *Inorg. Chem.* **2000**, *39*, 5747.

369. Messeri, D.; Lowe, M. P.; Parker, D.; Botta, M. *Chem. Commun.* **2001**, 2742.

370. Caravan, P.; Greenfield, M. T.; Li, X.; Sherry, A. D. *Inorg. Chem.* **2001**, *40*, 6580.

371. Bobba, G.; Dickins, R. S.; Kean, S. D.; Mathieu, C. E.; Parker, D.; Peacock, R. D.; Siligardi, G.; Smith, M. J.; Williams, J. A. G.; Geraldes, C. F. G. C. *J. Chem. Soc., Perkin Trans. 2* **2001**, 1729.

372. See pp. 320–323 of Ref. (*6*).

373. Saito, S.; Hoshino, H.; Yotsuyanagi, T. *Inorg. Chem.* **2001**, *40*, 3819.

374. Comblin, V.; Gilsoul, D.; Hermann, M.; Humblet, V.; Jacques, V.; Mesbahi, M.; Sauvage, C.; Desreux, J. F. *Coord. Chem. Rev.* **1999**, *185–186*, 451.

375. Sarka, L.; Burai, L.; Brücher, E. *Chem. Eur. J.* **2000**, *6*, 719.

376. Pulukkody, K. P.; Norman, T. J.; Parker, D.; Royle, L.; Broan, C. J. *J. Chem. Soc., Perkin Trans. 2* **1993**, 605.
377. Lukes, I.; Kotek, J.; Vojtísek, P.; Hermann, P. *Coord. Chem. Rev.* **2001**, *216–217*, 287.
378. Zheng, Z. *Chem. Commun.* **2001**, 2521.
379. Ma, B.-Q.; Zhang, D.-S.; Gao, S.; Jin, T.-Z.; Yan, C.-H. *New J. Chem.* **2000**, *24*, 251.
380. Zhang, D.-S.; Ma, B.-Q.; Jin, T.-Z.; Gao, S.; Yan, C.-H.; Mak, T. C. W. *New J. Chem.* **2000**, *24*, 61.
381. Wang, R.; Selby, H. D.; Liu, H.; Carducci, M. D.; Jin, T.; Zheng, Z.; Anthis, J. W.; Staples, R. J. *Inorg. Chem.* **2002**, *41*, 278.
382. Bourassa, J.; Lee, B.; Bernard, S.; Schoonover, J.; Ford, P. C. *Inorg. Chem.* **1999**, *38*, 2947.
383. Bourassa, J. L.; Ford, P. C. *Coord. Chem. Rev.* **2000**, *200–202*, 887.
384. Henderson, R. A.; Oglieve, K. E. *J. Chem. Soc., Dalton Trans.* **1999**, 3927.
385. Dunford, A. J.; Henderson, R. A. *Chem. Commun.* **2002**, 360.
386. Henderson, R. A. *J. Chem. Soc., Dalton Trans.* **1999**, 119.
387. Cui, Z.; Henderson, R. A. *Inorg. Chem.* **2002**, *41*, 4158.
388. Belyaev, A. N.; Simanova, S. A.; Matasov, V. B.; Shchukarev, A. V. *Zh. Prikl. Khim.* **1996**, *69*, 731.
389. Crimp, S. J.; Drljaca, A.; Smythe, D.; Spiccia, L. *J. Chem. Soc., Dalton Trans.* **1998**, 375.
390. Fujihara, T.; Yasui, M.; Ochikoshi, J.; Terasaki, Y.; Nagasawa, A. *Inorg. React. Mech.* **2000**, *2*, 119.
391. Lente, G.; Dobbing, A. M.; Richens, D. T. *Inorg. React. Mech.* **1998**, *1*, 3.
392. Hernandez-Molina, R.; Sykes, A. G. *Coord. Chem. Rev.* **1999**, *187*, 291.
393. Hernandez-Molina, R.; Sokolov, M. N.; Sykes, A. G. *Acc. Chem. Res.* **2001**, *34*, 223.
394. Malinak, S. M.; Coucouvanis, D. *Prog. Inorg. Chem.* **2001**, *49*, 599.
395. Fedin, V. P.; Sokolov, M.; Lamprecht, G. J.; Hernandez-Molina, R.; Seo, M.-S.; Virovets, A. V.; Clegg, W.; Sykes, A. G. *Inorg. Chem.* **2001**, *40*, 6598.
396. Gray, T. G.; Holm, R. H. *Inorg. Chem.* **2002**, *41*, 4211.
397. Gameiro, A. M. F.; Gillard, R. D.; Rees, N. H.; Schulte, J.; Sengül, A. *Croat. Chem. Acta* **2001**, *74*, 641.
398. Sengül, A.; Gillard, R. D. *Transition Met. Chem.* **1998**, *23*, 663.
399. Zhang, X.-M.; Tong, M.-L.; Chen, X.-M. *Angew. Chem. Int. Ed.* **2002**, *41*, 1029.
400. Wang, Y.; Rillema, D. P. *Inorg. Chem. Commun.* **1998**, *1*, 27.
401. Wee, S.; Grannas, M. J.; McFadyen, W. D.; O'Hair, R. A. J. *Aust. J. Chem.* **2001**, *54*, 245.
402. Hartshorn, R. M. *Aust. J. Chem.* **1996**, *49*, 905.
403. Eriksen, J.; Mønsted, O.; Mønsted, L. *Transition Met. Chem.* **1998**, *23*, 783.
404. Hay, R. W.; Govan, N. *Transition Met. Chem.* **1998**, *23*, 721.
405. Rawji, G. H.; Yamada, M.; Sadler, N. P.; Milburn, R. M. *Inorg. Chim. Acta* **2000**, *303*, 168.
406. Atkinson, I. M.; Lindoy, L. F. *Coord. Chem. Rev.* **2000**, *200–202*, 207.
407. Smoukov, S. K.; Quaroni, L.; Wang, X.; Doan, P. E.; Hoffman, B. M.; Que, L. *J. Am. Chem. Soc.* **2002**, *124*, 2595.
408. Fritsky, I. O.; Ott, R.; Krämer, R. *Angew. Chem. Int. Ed.* **2000**, *39*, 3255.
409. Deal, K. A.; Park, G.; Shao, J.; Chasteen, N. D.; Brechbiel, M. W.; Planalp, R. P. *Inorg. Chem.* **2001**, *40*, 4176.
410. Deck, K. M.; Tseng, T. A.; Burstyn, J. N. *Inorg. Chem.* **2002**, *41*, 669.
411. Ibrahim, M. M.; Shimomura, N.; Ichikawa, K.; Shiro, M. *Inorg. Chim. Acta* **2001**, *313*, 125.

412. Yamaguchi, K.; Akagi, F.; Fujinami, S.; Suzuki, M.; Shionoya, M.; Suzuki, S. *Chem. Commun.* **2001**, 375.
413. Hay, R. W.; Clifford, T.; Richens, D. T.; Lightfoot, P. *Polyhedron* **2000**, *19*, 1485.
414. Ercan, A.; Park, H. I.; Ming, L.-J. *Chem. Commun.* **2000**, 2501.
415. Forconi, M.; Williams, N. H. *Angew. Chem. Int. Ed.* **2002**, *41*, 849.
416. Hay, R. W.; Govan, N.; Parchment, K. E.; Kiss, E.; Clifford, T. *Inorg. React. Mech.* **1998**, *1*, 33.
417. Yamada, K.; Takahashi, Y.; Yamamura, H.; Araki, S.; Saito, K.; Kawai, M. *Chem. Commun.* **2000**, 1315.
418. Abe, K.; Izumi, J.; Ohba, M.; Yokoyama, T.; Ōkawa, H. *Bull. Chem. Soc. Jpn.* **2001**, *74*, 85.
419. Gómez-Tagle, P.; Yatsimirsky, A. K. *J. Chem. Soc., Dalton Trans.* **2001**, 2663.
420. Mejia-Radillo, Y.; Yatsimirsky, A. K. *Inorg. Chim. Acta* **2002**, *328*, 241.
421. Gómez-Tagle, P.; Yatsimirsky, A. K. *Inorg. Chem.* **2001**, *40*, 3786.
422. Jurek, P. E.; Jurek, A. M.; Martell, A. E. *Inorg. Chem.* **2000**, *39*, 1016.
423. Wang, C.; Choudhary, S.; Vink, C. B.; Secord, E. A.; Morrow, J. R. *Chem. Commun.* **2000**, 2509.
424. Gajda, T.; Krämer, R.; Jancsó, A. *Eur. J. Inorg. Chem.* **2000**, 1635.
425. Kimura, E.; Hashimoto, H.; Koike, T. *J. Am. Chem. Soc.* **1996**, *118*, 10963.
426. Koerner, T. B.; Brown, R. S. *Can. J. Chem.* **2002**, *80*, 183.
427. Hay, R. W.; Govan, N.; Parchment, K. E. *Inorg. Chem. Commun.* **1998**, *1*, 228.
428. Pasquato, L.; Rancan, F.; Scrimin, P.; Mancin, F.; Frigeri, C. *Chem. Commun.* **2000**, 2253.
429. Xia, J.; Li, S.; Shi, Y.; Yu, K.; Tang, W. *J. Chem. Soc., Dalton Trans.* **2001**, 2109.
430. Xia, J.; Xu, Y.; Li, S.; Sun, W.; Yu, K.; Tang, W. *Inorg. Chem.* **2001**, *40*, 2394.
431. Su, X.-C.; Sun, H.-W.; Zhou, Z.-F.; Lin, H.-K.; Chen, L.; Zhu, S.-R.; Chen, Y.-T. *Polyhedron* **2001**, *20*, 91.
432. Kurzecv, S. A.; Kazankov, G. M.; Ryabov, A. D. *Inorg. Chim. Acta* **2000**, *305*, 1.
433. Parr, J.; Ross, A. T.; Slawin, A. M. Z. *Inorg. Chem. Commun.* **1998**, *1*, 159.
434. Neverov, A. A.; Montoya-Pelaez, P. J.; Brown, R. S. *J. Am. Chem. Soc.* **2001**, *123*, 210.
435. Casas, J. S.; Castellano, E. E.; Couce, M. D.; Leis, J. R.; Sánchez, A.; Sordo, J.; Suárez-Gimeno, M. I.; Taboada, C.; Zukerman-Schpector, J. *Inorg. Chem. Commun.* **1998**, *1*, 93.
436. Kaminskaia, N. V.; Kostić, N. M. *Inorg. Chem.* **2001**, *40*, 2368.
437. Djuran, M.; Milinković, S. U. *Polyhedron* **2000**, *19*, 959.
438. Ren, R.; Yang, P.; Zheng, W.; Hua, Z. *Inorg. Chem.* **2000**, *39*, 5454.
439. Donnelly, P. S.; Harrowfield, J. M.; Skelton, B. W.; White, A. H. *Inorg. Chem.* **2001**, *40*, 5645.
440. Kaminskaia, N. V.; Spingler, B.; Lippard, S. J. *J. Am. Chem. Soc.* **2001**, *123*, 6555.
441. Montoya-Palaez, P. J; Brown, R. S. *Inorg. Chem.* **2002**, *41*, 309.
442. Barrios, A. M.; Lippard, S. J. *J. Am. Chem. Soc.* **2000**, *122*, 9172.
443. Pascaly, M.; Duda, M.; Schweppe, F.; Zurlinden, K.; Müller, F. K.; Krebs, B. *J. Chem. Soc., Dalton Trans.* **2001**, 828.
444. Leussing, D. L. *Transition Met. Chem.* **1998**, *23*, 771.
445. Rais, D.; Yau, J.; Mingos, D. M. P.; Vilar, R.; White, A. J. P.; Williams, D. J. *Angew. Chem. Int. Ed.* **2001**, *40*, 3464.
446. Nelson, J.; McKee, V.; Morgan, G. *Prog. Inorg. Chem.* **1998**, *47*, 167.
447. Curtis, N. F. *J. Chem. Soc.* **1960**, 4409; *Coord. Chem. Rev.* **1968**, *3*, 3.
448. Mendoza-Diaz, G.; Ruiz-Ramirez, L.; Moreno-Esparza, R. *Polyhedron* **2000**, *19*, 2149.
449. See, e.g., Gerbeleu, N. V.; Arion, V. B.; Burgess, J. *"Template Synthesis of Macrocyclic Compounds"*; Wiley-VCH: Weinheim, 1999; Sanders, J. K. M. *Pure Appl. Chem.* **2000**, *72*, 2265.

450. Bark, T.; Düggeli, M.; Stoeckli-Evans, H.; von Zelewsky, A. *Angew. Chem. Int. Ed.* **2001**, *40*, 2848.

451. Mimassi, L.; Guyard-Duhayon, C.; Raehm, L.; Amouri, H. *Eur. J. Inorg. Chem.* **2002**, 2453.

452. Hubin, T. J.; Busch, D. H. *Coord. Chem. Rev.* **2000**, *200–202*, 5.

453. Amendola, V.; Fabbrizzi, L.; Licchelli, M.; Mangano, C.; Pallavicini, P.; Parodi, L.; Poggi, A. *Coord. Chem. Rev.* **1999**, *190–192*, 649.

454. Wisner, J. A.; Beer, P. D.; Drew, M. G. B. *Angew. Chem. Int. Ed.* **2001**, *40*, 3606.

455. Reuter, C.; Schmieder, R.; Vögtle, F. *Pure Appl. Chem.* **2000**, *72*, 2233; Reuter, C.; Vögtle, F. *Org. Lett.* **2000**, *2*, 593.

456. Yamauchi, O.; Odani, A.; Hirota, S. *Bull. Chem. Soc. Jpn.* **2001**, *74*, 1525.

457. Aoyagi, M.; Biradha, K.; Fujita, M. *Bull. Chem. Soc. Jpn.* **2000**, *73*, 1369.

458. Rondelez, Y.; Bertho, G.; Reinaud, O. *Angew. Chem. Int. Ed.* **2002**, *41*, 1044.

459. Halpern, J., Ed. *Proc. Natl. Acad. Sci. (USA)* **2002**, *99*, 4762–5206.

460. Lalioti, N.; Raptopoulou, C. P.; Terzis, A.; Aliev, A. E.; Gerothanassis, I. P.; Manessi-Zoupa, E.; Perlepes, S. P. *Angew. Chem. Int. Ed.* **2001**, *40*, 3211.

461. Toyota, S.; Woods, C. R.; Benaglia, M.; Haldimann, R.; Wärnmark, K.; Hardcastle, K.; Siegel, J. S. *Angew. Chem. Int. Ed.* **2001**, *40*, 751.

462. Real, J. A.; Andrés, E.; Muñoz, M. C.; Julve, M.; Granier, T.; Bousseksou, A.; Varret, F. *Science* **1995**, *268*, 265; De Munno, G.; Cipriani, F.; Armentano, D.; Julve, M.; Real, J. A. *New J. Chem.* **2001**, *25*, 1031.

463. Umemoto, K.; Tsukui, H.; Kusukawa, T.; Biradha, K.; Fujita, M. *Angew. Chem. Int. Ed.* **2001**, *40*, 2620.

464. Kawano, T.; Kuwana, J.; Du, C.-X.; Ueda, I. *Inorg. Chem.* **2002**, *41*, 4078.

465. Bosch, E.; Barnes, C. L. *New J. Chem.* **2001**, *25*, 1376.

466. Schweiger, M.; Seidel, S. R.; Arif, A. M.; Stang, P. J. *Angew. Chem. Int. Ed.* **2001**, *40*, 3467.

467. Chang, C.-H.; Hwang, K. C.; Liu, C.-S.; Chi, Y.; Carty, A. J.; Scoles, L.; Peng, S.-M.; Lee, G.-H.; Reedijk, J. *Angew. Chem. Int. Ed.* **2001**, *40*, 4651.

468. Hubin, T. J.; Busch, D. H. *Coord. Chem. Rev.* **2000**, *200–202*, 5.

469. Skinner, P. J.; Blair, S.; Kataky, R.; Parker, D. *New J. Chem.* **2000**, *24*, 265.

470. Hunter, C. A.; Low, C. M. R.; Packer, M. J.; Spey, S. E.; Vinter, J. G.; Vysotsky, M. O.; Zonta, C. *Angew. Chem. Int. Ed.* **2001**, *40*, 2678.

471. Gunter, M. J.; Bampos, N.; Johnstone, K. D.; Sanders, J. K. M. *New J. Chem.* **2001**, *25*, 166.

472. Linke, M.; Fujita, N.; Chambron, J.-C.; Heitz, V.; Sauvage, J.-P. *New J. Chem.* **2001**, *25*, 790.

473. Maiya, B. G.; Bampos, N.; Kumar, A. A.; Feeder, N.; Sanders, J. K. M. *New J. Chem.* **2001**, *25*, 797.

474. Dietrich-Buchecker, C.; Rapenne, G.; Sauvage, J.-P. *Coord. Chem. Rev.* **1999**, *185–186*, 167.

475. Leigh, D. A.; Lusby, P. J.; Teat, S. J.; Wilson, A. J.; Wong, J. K. Y. *Angew. Chem. Int. Ed.* **2001**, *40*, 1538.

476. Cardenas, D. J.; Livoreil, A.; Sauvage, J.-P. *J. Am. Chem. Soc.* **1996**, *118*, 11 980.

477. Munakata, M.; Wu, L. P.; Kuroda-Sowa, T. *Adv. Inorg. Chem.* **1999**, *46*, 173.

478. Ziessel, R.; Harriman, A.; El-ghayoury, A.; Douce, L.; Leize, E.; Nierengarten, H.; Van Dorsselaer, A. *New J. Chem.* **2000**, *24*, 729.

479. Fatin-Rouge, N.; Blanc, S.; Pfeil, A.; Rigault, A.; Albrecht-Gary, A.-M.; Lehn, J.-M. *Helv. Chim. Acta* **2001**, *84*, 1694.

480. Mathieu, J.; Marsura, A.; Bouhmaida, N.; Ghermani, N. *Eur. J. Inorg. Chem.* **2002**, 2433.
481. Ziessel, R. *Coord. Chem. Rev.* **2001**, *216–217*, 195.
482. Hasenknopf, B.; Lehn, J.-M.; Boumediene, N.; Dupont-Gervais, A.; Van Dorssellaer, A.; Kneisel, B.; Fenske, D. *J. Am. Chem. Soc.* **1997**, *119*, 10 956.
483. Fatin-Rouge, N.; Blanc, S.; Leize, E.; Van Dorssellaer, A.; Baret, P.; Pierre, J.-L.; Albrecht-Gary, A. M. *Inorg. Chem.* **2000**, *39*, 5771.
484. Hasenknopf, B.; Lehn, J.-M.; Boumediene, N.; Leize, M.; Van Dorssellaer, A. *Angew. Chem. Int. Ed.* **1998**, *37*, 3265.
485. Tabellion, F. M.; Seidel, S. R.; Arif, A. M.; Stang, P. J. *Angew. Chem. Int. Ed.* **2001**, *40*, 1529.
486. Albrecht, M.; Blau, O.; Röttele, H. *New J. Chem.* **2000**, *24*, 619.
487. Woods, C. R.; Benaglia, M.; Toyota, S.; Hardcastle, K.; Siegel, J. S. *Angew. Chem. Int. Ed.* **2001**, *40*, 749.
488. Carina, R. F.; Dietrich-Buchecker, C.; Sauvage, J.-P. *J. Am. Chem. Soc.* **1996**, *118*, 9110.

OXYGEN TRANSFER REACTIONS: CATALYSIS BY RHENIUM COMPOUNDS

JAMES H. ESPENSON

Ames Laboratory and Department of Chemistry, Iowa State University,
Ames, IA 50011, USA

ADVANCES IN INORGANIC CHEMISTRY
VOLUME 54 ISSN 0898-8838

I. Rhenium Catalysts

This review deals with the transfer of an atom – usually oxygen, occasionally sulfur – from one species to another. Because the participants have closed electronic shells (they are octet-rule molecules for the most part), electronic interaction between them is not substantial. For that reason the intervention of a catalyst is nearly always required.

A. ATTRIBUTES OF A CATALYST

1. Steric Considerations

As one reflects on what the catalysts must offer, or in examining the findings in retrospect, two factors become evident. One is steric: the O-atom *donor* must be able to coordinate to an active site on the transition metal, which in this case is rhenium. This may entail coordination to an already-vacant position, or to a site made vacant by the departure of an existing (and presumably more weakly coordinating) ligand. Thus ligand addition and displacement reactions are clearly relevant to the subject, and will find a place in this narrative. The data point to the pre-existing oxo ligand as playing a substantial role in this respect. Multiple bonding from oxygen to rhenium will greatly weaken ligand binding to a position trans to the oxo group. For that reason most of the oxorhenium(V) catalysts that we have studied adopt an approximate square-pyramidal geometry, the position trans to the oxo group being unoccupied. If one forces the issue by use of a bidentate ligand, approximate octahedral geometry can be realized; as will be shown, such compounds appear to function as catalysts only to the extent that the sixth coordination site can open, or be opened, during the catalytic cycle. Otherwise, they have proved entirely inactive.

Stable structures are best supported by chelating dithiolate ligands. The monodentate PhS⁻ ligand gives some analogous compounds, but they have proved less stable throughout repeated catalytic cycles. The dithiols from which the chelates used in these studies have been derived are 2-(mercaptomethyl)thiophenol, 1,2-ethane dithiol and 1,3-propane dithiol.

2. Electronic Structure

The second attribute of the catalyst concerns its electronic structure, or more simply the valence electron count. Effective catalysts must, it seems, have < 18 VE, such that coordination of a substrate or the departure of a product does not itself pose a major kinetic barrier. Furthermore, it happens that the most stable valence states of the metal will differ by two units. Thus not only will the stoichiometry of atom transfer be supported, but also the mechanism. In the case of rhenium, the oxidation states are Re(V) and Re(VII); indeed scant indication of Re(VI) has been found in this chemistry, especially in a mononuclear species. Likewise, there is no indication of the involvement of free radical chemistry.

3. Structural Data

The tabulation of distances for metal–ligand multiple bonds compiled by Nugent and Mayer (1) has proved to be quite useful. To extend that effort to the new rhenium compounds synthesized during this research, Table I compiles data, which pertain principally to rhenium–oxygen bonds, but also provides data concerning imido- and thio-rhenium groups. The rhenium–oxygen distances are remarkably invariant among 28 compounds, the entire range of values lying between 166.2 and 169.8 pm. It is accurate to state that all the rhenium–oxygen distances are equivalent within the precision of the measurements. Even the one rhenium(VII) compound in the series has a distance of 168.1 pm, lying at an average value for rhenium(V)-oxo.

In the imido series, however, the rhenium–nitrogen distances for rhenium(V) average 180.1 pm, which is somewhat longer than the average of the values for several rhenium(VII) compounds, 174 pm. The one terminal thio-rhenium bond, 209.2 pm, is significantly shorter than any of the μ-S bonds, which average 231 pm.

TABLE I

COMPILATION OF RHENIUM–OXYGEN AND OTHER BOND DISTANCES IN OXO-, AND
IMIDO- AND THIO-RHENIUM COMPOUNDS

Compound	d(Re–O) (pm)
Part A. Dinuclear oxorhenium compounds	
I. {MeReO(dithiolate)}$_2$	
dithiolate = edt	167.6
pdt	167.2
mtp	167.4
II. {ReO}$_2$(dithiolate)$_3$	
dithiolate = edt	167.2, 167.9
mtp	167.6, 168.3
III. {MeReO(dithiolate)}$_2$L	
dithiolate = edt, L = dmso	167.8
IV. [{MeReO(mtp)}$_2$(μ-OH)]$^-$	168.0
V. {MeRe(O)$_2$}$_2$(dithioerythritol)^a	168.1
VI. MeReO(mtp)(μ-S)MeRe(mtp)	167.5
I. MeReO(mtp)L	
L = PPh$_3$	169.6
PMePh$_2$	168.4
NC$_5$H$_5$	168.5
NC$_5$H$_4$C(O)Me	168.6
2,2′-bipyridine	168.7
S$_2$CNEt$_2$	169.8
II. MeReO(edt)L	
L = PPh$_3$	168.1
bipyrimidine	169.2
ReO(edt)(SCH$_2$CH$_2$SMe)(1,3,5-phospha-adamantane)	170.0
III. Compounds containing PhS$^-$	
[ReO(SPh)$_4$]$^-$ [2-Pic$_2$H]$^+$	168.3
MeReO(SPh)(S$_2$CNEt$_2$)	166.4
[MeReO(edt)(SPh)]$^-$	169.2
IV. Other chelates	
MeReO(2-picolinate)$_2$	166.2
MeReO(SCH$_2$CH$_2$SCH$_2$CH$_2$S)	168.9
MeReO(SCH$_2$CH$_2$OCH$_2$CH$_2$S)	167.4
MeReO(SC(O)CH$_2$SCH$_2$C(O)S)	167.4
MeReO("NSN")	168.4
MeReO(8-quinolinethiol)	167.4
MeReO(8-hydroxyquinolinate)	167.4
MeReS(mtp)[P(OMe)Ph$_2$]	209.2 (Re-S$_T$)
MeReO(mtp)(μ-S)MeRe(mtp)	219.0, 230.7
{MeRe(NAr)$_2$(μ-S)}$_2$	234-242
{MeRe(NAr)$_2$}$_2$(μ-S)	228.9, 229.7

(*Continued*)

TABLE I

(*Continued*)

Compound	d(Re–O) (pm)
I. MeRe(NAr)$_2$L$_2$	
L = PMe$_2$Ph	180.9
κ^1-Ph$_2$PCH$_2$PPh$_2$	179.4
II. Imido compounds [b]	
MeRe(NAd)$_3$ [a]	173.9, 175.0
MeRe(edt)(NAr)$_2$ [a]	174.2, 175.0
III. Dinuclear compounds	
{MeRe(NAr)$_2$(μ-S)}$_2$ [a]	173.9
{MeRe(NAr)$_2$}$_2$(μ-S)	173.8–176.2
{MeRe(NAr)$_2$}$_2$(μ-O)	173.7, 174.2

[a] Rhenium(VII) compound.
[b] Ad = 1-adamantyl; Ar = 2,6-diisopropylphenyl.

4. Structural Motif

The multiple bond to the metal in a rhenium(V) compound is best represented by the MO description of it being a triple bond, consistent with the symmetry of the p$_x$ and p$_y$ orbitals on the ligand; see **1** (*1*). In keeping with that, these bonds are found to be relatively short and quite strong. The structure has also been presented as **1'**, a valence bond formulation that facilitates the counting of oxidation states but does not provide an accurate representation.

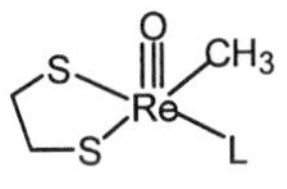

1 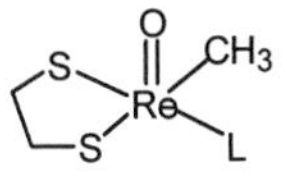   **1'**

In rhenium compounds that contain two metal-imido ligands, these tend not to occupy positions trans to one another for the same reason. A trigonal-bipyramidal geometry has been identified in **2** and **3**, even to the extent that the potentially chelating ligand Ph$_2$PCH$_2$PPh$_2$ acts as a monodentate ligand.

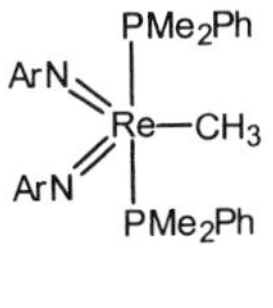

2 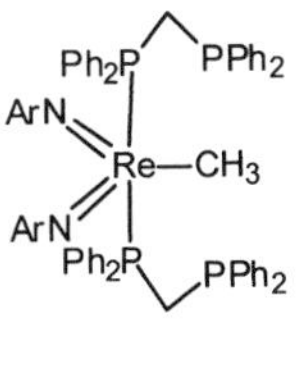   **3**

Indeed, to anticipate the findings from studies of the catalytic chemistry, the pre-existing oxo ligand is not a participant in the oxo-transfer reactivity. Its role appears to be the enforcement of a square-pyramidal geometry with an open position for substrate entry. The geometry-controlling role of the multiple bond to rhenium can be appreciated by comparing the structures of two closely-related compounds. One is **4** (*2*), a rhenium(V) compound that is fully analogous to **5** (*3*), its oxo analogue. The coordination geometry about rhenium is that of an approximate square pyramid in both compounds (*2*).

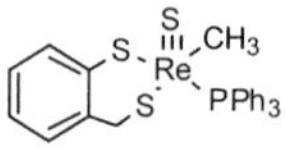

4 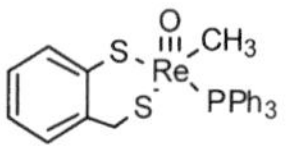   5 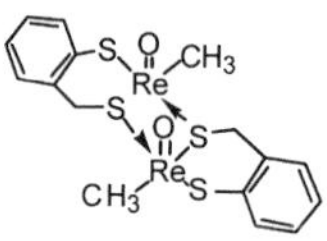   6 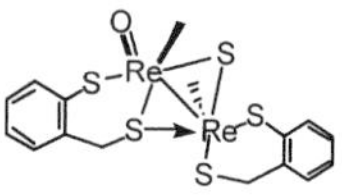   7

In contrast, there are two dimeric compounds to consider. One is the oxo,oxo derivative **6**, the other a mixed oxo,thio **7** (*3*). The oxo groups in all of **5–7** occupy terminal positions, as usual, whereas the thio group in **7** is in a bridging position. Each rhenium atom in **7** is therefore six-coordinate by virtue of a short rhenium–rhenium distance, 277.7 pm. Ignoring that bond, the local geometry about the one rhenium is a severely-distorted square pyramid, whereas the oxo-rhenium moiety has a geometry best described as a trigonal bipyramid.

5. Oxophilicity

Compounds and complexes of the early transition metals are *oxophilic* because the low d-electron count invites the stabilization of metal-oxo bonds by π-bond formation. To a substantial extent, their reactivity is typical of complexes of metals other than rhenium. That is particularly the case insofar as activation of hydrogen peroxide is concerned. Catalysis by d^0 metals – not only Re^{VII}, but also Cr^{VI}, W^{VI}, Mo^{VI}, V^V, Zr^{IV} and Hf^{IV} – has been noted. The parent forms of these compounds have at least one oxo group. Again the issue is the coordination of the oxygen *donating* substrate, HOOH, to the metal, usually by condensation:

$$M = O + HOOH \rightarrow M(\kappa^2 - O_2) + H_2O \qquad (1)$$

The electropositive metal center polarizes the peroxo group so that is much more electrophilic than free peroxide anion. Evidence has been accumulated, and it will be summarized later, that such polarization happens to the extent that nucleophilic reagents can attack a peroxo oxygen when it is coordinated to a d^0 metal. In other words, this situation provides *electrophilic* activation of peroxide (*4*), enabling attack by nucleophiles and other electron-rich substrates.

The d^0 electronic structure offers mutually advantageous situations for both groups; for an oxo group it invites π-back donation to the metal and the resulting stabilization; for $\kappa^2 - O_2^{2-}$ the d^0 structure offers strong σ-bond formation.

6. Syntheses and Characteristics of Oxorhenium Compounds

The easily-prepared compound (*5*) $MeReO_3$, known as MTO, forms the basis for synthesizing compounds such as **1** and of $MeReO(dithiolate)L$ compounds in general. The reaction by which the latter group of compounds can be prepared is:

$$MeReO_3 + 2\,dithiol + L \rightarrow MeReO(dithiolate)L + disulfide + 2H_2O$$

$$(2)$$

This preparation is carried out in an aprotic solvent (e.g. benzene, chloroform) with no special provision other than working in a well-ventilated fume hood to avoid ill-smelling sulfur compounds. Various ligands have proved successful: phosphines, pyridines, imidazoles, tetramethylthiourea, etc. When the same reaction is carried out in the absence of the Lewis base L, a dimer **6** is obtained, which is a useful catalyst in its own right and sometimes a much more active one; see Section VII.A. The chemical equation for that reaction is,

$$MeReO_3 + 2\,dithiol \rightarrow \{MeReO(dithiolate)\}_2\,(\mathbf{6}) + disulfide + 2H_2O$$

$$(3)$$

These rhenium compounds remain unchanged over at least several months. It is particularly noteworthy that their use as catalysts requires no special working conditions; benchtop reactions with provision to protect reactions from air or ordinary levels of atmospheric moisture are not required. This feature is the one that makes these systems of use in actual laboratory procedures that extend beyond their use as catalysts for the exploration of reaction mechanisms.

B. CATALYSIS BY MOLYBDENUM AND TUNGSTEN

An extensive literature deals with catalysis by molybdenum compounds (*6–15*), many of which are isoelectronic with the rhenium analogs of similar but not identical composition. Molybdenum chemistry will not be covered here except by way of comparison. The major oxidation states are the isoelectronic Mo(IV) and Mo(VI) compounds, except that Mo(V) intervenes more directly than Re(VI) does.

Tungsten catalysts are also known, and are of considerable interest because they are mimics for tungsten-based oxotransferase enzymes found in thermochemical bacteria. Studies on such compounds have been reported (16,17).

C. A PRELIMINARY EXAMPLE OF CATALYSIS

1. Catalyst Structure–Activity Correlations

Running the risk of getting ahead of the story, one example of a pertinent reaction will be examined. It is the reaction between 4-methylpyridine N-oxide and triphenylphosphine,

$$4\text{-MeC}_5\text{H}_4\text{NO} + \text{PPh}_3 \rightarrow 4\text{-MeC}_5\text{H}_4\text{N} + \text{Ph}_3\text{PO} \qquad (4)$$

The structural formulas of a group of related oxorhenium(V) compounds that has been examined as potential catalysts are:

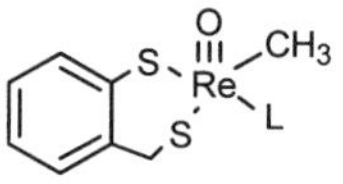

8 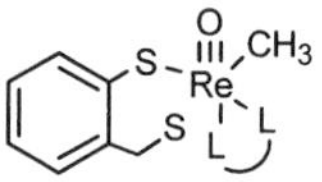   **9**

The relative catalytic efficiencies of **8** depend on the identity of the ligand L. With monodentate phosphines, $L = \text{PMePh}_2$ (rel. rate = 109), PPh_3 (100) and $\text{P(C}_6\text{H}_4\text{-4-Me})_3$ (16), the rate differences manifest only steric and inductive effects of the alkyl and aryl groups on phosphorus. When L is the chelating ligand bipyrimidine, the catalyst is less active, because the six-coordinate structure must open, or be opened, by picoline N-oxide for the process to begin (18). Compounds with strongly-coordinating chelates are completely inactive; this includes $L = 2,2'$-bipyridine, 1,2-bis(diphenylphosphino)benzene (19,20) and 1,10-phenanthroline (20). For them, the chelate effect is so strong as to render a coordination position unavailable to the substrate. The bipyrimidine compound, which has a ligand of lower Lewis basicity owing to the additional nitrogen atoms incorporated into its structure, can form a five-coordinate intermediate allowing entry of the substrate.

2. Thermochemical Considerations

It seems not entirely coincidental that the most effective catalysts are those with a pre-existing rhenium–oxygen bond; indeed, in our studies none that lack this feature are active. According to MO analysis, the rhenium–oxygen bond is a triple bond, Re≡O. There is every reason to believe, in the absence of a full set of thermochemical data, that the rhenium–oxygen bonds of the five compounds cited in the preceding section, and their relatives, are stronger than the phosphorus–oxygen bonds in R_3P–O. At least phosphines show no tendency to convert oxorhenium(V) compounds to rhenium(III) derivatives, whether for kinetic or thermodynamic reasons. The existence of the strong rhenium–oxygen bond, given the π-back donation to the metal, enforces a square-pyramidal geometry at the metal. This, in turn, facilitates substrate entry to the coordination sphere of rhenium.

II. Oxygen Atom Transfer: The Reactions Themselves

A. THE REDUCTION OF PYRIDINE N-OXIDE

Equation (4) depicts reduction of pyridine N-oxide to pyridine; the findings apply to RC_5H_4NO in general. The oxygen acceptor in that case is a phosphine, PPh_3 or PR_3 (R = alkyl, aryl) in general. This transformation holds significance in its own right, in that it represents a useful reaction for the deprotection of pyridines. The deprotection is not a facile process and often gives incomplete conversions or difficult workups (21–25). Rhenium compounds such as $MeReO(mtp)PPh_3$ are successful catalysts. On the subject of utility, it should be emphasized that these catalysts can be used in benchtop reactions.

However, interesting a case this makes for the examination of mechanism, the point must first be established that it represents a reaction worth conducting, and second that it can be carried out on a laboratory scale to give a useful amount of product. With a suitable rhenium catalyst, nearly quantitative conversions have been realized within a reasonable time. A total of 15 substrates were studied, such that the insensitivity to functional groups, steric and electronic variables were established. All the substrates examined gave essentially complete conversion to product on a scale of 1 g of PyO, irrespective of a wide range of functional groups present (26).

B. OTHER REACTION PARTNERS

Several other reactions of this general type are catalyzed by **1** and its relatives,

$$YO + X \rightarrow Y + XO \tag{5}$$

For example, satisfactory results have been realized from these {YO, X} combinations: {PyO, R$_2$S}, {PyO, R$_2$SO}, {R$'_2$SO, R$_2$S}, {Bu^{t}OOH, PR$_3$}, {Bu^{t}OOH, R$_2$S}, {Bu^{t}OOH, R$_2$SO} and a few others. One obvious requirement is that Eq. (5) be spontaneous as written. Except for the case {R$'_2$SO, R$_2$S}, this is obviously so from available bond energies (27). In the specific case indicated, the combination R'=Ph and R=Me together with a concentration imbalance provides the driving force; the solution thermodynamics of this combination contrasts with the gas-phase values (28).

A second requirement pertains to the inherent reactivity of the components which may pose a more serious limitation insofar as making this reaction fully general. For example, neither dinitrogen monoxide nor any epoxide has as yet shown the ability to serve as an oxygen donor. Weak coordination of the epoxide to rhenium may be at fault. Interestingly, an episulfide (thiirane) serves as a sulfur donor to phosphines with the same rhenium catalyst; see Section VIII.B (29). AsPh$_3$, although thermodynamically less favorable than PPh$_3$, should suffice but it has not always proved kinetically competent, perhaps because of its bulk.

III. Kinetics of Pyridine N-oxide Reduction

A. THE RATE LAW

Much of the kinetic data have been obtained with the specific components written in Eq. (4), after which determinations were extended to the generalized compounds RC$_5$H$_4$NO and PR$_3$ to characterize the specific electronic and steric effects of substituents. It was easy to demonstrate that the reaction went to completion by applying ^{1}H- and ^{31}P-NMR spectroscopies. Indeed, the reaction took place so rapidly with the concentrations used (44 mM of each reagent and 1 mM of catalyst **1**) that product formation was complete before the first NMR spectrum could be recorded.

It therefore became more convenient to monitor the reaction progress with UV/Visible spectrophotometry, because all the pyridine N-oxides have strong absorption bands near 330 nm, with $\varepsilon \sim 10^3$ L mol^{-1} cm^{-1}. Two approaches for the analysis of the kinetic data were used. In the first but much less precise method, the initial reaction rates were calculated from the objective method of fitting the experimental values of [PyO]$_t$ to this function (30):

$$[\text{PyO}]_t = [\text{PyO}]_0 - m_1 t - m_2 t^2 - m_3 t^3 - m_4 t^5 - m_5 t^5 \qquad (6)$$

The initial rate is given by the numerical value of m_1 from polynomial fitting. The rate proved to be a function of three concentration variables, [1], [PyO] and [PPh$_3$]. Values of the rate were determined in series with two variables maintained constant and the third varied. This led to this tentative rate equation:

$$v = -\frac{d[PyO]}{dt} = k_{cat}\frac{[1] \times [PyO]^2}{[PPh_3]} \tag{7}$$

Experiments were then designed in which absorbance readings were acquired over the full time course of the reaction. These experiments employed a low concentration of PyO as the limiting reagent and a large excess of phosphine. The data fit a precise pseudo-second-order analysis, and gave $k_{cat} = 1.5 \times 10^4 \, \mathrm{L\,mol^{-1}\,s^{-1}}$ in benzene at 298 K; a precision of $\pm 5\%$ was estimated.

Two unusual and complex features of this rate equation suggested additional studies would be informative. An inverse kinetic dependence on the concentration of one of the substrates was found. Also, the order with respect to the other substrate is two, despite which only a single PyO converts to Py for each cycle of catalysis.

B. A General Reaction Scheme

The form of the rate law points to PPh$_3$ being the product of an initial rapid reaction between 1 and PyO; it yields a new rhenium-containing species, designated **A**.

$$1 + PyO \; \rightleftarrows \; \mathbf{A} + PPh_3 \quad (k_1, \; k_{-1}) \tag{8}$$

The second-order dependence on PyO implicates that it is also involved in a second stage. One can write this reaction step, the significance of which will be considered in the following section.

$$\mathbf{A} + PyO \; \rightleftarrows \; \mathbf{B} \quad (k_2, \; k_{-2}) \tag{9}$$

Pyridine will be released from **B** in an irreversible step,

$$\mathbf{B} \longrightarrow \mathbf{C} + Py \quad (k_3) \tag{10}$$

More rapidly than these steps occur, **C** returns to 1 upon reaction with PPh$_3$, which will very likely occur in more than a single step with the overall result

$$\mathbf{C} + 2\,PPh_3 \; \xrightarrow{\text{fast}} \; 1 + Ph_3PO \tag{11}$$

The full Michaelis–Menten rate law that one can derive on this basis is,

$$-\frac{d[\mathrm{PyO}]}{dt} = \frac{k_1 k_2 k_3 [\mathbf{1}] \times [\mathrm{PyO}]^2}{(k_2 k_3)[\mathrm{PyO}] + (k_{-1} k_3 + k_{-1} k_{-2})[\mathrm{PPh_3}]} \tag{12}$$

The data indicate that there is no [PyO] term in the denominator, which allows simplification to the form

$$-\frac{d[\mathrm{PyO}]}{dt} = \frac{k_1 k_2 k_3 [\mathbf{1}] \times [\mathrm{PyO}]^2}{(k_{-1} k_3 + k_{-1} k_{-2})[\mathrm{PPh_3}]} \tag{13}$$

The data allow further consolidation and interpretation, but only after a consideration of the chemical steps, which is the subject of the next section.

IV. The Chemical Mechanism of Pyridine N-oxide Reduction

A. THE INITIAL STEP

Equation (8) appears to represent a reaction in which one ligand replaces another. Many examples of such reactions of these complexes have been found with ligands that are not capable of further steps, such as oxygen loss. In general, these reactions can be represented by the chemical equation,

$$\mathrm{MeReO(dithiolate)L}_i + \mathrm{L}_j \; \rightleftarrows \; \mathrm{MeReO(dithiolate)L}_j + \mathrm{L}_i \tag{14}$$

Both thermodynamic and kinetic issues come into play, as does the involvement of geometrical isomerization when a dithiolate with inequivalent thiolate donors, such as mtp, participates in the reaction; the ligand mtp is derived from mtpH$_2$, 2-(mercaptomethyl)thiophenol.

For purposes of pyridine N-oxide reduction, it suffices to say that many examples of Eq. (14) have been explored, including those in which PR$_3$ is displaced contra-thermodynamically (pK_{14} may be ca. 2) by PyO. Invoking such a step in the catalytic cycle is therefore perfectly acceptable. Actually, the ligand displacement chemistry is a rich and fascinating area in its own right, and for that reason it will be explored independently in Section V.

Experiments were carried out using a series of reagents, $\mathrm{P(C_6H_4R)_3}$. A Hammett correlation of k_{cat} against 3σ gave the reaction constant $\rho^{\mathrm{P}}_{\mathrm{cat}} = +1.03$. The positive value is consistent with the chemistry proposed at this stage, in that the most weakly-coordinated phosphine would be expected to be the most reactive catalyst.

B. PYRIDINE LIBERATION

It makes only sense that one difficult step, and most likely the most difficult, is scission of the N–O bond of PyO. The rate equation states that bond breaking is, in the next stage, not compensated by P–O bond making, in that there is not a positive rate dependence on [PR$_3$].

Conversion of PyO to Py is a reduction. It can be realized by the concomitant oxidation of rhenium(V) to rhenium(VII) in the step

$$MeRe^V O(dithiolate)OPy \rightarrow MeRe^{VII}(O)_2(dithiolate) + Py \qquad (15)$$

The data do not support a reaction as simple as this, however, the rate law implicates a *second* PyO at this stage. Similar experiments were extended with the use of a series of ring-substituted pyridine N-oxides, RC_5H_4NO, as the substrates. Correlation of the values of k_{cat} against σ_R gave a particularly large and negative Hammett reaction constant, $\rho_{cat}^N = -3.84$. This is so because PyO enters in three steps of the scheme, each of which is improved by electron donation.

C. NUCLEOPHILIC ASSISTANCE

Two transition states can be visualized, one as implied by Eq. (15), the other with an additional nucleophile (here, the second PyO) involved,

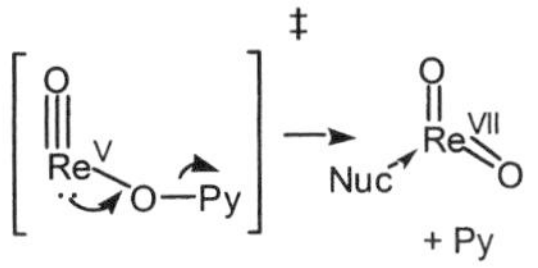

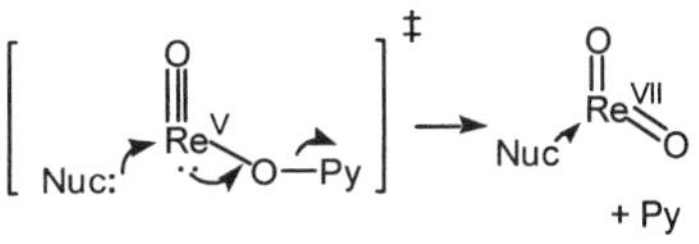

The role of PyO in the second of the proposals seems eminently reasonable. It assists oxidation of Re(V) to Re(VII) and reduction of PyO to Py. To test this hypothesis, three other nucleophiles were independently added to a system in which $2\text{-Me},4\text{-NO}_2C_5H_3NO$ was the substrate. They were 4-MeC_5H_4N, C_5H_4N, and $[Bu_4^nN]Br$. This substrate, which suffers from added steric and electronic barriers to reactivity, was selected so as to lower all of the rates into a more readily measured time frame.

Each of the nucleophiles exhibited the expected rate-acceleration effect. More to the point, the form of the rate equation was altered to:

$$v = -\frac{d[2-Me,4-NO_2C_5H_3NO]}{dt}$$
$$= k_{cat}\frac{[1]\times[2-Me,4-NO_2C_5H_3NO]\times[Nuc]}{[PPh_3]} \quad (16)$$

where Nuc represents the added nucleophile. The major point to note is that the reaction order with respect to [PyO] has been lowered to unity with the addition of a first-order term in [Nuc]. This implicates an alternative reaction,

$$\mathbf{A} + \text{Nuc} \rightleftharpoons \mathbf{B'} \quad (k_2, k_{-2}) \quad (9')$$

It should also be noted that the added nucleophilic reagents are not (cannot) themselves be transformed in any sense as a result of this chemistry; they serve as promoters of the rhenium catalyst.

D. THE PUTATIVE ROLE OF RHENIUM(VII)

According to the general postulates presented, a dioxorhenium(VII) species intervenes along the reaction pathway. Attempts to detect it under the conditions of the catalytic experiments were not successful. Indeed, species **A** and **B** were not seen either; **1** was the only rhenium species found by NMR spectroscopy. In an attempt to provide other validation for a rhenium(VII) intermediate, other experiments were carried out.

1. Kinetic Competition

A rapid reaction between dioxorhenium(VII) and PPh$_3$ was proposed in Eq. (11) for sake of consistency with the kinetic data. For reasons to be presented subsequently, this is shown as the addition of the phosphorus to an oxo oxygen. In effect, this is the reduction of rhenium(VII) with the concomitant creation of a rhenium(V) complex ligated by a phosphine oxide,

$$MeRe^{VII}(O)_2(dithiolate) + PPh_3 \rightarrow MeReO(dithiolate)(OPPh_3) \quad (17)$$

Although this reaction occurs too rapidly for direct observation, the relative rate constants for *pairs* of phosphines were determined by standard kinetic competition techniques. The rate constant ratios for P(C$_6$H$_4$R)$_3$ relative to PPh$_3$ correlate well with the Hammett constant 3σ, giving $\rho = -0.70$.

2. Direct Detection of Rhenium(VII) by Low-temperature NMR Spectroscopy

Reactions between MeReO(mtp)PPh$_3$ and four oxygen donors were carried out at 240 K in toluene-d$_8$. No phosphine other than the amount introduced with the starting reagent was added in these experiments, and ample 4-MeC$_5$H$_5$N was added instead. In each case, a new rhenium compound was detected in the ^{1}H-NMR spectrum. Most importantly, the same NMR spectrum was obtained when these O-atom donors were used: 4-MeC$_5$H$_4$NO, C$_5$H$_5$NO, Bu^{t}OOH and Me$_2$SO. The added 4-picoline stabilized the resulting dioxorhenium(VII), **12**, by coordination; this ligand, too, was found in the spectrum in the correct 1:1 ratio along with the dioxorhenium(VII). The fact that the newly-formed MeRe(O)$_2$(mtp)(NC$_5$H$_4$-4-Me), **10**, exchanges with 4-*tert*-butyl pyridine to produce **11** further substantiates these assignments (*31*).

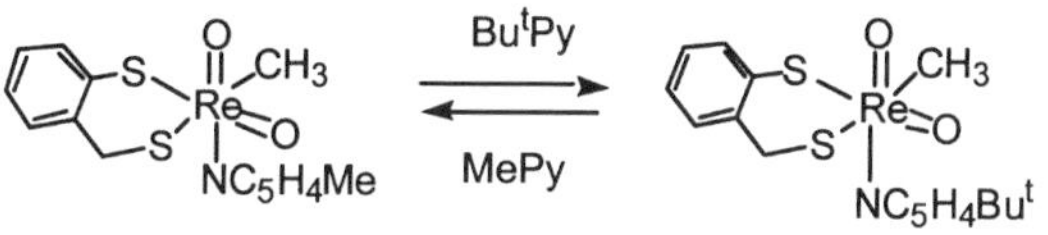

10   **11**

3. Synthetic Analogues of Dioxorhenium(VII) Dithiolates

Three ligand-stabilized dioxorhenium(VII) compounds were synthesized to allow direct evaluations of their reactivities (*32*). The strategy was to use an alkoxide-thiolate chelate rather than a dithiolate. Compounds **13**–**15** were obtained.

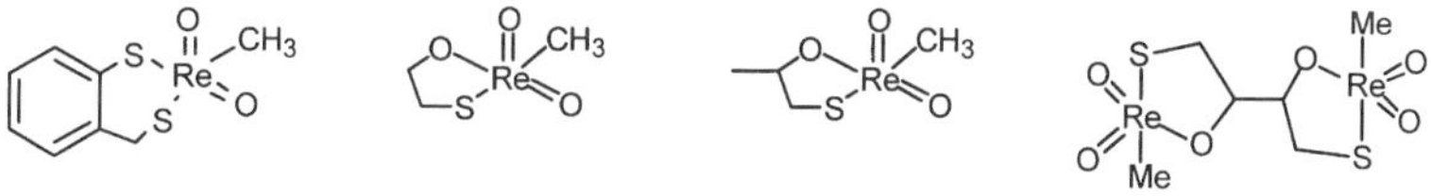

12   **13**   **14**   **15**

Only **15** was sufficiently stable for isolation and chemical and structural characterization (*33*). Compounds **12**–**13** persist for several hours in chloroform at room temperature and for 2–3 weeks at 251 K, particularly when water is carefully excluded. All three react readily with PR$_3$, forming oxorhenium(V) compounds that in these cases, unlike those with dithiolates, are metastable. The fastest reaction occurs between P(p-MeOC$_6$H$_4$)$_3$ and **13**, with a rate constant of 2.15×10^2 L mol^{-1} s^{-1} in chloroform at 298 K. Other reactions of PAr$_3$ upon applying the Hammett equation yield $\rho = -0.7$. This is the same value of the reaction constant that was obtained for $MeRe(O)_2(mtp)$ by kinetic competition, which supports the same assignment of mechanism.

E. The Chemical Mechanism at the PR_3 Stage

The results presented in Section IV.D.1 show that the net loss of the dioxorhenium(VII) species, and the ultimate formation of the phosphate R_3PO, must occur in two stages because the rate of reaction (17) shows a direct first-order phosphine dependence. That said, the chemical mechanism is still open to discussion: does the first step entail *abstraction* of an oxo oxygen or *addition* to it? If the former, the cycle is completed by PR_3 coordination to a four-coordinate rhenium intermediate; if the latter, the addition step is then followed by yet another ligand substitution reaction. The alternatives are presented in Schemes 1 and 2.

No data exist for rhenium reactions to support one chemical sequence over the other. The best that one can do is present arguments on the basis of precedent and plausibility. Nonetheless, an argument against the first assignment can be made, and another in favor of the second.

The argument against Scheme 1 is a negative one. Its basis derives from extensive studies carried out on the ligand replacement reactions of oxorhenium complexes of the family MeReO(dithiolate)L (*20,34–37*). Those studies (Sections V.B and V.C) show that all such processes studied to date proceed by direct displacement reactions without a recognizable intermediate from unassisted Re–L dissociation. (Indeed, in an early work, a dissociative step was written, but that formulation has since been revised; see Section V.D.)

One can prepare derivatives of **4** that contain a weakly-bound ligand L, such as a pyridine or a dialkyl sulfide. These compounds show no

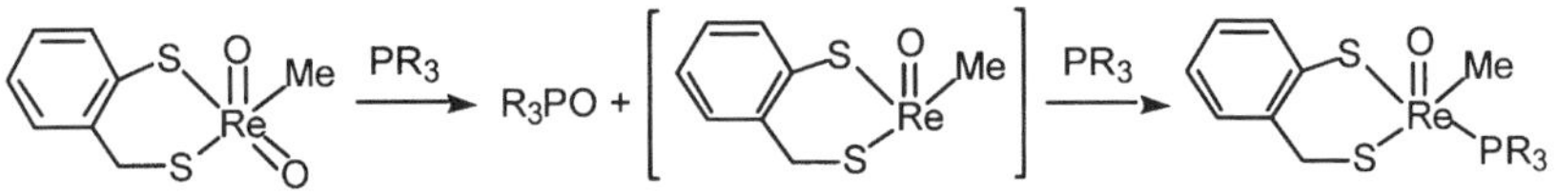

SCHEME 1. Abstraction mechanism for the reaction of phosphines with dioxorhenium(VII).

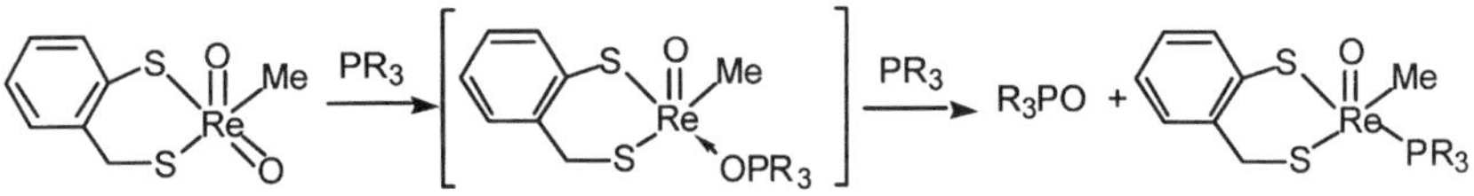

SCHEME 2. Addition mechanism for the reaction of phosphines with dioxorhenium (VII).

tendency to dissociate to produce MeReO(dithiolate), but instead equili-brate (again by an associative mechanism, Section V.D) with a dimer, **6**. The coordination number of rhenium(V) in dimer **6** is sustained at five by the formation of coordinate bonds with bridging thiolate sulfur atoms.

$$2\,\mathrm{MeReO(mtp)PPh_3}\,(\mathbf{4}) \rightleftarrows 2\,\mathrm{L} + \{\mathrm{MeReO(mtp)}\}_2\,(\mathbf{6}) \qquad (18)$$

The argument in favor of Scheme 2 is that a close model of the bracketed intermediate has been directly identified in molybdenum reactions *(38)*. The reaction between $\mathrm{PEt_3}$ and the dioxomolybdenum(VI) compound $\mathrm{L^{Pr}Mo(O)_2(OPh)}$, where $\mathrm{L^{Pr}}$ = hydrotris(3-isopropylpyraxol-1-yl)borate, yields a triethylphosphate complex, $\mathrm{L^{Pr}MoO(OPh)(OPEt_3)}$, which has been characterized fully. A relative of it is less stable: $\mathrm{L^{Me}MoO(Cl)(OPPh_3)}$ was detected by FABMS. Its decomposition occurs by replacement of $\mathrm{Ph_3PO}$ by water. Theoretical studies supported the for-mulation of that step as occurring by an associative mechanism. It is fully analogous to the second step of Scheme 2, for which we suggest the same assignment.

V. Digression to Ligand Exchange and Substitution

The first step in oxygen transfer is ligand substitution at an oxorhenium(V) center, Eq. (14). The final step (see Scheme 2, step 2) very likely is also ligand substitution. We have therefore examined the kinetics and mechanism of several reactions in which one monodentate ligand displaces another, represented in general as follows:

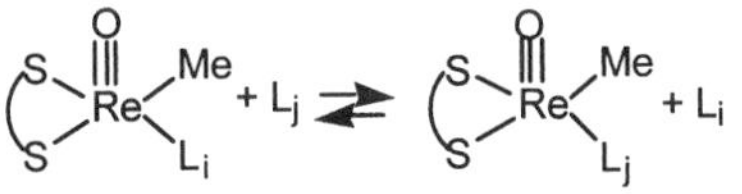

$$(19)$$

As a general statement, the reaction rate in each direction follows second-order kinetics for all the rhenium compounds studied. Moreover, the rate constants depend on the identities of L_i and L_j. Both findings argue for an associative (displacement) mechanism, which is also supported by the large and negative values of $\Delta S^{\ddagger}$, that often reach $-120\,\mathrm{J\,K^{-1}\,mol^{-1}}$ *(39)*.

A. Confronting an Issue of Microscopic Reversibility

It would be difficult to envisage entry of L_j into the coordination sphere of rhenium other than at the vacant position trans to the oxo group. It

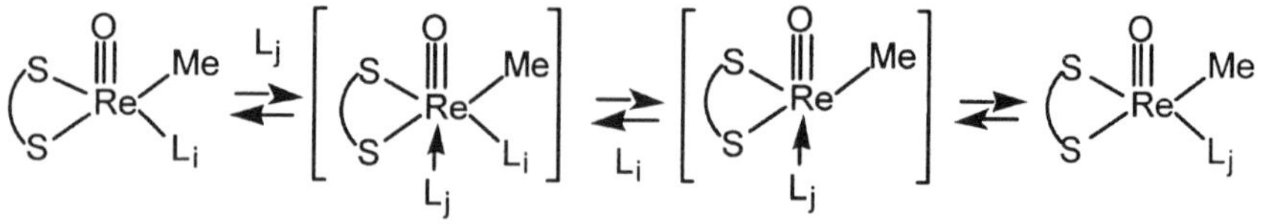

SCHEME 3. A disallowed mechanism that violates microscopic reversibility.

is not reasonable, however, to depict the leaving group as departing from its equatorial position, for the reaction coordinate would then not be symmetrical. In the limit where $L_i = L_j$, or nearly so, this would result in a clear violation of microscopic reversibility. Scheme 3 depicts the dilemma.

It is useful to note that the same issue does not arise in complexes that have a horizontal symmetry plane. There, a wedge geometry allows L_i and L_j to attain equivalent positions above and below that plane. These conceptual issues have been addressed in the case of oxorhenium(V) complexes by two experimental studies, each of which supports intervention of intermediate(s) that undergo turnstile or trigonal twist mechanisms. In so doing, L_i and L_j attain equivalent or at least inter-changeable positions. These studies are the subjects of the next two sections.

B. PYRIDINE EXCHANGE REACTIONS OF MeReO(edt)Py

On the basis of the preceding discussion, these complexes, where $Py = RC_5H_4N$, should undergo racemization, a process that will accompany pyridine exchange.

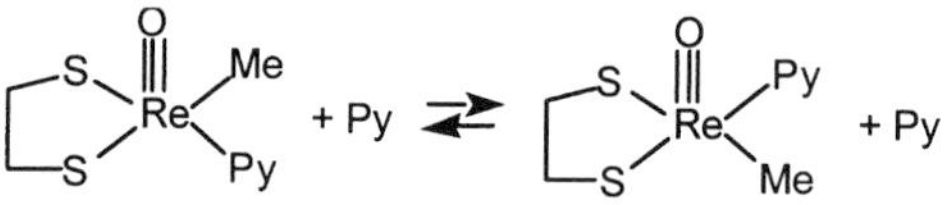

$$ (20) $$

Studies using ^{1}H-NMR spectroscopy in solutions containing excess Py were carried out. The width at half-height of the resonance peaks varied in proportion to [Py]. These experiments allowed determination of the rate constants for Eq. (20), which ranged systematically from 135 to $348 \, \text{L} \, \text{mol}^{-1} \, \text{s}^{-1}$ in C_6D_6 at 298 K as R was changed form 4-NC to 4-Me$_2$N. This is a small substituent effect, yielding $\rho = -0.4$. For R = 4-Me, $\Delta H^{\ddagger} = 28.9 \, \text{kJ} \, \text{mol}^{-1}$ and $\Delta S^{\ddagger} = -104 \, \text{J} \, \text{K}^{-1} \, \text{mol}^{-1}$ (20).

When MeReO(mtp)Py complexes were examined, yet another phenomenon was revealed. With this ligand, in which the two sulfur donor atoms are inequivalent, two geometric isomers of the compound were obtained. The more stable one, **8** in general, was always the compound isolated. It has the methyl group trans to the benzylic sulfur of mtp. In the case of L = Py, it was possible to detect a mixture of both isomers.

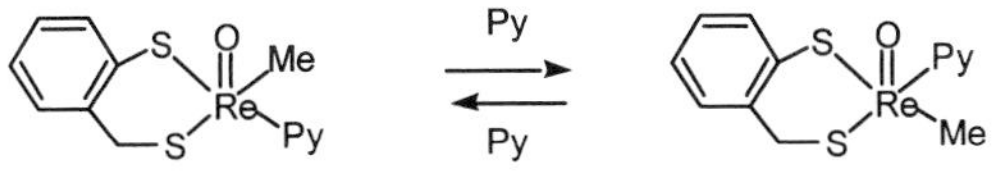

$$(21)$$

16   **16***

The less stable isomer is labeled with an asterisk. In this case, L = Py, both forms were detected in an equilibrium proportion with $K_{21} = 0.11$, a value that was barely perceptibly different among the different pyridines used (*20*).

The interconversion between **16** and **16*** does not take place intramolecularly. Just as in Eqs. (19)–(20) and in Scheme 3, reaction (21) takes place by a second-order associative mechanism. This is an interesting result to contemplate. Many five-coordinate complexes do rearrange by an intramolecular (fluxional or pseudorotation) process. In the oxorhenium complexes, the unimolecular mechanism may be disfavored owing to the oxo group, the presence of which disfavors a geometry other than square-pyramidal. In the case at hand, a suitable mechanism is provided by ligand addition. As the result of forming a *six*-coordinate intermediate, a trigonal rotational process can proceed. It will be depicted in the next section.

C. MeReO(mtp)PR₃: PHOSPHINE EXCHANGE AND SUBSTITUTION

Equations analogous to Eq. (21) will be written showing the metastable isomer as lying on the pathway for phosphine exchange (Scheme 4) and substitution (Scheme 5) in MeReO(mtp)PR₃ compounds.

The phosphine case stands out from that of pyridine in two ways. First, none of the MeReO(mtp)(P$_j$)* isomer can be detected at equilibrium. That is, the equilibrium constant analogous to K_{22} has always been found to be ≪ 1. Second, readily-detected concentrations of the metastable isomers are found by both UV/Visible, ^{1}H- and ^{31}P-NMR spectroscopies (*20,40*).

The overall reaction kinetics for substitution could easily be resolved into two stages, as described by these equations:

$$MeReO(mtp)P_i + P_j \rightarrow \{MeReO(mtp)P_j\}^* + P_i$$
$$v_{22} = k_{22} [MeReO(mtp)P_i] \times [P_j]$$

$$(22)$$

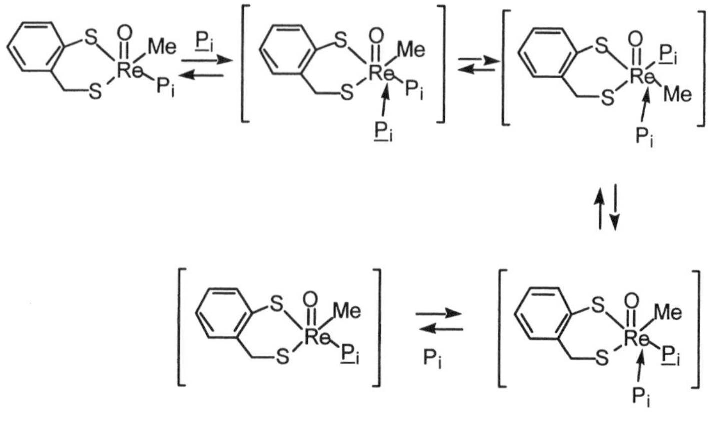

SCHEME 4. Proposed scheme for phosphine exchange.

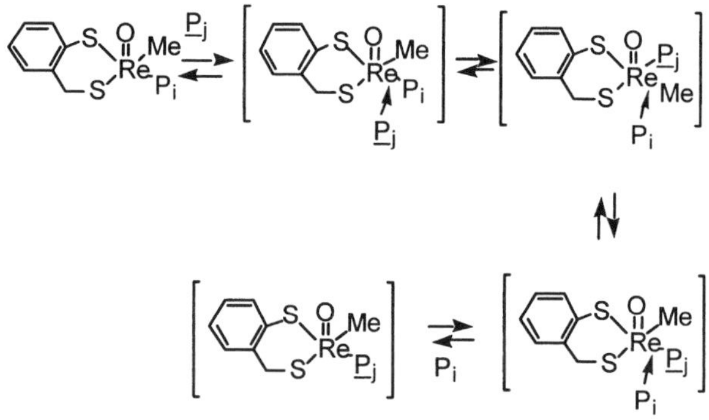

SCHEME 5. Proposed scheme for phosphine substitution.

$$\{\text{MeReO(mtp)}P_j\}^* + P_j \rightarrow \text{MeReO(mtp)}P_j + P_j$$
$$v_{23} = k_{23}\,[\{\text{MeReO(mtp)}P_j\}^*] \times [P_j] \tag{23}$$

The significant point is that *each stage proceeds at a rate directly proportional to* [P_j]. To make this quite explicit, the multi-step mechanism proposed is shown in Scheme 6. It is written as being unidirectional, so driven by the use of unbalanced concentrations, [P_j] ≫ [P_i], but of course each step is fully reversible except that the reversible arrows and reagents were omitted for sake of clarity.

The suggestion that equivalent phosphine positions are attained by a turnstile rotation is but one possibility, which is deemed to be the

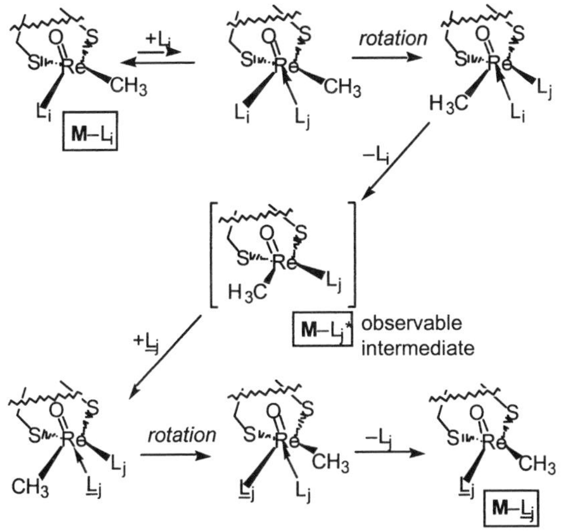

SCHEME 6. Turnstile mechanism for the ligand substitution reactions of oxo-rhenium(V) compounds.

most likely. One alternative imagines that a pentagonal pyramid is formed, by movement of L_j into the same basal plane as L_i. At least in the case where the ligands are phosphines, this scheme would suffer from severe steric strain, and seems rather less likely than the one depicted.

D. MONOMERIZATION OF DIMERIC OXORHENIUM(V) COMPOUNDS BY LIGATION

The mononuclear compound 8 and a related dimer 6 were character-ized in the early stages of this research (2,3,37). The two forms can be interconverted by the net reaction

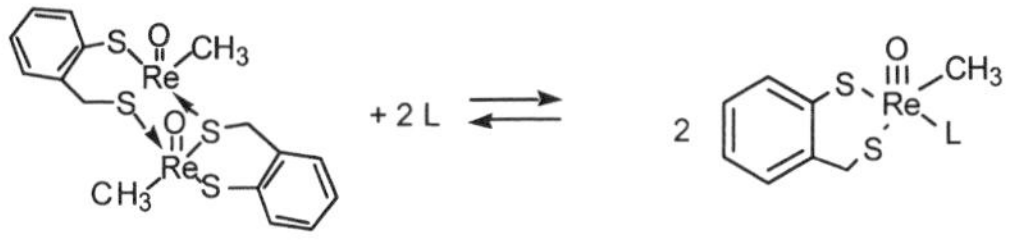

$$(24)$$

Both equilibrium and kinetic data have been obtained. The most exten-sive set of equilibrium constants was determined for pyridines, RC_5H_4N. Values of K_{24} lie in the range 2×10^{-2} (R = 2-Me) to $> 5\times10^4$ (R = 4-Me$_2$N); according to a Hammett correlation, $\rho = -7.5$ (35). The

TABLE II

[a]RATE CONSTANTS [b] AND HAMMETT REACTION CONSTANTS [c] FOR
MONOMERIZATION REACTIONS, Eq. (25)

	RC_5H_4N	$P(C_6H_4R)_3$
k_a (R = H)/L mol^{-1} s^{-1}	< 0.3	1.57×10^{-2}
k_b (R = H)/L^2 mol^{-2} s^{-1}	1.31×10^{-1}	5.1×10^{-2}
ρ_K [b]	−7.5	−
ρ_a [b]	−4.8	−1.3
ρ_b [b]	−6.1	−1.0

[a] Superscripts a–c refer to the footnotes; subscripts a and b refer to notation
used in Eq. (25).
[b] In benzene at 298 K.
[c] Correlation against (pyridines) or 3 (phosphines).

reactions between **6** and phosphines proceed essentially to completion
with but a modest excess of phosphine; i.e. $K_{24} \gg 1$. In two cases it
proved possible to determine quantitatively the Lewis basicity of PPh$_3$
relative to pyridine and dimethyl sulfide in the compounds
MeReO(mtp)L. The relative equilibrium constants lead to these rank-
ings: 900 (PPh$_3$) : 1.00 (Py) : 1.5 × 10^{-3}(Me$_2$S) (*35,41*).

The kinetic pattern for reaction (24) is more complex, and it is the same
for pyridines and phosphines. The general rate expression is,

$$\frac{d[MeReO(mtp)L]}{dt} = \{k_a[L] + k_b[L]^2\} \times [\{MeReO(mtp)L\}_2] \quad (25)$$

Values of k_a and k_b are correlated by the Hammett equation for pyridines
and for triaryl phosphines. Very large negative ρ values were found for
pyridine reactions especially (*35*). Key values are highlighted in Table II.

Formulating an acceptable chemical mechanism for each pathway is an
interesting challenge. In aid of that, in the reaction with L = Me$_2$SO,
intermediate **17** was isolated and fully characterized. It has a structure
in which one dmso molecule is coordinated to a rhenium atom of **6** trans
to its oxo group.

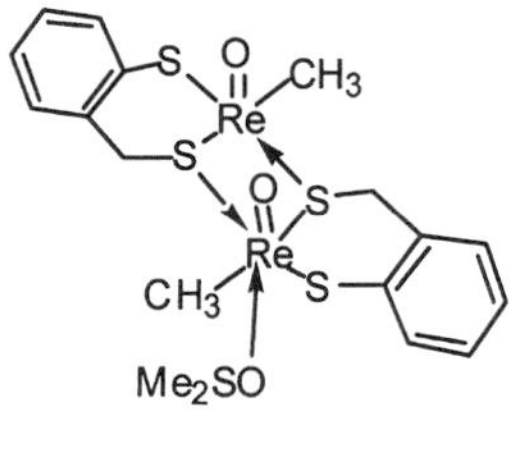

17

This structure validates a point made earlier, that ligand access occurs in the indicated position. At the point where this plausible but unproven assertion was first made, the reference was to reactions in which pyridine N-oxides were acting as oxygen donors. It remains pertinent for ligand substitution as well, but also for oxygen transfer reactions where sulfoxides are the oxygen donor atoms.

The pathway that is second-order with respect to [L] may entail a transition state analogous to **17**, save that two L molecules are coordinated; following that severance of both coordinated thiolate bonds would lead immediately to product **12**. Alternatively, following the point where the first L has been added, opening of a single coordinate bond facilitates entry of the second L. In any event, either sequence leads efficiently to **12**. Scheme 7 shows the preferred mechanism along both pathways. The pathway with a first-order dependence on [L] was first formulated as involving release of a four-coordinate rhenium intermediate from the ligated dimer, **17** (*34,35*). The revised representation is Scheme 7 thus appears preferable (*33*).

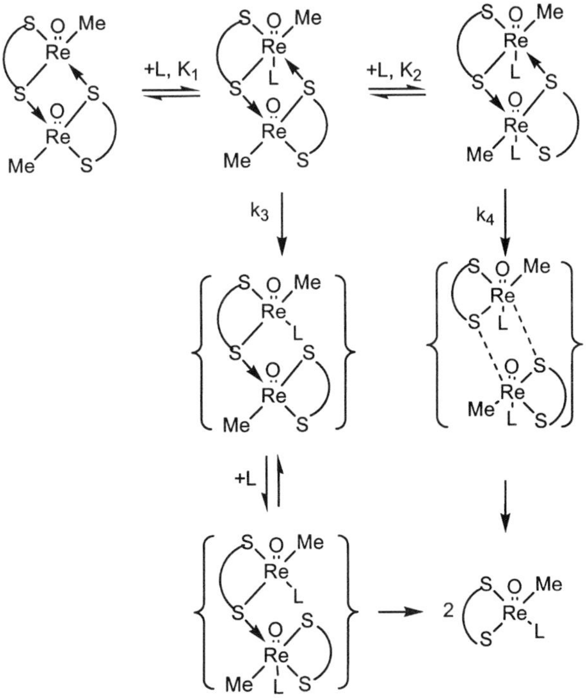

SCHEME 7. Proposed mechanism for monomerization.

VI. Additional Oxygen Atom Transfer Reactions

Additional reactions of the general type $YO + X \rightarrow Y + XO$ will now be described. The general conclusions arrived at from the thorough study of oxygen transfer from PyO to PR_3, presented in Sections III and IV, will be valid for these reactions as well. In terms of detail, however, each reaction offers further insight which is the reason such research is rewarding.

A. OXYGEN TRANSFER FROM *tert*-BUTYL HYDROPEROXIDE TO SULFIDES

1. Substrates, Procedures, Conversions, Yields

The same oxorhenium(V) compounds as used previously, such as **1**, catalyze this transformation under mild conditions

$$RSR' + Bu^tOOH \xrightarrow[\text{CHCl}_3,\ 25^\circ C]{\text{cat.MeReO(mtp)PPh}_3} RS(O)R' + Bu^tOH \qquad (26)$$

Provided an excess of the hydroperoxide is not used, sulfoxides are obtained in essentially quantitative yields in short reactions times, usually 0.7–2.5 h (*42*). The method is uncomplicated and can be carried out on the benchtop. The long shelf-life of **1** (>3 months) adds to the convenience of this procedure. A wide variety of functional groups is tolerated on R and R'. The reaction affords nearly pure sulfoxides without contamination from sulfones. The product is obtained simply be evaporating the solvent and *tert*-butyl alcohol. This method avoids aqueous workup, which is often required when peracids are used (*43*), and is thus convenient for water-soluble sulfoxides.

Dibenzothiophene is among the sulfides oxidized, and its monoxide was obtained in 89% yield albeit in a longer (6 h) time. Thiophene itself was also oxidized, but its monoxide is known to be too labile for isolation (*44–46*). Instead, it was trapped by a Diels-Alder reaction, as shown in Scheme 8.

2. Kinetics of Sulfoxide Formation

The oxidation of methyl tolyl sulfoxide, a representative substrate, was monitored by the buildup of the sulfoxide as a function of time under many sets of conditions (*42*). A representative concentration–time plot is presented in Fig. 1. In this case, and in all the others, the buildup curves showed similar features, the most noticeable of which is a distinct induction period. Its length depends on the concentrations, decreasing with increasing [Bu^{t}OOH] and [**1**]. At the same time the rate itself was

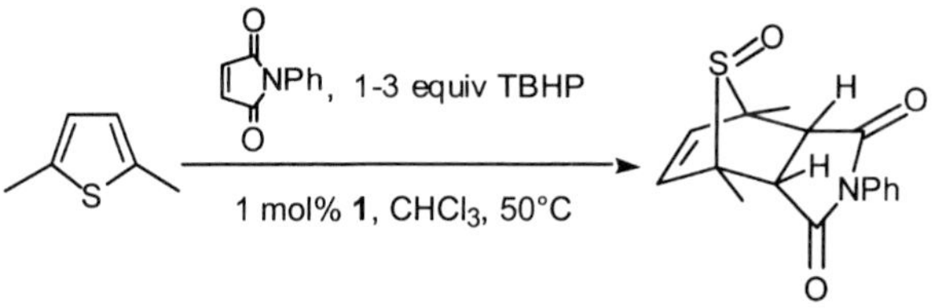

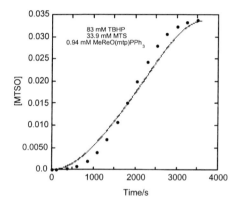

SCHEME 8. Oxidation and trapping of 2,5-dimethylthiophene monoxide.

FIG. 1. Formation of MeS(O)Tol, showing fitted (solid line) versus experimental data during the reaction of 33.9 mM MeSTol and 83 mM *tert*-butyl hydroperoxide in the presence of 0.31 mM MeReO(mtp)PPh₃, **1**. Reactions were conducted in benzene at 298 K. The experimental progress curve was modeled by a kinetics simulation routine that gave the optimum fit to data from six such experiments.

higher. With increasing [MeSTol], however, the reaction slowed down dramatically and the induction time grew longer.

Independent of this study, it has been shown that PPh₃ is oxidized by Bu^{t}OOH more rapidly than MeSTol, when **1** catalyzes both reactions. Nonetheless, addition of 1.0 mM PPh₃, along with 34 mM MeSTol, reduces $t_{1/2}$ from 3900 to 1900 s. Owing to these complications it was not possible to present a closed-form expression for the reaction rate. We do note the strong inhibition by phosphine, which inevitably remains coordinated to rhenium in a generally less active form, at least until oxidized.

3. The Reaction Scheme and Chemical Mechanism

On the basis of the qualitative observations presented in the preceding section, we have formulated a scheme that explains all of the qualitative aspects. Scheme 9 presents the proposal. It includes the now

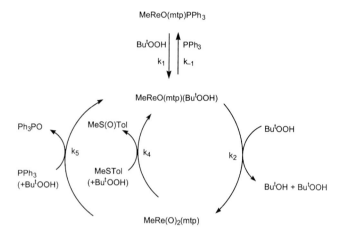

SCHEME 9. Catalytic cycle used for kinetic modeling.

well-precedented ligand substitution reactions (the steps with rate constants k_1, k_4 and k_5) and the formation of a dioxorhenium(VII) intermediate. It was independently observed in the absence of PPh$_3$ by low-temperature NMR experiments, Section IV.D.2.

This model was adequately but imperfectly modeled by kinetics simulation and fitting methods (42). The scheme itself suggests why matching a model to the observations is a difficult problem. First, PPh$_3$ is slow to be displaced, as it is by far the strongest Lewis base in the system. Then, as PPh$_3$ is released, it is subject to oxidation along with, and in competition with, MeSTol. For those reasons the reaction starts slowly and then accelerates. As PPh$_3$ is oxidized, however, **1** cannot remain in its original form. Thus the predominant form of the catalyst changes over the reaction period.

B. OXYGEN TRANSFER FROM *tert*-BUTYL HYDROPEROXIDE TO SULFOXIDES

1. Reactions Forming Sulfones

The same combination of reagents, now with a 3 : 1 ratio of hydroperoxide to sulfide, gives a quantitative yield of RS(O)$_2$R'. The reactions proceed to completion in about 2 h at 323 K in chloroform. Only 0.3–0.5 mol% of **1** relative to RSR' was needed. It was shown that oxidation to sulfoxide precedes sulfone formation. Again, the reaction is tolerant of many functional groups (42).

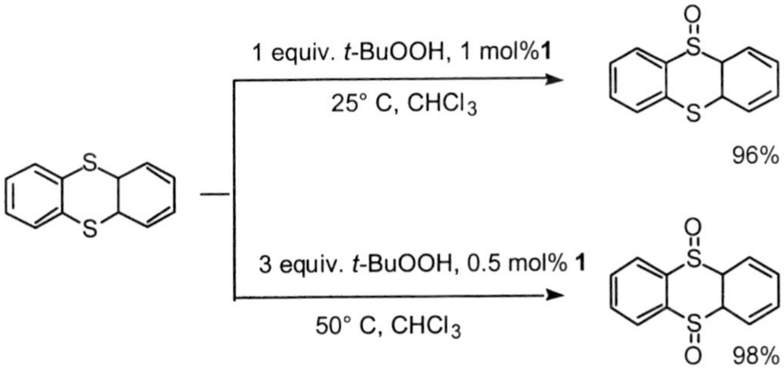

SCHEME 10. Oxidation of thianthrene.

2. The Oxidation of Thianthrene

Scheme 10 shows the course taken by these reactions. No trace was found of the 5,5-dioxide, the 5,5,10-trioxide, or the 5,5,10,10-tetraoxide. This reaffirms that sulfide oxidation precedes sulfone formation (42).

C. THE SPECIAL CASE OF 4,6-DIMETHYLDIBENZOTHIOPHENE

This compound is notorious as a "hard" sulfide that cannot be removed from petroleum by current hydrodesulfurization processing. Oxidation by tert-butyl hydroperoxide occurs readily when 1 is used as the catalyst. After trying several combinations, this was the most effective: 0.5 mmol of DMDBT, 1.75 mmol Bu^{t}OOH, and 0.05 mol% of 1 were placed in refluxing toluene (384 K). A quantitative yield of the dioxide was obtained in 2 h. The oxidation product is insoluble in toluene and can readily be removed by filtration (42).

D. THE OXIDATION OF PHOSPHINES BY tert-BUTYL HYDROPEROXIDE

A further test of whether MeReO(mtp)PAr$_3$ (5) catalyzes other oxygen atom transfer reactions was made through studies of this reaction,

$$Bu^tOOH + PAr_3 \rightarrow Bu^tOH + Ar_3PO \qquad (27)$$

The rate law indicates that the uncatalyzed pathway makes a contribution in this case,

$$v = [PAr_3] \times [Bu^tOOH] \times \{k_u + k_{cat}[5]\} \qquad (28)$$

For $P(C_6H_4-4-Me)_3$ at 298 K in chloroform, $k_u = 1.1 \, \text{L mol}^{-1} \text{s}^{-1}$ and $k_{cat} = 4.0 \times 10^4 \, \text{L}^2 \text{mol}^{-2} \text{s}^{-1}$. A plausible scheme for this reaction is

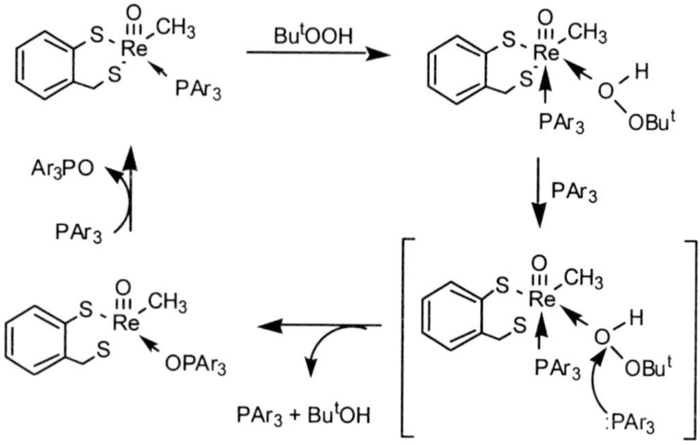

SCHEME 11. Suggested mechanism for rhenium-catalyzed phosphine oxidation.

presented in Scheme 11. It resembles what was shown before, except in the following respects (19). First, the first intermediate evidently has both hydroperoxide and phosphine coordinated, in that no inverse phosphine concentration dependence was found. Second, external phosphine attack on the coordinated oxygen of the hydroperoxide is, according to the rate law, needed to complete the catalytic cycle.

VII. Other Oxorhenium(V) Compounds as Catalysts

A. OXIDATION OF PAr_3 CATALYZED BY A RHENIUM(V) DIMER

The oxorhenium(V) dimer, $\{MeReO(mtp)\}_2$ (6) is remarkably more active than its mononuclear phosphine form, $MeReO(mtp)PPh_3$ (1) in certain reactions. Examples of this effect include reactions in which PyO (47) and MeS(O)Ph (48) transfer their oxygen atoms to a triarylphosphine. Figure 2 illustrates how much more effective 6 is compared to 1 in the PyO reaction.

In the final analysis, the matter comes down to the fact that avoidance of phosphine coordination eliminates a substantial kinetic barrier. Because the monomerization reaction between 6 and PAr_3 takes place so slowly, it amounts to but a minor correction during the reaction time. Without phosphine coordination, the system appears to involve "half-opened" derivatives of 6 as intermediates. Scheme 12 depicts the suggested mechanism.

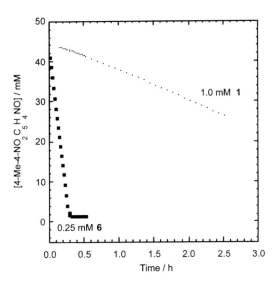

F $_{\text{IG}}$. 2. Concentration–time plot for the reaction between a PyO (44.4 mM) and PPh$_3$ (44.4 mM), showing the great effectiveness of the dimeric catalyst **6** over its monomer-phosphine form. Reactions were conducted in benzene at 298 K.

S $_{\text{CHEME}}$ 12. Proposed mechanism for reactions catalyzed by {MeReO(mtp)}$_2$.

B. Two Oxorhenium(V) Compounds with Tridentate Ligands

Two different structural types were prepared from $MeReO_3$ by using bis(2-mercaptoethyl) ether, which gives rise to **18**, and thiodiglycolic acid, from which **19** was prepared.

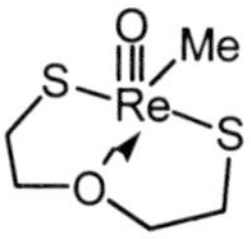

18 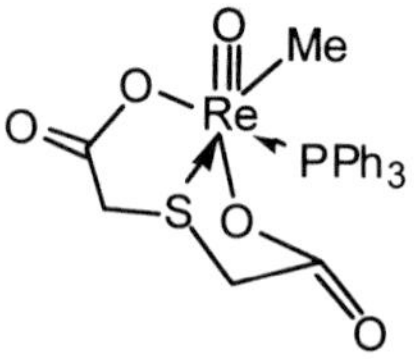   19

1. An Oxorhenium(V) Compound with an "SOS" Tridentate Ligand

Compound **18** adopts the now-familiar square-pyramidal structure. Reaction between it and PyO causes its decomposition to MTO and a disulfide

$$MeReO(SOS)\,(18) + 2\,PyO \rightarrow MeReO_3 + O(CH_2CH_2S)_2 + 2\,Py \quad (29)$$

This reaction follows second-order kinetics ($k_{298} = 1.6 \times 10^{-3}\,L\,mol^{-1}\,s^{-1}$ in chloroform). Electron donating R groups on RC_5H_4NO accelerate the reaction sharply; $\rho = -5$ (*41*).

2. An Oxorhenium(V) Compound with an "OSO" Tridentate Ligand

The rhenium atom in compound **19** is centered in a distorted octahedron. Compound **19** also reacts with PyO in a step that uses a molecule of water as well.

$$MeReO(OSO)PPh_3\,(19) + 2\,PyO + H_2O$$
$$\rightarrow MeReO_3 + S[CH_2CO_2H]_2 + 2\,Py \quad (30)$$

The rate of this reaction is first-order with respect to [**19**] and [H_2O]; there is no kinetic dependence on the concentration or identity of the pyridine N-oxide used. Electron-attracting substituents on $P(C_6H_4R)_3$ do, however, accelerate this reaction. The full details of the mechanism have not been defined at this time.

VIII. Sulfur Atom Abstraction

A. REACTIONS OF IMIDORHENIUM μ-S COMPOUNDS

1. Structural Issues and Isomers

Two μ-thio compounds are noteworthy, $\{MeRe(NAr)_2\}_2(\mu\text{-S})_2$, **20**, and $\{MeRe(NAr)_2\}_2(\mu\text{-S})$, **21**. For comparison purposes reference is also made to **22**, the μ-oxo analogue of **21**.

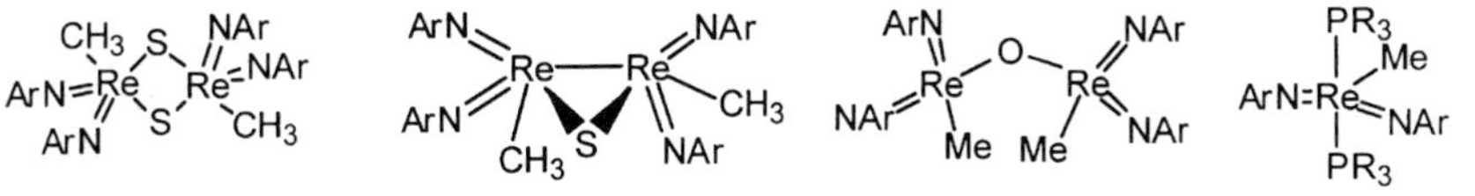

20	**21**	**22**	**23**

Compounds **21** and **22** are not isostructural: **22** has a syn,syn arrangement, whereas **21** exists as the syn,anti form with C_s symmetry. There was no indication that **21** rearranges over time at 243–323 K. Even at 263 K, **21** shows the two CH_3 resonances merged into one, but two resonances for the Me_2CH proton of the isopropyl groups were observed. In contrast, one CH_3 and one Me_2CH resonance were observed for **22**.

2. Reaction of **20** with Phosphines (49)

The net reaction with excess phosphine is,

$$\{MeRe(NAr)_2\}_2(\mu\text{-S})_2 \,(\mathbf{20}) + 4\,PR_3$$
$$\rightarrow 2\,MeRe(NAr)_2(PR_3)_2 \,(\mathbf{23}) + R_3PS \tag{31}$$

With mono- and di-phosphines the rate laws take different forms,

$$-\frac{d[\mathbf{20}]}{dt} = k_m[\mathbf{20}] \times [PR_3]^2 \quad \text{(monodentatephosphine)} \tag{32}$$

$$-\frac{d[\mathbf{20}]}{dt} = k_m[\mathbf{20}] \times [Ph_2P(CH_2)_2PPh_2] \quad \text{(bidentatephosphine)} \tag{33}$$

With insufficient or equimolar phosphine, the net reaction is,

$$\{MeRe(NAr)_2\}_2(\mu\text{-S})_2 \,(\mathbf{20}) + PR_3 \rightarrow \{MeRe(NAr)_2\}_2 \,(\mathbf{21}) + R_3PS \tag{34}$$

This poses a complication for the mechanism, because a direct study of the kinetics shows that **21** reacts with PR_3 much more slowly than **20** does, Eq. (31) (49). The rate law for **21** takes this form for both mono- and di-phosphines

$$-\frac{d[21]}{dt} = k_{\mathrm{m}}[21] \times [\text{Any Phosphine}] \qquad (35)$$

To explore the origin of the term in Eq. (32) with the $[PR_3]^2$ dependence, experiments were carried out with added pyridine. The rate law now includes a second term

$$k_\psi = k_{\mathrm{m}}[PMe_2Ph]^2 + k'_m[PMe_2Ph][Py] \qquad (36)$$

The key issues reduce to these: (a) given that a phosphine sulfide is the product, at what positions do phosphine atoms attack in the transition state? (b) What is the role for pyridine? (c) Although **21** is formed from **20**, with excess phosphine **23** results; given that **21→23** is slower than **20→23** and **23→21**, is **21** an intermediate along the pathway of the **20→23** reaction?

We conclude that only one phosphine attacks a μ-S group of **20**. Indeed, the phosphine sulfide product from dmpm is $Me_2P(S)CH_2PMe_2$, not a bis(sulfide). The other phosphine implicated by Eqs. (32) and (33), whether from a second PR_3 or from a di-phosphine, adds as a Lewis base to rhenium. The basis for this is that a pyridine easily replaced one of the phosphines in the mechanism. This proposed mechanism should steer the reaction directly towards formation of **21**, but owing to the dilemma posed in point (c), an intermediate likely intervenes. It can partition to the slowly-reacting **21** or to **23**, the latter step requiring additional phosphine; Scheme 13 shows the postulated mechanism. The dual role for phosphine and the use of Py in its place and the proposal for partitioning are shown in Scheme 14.

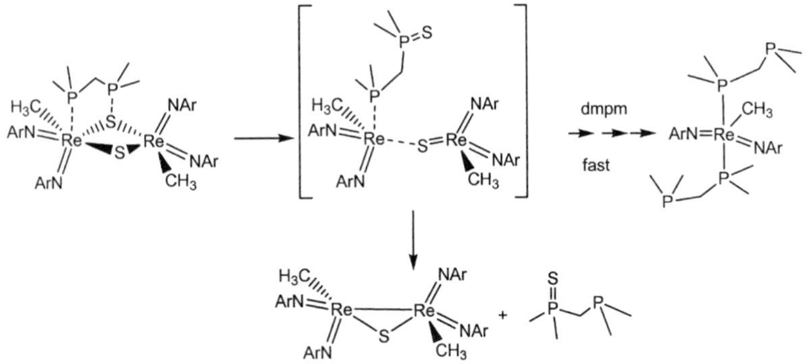

SCHEME 13. Proposed mechanism for reactions of a using dmpm as an example.

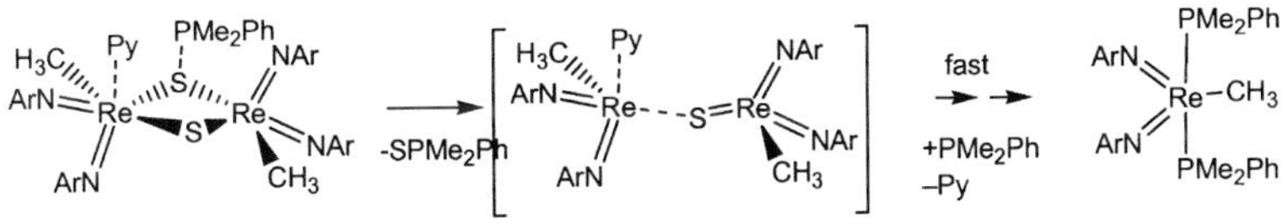

SCHEME 14. Proposed mechanism for the pyridine effect.

B. ABSTRACTION FROM THIIRANES (EPISULFIDES)

Phosphines undergo this reaction very slowly, $k \sim 10^{-5}\,\mathrm{L\,mol^{-1}\,s^{-1}}$ (50–52). An ill-defined catalyst for it was obtained by passing hydrogen sulfide through an acetonitrile solution of $MeReO_3$, followed by vigorous purging with argon. This allowed nearly every substrate to be quantitatively converted to alkene in a few hours or less (29).

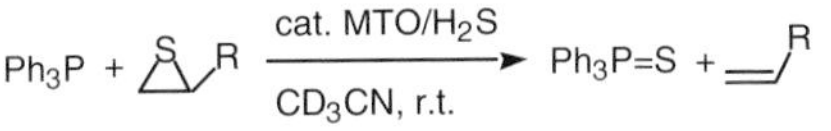

$$Ph_3P + \underset{}{\overset{S}{\triangle}}\!\!\diagup^R \xrightarrow[\text{CD}_3\text{CN, r.t.}]{\text{cat. MTO/H}_2\text{S}} Ph_3P{=}S + \underset{}{\diagup}\!\!\diagup^R \qquad (37)$$

The well-characterized molecular compound, dimer **6**, also catalyzes the same transformation. A slow, stoichiometric reaction occurs between propylene sulfide and **6**. With added PPh_3, and only a catalytic amount of **6**, a more rapid reaction occurs. The mechanism is currently under investigation.

C. OXYGEN ATOM TRANSFER FROM SULFOXIDE TO SULFIDE

The indicated transformation in dichloromethane solution is spontaneous (28)

$$Ph_2SO + Me_2S \rightarrow Ph_2S + Me_2SO \quad \Delta G° = -12\,\mathrm{kJ\,mol^{-1}} \qquad (38)$$

eventhough it is characterized by $\Delta G° = +13\,\mathrm{kJ\,mol^{-1}}$ in the gas-phase (27). One catalyst for the reaction is **24**, which loses a trifluorosulfonate ion to give rise to a cationic form. The reaction kinetics are complex, however, because both substrates compete for binding sites on the catalyst. The catalytic pathway is carried forward by coordination of Ph_2SO, release of Ph_2S as a relatively long-lived dioxorhenium(VII) is formed. It persists for days at 263 K in acetonitrile (28).

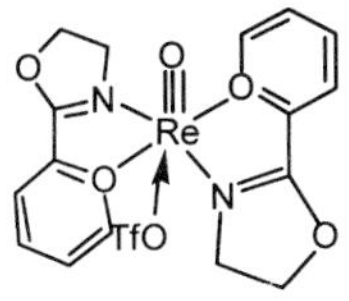

24 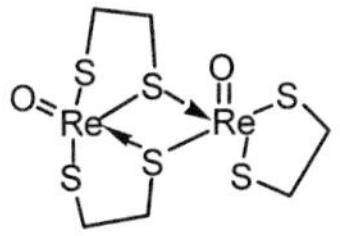   25

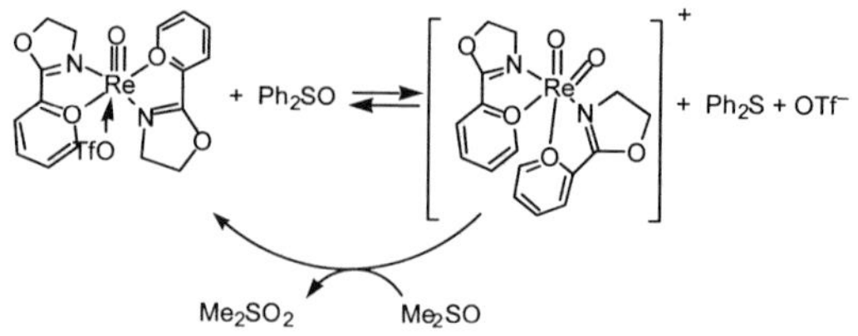

SCHEME 15. Mechanism suggested for sulfoxide disproportionation.

The dinuclear compound $\{ReO\}_2(edt)_3$, **25**, catalyzes the same reaction (*41*). The complex kinetics are much the same as for catalyst **24**. There is an inverse dependence on $[Me_2S]$. The rate increases linearly with $[Ph_2SO]$ to a maximum and then decreases to a steady, non-zero value.

D. OXYGEN ATOM TRANSFER BETWEEN SULFOXIDES

The net result is the disproportionation of the sulfoxides, e.g. (*28*)

$$2\,Me_2SO \rightarrow Me_2SO_2 + Me_2S \quad \Delta G^\circ = -105\,kJ\,mol^{-1} \qquad (39)$$

Complex **24** catalyzes this reaction, although the process occurs much more slowly than does Eq. (38). An extensive set of kinetic data lead to the mechanism presented in Scheme 15. It features two sequential sulfoxide reactions, one to form dioxorhenium(VII), the next to reconvert it to **24**.

IX. The Activation of Molecular Oxygen

As noted previously, the rhenium-catalyzed reactions dealt with to this point can be carried out in vessels open to the atmosphere. This is advantageous because of the convenient working procedures it allows. On the other hand, it means that molecular oxygen is not available as the stoichiometric oxidizing reagent (*53–55*). Three rhenium(V) compounds, **24–26**, have been prepared that do activate O_2.

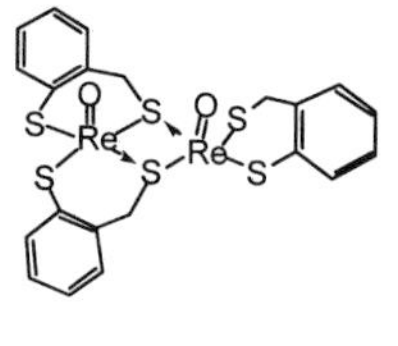

26 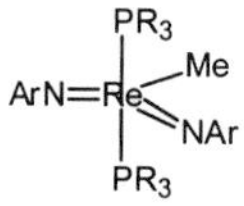   27

A. PHOSPHINE OXIDATION

1. Rhenium-Phosphine Coordination

These compounds, unlike their near-neighbors $\{MeReO(dithiolate)\}_2$ and MeRe(dithiolate)L, catalyze the oxidation of selected substrates by O_2. The first indication came with the finding that **25** is air-sensitive, unlike the others. The preparation of **25** is represented by the chemical reaction:

$$Re_2O_7 + 5\,H_2mtp \rightarrow \{ReO\}_2(mtp)_3 + 2\,RS\text{--}SR + 5\,H_2O \qquad (40)$$

where RS–SR is the cyclic disulfide of mtp. Unlike many other compounds, **25** and **26** lack a methyl group bound to rhenium. It appears that this is probably not a crucial issue, at least as far as oxygen reactivity is concerned.

In the absence of O_2, a very rapid (< 0.5 s) reaction occurs between PAr_3 and **25**, leading to the quantitative formation of the adduct **25**–PAr_3. Under these conditions, and indeed throughout oxygen catalysis, it is the only form of the catalyst that can be detected by NMR spectroscopy. Smaller phosphines, e.g. $PMePh_2$ and PMe_2Ph, engage in more complex interactions such that three species were seen. They are the two isomers of **25**–PR_3, featuring coordination at either rhenium atom as well as at the bis-phosphine compound **25**–$(PR_3)_2$ (*56*).

2. Kinetics of Phosphine Oxidation

The findings will be illustrated for a typical substrate, tri(*para*-tolyl)-phosphine. Typical conditions are 1–30 mM $PTol_3$, 0.1–9.1 mM O_2 and 5–$500\,\mu$M **25**. The major product was Tol_3PO ($> 90\%$), accompanied by a small amount of Tol_3PS, evidently from side reactions between $PTol_3$ and a dithiolate ligand of **25**.

The kinetic pattern follows the Michaelis-Menten pattern, the rate law for which is,

$$v = \frac{k_1 k_2 [\textbf{25}\text{--}PTol_3] \times [O_2] \times [PTol_3]}{k_1[O_2] + k_2[PTol_3] + k_{-1}} \qquad (41)$$

The value of k_1 proved to be so small that it could be omitted from the denominator. The initial rate increased with $PTol_3$ towards saturation limit in accord with Eqn. 41. The series of reagents $P(C_6H_4R)_3$ were studied, but the rate constant pattern is not uniform because the active catalyst, **25**–PR_3, as well as the substrate is changing (*56*).

Scheme 16 shows a mechanism for the catalysis that is consistent with the kinetics and other data. ^{1}H-NMR experiments in toluene-d_8 at 250 K

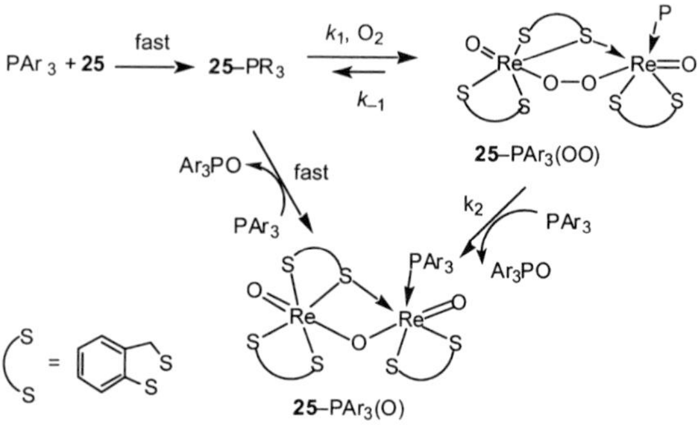

SCHEME 16. Mechanism for activation of oxygen by $\{ReO\}_2(mtp)_3$.

gave evidence of a new species. Its resonance reached a maximum intensity in ca. 30 min and had disappeared by 5–6 h. The spectrum is consistent with, but not proof of, the peroxo intermediate shown in the scheme.

B. OTHER SUBSTRATES FOR $\{ReO\}_2(mtp)_3$, 25

Phosphines offer the advantage of forming the strongest bonds to oxygen. In that sense their reactions should be most preferred. The sterically-encumbered reagent $P(o\text{-Tol})_3$ was oxidized slowly by this method. With 25 mM of the phosphine in air and 5% 25, only 10% $(o\text{-Tol})_3PO$ was formed after 16 h. When 1.0 mM $P(p\text{-Tol})_3$ was also added at the start, then the yield of $(o\text{-Tol})_3PO$ was 6.3 mM (25%); at the same time 1.0 mM $(p\text{-Tol}_3)PO$ (100%) was obtained. It is likely that the para isomer initiates formation of the peroxo intermediate more efficiently.

Experiments were also carried out with selected sulfides, dienes and cyclohexene. Essentially no oxidation occurred. Addition of $(p\text{-Tol}_3)_3P$ marginally improved the extent of oxidation, but yields were still only in the range 0.5–14% (56).

C. $\{ReO\}_2(mtp)_3$, 25, AS A CATALYST FOR OTHER OXYGEN-TRANSFER REACTIONS

In the absence of oxygen, 25 catalyzes the oxidation of PR_3 by Me_2SO and PyO

$$PR_3 + YO \rightarrow R_3PO + Y \tag{42}$$

These reactions require a modest excess of the sulfoxide and 5 mol% of 25, and result in 65–90% conversions to R_3PO (56).

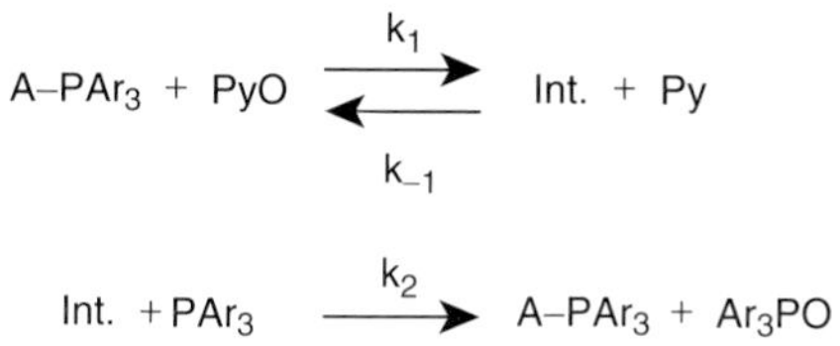

$$A{-}PAr_3 + PyO \underset{k_{-1}}{\overset{k_1}{\rightleftarrows}} Int. + Py$$

$$Int. + PAr_3 \xrightarrow{k_2} A{-}PAr_3 + Ar_3PO$$

SCHEME 17. Reaction scheme of oxygen transfer.

1. Competition Experiments for Oxygen Transfer from PyO

In one family of experiments, a pair of phosphine reagents was used to evaluate the relative reactivities of PyO towards the peroxo intermediate. The reactivity ratio proved to be independent of the PyO used, consistent with the absence of the Py group in the intermediate. The most reactive phosphines were those with electron-donating substituents (57).

2. Kinetics of Oxygen Transfer from PyO

$$v = \frac{k_1[\mathbf{25}{-}PPh_3] \times [PyO] \times [PPh_3]}{(k_{-1}/k_2)[PyO] + [PPh_3]} \tag{43}$$

With more reactive forms, a simplified rate equation could be attained,

$$v \cong k_1[\mathbf{25}{-}PPh_3] \times [PyO] \tag{44}$$

A scheme with two reaction steps can be written, a presented in Scheme 17.

Direct evidence was obtained that one of the two rhenium atoms is present as dioxorhenium(VII) in the active intermediate (57).

D. PHOSPHINE AUTOXIDATION CATALYZED BY $MeRe(NAr)_2(PR_3)_2$

1. Phosphine Dissociation and Substitution

The following dynamic process has been established for **27** by variable-temperature ^{1}H- and ^{31}P-NMR studies

$$MeRe(NAr)_2(PR_3)_2\ (\mathbf{27}) \rightleftarrows MeRe(NAr)_2PR_3 + PR_3 \tag{45}$$

The rates vary widely with the identity of the phosphine. PMe_3 and $Me_2P(CH_2)_2PMe_2$ (dmpe) equilibrate slowly as compared with the NMR time scale, whereas low temperatures are needed to observe a sharp resonance for others: PMe_2Ph, 283 K; $P(OMe)_2Ph$, 253 K; $P(OMe)_3$, 243 K and $P(OEt)Ph_2$, 223 K (58).

Equilibrium constants have been reported for the stepwise conversion of $MeRe(NAr)_2(PR_3)_2$ to the mixed-phosphine and then the $(PR_3')_2$ derivative. Both steric and electronic factors come into play, typical of phosphines. The stability ordering is $PMe_3 > dmpe > PMe_2Ph > P(OMe)_2Ph > PEt_3 > P(OEt)_3 > PMePh_2 > P(OEt)Ph_2 > PPh_3$ (58).

2. Reactions with Oxygen

The various $MeRe(NAr)_2(PR_3)_2$ compounds react with O_2 directly, in two steps. The rate law for phosphine oxidation is,

$$v = k_{cat} \times \frac{[27] \times [O_2]}{K + [PR_3]} \cong k_{cat}[27] \times [O_2] \qquad (46)$$

With a limiting amount of oxygen and excess $P(OMe)_3$, the absorbance of $MeRe(NAr)_2[P(OMe)_3]$ at 580 nm decreased very rapidly, but gradually returned to its original value. This phenomenon repeats itself as further increments of oxygen are added. Two reactions occur,

$$MeRe(NAr)_2(PR_3)_2 \, (\mathbf{27}) + O_2 \rightarrow MeRe(NAr)_2O \, (\mathbf{22}) + R_3PO + PR_3 \quad (47)$$

$$MeRe(NAr)_2O \, (\mathbf{22}) + 3\,PR_3 \rightarrow MeRe(NAr)_2(PR_3)_2 \, (\mathbf{27}) + R_3PO \qquad (48)$$

From these equations one can predict that $MeRe(NAr)_2O$, **22**, will be an efficient catalyst for reactions between PR_3 and O_2. Scheme 18 shows the proposed mechanism.

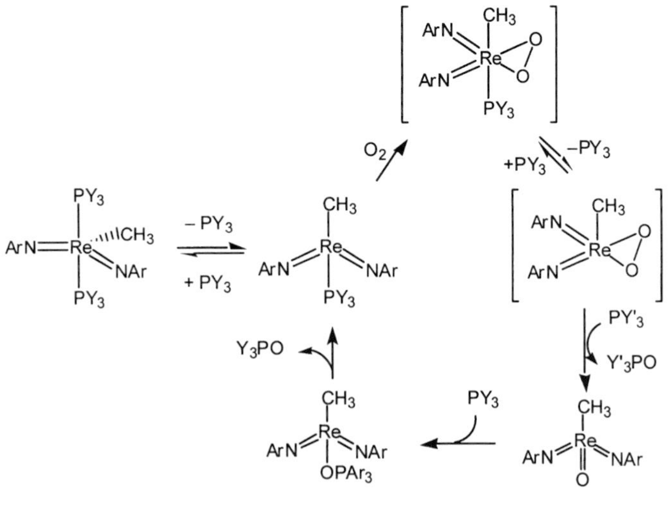

SCHEME 18. Reactions of $MeRe(NAr)_2(PY_3)_2$ with oxygen.

The catalytic cycle can be considered in two parts. First, a rhenium(V) species is generated from $MeRe(NAr)_2O$ as $(MeO)_3PO$ is formed. This step has been studied independently, because **22** has been independently isolated and characterized. A reaction occurs between **22** and PMe_2Ph

$$MeRe(NAr)_2O + 3\,PR'_3 \rightarrow MeRe(NAr)_2(PR'_3)_2\,(\textbf{27}) + R'_3PO \qquad (49)$$

This reaction is first-order with respect to each rate constant, the values of which reflect a combination of steric and electronic effects. The values are summarized in Table III. The values can be interpreted in terms of Giering's QALE procedure (59).

When only 0.5 equiv. of $P(OMe)_3$ was added to $MeRe(NAr)_2O$, the solution changed first to brownish-green and then slowly to orange–red. These analytical and structural data confirmed the formation of $\{MeRe(NAr)_2\}_2O$, **22**, under these conditions,

$$2\,MeRe(NAr)_2O + P(OMe)_3$$
$$\rightarrow \{MeRe(NAr)_2\}_2(\mu\text{-O})\,(\textbf{22}) + (MeO)_3PO \qquad (50)$$

A reaction scheme is presented in Scheme 19, showing the different pathways for the reaction.

TABLE III

RATE CONSTANTS [a] FOR THE REACTIONS OF PHOSPHINES
WITH $MeRe(NAr)_2O$ (**22**)

L	k_{298} $(L\,mol^{-1}\,s^{-1})$
$P(OMe)_3$	0.72(1)
$P(OMe)_2Ph$	72(3)
$P(OMe)Ph_2$	21.7(6)
$P(OEt)_3$	0.57(1)
$P(OEt)_2Ph$	37.0(5)
$P(OEt)Ph_2$	9.4(1)
PMe_3	$1.46(4)\times10^{-3}$
PMe_2Ph	$2.15(2)\times10^{-4}$
PEt_3	$1.01(5)\times10^{-3}$
Dmpe	$1.85(3)\times10^{-3}$
PCy_3	$<5\times10^{-6}$

[a] In C_6H_6 at 298 K; errors represent one standard deviation.

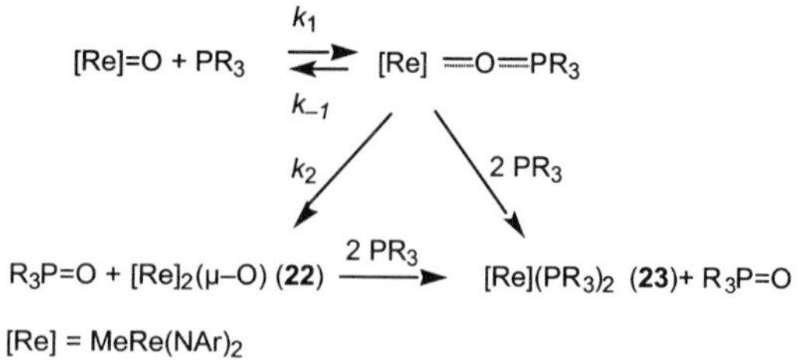

$$[Re]\!=\!O + PR_3 \; \underset{k_{-1}}{\overset{k_1}{\rightleftharpoons}} \; [Re] \!=\!O\!=\!PR_3$$

$$R_3P\!=\!O + [Re]_2(\mu\text{--}O) \; \textbf{(22)} \; \xrightarrow{\;2\,PR_3\;} \; [Re](PR_3)_2 \; \textbf{(23)} + R_3P\!=\!O$$

with k_2 pathway and 2 PR$_3$ pathway branching from the intermediate.

$[Re] = MeRe(NAr)_2$

SCHEME 19. Reactions of MeRe(NAr)$_2$O with phosphines.

X. Imido-rhenium Compounds

A. TRIS(IMIDO)RHENIUM(VII) COMPOUNDS

Three compounds with the formula MeRe(NR)$_3$ have been reported, with R = *tert*-butyl (*60*), 2,6-diisopropylphenyl, **28** (*60*), and 1-adamantyl, **29** (*61*). They are prepared by the reaction:

$$MeReO_3 + 3\,RNCO \rightarrow MeRe(NR)_3 + 3\,CO_2 \qquad (51)$$

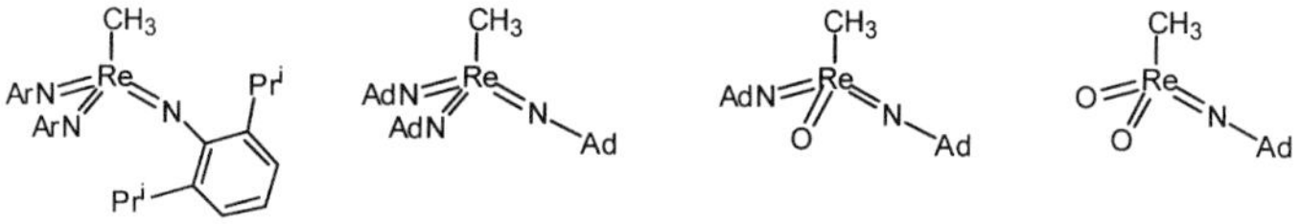

      28   **29**   **30**   **31**

B. PAIRWISE OXO–IMIDO EXCHANGE REACTIONS

Redistribution reactions between **28** and **29** to form mixed-imido compounds were not detected. On the other hand, oxo–imido interchange reactions readily take place between MeReO$_3$ and **29**. As a consequence, new compounds are formed, MeReO(NR)$_2$, **30**, and MeReO$_2$NR, **31**. The product obtained is largely under stoichiometric control, as represented by these equations (*62*):

$$MeReO_3 + 2\,MeRe(NR)_3 \rightarrow 3\,MeReO(NR)_2 \qquad (52)$$

$$2\,MeReO_3 + MeRe(NR)_3 \rightarrow 3\,MeReO_2(NR) \qquad (53)$$

The isolated products are not perfectly pure owing to further redistribution reactions. Several reactions steps in the sequence do not give rise to observable net reaction, e.g.,

$$MeReO(NR)_2 + MeReO_2(NR) \rightarrow MeReO_2(NR) + MeReO(NR)_2 \qquad (54)$$

The three exchange reactions that do lead to net observable chemical change are:

$$MeReO_3 + MeRe(NR)_3 \rightleftharpoons MeReO_2(NR) + MeReO(NR)_2 \quad (55)$$

$$MeReO_2(NR) + MeRe(NR)_3 \rightleftharpoons 2\,MeReO(NR)_2 \quad (56)$$

$$MeReO_3 + MeReO(NR)_2 \rightleftharpoons 2\,MeReO_2(NR) \quad (57)$$

Only two of them are independent, and the following thermodynamic and kinetic relationships are required to be met,

$$K_{55} = K_{56} \times K_{57}; \quad \frac{k_{55}}{k_{-55}} = \frac{k_{56}}{k_{-56}} \times \frac{k_{57}}{k_{-57}} \quad (58)$$

The equilibrium constants were evaluated for R=1-Ad and Ar by integration of NMR signals. Kinetic data were obtained by ^{1}H-NMR and UV/Visible spectroscopies. The results are presented in Table IV.

The tabulation shows that the reaction favors the mixed oxo–imido complex in each case. The rate constants span a large range. The values of k_f for Eqs. (55)–(57) are greater than those of k_r, often by a large margin. Where determined, $\Delta S^{\ddagger}$ has large negative values, -134 and $-175\,J\,K^{-1}\,mol^{-1}$ (62).

A mechanism involving a 4-centered transition state has been well documented in other instances (1). The data in Table IV show that the rate constants decrease with the total number of 2,6-diisopropylimido ligands, suggesting steric crowding in the transition state plays a significant role. The diagram in Scheme 20 shows how that effect is manifest, using a small circle for oxo and a larger one for imido.

TABLE IV

EQUILIBRIUM AND RATE CONSTANTS FOR OXO–IMIDO EXCHANGE BETWEEN Re(VII) COMPOUNDS ^{a}

	K_{298}	$k_{f,298}$ $(L\,mol^{-1}\,s^{-1})$	$k_{r,298}$ $(L\,mol^{-1}\,s^{-1})$
Reaction (R = Ad)			
(55)	$10(1) \times 10^5$	$\sim 10^2$	$\sim 10^{-3}$
(56)	$3.7(3) \times 10^2$	~ 0.2	$\sim 5 \times 10^{-4}$
(57)	$2.60(25) \times 10^2$	$7.4(3) \times 10^{-1}$	$2.8(3) \times 10^{-3}$
Reaction (R = Ar)			
(55)	$1.8(2) \times 10^3$	$1.7(1) \times 10^{-2}$	$9(1) \times 10^{-6}$
(56)	$2.2(2) \times 10^1$	$\sim 2._4 \times 10^{-3}$	$\sim 10^{-4}$
(57)	$8.4(7) \times 10^1$	$2.7(1) \times 10^{-1}$	$3.0(3) \times 10^{-3}$

^{a} In C_6H_6 at 298 K; errors represent one standard deviation.

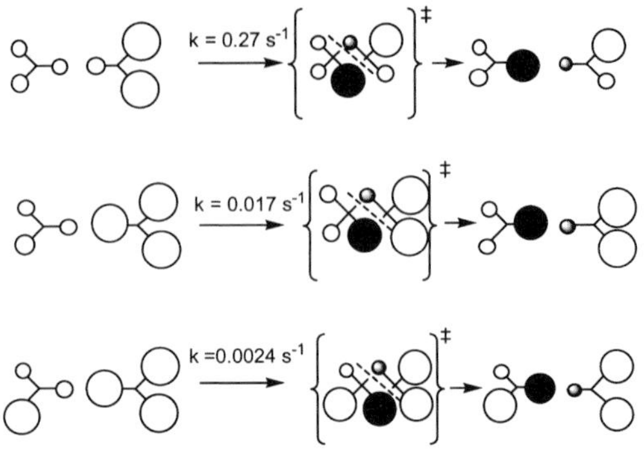

$$k = 0.27 \ s^{-1}$$

$$k = 0.017 \ s^{-1}$$

$$k = 0.0024 \ s^{-1}$$

SCHEME 20. Transition state diagrams for oxo–imido reactions, illustrating steric effects.

C. REACTIONS OF TRIS(IMIDO)RHENIUM(VII) WITH ALDEHYDES

A reaction between **29** and ArCHO occurs,

$$MeRe(NAd)_3 \ (\textbf{29}) + 3\,ArCHO \rightarrow MeReO_3 + 3\,ArCH = NAd \qquad (59)$$

This is a stepwise process during which two mixed oxo–imido compounds were detected, MeRe(O)$_2$NAr, **31**, and MeReO(NAr)$_2$, **32**. These intermediates could be prepared independently, as described in the preceding section. The first stage is,

$$MeRe(NAd)_3 \ (\textbf{29}) + ArCHO$$
$$\rightarrow MeReO(NAd)_2 \ (\textbf{30}) + ArCH = NAd \qquad (60)$$

The rate constants are presented in Table V. It is clear that a change in substituent X for the substrates XC_6H_4CHO gives rise to a substantial kinetic effect. Electron-donating groups lower the rate; the Hammett reaction constant is $\rho = +0.9$ (61).

The mechanism suggested for this reaction is represented in Scheme 21. The involvement of a four-centered transition state is generally accepted for many related (2+2) exchange reactions (1,63).

The conceptual combination of Eqs. (59) and (60) suggests that $MeReO_3$ will catalyze this reaction,

$$ArNCO + ArCHO \rightarrow ArCH = NAd + CO_2 \qquad (61)$$

Indeed, 4-NO$_2$C$_6$H$_4$CH=NAd was formed from these reagents in the presence of MTO in refluxing toluene (61).

TABLE V

RATE CONSTANTS FOR THE REACTION OF
AROMATIC ALDEHYDES WITH $CH_3Re(NAd)_3$ IN
C_6D_6 AT 298 K

4-XC_6H_4CHO X=	k_1 (10^{-4} L mol^{-1} s^{-1})
4-NO_2	7.0 [a]
CHO	5.6 [b]
MeOC(O)	3.2 [b]
H	1.4 [b]
MeO	0.4$_2$ [b]
Me$_2$N	0.2$_8$ [b]

[a] By direct kinetic determinations.
[b] By competition kinetics against the first compound.

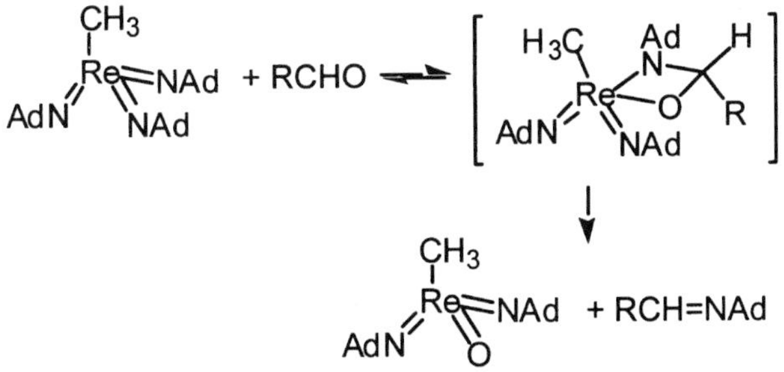

SCHEME 21. Methathesis reactions with aldehydes and MeRe(NAd)$_3$.

D. HYDROLYSIS, HYDROSULFIDOLYSIS AND AMINOLYSIS OF IMIDO-RHENIUM COMPOUNDS

The compound MeRe(NAd)$_3$, **29**, reacts with acidic protons. Water produces AdNH$_2$ and additionally MeReO(NAd)$_2$, **30**, which can undergo further reaction to **31**. In reaction with H$_2$S both **30** and **31** form {MeRe(NR)$_2$}$_2$(μ-S)$_2$. The rate constants for these reactions are summarized in Table VI.

Scheme 22 summarizes the mechanism proposed. As in the case of oxygen exchange between MeReO$_3$ and H$_2$O, the proposed intermediate was not detected. Examples of analogous intermediates that carry out the hydrolysis of imines and the aminolysis of ketones are well documented (*64–66*).

TABLE VI

SECOND-ORDER RATE CONSTANTS ($L\,mol^{-1}\,s^{-1}$) FOR REACTIONS OF
$CH_3Re(NR)_3$ AND $CH_3Re(NR)_2O$ WITH H_2O AND H_2S IN C_6H_6 AT 298 K

	H_2O	H_2S
$CH_3Re(NAd)_3$	3.3 ± 0.8	$17(2)$
$CH_3Re(NAr)_3$	$1.0(2) \times 10^{-4\ a,b}$	$1.6(3) \times 10^{-4\ a}$
$CH_3Re(NAd)_2O$	$9(1) \times 10^{-3}$	$95(10)$
$CH_3Re(NAr)_2O$	$4(1) \times 10^{-6\ a,b}$	$3.3(3) \times 10^{-3\ a,c}$

[a] In CH_3CN.
[b] At 313 K.
[c] $0.28\ L\,mol^{-1}\,s^{-1}$ in the presence of 1.4 mM $ArNH_2$.

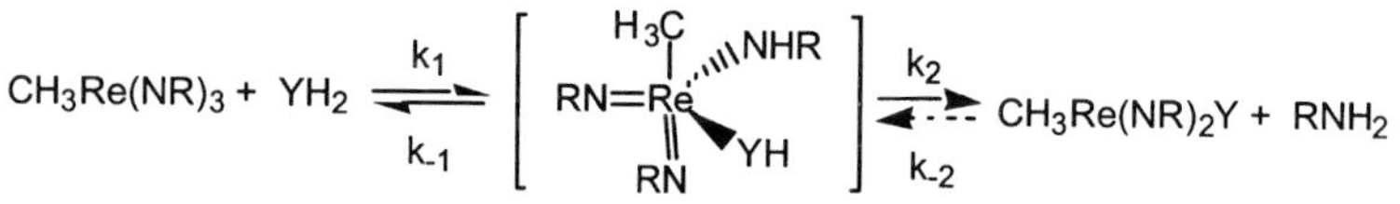

SCHEME 22. Mechanism proposed for hydrolysis (etc.) reactions of MeRe(NAd)$_3$.

XI. Conclusions

The catalytic capabilities of rhenium compounds burst on the scene about one decade ago, featuring $MeReO_3$ as a catalyst for reactions of hydrogen peroxide. It was quickly verified that peroxorhenium(VII) compounds were the active intermediates. With them, practical reactions and fundamental questions of mechanism could then be resolved.

A new generation of oxorhenium compounds has now been prepared. They catalyze oxidation reactions of a different type, and appear to function by a different mechanism. They are oxorhenium(V) compounds that form usually metastable dioxorhenium(VII) intermediates. The mechanisms feature $Re^V(O)$-to-$Re^{VII}(O)_2$ interconversions and catalyze oxygen atom transfer reactions. The mechanisms show a certain diversity as to the steps that enter in a kinetic sense. Yet the schemes presented in this review show a great deal of similarity in their overall mode of action.

REFERENCES

1. Nugent, W. A.; Mayer, J. M. *Metal–Ligand Multiple Bonds*; Wiley-Interscience: New York, 1988.
2. Jacob, J.; Guzei, I. A.; Espenson, J. H. *Inorg. Chem.* **1999**, *38*, 3266–3267.
3. Jacob, J.; Guzei, I. A.; Espenson, J. H. *Inorg. Chem.* **1999**, *38*, 1040–1041.

4. Sheldon, R. A.; Kochi, J. K. *"Metal-Catalyzed Oxidations of Organic Compounds"*, 1981, Chapter 4, p. 81.

5. Herrmann, W. A.; Kratzer, R. M.; Fischer, R. W. *Angew. Chem., Int. Ed. Engl.* **1997**, *36*, 2652–2654.

6. Hille, R. *Chem. Rev.* **1996**, *96*, 2757–2816.

7. Berg, J. M.; Holm, R. H. *J. Am. Chem. Soc.* **1985**, *107*, 917–925.

8. Caradonna, J. P.; Reddy, P. R.; Holm, R. H. *J. Am. Chem. Soc.* **1988**, *110*, 2139–2144.

9. Holm, R. H.; Berg, J. M. *Acc. Chem. Res.* **1986**, *19*, 363–370.

10. Holm, R. H. *Chem. Rev.* **1987**, *87*, 1401–1449.

11. Sung, K.-M.; Holm, R. H. *J. Am. Chem. Soc.* **2001**, *123*, 1931–1943.

12. Baird, D. M.; Aburri, C.; Barron, L. S.; Rodriguez, S. A. *Inorg. Chim. Acta* **1995**, *237*, 117–122.

13. Li, H.; Palanca, P.; Sanz, V.; Lahoz, L. *Inorg. Chim. Acta* **1999**, *285*, 25–30.

14. Laughlin, L. J.; Young, C. G. *Inorg. Chem.* **1996**, *35*, 1050–1058.

15. Smith, P. D.; Slizys, D. A.; George, G. N.; Young, C. G. *J. Am. Chem. Soc.* **2000**, *122*, 2946–2947.

16. Johnson, M. K.; Rees, D. C.; Adams, M. M. W. *Chem. Rev.* **1996**, *96*, 2817.

17. Sung, K.-M.; Holm, R. H. *Inorg. Chem.* **2000**, *39*, 1275–1281.

18. Cai, Y.; Espenson, J. H. Unpublished results.

19. Saha, B.; Espenson, J. H. Unpublished results.

20. Espenson, J. H.; Shan, X.; Lahti, D. W.; Rockey, T. M.; Saha, B.; Ellern, A. *Inorg. Chem.* **2001**, 6717–6724.

21. Konwar, D.; Boruah, R. C.; Sandhu, J. S. *Synthesis* **1990**, 337–339.

22. Ochiai, E. *Aromatic Amine Oxides*; Elsevier: New York, 1967.

23. Rosenau, T.; Potthast, A.; Ebner, G.; Kosma, P. *Synlett* **1999**, *6*, 623.

24. Sim, T. B.; Ahn, J. H.; Yoon, N. M. *Synthesis* **1996**, 324–326.

25. Trost, B. M.; Fleming, L. *"Comprehensive Organic Synthesis"*, vol. 8; 1991; p. 390.

26. Wang, Y.; Espenson, J. H. *Org. Lett.* **2000**, *2*, 3525–3526.

27. Holm, R. H.; Donahue, J. P. *Polyhedron* **1993**, *12*, 571–589.

28. Arias, J.; Newlands, C. R.; Abu-Omar, M. M. *Inorg. Chem.* **2001**, *40*, 2185–2192.

29. Jacob, J.; Espenson, J. H. *Chem. Commun.* **1999**, 1003–1004.

30. Hall, K. J.; Quickenden, T. I.; Watts, D. W. *J. Chem. Educ.* **1976**, *53*, 493.

31. Wang, Y.; Espenson, J. H. *Inorg. Chem.* **2001**, *41*, 2226.

32. Dixon, J.; Espenson, J. H. *Inorg. Chem.* **2002**, *41*, 4727.

33. Espenson, J. H.; Shan, X.; Wang, Y.; Huang, R.; Lahti, D. W.; Dixon, J.; Lente, G.; Ellern, A.; Guzei, I. A. *Inorg. Chem.* **2002**, *41*, 2583–2591.

34. Lente, G.; Jacob, J.; Guzei, I. A.; Espenson, J. H. *Inorg. React. Mech.* **2000**, *2*, 169–177.

35. Lente, G.; Guzei, I. A.; Espenson, J. H. *Inorg. Chem.* **2000**, *39*, 1311–1319.

36. Lente, G.; Espenson, J. H. *Inorg. Chem.* **2000**, *39*, 4809–4814.

37. Jacob, J.; Lente, G.; Guzei, I. A.; Espenson, J. H. *Inorg. Chem.* **1999**, *38*, 3762–3763.

38. Smith, P. D.; Millar, A. J.; Young, C. G.; Ghosh, A.; Basu, P. *J. Am. Chem. Soc.* **2000**, *122*, 9298–9299.

39. Basolo, F.; Pearson, R. G. *Mechanisms of Inorganic Reactions*; Wiley: New York, 1967.

40. Lahti, D. W.; Espenson, J. H. *J. Am. Chem. Soc.* **2001**, *123*, 6014–6024.

41. Shan, X.; Espenson, J. H. Unpublished results.

42. Wang, Y.; Lente, G.; Espenson, J. H. *Inorg. Chem.* **2002**, *41*, 1272–1280.

43. Madesclaire, M. *Tetrahedron* **1986**, *42*, 5459–5495.

44. Nagasawa, H.; Sugihara, Y.; Ishii, A.; Nakayama, J. *Bull. Chem. Soc. Jpn.* **1999**, *72*, 1919–1926.

45. Nakayama, J.; Nagasawa, H.; Sugihara, Y.; Ishii, A. *J. Am. Chem. Soc.* **1997**, *119*, 9077–9078.

46. Nakayama, J. *Bull. Chem. Soc. Jpn.* **2000**, *73*, 1–17.

47. Wang, Y.; Espenson, J. H. *Inorg. Chem* **2002**, *41*, 2266–2274.

48. Koshino, N.; Espenson, J. H. Unpublished results.

49. Wang, W.-D.; Guzei, I. A.; Espenson, J. H. Submitted for publication.

50. Espenson, J. H.; Abu-Omar, M. M. *Adv. Chem. Ser.* **1997**, *253*, 99–134.

51. Espenson, J. H. *Chem. Commun.* **1999**, 479–488.

52. Romão, C. C.; Kühn, F. E.; Herrmann, W. A. *Chem. Rev.* **1997**, *97*, 3197–3246.

53. Simandi, L. I. *Catalytic Activation of Dioxygen by Metal Complexes*; Kluwer Academic Publishers: Dordrecht, 1992.

54. Barton, D. H. R.; Martell, A. E.; Sawyer, D. T. *The Activation of Dioxygen and Homogeneous Catalytic Oxidation*; Plenum: New York, 1993.

55. Foote, C. S.; Valentine, J. S.; Greenberg, A.; Liebman, J. F. In *Active Oxygen in Chemistry*, Eds.; Liebman, J. F., Greenberg, A.; Blackie Academic & Professional: New york, 1995.

56. Huang, R.; Espenson, J. H. *J. Mol. Catal.* **2001**, *168*, 39–46.

57. Huang, R.; Espenson, J. H. *Inorg. Chem.* **2001**, *40*, 994–999.

58. Wang, W.; Espenson, J. H. *Inorg. Chem.* **2001**, *40*, 1323–1328.

59. Wang, W.-D.; Guzei, I. A.; Espenson, J. H. *Organometallics* **2001**, *20*, 148–156.

60. Cook, M. R.; Herrmann, W. A.; Kiprof, P.; Takacs, J. *J. Chem. Soc., Dalton Trans.* **1991**, 797.

61. Wang, W.-D.; Espenson, J. H. *Organometallics* **1999**, *18*, 5170–5175.

62. Wang, W.-D.; Espenson, J. H. *Inorg. Chem.* **2002**, *41*, 1782–1787.

63. Wigley, D. E. *Prog. Inorg. Chem.* **1994**, *42*, 239.

64. Howard, W. A.; Parkin, G. *Organometallics* **1993**, *12*, 2363.

65. Bottomley, F.; Drummond, D. F.; Egharevba, G. O.; White, P. S. *Organometallics* **1986**, *5*, 1620.

66. Bortolin, R.; Patel, V.; Munday, I.; Taylor, N. J.; Carty, A. J. *J. Chem. Soc., Chem. Commun.* **1985**, 456.

REACTION MECHANISMS OF NITRIC OXIDE WITH BIOLOGICALLY RELEVANT METAL CENTERS

PETER C. FORD, LEROY E. LAVERMAN and IVAN M. LORKOVIC

Department of Chemistry and Biochemistry, University of California,
Santa Barbara, CA 93106, USA

I. Introduction

Nitric oxide is important to a wide variety of mammalian physiological processes *(1,2)*, beyond being a constituent of air pollution *(3)*. Natural physiological activities are now known to include roles in blood pressure control, neurotransmission and immune response, and a number of disease states involving NO imbalances have been reported *(2,4)* as the result of extensive research activity into the chemistry, biology and pharmacology of NO. Understanding the fundamental reaction mechanisms

ADVANCES IN INORGANIC CHEMISTRY
VOLUME 54 ISSN 0898-8838

of NO chemistry provides the basis for understanding the physiological functions of this "simple" molecule.

The reactions and interactions of NO with metal centers are of particular interest since transition metal centers such as hemes are well established as targets for NO reactions in mammalian biology. In this context, we present an overview of developments in mechanistic chemistry involving the formation and reactions of selected metal nitrosyl complexes. Such studies provide quantitative understanding of pathways in which NO may participate and allow one to evaluate those which may be the most significant among the multitude of chemical trajectories that must be considered in interpreting biological systems. Notably, nitric oxide complexes have long been of interest to transition metal chemists, and numerous reviews have appeared over the past several decades (5). The present article will not duplicate these previous efforts but will focus on relatively recent mechanistic investigations of metal nitrosyl complexes, especially in the context of possible biomedical roles.

The solubility and transport properties of NO are similar to those of dioxygen (6,7). The aqueous solution solubility of NO is $1.9 \, \text{mM atm}^{-1}$ at 298 K and $1.4 \, \text{mM atm}^{-1}$ at 310 K (6). In organic solvents, the solubility is higher, ranging from $\sim 3 \, \text{mM atm}^{-1}$ in DMSO to $15.0 \, \text{mM atm}^{-1}$ in cyclohexane at 298 K. NO is readily diffusible, and a diffusion constant in solution of $3300 \, \mu\text{m}^2 \text{s}^{-1}$ has been reported under physiological conditions (7). Diffusion of NO in cellular and vascular systems has been modeled quantitatively by Lancaster (7b).

NO is a stable free radical with an electronic structure analogous to the dioxygen cation O_2^+, and this is understandably a dominant theme in its chemistry. It reacts rapidly with other free radicals and with substitution labile, redox active metals, but is not a strong one-electron oxidant or reductant. The character of the NO ligand in a complex with a metal center can range from that of a nitrosyl cation (NO^+), isoelectronic with CO with approximately linear M–N–O bonds, to that of a nitroxyl anion (NO^-) for which a bond angle of $\sim 120°$ is expected. The former case involves considerable charge transfer to the metal center, while in the latter, charge transfer is in the opposite direction. A generalized description of the metal–NO interaction was offered some years ago by Feltham and Enemark (8), who proposed the $\{MNO\}^n$ formulation, where n is the sum of the metal d-electrons and the nitrosyl π^* electrons. Walsh-type diagrams were used to predict the bond angles of this unit. When the other ligands on the metal include a strong C_{4v} perturbation, as is the case with metallo-porphyrins and trans-cyclam metal complexes, the M–N–O angle is predicted to be linear for $n \leq 6$ but bent for $n > 6$. Much less common structures are certain metastable complexes

generated photochemically in low temperature solids that have either an
η^1-NO coordinated at the oxygen or an η^2-NO coordinated with the NO
bond perpendicular to the metal–ligand axis (9,10). In some polynuclear
complexes NO has been noted to bridge two metal ions via the
nitrogen (11).

The redox chemistry of the nitrogen oxides is complex and extremely
medium dependent. For example, the aqueous redox chemistry of NO is
highly sensitive to pH, as both nitrite reduction to NO (2 H^+/e^-) and
NO reduction via HNO to N_2O and H_2O (1 H^+/e^-) (12,13) are proton
coupled reductions (Eqs. (1) and (2)). As a result NO is a relatively
strong reducing agent at high pH, and nitrous acid a strong oxidant at
low pH ($E_{1/2}$ (H^+)$NO_2^-/NO = -0.46$ (+1.00) V vs. NHE, pH 14 (0)), but NO
is easy to reduce at low pH ($E_{1/2}$ $NO/N_2O = +0.76$ V vs. NHE at pH $= 0$)
(Eq. (1)) (14). Furthermore, the prompt products of NO oxidation and
reduction, NO^+ and NO^-, are unstable in aqueous media, and exact
potentials for these one electron steps have been difficult to obtain.

$$NO_2^- + 2H^+ + e^- \longrightarrow NO + H_2O \quad E^\circ = 0.37\,V \text{ (vs. NHE, pH 7)} \quad (1)$$

$$NO + H^+ + e^- \longrightarrow HNO\,(\text{singlet}) \quad E^\circ = -0.55\,V \text{ (vs. NHE, pH 7)} \quad (2)$$

$$NO^+ + e^- \longrightarrow NO \qquad E^\circ = \sim 1.2\,V \text{ (vs. NHE)} \quad (3)$$

Nitrosonium (NO^+) is a strong oxidant and the reduction potential to
NO has been measured in non-aqueous media (1.67 V vs. SCE in
CH_3CN), and estimated for water (Eq. (3)) (12,15). NO^+ is subject to rapid
hydrolysis to nitrite ($2H^+ + NO_2^-$), and therefore if formed in biological
media would be short-lived. However, other less water-sensitive chemical
species can act as NO^+ donors in reactions leading to the nitrosation of
various substrates. For example, the reactions of certain metal nitrosyl
complexes with nucleophiles such as R–SH can lead to the transfer of
NO^+ as illustrated in Eq. (4). Such reactions will be discussed in greater
detail below.

$$R-SH + L_xM-NO \longrightarrow R-SNO + L_xM^- + H^+ \quad (4)$$

As mentioned above, NO can also be reduced to the nitroxyl anion
(NO^-) (Eq. (5)), which is isoelectronic with O_2, and like dioxygen has a
triplet ground state (15).

$$NO + e^- \longrightarrow {}^3NO^- \quad (5)$$

There is increasing interest in possible biological roles of the nitroxyl
anion in both singlet and triplet forms as well as of the conjugate

acid HNO (*16*). The standard potential for NO reduction (Eq. (5)) has been estimated to be -0.33 V vs. NHE (*15*), while other estimates fall in the range -0.5 to -0.8 V vs. NHE (16j). Another recent study (*13*) has concluded that the data used in these estimates were incomplete and that the reduction of NO is even less favorable. This more recent work shows that while $^3NO^-$ ($pK_a = -1.8$) is more stable than $^1NO^-$ ($pK_a = 23$), ^1HNO (-0.15 V, pH $= 0$) is more stable than either ^3HNO or $^3NO^-$ (-0.81 V, pH $= 0$). It can be shown from these data that the effective pK_a for the transformation $^1HNO \rightleftharpoons {}^3NO^- + H^+$ is relatively high (~ 11.3) (*13*), indicating that if an uncoupled nitrogen (*1*) species were formed in biological media, it would be present as ^1HNO. This is an important distinction because $^3NO^-$ is reported to react at a near diffusion-limited rate with $^3O_2$ to form the peroxynitrite ion $ONOO^-$, while the corresponding reaction with ^1HNO is spin-forbidden and slow (*13a*). Over a longer timescale, ^1HNO may dimerize and dehydrate to nitrous oxide (N_2O), or add two more equivalents of NO to give $N_3O_3^-$, which also decomposes to give N_2O and NO_2^-.

II. Formation of Metal Nitrosyl Complexes

When exploring the formation of a metal nitrosyl complex (e.g. Eq. (6)), it is instructive to consider whether or not the free radical nature of NO leads to different reaction mechanisms than seen for other small diatomic ligands such as CO. In many cases, the reactivity pattern for NO appears similar to that observed for other small Lewis bases. This might be rationalized on the basis that, since the odd electron of NO resides in the π^* orbital, it does not become involved until the metal–ligand bond is largely formed. Thus the kinetics of a bimolecular reaction, as illustrated by Eq. (6), are dominated by the lability of the metal complex ML_nX. On the other hand, back reactions of geminate pairs $\{ML_n, AB\}$ formed by flash photolysis of a ML_n–AB complex or M–AB formation from an analogous encounter pair formed by the diffusion of M and AB together, show significant reactivity differences between cases when AB is NO vs. when AB is CO. Furthermore, in the example described immediately below, kinetics data suggest that the radical nature of NO leads to an associative substitution pathway, in at least one case involving a paramagnetic metal ion complex.

$$ML_nX + NO \rightleftharpoons ML_n(NO) + X \qquad (6)$$

A. Ruthenium Ammine Complexes

Until very recently, there has been little systematic study of the reaction mechanism(s) of metal–NO bond formation. An exception was an early study of the kinetics of nitrosylation of the ruthenium(III) complex, $Ru(NH_3)_6^{3+}$, in aqueous solution (Eq. (7)) (17). Armor, Scheidegger and Taube (17a) found the rate for this reaction ($k_{NO} = 0.2\,M^{-1}s^{-1}$ at 298K) to be much faster than the replacement of NH_3 by other ligands. They concluded that the reaction probably proceeds by an associative mechanism, where the paramagnetic d^5 Ru(III) center interacts with the odd electron of NO to give a seven-coordinate intermediate $Ru(NH_3)_6(NO)^{3+}$. The associative mechanism gained further strong support from subsequent activation parameter studies by Pell and Armor (17b), who found a small $\Delta H^{\ddagger}$ ($36\,kJ\,mol^{-1}$) but a large and negative $\Delta S^{\ddagger}$ ($-138\,J\,K^{-1}mol^{-1}$) for the reaction described by Eq. (7) in acidic solution. A recent study by Czap and van Eldik confirmed these activation parameters and an activation volume of $-13.6\,cm^3\,mol^{-1}$ was determined, in support of an associative mechanism (18).

$$Ru(NH_3)_6^{3+} + NO + H^+ \longrightarrow Ru(NH_3)_5(NO)^{3+} + NH_4^+ \qquad (7)$$

Pell and Armor found entirely different products in alkaline solution. Above pH 8.3, the sole ruthenium product of the reaction of $Ru(NH_3)_6^{3+}$ with NO was the dinitrogen complex $Ru(NH_3)_5(N_2)^{2+}$. Under these conditions the rate law proved to be first-order in $[Ru(NH_3)_6^{3+}]$, [NO] and [OH⁻]. A likely mechanism is the reversible reaction of $Ru(NH_3)_6^{3+}$ with OH⁻ to give the intermediate $Ru(NH_3)_5(NH_2)^{2+}$, followed by electrophilic NO attack at the amide ligand and release of water. However, the kinetic evidence does not exclude other sequences.

B. Ruthenium(III) Salen Complexes

Several Ru(III) salen complexes of the type $Ru^{III}(salen)(X)(NO)$ (X = Cl⁻, ONO⁻, H_2O; salen = N,N'-bis(salicylidene)-ethylenediamine dianion) have been examined as possible photochemical NO precursors (19). Photo-excitation of the $Ru^{III}(salen)(NO)(X)$ complex labilizes NO to form the respective solvento species $Ru^{III}(salen)(X)(Sol)$. The kinetics of the subsequent back reactions to reform the nitrosyl complexes (e.g. Eq. (8)) were studied as a function of the nature of the solvent (Sol) and reaction conditions. The reaction rates are dramatically dependent on the identity of Sol, with values of k_{NO} (298 K, X = Cl⁻) varying from $5 \times 10^{-4}\,M^{-1}s^{-1}$ in acetonitrile to $4 \times 10^7\,M^{-1}s^{-1}$ in toluene, a much weaker electron donor. In this case, Ru^{III}–Sol bond breaking clearly plays an important role in the rate-limiting step for NO substitution of Sol.

$$
\text{(structure with Sol ligand)} \quad +NO \longrightarrow \quad \text{(structure with NO ligand)} \quad +Sol \qquad (8)
$$

Activation parameters were also measured for the "on" reaction for two Ru^{III}(salen) complexes in toluene. Temperature dependent kinetic studies yielded values for $\Delta H^{\ddagger}$ of 34 and 20 kJ mol^{-1} and $\Delta S^{\ddagger}$ of +10 and -46 J mol^{-1} K^{-1} for Ru(^{t}Bu$_4$salen) and Ru(^{t}Bu$_4$salophen), respectively (^{t}Bu$_4$salen = N,N'-ethylenebis(3,5-di-t-butylsalicylideneiminato) dianion, ^{t}Bu$_4$salophen = N,N'-1,2-phenylenediaminebis(3-t-butylsalicylideneimi-nato) dianion). Kinetics for the "on" reaction were also explored as a function of hydrostatic pressure yielding $\Delta V^{\ddagger}$ values of +22 and +16 cm^3 mol^{-1} for Ru(^{t}Bu$_4$salen) and Ru(^{t}Bu$_4$salophen), respectively. These observations are consistent with k_{on} rate constants that vary over 12 orders of magnitude depending upon the nature of the leaving group (Sol in Eq. (8)). Reactions with NO in highly coordinating solvents occur at substantially slower rates than in weakly coordinating solvents. The large and positive values for the activation volumes are indicative of a dissociative ligand exchange mechanism in which a solvent ligand must first dissociate from the ruthenium center prior to reaction with NO. However, it should be noted that these positive $\Delta V^{\ddagger}$ values were measured for NO displacement of a labile ligand, toluene. It remains to be seen whether this mechanistic argument holds true for a much less labile ligand such as CH_3CN.

C. METALLOPORPHYRINS

Ligand substitution reactions of NO leading to metal–nitrosyl bond formation were first quantitatively studied for metalloporphyrins, (M(Por)), and heme proteins a few decades ago (20), and have been the subject of a recent review (20d). Despite the large volume of work, systematic mechanistic studies have been limited. As with the Ru^{III}(salen) complexes discussed above, photoexcitation of metalloporphyrin nitrosyls results in labilization of NO. In such studies, laser flash photolysis is used to labilize NO from a M(Por)(NO) precursor, and subsequent relaxation of the non-steady state system back to equilibrium (Eq. (9)) is monitored spectroscopically.

$$
M(Por) + NO \underset{k_{off}}{\overset{k_{on}}{\rightleftharpoons}} M(Por)(NO) \qquad (9)
$$

In the presence of excess NO, the observed transient spectra are expected to decay exponentially with an observed rate constant (k_{obs}) equal to:

$$k_{obs} = k_{on}[NO] + k_{off} \tag{10}$$

Consequently, plots of k_{obs} vs. [NO] should be linear with a slope equal to k_{on} and an intercept equal to k_{off} as illustrated in Fig. 1 for the relaxation kinetics for reaction of the iron(III) heme protein met-myoglobin (metMb), in the presence of excess NO (21–23). In this case, thermal ligand dissociation (k_{off}) is sufficiently fast relative to k_{on} to allow determination of an accurate value of the intercept; thus the ratio k_{on}/k_{off} gives a reasonable value for the equilibrium constant (K) for nitrosyl complex formation. However, for many systems, especially those of ferro-heme complexes and certain proteins, the "off" reaction is too slow to give acceptably accurate intercepts, and it is difficult to determine either k_{off} or K from these plots. In certain cases, rates of NO labilization have been determined directly by following the thermal disappearance of the nitrosyl complex in the presence of an efficient trapping agent for free NO, an example being the Ru(III) complex Ru(edta)$^-$ (23).

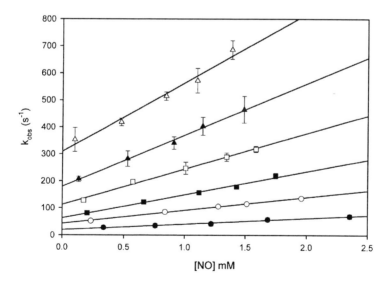

FIG. 1. Plot of k_{obs} vs. [NO] for the reaction of metMb with NO as measured by laser flash photolysis at different temperatures in pH 7.0 phosphate buffer solution (50 mM): 15 °C (filled circles), 25 °C (open circles), 30 °C (filled squares), 35 °C (open squares), 40 °C (filled triangles), and 45 °C (open triangles) (21c).

D. KINETICS OF NO REACTIONS WITH HEME PROTEINS

The kinetics of reactions of NO with ferri- and ferro-heme proteins and models under ambient conditions have been studied by time-resolved spectroscopic techniques. Representative results are summarized in Table I (*22–28*). Equilibrium constants determined for the formation of nitrosyl complexes of met-myoglobin (metMb), ferri-cytochrome-*c* (Cyt^{III}) and catalase (Cat) are in reasonable agreement when measured both by flash photolysis techniques ($K = k_{on}/k_{off}$) and by spectroscopic titration in aqueous media (*22*). Table I summarizes the several orders of magnitude range of k_{on} and k_{off} values obtained for ferri- and ferro-heme proteins. Many k_{off} values were too small to determine by flash photolysis methods and were determined by other means. The small values of k_{off} result in very large equilibrium constants K for the

TABLE I

RATE CONSTANTS k_{on} AND k_{off} FOR NITROSYLATIONS OF REPRESENTATIVE FERRO- AND FERRI-HEME PROTEINS

Conditions [a]		k_{on} (M^{-1}s^{-1})	k_{off} (s^{-1})	Reference
Ferric proteins [a]				
metMb [b]	H$_2$O, pH 6.5	1.9×10^5	13.6	(*22*)
metMb [c]	50 mM phosphate, pH 7.0, 298 K	4.8×10^4	43	(*23*)
Cyt^{III}	H$_2$O, pH 6.5, 293 K	7.2×10^2	4.4×10^{-2}	(*22*)
Cat	H$_2$O, pH 6.5, 293 K	3.0×10^7	1.7×10^2	(*23*)
Enos	283 K, 1 mM arginine	8.2×10^5	70	(*24*)
nNOS	pH 7.8, 293 K, heme domain	2.1×10^7	40	(*24*)
NP [d]	298 K	$1.5–2.2 \times 10^6$	0.006–2.2	(*25*)
MPO	pH 7.0, 283 K	1.07×10^6	10.8	(*29*)
Ferrous proteins [a]				
Hb$_4^{T}$ [e]	pH 7.0, 293 K	2.6×10^7	3.0×10^{-3}	(*26*)
Hb$_4^{R}$ [e]	pH 7.0, 293 K	2.6×10^7	1.5×10^{-4}	(*26*)
sGC	pH 7.4, 293 K	1.4×10^8	6.8×10^{-4}	(*27*)
sGC	pH 7.4, 293 K, 3 mM Mg^{2+}, 0.5 mM GTP	–	5.0×10^{-2}	(*27*)
Mb	phosphate buffer pH 7.0, 293 K	1.7×10^7	1.2×10^{-4}	(*26*)
Cyt^{II}	H$_2$O, pH 6.5	8.3	2.9×10^{-5}	(*22*)
eNOS	283 K, 1 mM arginine	1.1×10^6	70	(*24*)
nNOS	pH 7.8, 293 K, heme domain	1.1×10^7	~ 0	(*24*)
MPO	pH 7.0, 283 K	1.0×10^5	4.6	(*29*)

[a] Abbreviations listed in Section VI.

[b] Sperm whale skeletal metMb.

[c] Horse heart metMb.

[d] Rate constants are the range for NP1, NP2, NP3 and NP4, pH 5.0 and pH 8.0, the k_{off} displays two phases.

[e] Two phases are observed for NO binding.

ferrous-hemes with the exception of Cyt^{II}, which also displays a very small k_{on} value.

The low reactivity of both Cyt^{III} and Cyt^{II} toward NO can be attributed to occupation of the heme iron axial coordination sites by an imidazole nitrogen and by a methionine sulfur of the protein (28). Thus, unlike other heme proteins where one axial site is empty or occupied by H_2O, formation of the nitrosyl complex not only involves ligand displacement but also significant protein conformational changes which inhibit the reaction with NO. However, the protein does not always inhibit reactivity given that Cat and nNOS are more reactive toward NO than is the model complex $Fe^{III}(TPPS)(H_2O)_2$ (Table II). Conversely, the k_{off} values

TABLE II

THE RATE CONSTANTS (298 K) FOR THE "ON" AND "OFF" REACTIONS OF NO AND CO WITH CERTAIN WATER SOLUBLE METAL COMPLEXES IN AQUEOUS SOLUTION

"on" Reactions [a]	k_{on} $(M^{-1}s^{-1})$	$\Delta H^{\ddagger}$ $(kJ\,mol^{-1})$	$\Delta S^{\ddagger}$ $(J\,mol^{-1}\,K^{-1})$	$\Delta V^{\ddagger}$ $(cm^3\,mol^{-1})$	Reference
$Fe^{III}(TPPS)+NO$	4.5×10^5	69 ± 3	95 ± 10	9 ± 1	(21c)
$Fe^{III}(TMPS)+NO$	2.8×10^6	57 ± 3	69 ± 11	13 ± 1	(21c)
metMb+NO	4.8×10^4	63 ± 2	55 ± 8	20 ± 6	(23)
$Fe^{II}(H_2O)_6+NO$	1.4×10^6	37 ± 0.5	-3 ± 2	6.1 ± 0.4	(30)
$Fe^{II}(Hedtra)+NO$	6.1×10^7	26 ± 1	-12 ± 3	2.8 ± 0.1	(31b)
$Fe^{II}(edta)+NO$	2.4×10^8	24 ± 1	-4 ± 3	4.1 ± 0.2	(31b)
$Fe^{II}(nta)+NO$	2.1×10^7	24 ± 1	-22 ± 3	-1.5 ± 0.1	(31b)
$Fe^{II}(TPPS)+NO$	1.5×10^9	24 ± 3	12 ± 10	5 ± 1	(21c)
$Fe^{II}(TMPS)+NO$	1.0×10^9	26 ± 6	16 ± 21	2 ± 2	(21c)
$Fe^{II}(TPPS)+CO$	3.6×10^7	11 ± 6	-64 ± 2	-6.6 ± 0.6	(21c)
$Fe^{II}(TMPS)+CO$	6.0×10^7	31 ± 4	6 ± 13	-4.0 ± 0.7 [b]	(21c)
$Co^{II}(TPPS)+NO$	1.9×10^9	28 ± 2	26 ± 7	[b]	(21c)

"off" Reactions	k_{off} (s^{-1})	$\Delta H^{\ddagger}$ $(kJ\,mol^{-1})$	$\Delta S^{\ddagger}$ $(J\,mol^{-1}\,K^{-1})$	$\Delta V^{\ddagger}$ $(cm^3\,mol^{-1})$	
$Fe^{III}(TPPS)(NO)$	0.5×10^3	76 ± 6	60 ± 11	18 ± 2	(21c)
$Fe^{III}(TMPS)(NO)$	0.9×10^3	84 ± 3	94 ± 10	17 ± 3	(21c)
metMb(NO)	42	68 ± 4	14 ± 13	18 ± 3	(23)
$Fe^{II}(H_2O)_5(NO^-)$	3.2×10^3	48 ± 1	-15 ± 5	1.3 ± 0.2	(30)
$Fe^{II}(Hedtra)(NO)$	4.2	73 ± 1	11 ± 4	4.4 ± 0.8	(31b)
$Fe^{II}(edta)(NO)$	91	61 ± 2	-5 ± 7	7.6 ± 0.6 [b]	(31b)
$Fe^{II}(nta)(NO)$	9.3	66 ± 1	-5 ± 4	-3.5 ± 0.7	(31b)
$Fe^{II}(TPPS)(NO)$	6.4×10^{-4}	[b]	[b]	[b]	(21c)
$Co^{II}(TPPS)(NO)$	1.5×10^{-4}	[b]	[b]	[b]	(21c)

[a] Abbreviations given in Section VI.
[b] Not determined.

for metMb, Cyt^{III}, and Cat are all smaller than for $Fe^{III}(TPPS)$ suggesting retardation of NO dissociation from the iron nitrosyl complex by these proteins.

E. Mechanism Studies with Iron Porphyrin Systems

Laverman and coworkers have reported activation parameters for the aqueous solution reactions of NO with the iron(II) and iron(III) complexes of the water soluble porphyrins TPPS and TMPS (21). These studies involved systematic measurements to determine k_{on} and k_{off} as functions of temperature (298–318 K) and hydrostatic pressure (0.1–250 MPa) to determine values of $\Delta H^{\ddagger}$, $\Delta S^{\ddagger}$ and $\Delta V^{\ddagger}$ for the "on" and "off" reactions of the ferri-heme models and for the "on" reactions of the ferro-heme models (Table II). Figure 2 illustrates hydrostatic pressure effects on k_{on} and k_{off} for $Fe^{III}(TPPS)$.

The large and positive $\Delta S^{\ddagger}$ values and, particularly the large and positive $\Delta V^{\ddagger}$ values obtained (21) for k_{on} and k_{off} represent signatures for a substitution mechanism dominated by ligand dissociation, for the ferri-heme complexes, i.e.,

$$Fe^{III}(Por)(H_2O)_2 \underset{k_{-1}}{\overset{k_1}{\rightleftharpoons}} Fe^{III}(Por)(H_2O) + H_2O \qquad (11)$$

$$Fe^{III}(Por)(H_2O) + NO \underset{k_{-2}}{\overset{k_2}{\rightleftharpoons}} Fe^{III}(Por)(H_2O)(NO) \qquad (12)$$

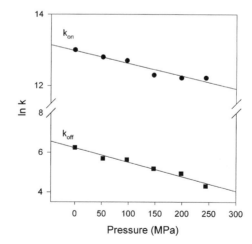

Fig. 2. Plots of $\ln(k_{on})$ (circles) and $\ln(k_{off})$ (squares) vs. hydrostatic pressure to determine activation volume values $\Delta V^{\ddagger}_{on}$ and $\Delta V^{\ddagger}_{off}$ for the reaction of NO with $Fe^{III}(TPPS)$ in aqueous solution at 298 K (21a).

The H_2O exchange mechanism was studied by Hunt et al. (32) who reported that exchange between aqueous solvent and $Fe^{III}(TPPS)(H_2O)_2$ occurs with a first-order rate constant ($k_{ex} = 1.4 \times 10^7 \, s^{-1}$ in water at 298 K) far exceeding the k_{obs} values determined at any [NO]. If the steady state approximation was applied with regard to the intermediate $Fe^{III}(Por)(H_2O)$, the k_{obs} for the exponential relaxation of the non-equilibrium mixture generated by the flash photolysis experiment would be,

$$k_{obs} = \frac{k_1 k_2 [NO] + k_{-1} k_{-2} [H_2O]}{k_{-1}[H_2O] + k_2[NO]} \tag{13}$$

It may be assumed that $k_{-1}[H_2O] \gg k_2[NO]$ since both steps involve trapping of an unsaturated metal center and $[H_2O] \gg [NO]$. Accordingly, $k_{on} = k_1 k_2 / k_{-1}[H_2O]$ and $k_{off} = k_{-2}$. In this context, the apparent activation parameters for k_{on} would be:

$$\Delta S_{on}^{\ddagger} = \Delta S_1^{\ddagger} + \Delta S_2^{\ddagger} - \Delta S_{-1}^{\ddagger} \quad \text{and} \quad \Delta V_{on}^{\ddagger} = \Delta V_1^{\ddagger} + \Delta V_2^{\ddagger} - \Delta V_{-1}^{\ddagger} \tag{14}$$

The k_2 and the k_{-1} steps represent similar (very fast) reactions of the unsaturated intermediate $Fe^{III}(Por)(H_2O)$ with an incoming ligand (NO and H_2O, respectively), so the differences in their activation parameters (e.g. $\Delta S_2^{\ddagger} - \Delta S_{-1}^{\ddagger}$ and $\Delta V_2^{\ddagger} - \Delta V_{-1}^{\ddagger}$) should be small. In such a case the principal contributor to $\Delta S_{on}^{\ddagger}$ would be $\Delta S_1^{\ddagger}$, the activation entropy for the H_2O dissociative step. The k_1 step should thus display a positive $\Delta H_1^{\ddagger}$ consistent with the energy necessary to break the $Fe^{III}-OH_2$ bond, a large, positive $\Delta S_1^{\ddagger}$ owing to formation of two species from one, and a substantially positive $\Delta V_1^{\ddagger}$ for the same reason. These conditions are met for the k_{on} activation parameters for both model complexes (Table II). Furthermore, the values measured by Hunt et al. for $\Delta H_{ex}^{\ddagger}$ ($57 \, kJ \, mol^{-1}$) and $\Delta S_{ex}^{\ddagger}$ ($84 \, J \, K^{-1} \, mol^{-1}$) for the H_2O exchange on $Fe^{III}(TPPS)(H_2O)_2$ (32) are very similar to the respective activation parameters relating to k_{on} obtained for the reaction with NO. The H_2O exchange was reexamined in a recent study by van Eldik et al. (33) using variable temperature/pressure NMR techniques. These workers reported $\Delta H_{ex}^{\ddagger} = 67 \, kJ \, mol^{-1}$ and $\Delta S_{ex}^{\ddagger} = 99 \, J \, mol^{-1} \, K^{-1}$ and $\Delta V_{ex}^{\ddagger} = 7.9 \, cm^3 \, mol^{-1}$, for $Fe^{III}(TPPS)(H_2O)_2$, which are in even better agreement with those measured by flash photolysis for the k_{on} pathway with NO (21). Thus, the factors that determine the exchange kinetics for $Fe^{III}(TPPS)(H_2O)_2$ with solvent H_2O dominate the reaction of NO with the same species. The activation parameters relevant to k_{on} for this iron(III) heme model are largely defined by a dissociative mechanism, the limiting step being k_1 in Eq. (11).

Based on the principle of microscopic reversibility one may conclude that the intermediate(s) in the "off" step will be the same as those generated during the k_{on} pathway, thus iron–nitrosyl bond breakage (k_{-2})

would be the energetically dominant step. Coordination of NO to $Fe^{III}(Por)$ is accompanied by considerable charge transfer to give a linearly bonded, diamagnetic complex that can be formally represented as $Fe^{II}(Por)-(NO^+)$. Thus, the activation parameters of the "off" reaction must reflect the intrinsic entropy and volume changes associated with the spin change and solvent reorganization as the charge localizes on the metal. The large positive $\Delta V_{ex}^{\ddagger}$ values for $Fe^{III}(TPPS)(H_2O)(NO)$ are consistent with the limiting dissociative mechanism as outlined in Eqs. (11) and (12).

The specific solvation of NO coordinated to Fe(III) and the resulting solvent reorganization upon NO dissociation (Fig. 3) finds some analogy with the nitrophorins, which are heme protein systems for NO transfer found in certain blood sucking insects. The crystal structure of one nitrophorin, NP4, shows that binding of NO to the Fe(III) center leads to a collapse of the protein around the coordinated NO. The distal heme-binding pocket in nitrophorin NP4 is quite open to solvent in the absence of NO. It was postulated that collapse of the protein around the heme nitrosyl led to increased retention of bound NO at low pH (25).

Activation parameters for the reaction of NO with metMb, Eq. (15), were determined in this laboratory and in collaboration with van Eldik and Stochel (Table II) (23). Comparison of these activation parameters with those determined for reactions of NO with the water soluble ferri-heme complexes $Fe^{III}(TPPS)(H_2O)_2$ and $Fe^{III}(TMPS)(H_2O)_2$ (Table II) demonstrate that the latter compounds represent reasonable models for the kinetics for the analogous reaction with metMb. For example, the k_{on} step would appear to be defined largely by the H_2O lability of $metMb(H_2O)$, although it is clear that the diffusion through protein channels, the distal residues and the proximal histidine binding to the Fe(III) center must all influence the NO binding kinetics (23,24). These properties may indeed be reflected in the lower $\Delta S^{\ddagger}$ values for both the "on" and "off" reactions on metMb. In a related study, Cao et al. recently

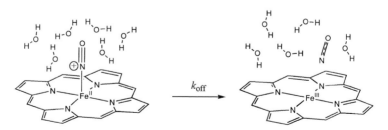

FIG. 3. Schematic representation of the solvent reorganization that may occur upon the dissociation of nitric oxide from $Fe^{III}(Por)$ in aqueous solution.

reported the observation of the 5-coordinate intermediate metMb generated by flash photolysis of metMb(NO) and trapping of that species by H_2O (*34b*). They further showed that the k_{on} step is several orders of magnitude faster for the metMb mutant H64G than for wild type horse heart metMb. Since in the H64G mutant a glycine is substituted for the distal histidine, the greater reactivity of the mutant was interpreted in terms of hydrogen bonding from His-64 stabilizing the coordinated H_2O of the wild type protein. One might expect such stabilization to be reflected by a higher $\Delta H_{on}^{\ddagger}$ value for the wild type protein, but activation parameters were not reported for the mutant protein.

Ferri-heme proteins for which k_{on} and k_{off} values have been reported include two forms of nitric oxide synthase eNOS and nNOS as well as several forms of nitrophorin (Table I, Eq. (15)).

$$\text{metMb(H}_2\text{O)} + \text{NO} \underset{k_{off}}{\overset{k_{on}}{\rightleftharpoons}} \text{metMb(NO)} + \text{H}_2\text{O} \qquad (15)$$

Hoshino *et al.* (*35*) have reported rates for the nitrosylation of model compounds M(Por) (M = Fe(II), Co(II), and Mn(II), Por = TPP or OEP) in toluene solutions by flash photolysis of $M^{II}(Por)(NO)$ in the absence of excess NO. The bimolecular rate constants, k_{on}, were obtained directly from analysis of the decay of the transient at various temperatures to give $\Delta H^{\ddagger}$ and $\Delta S^{\ddagger}$ values (*35,36*). The general pattern observed was that all the "on" rates are fast, the activation enthalpies are quite small and activation entropies are negative. This pattern is consistent with bimolecular NO trapping by a M(Por) species that is, at most, weakly coordinated by the toluene solvent. The trapping rates follow the order $Mn^{II}(Por) < Fe^{II}(Por) < Co^{II}(Por)$, and it was noted that, since nitrosylation changes the spin state (S = $\sum m_s$ from 5/2, 2 and 1/2, respectively, to 0, 1/2 and 0) the slower rates were observed for those complexes showing the largest reorganization of spin multiplicity.

The water-soluble ferrous complexes $Fe^{II}(TPPS)$ and $Fe^{II}(TMPS)$, react with NO about three orders of magnitude faster than for the iron(III) analogs (Table II). Correspondingly, the activation parameters show much lower values of $\Delta H_{on}^{\ddagger}$ and $\Delta S_{on}^{\ddagger}$. The magnitude of the latter is consistent with rates largely defined by diffusional factors, although the k_{on} values reported are nearly an order of magnitude less than diffusion-limited rate constants in water. High spin $Fe^{II}(Por)$ complexes are considerably more labile than the $Fe^{III}(Por)$ analogs not only for the model compounds but also for most heme proteins (Table I). Since the ferro-heme center may be 5-coordinate in such cases, formation of a metal–NO bond does not require initial displacement of another ligand, and thus is not limited by the rate of ligand labilization.

Earlier kinetics studies (20c) of ferro-heme proteins and model compounds have led to the proposal of a mechanism in which an encounter complex, $\{Fe^{II}(Por)\|L\}$, is formed prior to ligand bond formation according to Eq. (16).

$$Fe^{II}(Por) + L \underset{k_{-D}}{\overset{k_D}{\rightleftharpoons}} \{Fe^{II}(Por)\|L\} \xrightarrow{k_a} Fe^{II}(Por)L \qquad (16)$$

In this equation, k_D is the rate constant for the diffusion-limited formation of the encounter complex, k_{-D} is the rate constant for diffusion apart, and k_a is that for the "activation" step, i.e. M–L bond formation. Based on the steady-state approximation for the encounter complex concentration, the apparent rate constant for the "on" reaction is $k_{on} = k_D k_a / (k_{-D} + k_a)$, and the activation volume is defined as

$$\Delta V_{on}^{\ddagger} = \Delta V_D^{\ddagger} + \Delta V_a^{\ddagger} - RT\left(\frac{d\ln(k_a + k_{-D})}{dP}\right) \qquad (17)$$

There are two limiting cases in this model, one in which the reaction is diffusion limited ($k_a \gg k_{-D}$), the other in which the reaction is activation limited ($k_{-D} \gg k_a$). In the activation limited process, Eq. (17) simplifies to

$$\Delta V_{on}^{\ddagger} = \Delta V_D^{\ddagger} + \Delta V_a^{\ddagger} - \Delta V_{-D}^{\ddagger} \qquad (18)$$

where $\Delta V_D^{\ddagger} - \Delta V_{-D}^{\ddagger}$ is the volume difference between the encounter complex and the solvent separated species. Although unknown, this difference is likely to be small for a neutral ligand such as NO, since the encounter complex does not involve the formation or breakage of bonds and should have only modest impact on solvation. The dominant term would be $\Delta V_a^{\ddagger}$ which should be negative owing to the formation of a Fe^{II}–L bond and the concomitant change of the spin state from high (quintet $Fe^{II}(Por)$ plus doublet NO) to low spin (doublet $Fe^{II}(Por)(NO)$).

For the condition $k_a \gg k_{-D}$, the reaction would become diffusion-limited and Eq. (17) reduces to $\Delta V_{on}^{\ddagger} = \Delta V_D^{\ddagger}$. Activation volumes for diffusion in various solvents are positive owing to viscosity increases at higher pressure (+7.5, +9.5 and +0.8 $cm^3\,mol^{-1}$ in CH_3CN, C_2H_5OH, and H_2O, respectively) (38). For $Fe^{II}(TPPS)(H_2O)_2$ and $Fe^{II}(TMPS)(H_2O)_2$, the positive $\Delta V_{on}^{\ddagger}$ values are somewhat larger than expected for a diffusion-limited process in aqueous solution but are significantly smaller than those measured for the iron(III) analogs. The relevant kinetic data from high pressure experiments are consistent with a process having a k_{on} value within an order of magnitude of the diffusion limit in water ($k_D \sim 10^{10}\,M^{-1}\,s^{-1}$ at 298 K) (39). If a similar analysis were made with respect to $\Delta S_{on}^{\ddagger}$, then $\Delta S_{on}^{\ddagger} = \Delta S_D^{\ddagger}$ in the diffusion limited case. The activation entropy for diffusion in aqueous solution can be calculated as $\sim 34\,\mathrm{J\,mol^{-1}\,K^{-1}}$ (40); thus the measured $\Delta S_{on}^{\ddagger}$ values for $Fe^{II}(TPPS)$ $(H_2O)_2$ and $Fe^{II}(TMPS)(H_2O)_2$ (12 ± 10 and $16 \pm 21\,\mathrm{J\,mol^{-1}\,K^{-1}}$, respectively) are not entirely inconsistent with a process limited by diffusion. Similar arguments can be made for the aqueous solution reaction of NO with $Co^{II}(TPPS)$ (21).

In order for NO to act as an intracellular signaling agent at sub-micromolar concentrations, it must be generated near the target, and the reactions with ferro-hemes must be very fast to compete with other chemical and physiological processes leading to NO depletion. The above study is consistent with the intuitive notion that the fast reactions of ferro-heme proteins with NO are due to a vacant or exceedingly labile coordination site.

A model parallel to that described by Eq. (16) applies to the analogous reactions with CO. The second-order rate constant for the reaction of $Fe^{II}(TPPS)(H_2O)_2$ with CO is several orders of magnitude below the diffusion limit. As a consequence, the rate of this reaction must be activation controlled. In contrast to the reaction with NO, the $\Delta V_{on}^{\ddagger}$ values for CO are negative. These results parallel other studies of ferro-heme complexes that found reaction with NO to be diffusion-limited while reaction with CO is activation limited. This model was confirmed by a study of the reaction of CO with $Fe^{II}(MCPH)$ (MCPH = monochelated protoheme, or protohemin 3-(1-imadazoyl) propylamide stearyl ester) in toluene/mineral oil solutions. By exploiting pressure effects it was possible to tune the reaction mechanism from an activation-limited process to a diffusion-limited process (37). The viscosity of this solvent mixture is very sensitive to applied pressure and pressure increases led to slower limiting diffusion rates (i.e. smaller values of k_D and k_{-D}) to the point where diffusion becomes rate-limiting at higher pressures.

Flash photolysis techniques were unsuitable for measuring the slow "off" reactions for the iron(II) model complexes such as $Fe^{II}(TPPS)(NO)$, since the experimental uncertainties in the extrapolated intercepts of k_{obs} vs. [NO] plots were larger than the values of the intercepts themselves. When trapping methods were used to evaluate NO labilization from $Fe^{II}(TPPS)(NO)$, k_{off} values were found to be quite small but were sensitive to the nature of the trapping agents used. Lewis bases that could coordinate the metal center appeared to accelerate NO loss. More reliable estimates for the uncatalyzed "off" reaction were obtained by using $Ru(edta)^-$ as an NO scavenger, and the k_{off} values listed in Table I were obtained in this manner (21c). The small k_{off} values found for Fe(II) models are consistent with the trend observed for the ferro-heme proteins discussed above.

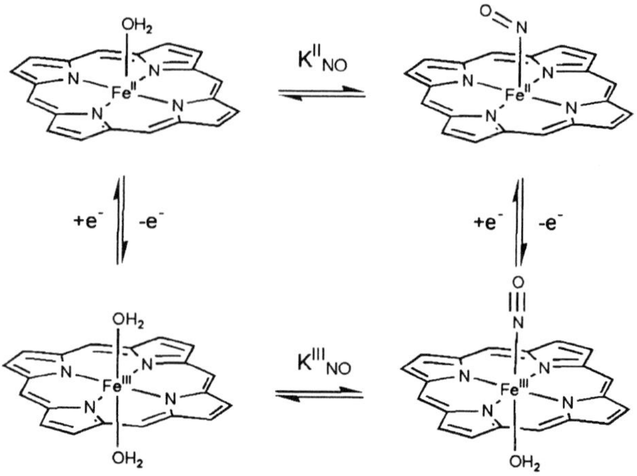

SCHEME 1. Porphyrin ligand substituents have been omitted for clarity.

Another method for estimating k_{off} is shown in Scheme 1. This involves using a Born-Haber type cycle to calculate the equilibrium constant K_{NO}^{II} for formation of $Fe^{II}(TPPS)(NO)$ from the value of K_{NO}^{III} $(1.1 \pm 0.1 \times 10^3 \, M^{-1})$ (21) and the reduction potentials for $Fe^{III}(TPPS)(NO)$ (+0.35 V vs. SCE) and $Fe^{III}(TPPS)$ (−0.23 V vs. SCE) in aqueous solution (41). The value $k_{off}^{II} = 2 \times 10^{-4} \, s^{-1}$ was estimated using the expression in Eq. (19) below, where $k_{on}^{II} = 1.5 \times 10^9 \, M^{-1} \, s^{-1}$. This is about threefold smaller than the value measured by NO scavenging, but given the uncertainties in the electrochemical values used in the estimate, the agreement is quite reasonable (21c).

$$k_{off}^{II} = \frac{k_{on}^{II}}{K_{NO}^{III} \, e^{nF\Delta E/RT}} \tag{19}$$

Kinetics and activation parameters for NO reactions with a series of iron(II) aminocarboxylato complexes have been obtained (Table II) in aqueous solution (31). Rate constants for these reactions ranged from 10^5 to $10^8 \, M^{-1} \, s^{-1}$ for the series of iron(II) complexes studied. The reactions of NO with $Fe^{II}(edta)$ (edta = ethylenediaminetetraacetate) and $Fe^{II}(Hedtra)$ (Hedtra = hydroxyethylenediaminetriacetate) yielded activation volumes of +4.1 and +2.8 cm³ mol⁻¹, respectively and were assigned to a dissociative interchange (I_d) mechanism (31b). All of the iron(II) aminocarboxylato complexes studied followed a similar pattern with the exception of the $Fe^{II}(nta)$ (Nta = nitriloacetic acid) complex which gave a $\Delta V^{\ddagger}$ value of −1.5 cm³ mol⁻¹. The reaction of this complex with

NO was proposed to occur through an associative interchange mechanism (I_a). A recent study of the formation of $[Fe(H_2O)_5(NO)]^{2+}$ from aquated ferrous ion (30) resulted in activation parameters similar to those for chelated ferrous ion (Table II). The small and positive activation volumes were used to assign the reaction mechanism as dissociative interchange in character.

III. Reactions of Metal Nitrosyl Complexes

The versatility of NO as a ligand is illustrated by the linear and bent coordination modes of metal–NO bonding (Fig. 4). Linear coordination is often viewed in terms of charge transfer to the metal giving (formally) the nitrosyl (NO^+) ligand, which is isoelectronic with carbon monoxide. Such charge transfer from the π^*_{NO} orbital to the metal is qualitatively consistent with the relatively high ν_{NO} stretching frequencies (~ 1800–$1950\,cm^{-1}$) and reflects the nitrosyl triple bond character in these cases. Bent M–NO coordination implies less electronic charge transfer from NO to M, and consequently the ν_{NO} values are lower. Indeed, as the angle approaches $120°$, the polarity of the charge transfer is reversed, and the ligand is formally a nitroxyl anion (NO^-). In this context one can easily speculate that a NO molecule coordinated linearly to a cationic metal center may be susceptible to nucleophilic attack, while the bent nitroxyl complexes would be more inclined to react with electrophiles such as H^+. This qualitative picture has indeed been realized for each limiting case. Such reactivity patterns of coordinated NO have been reviewed by McCleverty (1977) (16b) and Bottomley (1989) (16d), so the present article will largely focus on more recent examples.

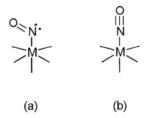

(a)   (b)

Fig. 4. Illustration of limiting cases of NO binding to a metalloporphyrin center as: (a) the nitroxyl anion (NO^-) with a M–N–O bond angle of $\sim 120°$; or as (b) the nitrosyl cation (NO^+) with a M–N–O bond angle of $\sim 180°$.

A. ELECTRON TRANSFER REACTIONS

The NO/NO^+ and NO/NO^- self-exchange rates are quite slow (42). Therefore, the kinetics of nitric oxide electron transfer reactions are strongly affected by transition metal complexes, particularly by those that are labile and redox active which can serve to promote these reactions. Although iron is the most important metal target for nitric oxide in mammalian biology, other metal centers might also react with NO. For example, both cobalt (in the form of cobalamin) (43,44) and copper (in the form of different types of copper proteins) (45) have been identified as potential NO targets. In addition, a substantial fraction of the bacterial nitrite reductases (which catalyze reduction of NO_2^- to NO) are copper enzymes (46). The interactions of NO with such metal centers continue to be rich for further exploration.

There have been several studies with vitamin B_{12} derivatives. For example, it was claimed that the Co(III) complex aquacobalamin (Vitamin B_{12a}) reacts with NO to form a stable complex, and this reaction was attributed to biological roles (43), in contradiction to an earlier conclusion by Williams and coworkers (47) that B_{12a} is unreactive with nitric oxide. However, van Eldik and coworkers (44a) recently concluded that the claimed reactivity with B_{12a} is due instead to the common aqueous solution impurity NO_2^- which forms the B_{12a}–NO_2 complex. They also demonstrated (44c) that NO reacts rapidly with the reduced form of B_{12a} (Co^{II}) to give the corresponding nitrosyl complex with a second-order rate constant $k_{on} = 7.4 \times 10^8 \, M^{-1} s^{-1}$ at pH 7.4 (298 K). The water soluble cobalt(II) porphyrin complex $Co^{II}(TPPS)$ has been shown to react with NO to give the nitrosyl adduct with comparable rates (Table II) (21).

The NO reduction of the Cu(II) complex $Cu(dmp)_2(H_2O)^{2+}$ (dmp = 2,9-dimethyl-1,10-phenanthroline) to give $Cu(dmp)_2^+$ plus nitrite ion (Eq. (20)) has been studied in aqueous solution and various mixed solvents (42a). The reduction potential for $Cu(dmp)_2(H_2O)^{2+}$ (0.58 V vs. NHE in water) (48) is substantially more positive than those for most cupric complexes owing to steric repulsion between the 2,9-methyl substituents that provide a bias toward the tetrahedral coordination of Cu(I). The less crowded bis(1,10-phenanthroline) complex $Cu(phen)_2(H_2O)^{2+}$ is a weaker oxidant (0.18 V) (48).

In methanol, the product of the $Cu(dmp)_2(H_2O)^{2+}$ oxidation of NO is CH_3ONO; in water, it is NO_2^- (Eq. (20)). The reaction did not occur in CH_2Cl_2 unless methanol was added.

$$Cu(dmp)_2(H_2O)^{2+} + NO + ROH \longrightarrow Cu(dmp)_2^+ + RONO + H^+ + H_2O$$

$$(20)$$

The kinetics of this reaction were followed by tracking the appearance of $Cu(dmp)_2^+$ at 455 nm, the λ_{max} of the metal to ligand charge transfer (MLCT) absorption band. At a fixed pH, the kinetics in aqueous solution followed the rate law.

$$\frac{d[Cu(dmp)_2^+]}{dt} = k_{NO}[NO][Cu(dmp)_2^{2+}] \tag{21}$$

Addition of $NaNO_2$ (50 µM) had no effect on the reaction profile with NO present, and no reaction was observed (on the time scale of the stopped-flow experiment) when NO was absent. However, at higher concentrations, anions, including the conjugate bases of various buffers (B^-), slowed down the reaction. This was attributed to the competition between water and the anions for the labile 5th coordination site of $Cu(dmp)_2(H_2O)^{2+}$.

These results can be analyzed in the context of two different mechanisms. The first would be simple outer-sphere electron transfer to give Cu(I) plus NO^+ followed by hydrolysis of the latter (Eqs. (22) and (23)).

$$Cu(dmp)_2^{2+} + NO \underset{k_{-OS}}{\overset{k_{OS}}{\rightleftharpoons}} Cu(dmp)_2^+ + NO^+ \tag{22}$$

$$NO^+ + H_2O \overset{k_{hyd}}{\longrightarrow} H^+ + HNO_2 \tag{23}$$

The kinetic scheme of the outer-sphere mechanism can yield two limiting cases, both first-order in [NO]. One would involve a reversible equilibrium (Eq. (22)) followed by hydrolysis of the nitrosonium ion. In this case the second-order rate constant would be $k_{NO} = K_{OS} k_{hyd}/[Cu(dmp)_2^+]$ where $K_{OS} = k_{OS}/k_{-OS}$ and k_{hyd} is the rate constant for NO^+ hydrolysis. At the other extreme, k_{OS} would be rate limiting ($k_{NO} = k_{OS}$), and electron transfer is effectively irreversible owing to rapid hydrolysis of NO^+. The latter would appear more likely given the expectation that k_{hyd} is quite large and that the $Cu(dmp)_2^+$ concentration is initially very small (49). In either case, the reaction rate would be expected to be lower when an anion is coordinated to the Cu(II) instead of H_2O owing to the (likely) lower reduction potential of a $Cu(dmp)_2(B)^+$ species. For either limiting case, k_{OS} is the maximum rate constant by which NO reduction of Cu(II) would occur, and a value for this can be estimated from the Marcus cross relationship (50), i.e. $k_{OS} \sim (k_{11} k_{ex} K_{OS})^{1/2}$, where k_{11} is the $Cu(dmp)_2^{2+}/Cu(dmp)_2^+$ self exchange rate constant and k_{ex} is the self-exchange rate constant, for NO^+/NO. This treatment gave $\sim 3 \times 10^{-3}$ $M^{-1}s^{-1}$ as an estimate for k_{OS}, a value five orders of magnitude smaller than the value of k_{NO} measured for Eq. (20) at lower pH values.

On this basis, the outer-sphere reaction mechanism was concluded to be unlikely.

An inner-sphere mechanism can be offered as an alternative interpretation of the kinetics for NO reduction of aqueous $Cu(dmp)_2(H_2O)^{2+}$.

$$L = dmp$$

$$L_2Cu^{II}(ROH)^{2+} + NO \underset{}{\overset{K_{NO}}{\rightleftharpoons}} [L_2Cu^I-NO^+]^{2+} + ROH \qquad (i)$$

$$OR$$

$$[L_2Cu^I-NO^+]^{2-} + ROH \xrightarrow{k_{ROH}[ROH]} [L_2Cu^I-\overset{|}{N}=O]^+ + H^+] \qquad (ii)$$

$$[L_2Cu^I-(N(O)OR]^+ \xrightarrow{fast} [L_2Cu^+ + RONO \qquad (iii)$$

$$rate = k_{ROH}K_{NO}[Cu(dmp)_2^{2+}][NO][ROH] \qquad (24)$$

The three steps would be: (i) the reversible equilibrium displacement of solvent (H_2O or ROH) by NO to form an inner-sphere Cu(II) nitric oxide complex, which is activated toward nucleophilic attack by ROH (step ii) owing to charge transfer from NO to the metal ($Cu^{II}-NO\leftrightarrow Cu^I-NO^+$). Dissociation of the RONO complex (step iii) would be rapid owing to the preference of cuprous complexes for tetrahedral coordination. This inner-sphere pathway parallels the reductive nitrosylation mechanisms discussed below, with the exception that the $Cu^{II}-NO$ complex is formed with a very low K_{NO} (Attempts to observe formation of the putative inner-sphere complex $[Cu(dmp)_2(NO)]^{2+}$ gave no UV–Visible spectroscopic evidence for a new species). In this context the rate law predicted for Eq. (24) would also be second-order ($k_{NO} = K_{NO}k_{ROH}$) if ROH is the solvent. While the ROH concentration is not a variable in aqueous or methanolic solution, kinetic dependence on [MeOH] in methanolic dichloromethane is in agreement with this model.

Even though the rate law for Eq. (20) might suggest a simple outer-sphere electron transfer mechanism from NO to the copper complex, the evidence in this case points to an inner-sphere pathway, involving NO coordination *followed* by the reaction with a solution nucleophile. Perhaps this is not surprising given the relatively high potential required and slow self-exchange rate for the simple one electron oxidation of NO.

B. NUCLEOPHILIC REACTIONS WITH THE COORDINATED NITROSYLS

Nucleophilic reactions with coordinated NO are illustrated by the well known reversible reaction of hydroxide ion with the nitrosyl ligand of the nitroprusside ion (NP) to give the nitro analog $Fe(CN)_5(\underline{N}O_2)^{4-}$ (Eq. (25)). The rate of this reaction is first-order in $[OH^-]$ and in $[Fe(CN)_5(NO)^{2-}]$ (51), so the likely reactive intermediate is the adduct $Fe(CN)_5(N(O)OH)^{3-}$. The reaction is reversed in strongly acidic solution. Similar reactions are seen with the ruthenium and osmium analogs (51c,d) as well with numerous other simple coordination compounds of NO (Table III) (52).

$$Fe(CN)_5(NO)^{2-} + 2\,OH^- \rightleftharpoons Fe(CN)_5(\underline{N}O_2)^{4-} + H_2O \qquad (25)$$

Olabe and coworkers (51c,d) have systematically studied the mechanism of reaction (25) and the analogous transformations using $Ru(CN)_5(NO)^{2-}$ and $Os(CN)_5(NO)^{2-}$. The equilibrium constant K of reaction (25) depends on the nature of the metal center with values of 1.5×10^5, 4.4×10^6, and $42\,M^{-2}$ for the Fe^{II}, Ru^{II} and Os^{II} species, respectively (51c). The much lower K for the osmium complex is consistent with a relatively low ν_{NO} value for this species ($\nu_{NO} = 1897\,cm^{-1}$ for the Os complex vs. $1945\,cm^{-1}$ for the Fe analog). The rationale for this correlation is the higher frequency ν_{NO} values reflect the more electron accepting metal centers. A similar correlation has been described for activation of coordinated CO by nucleophiles (58).

Numerous reports describe the reactions of nucleophiles with nitroprusside (NP). For example, HS^- reacts with NP to give a strongly colored species thought to be a thiol analog of the nitro product, namely, $Fe(CN)_5(N(O)S)^{4-}$. However, this is not stable and undergoes oligomerization, possibly via the formation of bridging disulfide bonds (54). NP reacts with mercaptans (RSH) and mercaptides (RS^-) apparently to form metal nitrosothiolato intermediates with deep red or

TABLE III

SELECTED REACTIONS OF METAL NITROSYL COMPOUNDS WITH NUCLEOPHILES

Nucleophile	Substrate	Product	Reference
OH^-	$Ru(hedta)(NO)^{+3}$	$Ru(hedta)(NO_2)^+$	(52a,52c)
ROH, OR^-	$IrCl_3(PPh_3)_2(NO)^+$	$IrCl_3(PPh_3)_2(N(O)OR)$	(52b)
RSH, SR^-	$Fe(CN)_5(NO)^{-2}$	$Fe(CN)(OH_2)^{-3}+RSSR+NO$	(53)
S_2^-	$Fe(CN)_5(NO)^{-2}$	$[Fe(CN)_5N(O)S]_2^{6-}$	(54)
NH_3	$Fe(CN)_5(NO)^{-2}$	$Fe(CN)_5(OH_2)+N_2$	(52c)
RNH_2	$Fe(CN)_5(NO)^{-2}$	$Fe(CN)_5(H_2O)^{3-}+N_2+ROH$	(55)
N_2H_4	$Fe(CN)_5(NO)^{-2}$	$Fe(CN)_2(H_2O)+N_2O+NH_3+H^+$	(56)
N_3^-	$Ru(das)_2(NO)Cl^{+2}$	$Ru(Cl)(das)_2N_3+N_2+N_2O$	(52d)
$O=C-CHR^-$	$Ru(py)_4Cl(NO)^{+2}$	$Ru(py)_4Cl(N(OH)CCNR$	(52e)
$ArNH_2$	$Ru(bpy)_2(NO)Cl^{+2}$	$Ru((bpy)_2(N_2Ar)Cl$	(57)

purple colors (*53*). These intermediates are unstable and decay via formation of disulfides and reduced nitroprusside, and the latter subsequently
decomposes by dissociation to cyanide and NO. Similar processes may
be responsible for the biological activity of sodium nitroprusside, which
is used as an intravenously administered vasodilator drug (*59*). The reactivity of thiols with metal nitrosyls continues to be a fertile field for discovery, particularly in the context of the high in vivo concentrations of
reduced sulfur species such as glutathione.

In solutions with relatively high ammonia concentrations, NH_3 reacts
with NP to give $Fe(CN)_5(H_2O)^{3-}$ plus N_2, effecting comproportionation
of NH_3 and NO^+ to N_2 (*60,52c*). Likewise, primary amines (RNH_2) are
diazotized by aqueous NP to give the alcohols (ROH) plus N_2, with the
maximum rate occurring about pH 10.5 (*55*). The rates of these reactions
are first-order in [NP] and [RNH_2] and increase with the basicity of the
amines.

NP reacts with hydrazine to form NH_3 and nitrous oxide (Eq. (26)) with
the rate law: $-d[NP]/dt = k[NP][NH_2NH_2]$ (*61*). The hydrazinium ion
$N_2H_5^+$ was inactive, so the rate decreased to near zero at pH 6 consistent
with the pK_a value of hydrazine.

$$Fe(CN)_5(NO)^{2-} + NH_2NH_2 \longrightarrow Fe(CN)_5(H_2O)^{3-} + NH_3 + N_2O + H^+$$

$$(26)$$

Reaction of metal nitrosyls with azide ion proceeds with formation of N_2
and N_2O (*56*). This can be viewed as the result of a nitrene transfer reaction in analogy with the Curtius rearrangement (*62*) and its organometallic counterpart (*63*).

$$[(NC)_5FeNO]^{2-} + N_3^- \rightleftharpoons \left\{ (CN)_5Fe-N \begin{matrix} \diagup O \\ \diagdown N=N=N \end{matrix} \right\}^{3-}$$

$$\xrightarrow{H_2O} Fe(CN)_5(H_2O)^{3-} + N_2O + N_2 \qquad (27)$$

Some other reactions of metal nitrosyls $L_xM(NO)$ with various nucleophiles (Nuc) are summarized in Table III. The pattern indicated by the
studies described above is repeated; simple adduct formation occurs
when the coordinated nitrosyls are sufficiently electrophilic and the
nucleophiles sufficiently basic. The first species formed is probably the
N-coordinated nucleophile nitrosyl adduct $L_xM(N(O)Nuc)$, e.g. Eq. (27).
Subsequent reactions depend on the substitution lability of these species,
as well as on the redox stability of the complex and of the ligand.

For example, the substituted aniline $Ar-NH_2$ ($Ar = p$-$CH_3OC_6H_4^-$) reacts with the ruthenium nitrosyl complex $Ru(bpy)_2(Cl)(NO)^{2+}$ (bpy = 2,2'-bipyridine) to give a complex of the diazo ligand, namely $Ru(bpy)_2(Cl)(NNAr)^{2+}$ (57). Upon employing the ^{15}N labeled nitrosyl complex $Ru(bpy)_2Cl(^{15}NO)^{2+}$ this reaction resulted in the ^{15}N coordinated product, $Ru(bpy)_2Cl(^{15}NNAr)^{2+}$, demonstrating that the reaction occurs within the metal complex coordination sphere. When the reactions were conducted in non-protic solvents, these nucleophile-nitrosyl adducts could be isolated.

C. REDUCTIVE NITROSYLATION REACTIONS

As described above for $Cu(dmp)_2(H_2O)^{2+}$, nucleophilic attack on a coordinated nitrosyl is a logical mechanism for metal reduction by NO which in turn is oxidized to nitrite or another N(III) species. Ferric porphyrins and other redox active metal centers have long been known to undergo reductive nitrosylation in the presence of excess NO (64–66). For example, the iron(III) complex, $Fe^{III}(TPP)(Cl)$, reacts with NO in toluene solution containing a small amount of methanol to give $Fe^{II}(TPP)(NO)$, consistent with the scheme shown in Eqs. (28)–(30) (64,65c). Analogously, aqueous ferri-hemoglobin, (metHb) reacts with NO to give the ferro-hemoglobin NO adduct, Hb(NO) (66).

$$Fe^{III}(TPP)Cl + NO \rightleftharpoons Fe^{III}(TPP)(Cl)(NO) \qquad (28)$$

$$Fe^{III}(TPP)(Cl)(NO) + CH_3OH \longrightarrow Fe^{II}(TPP) + CH_3ONO + HCl \qquad (29)$$

$$Fe^{II}(TPP) + NO \longrightarrow Fe^{II}(TPP)(NO) \qquad (30)$$

Additional mechanistic insight into the reductive nitrosylation of ferri-heme proteins was obtained from kinetic studies carried out on aqueous solutions of Cyt^{III}, metMb, and metHb at various pH values (67). For example, Cyt^{III} undergoes reduction by NO to Cyt^{II} in aqueous solutions at pH values >6.5. A hypothetical reaction mechanism is shown in Scheme 2 which would predict the rate law presented in Eq. (31) (67).

$$\frac{d[Fe^{II}]}{dt} = k_d[Fe^{III}(Por)]\left(\frac{K_{NO}[NO]}{1 + K_{NO}[NO]}\right)\left(\frac{K_{OH}[OH^-]}{1 + K_{OH}[OH^-]}\right) \qquad (31)$$

Because the reaction of NO with Cyt^{II} to form $Cyt^{II}(NO)$ is very slow (see Section II), the formation of Cyt^{II} could be observed directly. The observed rates are functions of [NO] and $[OH^-]$ as predicted by Eq. (31),

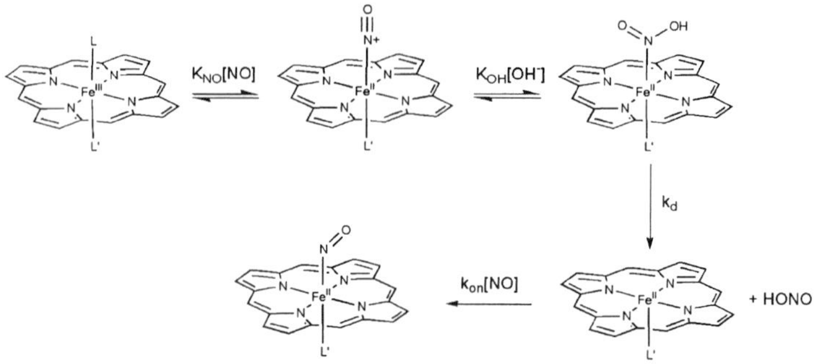

SCHEME 2. Porphyrin ligand substituents have been omitted for clarity.

TABLE IV

REDUCTIVE NITROSYLATION OF FERRIHEMOPROTEINS AT 25.0 °C. VALUES OF
CONSTANTS DETERMINED (67) [a]

	Cyt^{III}	metMb	metHb [b]
K_1 (M^{-1})	1.4×10^4	$(1.3\text{–}0.62) \times 10^{3\ c}$	1.3×10^4
k_{OH} (M^{-1}s^{-1})	1.5×10^3	3.2×10^2	3.2×10^3
k_{NO} (M^{-1}s^{-1})	8.3	1.7×10^7	2.5×10^7
pH	6.1–8.45	6.0–7.2	5.6–7.4

[a] Reprinted with permission from Ref. (5k). Copyright 2002 American Chemical Society.
[b] metHb(NO) reacts with H_2O in pH 6 water with a rate constant $k_{H_2O} = 1.1 \times 10^{-3}$ s^{-1}.
[c] K_1 for metMb is pH dependent, decreasing to half its value at higher pH.

that is, $k_{obs} = k_{OH} \times K_{NO}[NO][OH^-]/(1 + K_{NO}[NO])$ at low pH (where $k_{OH} = k_d \times K_{OH}$) and $k_{obs} = k_{OH}[OH^-]$ at high [NO]. No evidence for the N-bound nitrous acid complex $Fe^{II}(\underline{N}(O)OH)$ was found for the three ferri-heme proteins studied. Thus, either the formation of this intermediate is rate limiting, or K_{OH} is very small in each case. Values of K_{NO} were determined from the spectrophotometric titration of the respective ferri-heme protein by NO, and kinetics studies gave rise to the values for k_{OH} listed in Table IV.

The mechanisms for metMb and metHb reductive nitrosylation are thought to be similar to that for Cyt^{III}. However, given that both Mb and Hb readily react with NO, the only observable products are Mb(NO) and Hb(NO) (67). For metMb, K_{NO} values decreased at higher pH values, suggesting that pH change may induce protein conformational changes. Reductive nitrosylation of metHb also occurs at lower pH values (< 6), implying that metHb(NO) reacts with not only OH$^-$ but also with H_2O (Eq. (32)), with a rate constant, $k_{H_2O} = 1.1 \times 10^{-3}\,s^{-1}$ in aqueous solution at 298 K. In contrast reductive nitrosylations of metMb and Cyt^{III} were not observed at low pH; thus, direct reactions of metMb(NO) and Cyt^{III}(NO) with H_2O appear to be much slower than for metHb (67).

$$Hb^{III}(NO) + H_2O \xrightarrow{k_{H_2O}} Hb^{II} + ONO^- + 2H^+ \qquad (32)$$

Recently, NO reduction of metMb in pH 7.4 phosphate buffer, in the presence of the biological antioxidant glutathione, GSH, was demonstrated (68). Absorbance changes in the porphyrin Q band region showed Mb(NO) to be one product while amperometric sensor experiments were interpreted in terms of the nitrosoglutathione, GSNO, being the other product. The second-order rate constant for reaction of GSH with metMb(NO) was surprisingly large ($47\,M^{-1}s^{-1}$) given that k_{OH} for the smaller and more basic OH^- ion (67) is only an order of magnitude higher (Table IV).

Studies in this laboratory (69) of the water soluble ferri-heme model Fe^{III}(TPPS) in aqueous solution have shown that this species also undergoes reductive nitrosylation in solutions that are moderately acidic (pH 4–6) (Eq. (32)). The rate of this reaction includes a buffer dependent term indicating that the reaction of the Fe^{III}(TPPS)(NO) complex with H_2O is subject to general base catalysis. The reaction depicted in Eq. (33) is not observable at pH values < 3, since the half-cell reduction potential for the nitrite anion (Eq. (1)) is pH dependent, and Eq. (33) is no longer thermodynamically favorable.

$$Fe^{III}(TPPS)(NO) + H_2O + 2\,NO \xrightarrow{k_{H_2O}} Fe^{II}(TPPS)(NO) + ONO^- + 2H^+ \qquad (33)$$

It was also found that NO_2^- accelerated the observed rates of reductive nitrosylation ($k_{NO_2} = 3.1 \pm 0.1\,M^{-1}s^{-1}$ in 16 mM acetate at pH 4.96) (69). Since nitrite is a product of the reductive nitrosylation reaction in aqueous solution, the system is, in principle, autocatalytic.

There are two mechanisms that could explain the catalytic effects of nitrite; an inner-sphere mechanism in which nitrite acts as a nucleophile toward the {$Fe^{II}NO^+$} moiety (Scheme 3, pathway A) and an outer-sphere path in which nitrite is oxidized to NO_2 which then reacts with excess NO to form N_2O_3 (Scheme 3, pathway B). Although the initial electron transfer step in pathway B is thermodynamically uphill ($\Delta E = -0.3\,V$) (41,70), one cannot rule out pathway B since N_2O_3 is rapidly hydrolyzed, once formed (71).

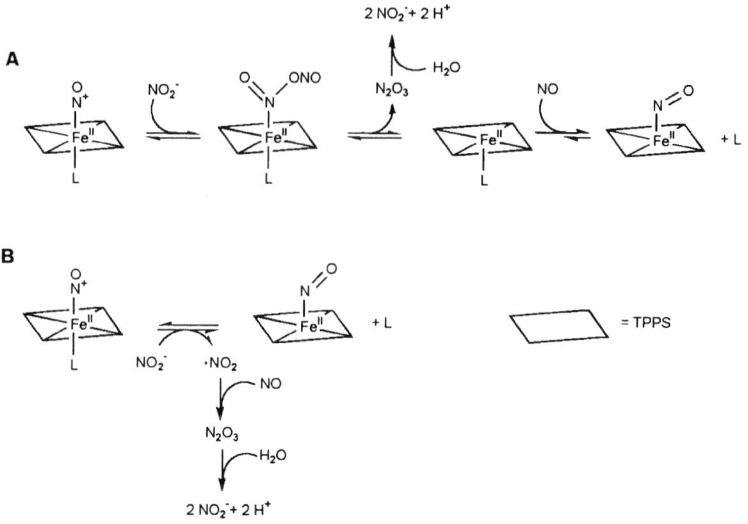

SCHEME 3.

D. ELECTROPHILIC REACTIONS

Nucleophilic attack on coordinated nitrosyl generally occurs at the nitrogen atom bound to the metal, although there are examples in the organometallic literature of ligand addition to an ML_xNO complex with a linear (3 electron donor) nitrosyl giving a $L_{x+1}M–NO$ species with a bent (or one electron donor) NO. Electrophilic attack may be less selective. For example, protonation may occur at the metal center, at the nitrosyl nitrogen or at the nitrosyl oxygen (Scheme 4), an important facet being the choice of the conjugate base counter-ion. If anion coordination occurs at the metal, this promotes electron release to the nitrosyl making the latter more basic. Thus, a strong acid with a non-coordinating counter-ion might not protonate a coordinated nitrosyl, while a weaker acid with a more strongly binding counter-ion would generate an HNO complex. The result would effectively be HX oxidative addition across M–NO (72). Protonation at the nitrosyl oxygen or the metal center is more likely when a strong acid with a non-coordinating counter-ion is employed.

An additional path to protonation of coordinated NO is the reduction of the M–NO unit electrochemically (73). Such coupled reduction/protonation schemes have been argued to be relevant to enzymatic nitrogen oxide reductases (41). Farmer and coworkers (74) accomplished such reductions by using graphite electrodes modified by depositing

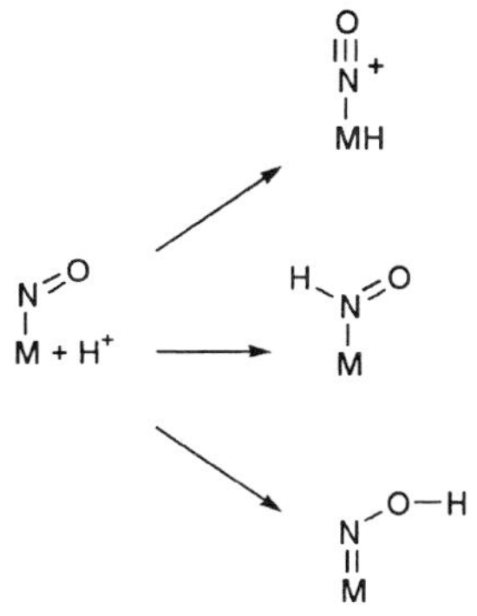

$$
\begin{array}{c}
\text{O} \\
\||| \\
\text{N}^+ \\
| \\
\text{MH}
\end{array}
\qquad\nearrow
$$

$$
\begin{array}{ccc}
\text{N}{=}\text{O} & & \text{H}{\sim}\text{N}{=}\text{O} \\
| & \longrightarrow & | \\
\text{M} + \text{H}^+ & & \text{M}
\end{array}
$$

$$
\searrow\qquad
\begin{array}{c}
\text{N}\;{\sim}\text{O}{-}\text{H} \\
\|| \\
\text{M}
\end{array}
$$

Scheme 4.

surfactant films of Mb(NO) on the surface. Electrochemical reduction of Mb(NO)$_{\text{surface}}$ ($E_{1/2}=-0.63$ V vs. NHE) was accompanied by protonation to provide a Mb(H$\underline{\text{N}}$O)$_{\text{surface}}$ complex. The Mb(NO$^-$)$_{\text{surface}}$ underwent catalytic reaction with excess NO in solution at more negative potentials to give N$_2$O, suggesting that a N–N coupling reaction occurs between the bound nitroxyl ion and free NO. Mb(H$\underline{\text{N}}$O) was prepared independently in aqueous solution by reacting Mb(NO) with Cr^{2+} (Eq. (34)) and proved to be surprisingly stable (74b). The ^{1}H NMR displayed a singlet at 14.8 ppm; however, the pK_a of the coordinated HNO was not reported.

$$\text{Mb(NO)} + \text{Cr}^{2+} + \text{H}^+ \longrightarrow \text{Mb(H}\underline{\text{N}}\text{O)} + \text{Cr}^{3+} \qquad (34)$$

Li$^+$ and BF$_3$ have been shown to bind weakly to Co(salophen)(NO) complexes (salophen $= N,N'$-1,2-phenylenediamine-bis(salicylidenimato)), presumably at the oxygen atom, causing ν_{NO} to shift ~ 20 cm^{-1} to higher frequency (75). This also serves to labilize the nitrosyl. Dioxygen and other oxidants have also been examined as potential electrophiles in reactions with coordinated NO. Since these reactions are generally accompanied by subsequent processes leading to nitro or nitrito ligands or to dissociated NO$_x$ products as well as other transformations of the metal complex, they will be discussed in a following section on reactions with dioxygen.

E. NO Disproportionation

In homogeneous solutions NO disproportionation may be promoted by transition metal complexes, and a variety of mechanisms seem to be available owing to the many possible modes of coordination. One example is the reaction of NO with nickel carbonyl shown in Eq. (35) (76), where the nitrogen containing products are N_2O (oxidation state formally $N(I)$) and coordinated NO^+ and ONO^- (both $N(III)$). A number of other metal complexes have been shown to promote similar transformations (77).

$$Ni(CO)_4 + 4\ NO \longrightarrow [Ni(NO)(NO_2)]_n + 4\ CO + N_2O \qquad (35)$$

One example was reported by Tolman and coworkers (78) who found that the copper(I) complex Cu^ITp^{R2} (Tp^{R2}=tris(3-(R^2)-5-methylpyrazol-1-yl)hydroborate) promotes NO disproportionation via a weakly bound $Cu^ITp^{R2}(NO)$ intermediate (formally a $\{MNO\}^{11}$ species). The products are N_2O and a copper(II) nitrito complex (Eq. (36)). The rate law established the reaction to be first-order in copper complex concentration and second-order in [NO], and this was interpreted in terms of establishment of a pre-equilibrium between NO and the Cu(I) precursor and the $Cu^I(NO)$ adduct, followed by rate-limiting electrophilic attack of a second NO molecule (mechanism B of Scheme 5) (78b).

$$Cu^ITp^{R2} + 3\ NO \longrightarrow Cu^{II}Tp^{R2}(NO_2) + N_2O \qquad (36)$$

Complexes of N–N bonded dinitrogen dioxide, such as depicted in pathway B of Scheme 5, would appear to be necessary in order to effect the formation of the N–N bond. This has been treated theoretically as a metal promoted reductive coupling of 2 NO to form a hyponitrite complex (79). The $Cu^I(Tp^{R2})$ system was also shown to catalyze NO oxidations of benzyl and isopropyl alcohol to benzaldehyde and acetone (Eq. (37)). Electrospray mass spectrometry indicated that higher

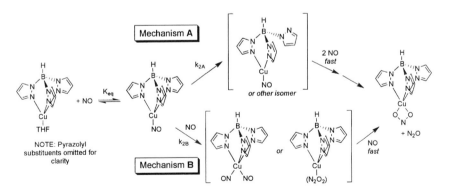

SCHEME 5. Reprinted with permission from Ref. (78b). Copyright 1998 American Chemical Society.

oligomers of copper are involved in the transfer of oxidizing equivalents
to substrate (80).

$$2 \text{ NO} + R_2\text{CHOH} \xrightarrow{\text{Cu}^{\text{I}}(\text{Tp}^{R2})} N_2O + H_2O + R_2C=O \qquad (37)$$

In a related observation, reported by Tanaka et al. (81), the copper(II)
complex Cu(tpa)^{2+} (tpa = tris[(2-pyridyl)methyl]amine) was shown to
serve as a catalyst for the electrochemical reduction of nitrite to N_2O
and traces of NO in aqueous solution. NO and/or a copper nitrosyl complex would appear to be the likely intermediates in this process (81a).

NO disproportionation has been shown to be promoted by the Mn(II)
tropocoronand complex Mn(TC-5,5) (82) (Eq. (38)), and the reaction was
found to involve three equivalents of NO leading to formation of N_2O
and O-coordinated nitrito ligand. The electron balance is provided by
oxidation of Mn(II) to Mn(III). The mononitrosyl complex Mn(TC-
5,5)(NO) was proposed to react with NO to produce an unstable cis-dini-
trosyl, Mn(TC-5,5)(NO)_2, which is then poised to form an N-coordinated
hyponitrito (O=N–N=O) ligand from which oxygen transfer occurs to
another NO (82a). The intermediacy of a hyponitrito ligand parallels
other proposed mechanisms for metal complex promoted NO disproportionation (5a–d).

$$\text{Mn(TC-5,5)} + 3 \text{ NO} \longrightarrow \text{Mn(TC-5,5)(\underline{O}NO)} + N_2O \qquad (38)$$

The mononitrosyl complex Fe(TC-5,5)(NO) was suggested to be a logical
intermediate in the disproportionation promoted by the Fe(II) system
(82b). However, the NO_2 released during the reaction (Eqs. (39) and (40)),
nitrates the aromatic rings of the tropocoronand ligand and renders the
resulting complex inactive as a disproportionation catalyst.

$$\text{Fe(TC-5,5)} + 4 \text{ NO} \longrightarrow \text{Fe(TC-5,5)(NO)} + NO_2 + N_2O \qquad (39)$$

$$\text{Fe(TC-5,5)(NO)} + NO_2 \longrightarrow \text{Fe(TC-5,5-NO}_2\text{)(NO)} \qquad (40)$$

Ruthenium(II) porphyrin complexes, $\text{Ru}^{\text{II}}(\text{Por})\text{CO}$, react with NO to
give the nitrosyl nitrito complex Ru(Por)(NO)(ONO) (Por = TPP, OEP
and related porphyrins) (83,84). Stoichiometric quantities of N_2O and
CO were found to be released (Eq. (41)) (83).

$$\text{Ru}^{\text{II}}(\text{Por})\text{CO} + 4 \text{ NO} \longrightarrow \text{Ru}^{\text{II}}(\text{Por})(\text{NO})(\underline{O}NO) + N_2O + CO \qquad (41)$$

A detailed stopped-flow kinetics investigation (85) of this reaction
(Por = TmTP and OEP) demonstrated that the mechanism occurs in two
stages. The first was quite fast and was suppressed by the presence
of excess CO and other coordinating ligands. Time resolved infrared

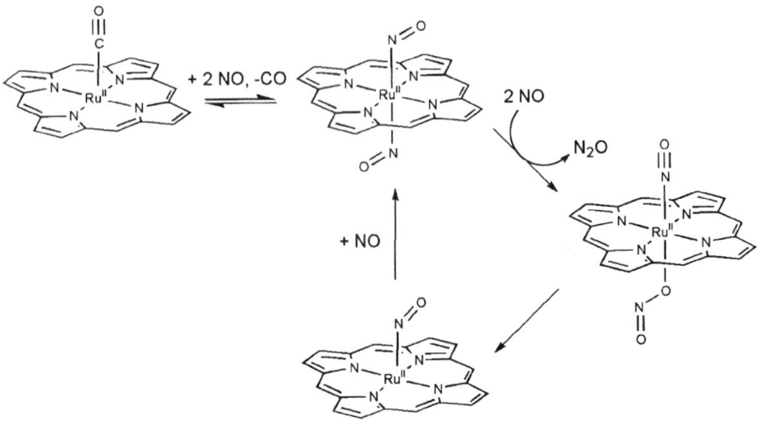

SCHEME 6. Porphyrin ligand substituents have been omitted for clarity.

spectral studies indicate that the intermediate formed is the *trans*-dinitrosyl complex $Ru(Por)(NO)_2$, characterized by a strong, single ν_{NO} band at $1642\,cm^{-1}$ (for $Por = TmTP$ in cyclohexane solution) (86). The rate of the second stage leading to formation of $Ru(Por)(NO)(\underline{O}NO)$ proved to be second-order in [NO] (Scheme 6).

The analogous dinitrosyl intermediate was also observed in time resolved UV–Vis and IR studies of the species generated via the 355 nm laser flash photolysis of $Ru(Por)(NO)(ONO)$ in the presence of excess NO (87). Photo-induced dissociation of NO_2 from $Ru(Por)(NO)(ONO)$ followed by trapping with NO yielded $Ru(Por)(NO)_2$ as observed by flash photolysis experiments using stepped scan FTIR detection (87). The $Ru(Por)(NO)_2$ species reacted with additional NO to regenerate $Ru(Por)(NO)(ONO)$ via the disproportionation reaction shown above. When flash photolysis was carried out upon unlabeled $Ru(P)(NO)(ONO)$ and doubly labeled $^{15}N^{18}O$ in solution, only singly and triply labeled nitrite ligand and fully labeled $^{15}N_2{}^{18}O$ were formed. This indicates that the nitrito ligand is formed by an oxygen atom transfer from two $^{15}N^{18}O$ molecules to one of the two coordinated NO molecules of the dinitrosyl, one of which must be an unlabeled NO from the original substrate.

Literature reports of NO disproportionation reactions with Fe(II) porphyrins contain many mutually inconsistent observations. Although facile NO disproportionation is promoted by Ru(II) and Os(II) (88) porphyrins to yield N_2O and the respective $M(Por)(NO)(\underline{O}NO)$ complexes, the reactivity appears to be quite different with analogous Fe(II) complexes. Ferrous porphyrins such as $Fe^{II}(TPP)$ undergo NO addition in ambient temperature solution to give the relatively stable paramagnetic mononitrosyl complex, e.g. Fe(TPP)(NO), although there have been significant disagreements regarding the subsequent reaction with excess NO. Fe(TPP)(NO) has been reported to promote NO disproportionation to give the N-bonded nitro species Fe(TPP)(NO)($\underline{N}O_2$) (*89*). However, this observed reactivity is apparently due to the presence of NO_x impurities (Eq. (42)). Recent studies (*90*) in this laboratory have demonstrated that ambient temperature solutions of Fe(TPP)(NO) display no changes in IR or optical spectra when treated with NO *carefully cleaned of higher NO_x impurities*, consistent with an early ESR study by Wayland and Olson (*91*).

$$\text{Fe(TPP)(NO)} + \text{N}_2\text{O}_3 \longrightarrow \text{Fe(TPP)(NO)(NO}_2) + \text{NO} \qquad (42)$$

This reaction has been reexamined using optical, IR and NMR spectroscopic methods to probe NO reactions with Fe(TPP)(NO) and the more soluble Fe(TmTP)(NO) (*92*). These studies confirmed the formation of Fe(Por)(NO)$_2$ in toluene-d_8 at low temperature (Eq. (43)). NMR line shape analysis was used to calculate $K_{43} = 23\,\text{M}^{-1}$ at 253 K (3100 M^{-1} at 179 K, $\Delta H^\circ = -28\,\text{kJ mol}^{-1}$) (*92*). The failure of the Fe^{II}(Por) complexes to promote NO disproportionation, in contrast to the behavior of the respective Ru(II) and Os(II) analogs, may find its origin partly in the relatively low stability of the dinitrosyl intermediate (K_{52} estimated to be 2.8 M^{-1} at 298 K) and unfavorable kinetics of subsequent reaction of this species with NO.

$$\text{Fe}^{II}(\text{TmTP})(\text{NO}) + \text{NO} \overset{K_{43}}{\rightleftharpoons} \text{Fe}^{II}(\text{TmTP})(\text{NO})_2 \qquad (43)$$

Kadish *et al.* (*93*) have reported UV–Vis and EPR spectra of the 19-electron complex Fe(P)(NO)$_2^+$, at room temperature. Recently the 20 electron species M(Pc)(NO)$_2^-$ (M = Re, Mn, Pc = phthalocyaninato), electronically analogous to Fe(P)(NO)$_2$, was also described (*94*).

F. REACTIONS WITH DIOXYGEN

Redox reactions involving nitric oxide have important implications beyond their fundamental chemistry as demonstrated by the controversy in the biomedical literature regarding conditions under which generation of NO leads to the amelioration or the exacerbation of oxidative stress in mammalian systems (*95*). "Oxidative stress" is defined as a disturbance in the balance between production of reactive oxygen species (pro-oxidants) and antioxidant defenses (*96*). Reactive oxygen species include free radicals and peroxides as well as other reactants such as oxidative enzymes with metal ion sites in high oxidation states. The physiological damage to an organism by oxidative stress during cardio-vascular events, for example, or by less acute problems such as chronic autoimmune disease or infection, is an important issue in many disease states.

The biological chemistry of NO is ultimately defined by its activity at the molecular level. For example, NO readily reacts with other free radicals such as $HO^{\bullet}$ to give nitrite or with O_2^- to give peroxynitrite (Eq. (44)) at near diffusion-limited second-order rate constants (e.g. $k_2 \sim 10^{10}\,\mathrm{M^{-1}\,s^{-1}}$ for O_2^-) (97).

$$NO + O_2^- \longrightarrow ONOO^- \tag{44}$$

In contrast, processes requiring multiple electron changes, such as the reaction of NO with O_2 in aqueous media to give nitrite ion (Eq. (45)) are generally much slower under physiological conditions (98).

$$4NO + O_2 + 2H_2O \longrightarrow 4H^+ + 4NO_2^- \tag{45}$$

The explanation lies in the rate law for the autoxidation of NO in aqueous solution, which follows third-order kinetics (Eq. (46) where $4k_{aq} = 9 \times 10^6\,\mathrm{M^{-2}\,s^{-1}}$) (98,99).

$$-\frac{d[NO]}{dt} = 4k_{aq}[NO]^2\,[O_2] \tag{46}$$

At the low [NO] relevant to bioregulatory processes, autoxidation is slow relative to other depletion pathways, and lifetimes are sufficient to allow for fast reactions with ferro-heme proteins in close proximity such as guanylyl cyclase (99). However, when much higher NO levels are produced, e.g. by stimulated macrophages under immune response, auto-xidation is faster and has potentially greater biological significance. Autoxidation intermediates, most prominently a species with the stoi-chiometry N_2O_3, are apparently responsible for oxidative and nitrosative reactions that contribute to cytotoxic and mutagenic activities under these conditions (100,101). Thus, third-order kinetics defines how this reactive molecule can play important bioregulatory roles in oxygenated media, yet participate in cytotoxic action when generated at higher concentration.

Another oxidant, peroxynitrite, (formed from NO plus O_2^-, Eq. (44)) has received considerable attention as a possible toxic/mutagenic agent formed during immune response (102). The role of peroxynitrite in this regard is a matter of continuing debate given that another school of thought argues that $ONOO^-$ is less damaging than the superoxide ion. Hence, reaction of the latter with NO is actually a cytoprotective mechanism (100a).

In aprotic solvents NO autoxidation also follows a third-order rate law but the product under these conditions is nitrogen dioxide. NO_2 is much more reactive than nitrite ion, the autoxidation product in aqueous solution, especially as a nitrosating reagent (100,103). Autoxidation in aprotic media may have biological relevance owing to the higher solubility of both NO and O_2 in hydrophobic media. As a consequence of reactant partitioning between cellular hydrophobic and hydrophilic regions and the third-order nature of this reaction, a disproportionately large fraction of autoxidation may occur in hydrophobic regions to yield nitrogen dioxide as the key intermediate under these conditions (104).

The reactivity of NO with O_2 is dramatically affected upon coordination of one of the diatomic components to a metal center. For example, the second-order reactions of NO with oxyhemoglobin, $Hb(O_2)$ and oxymyoglobin, $Mb(O_2)$ (e.g. Eq. (47)) are quite fast and have been used as colorimetric tests for NO (105). The nitrogen product is NO_3^- rather than NO_2^- that is the product of aqueous autoxidation (106). While the reaction of O_2 with nitrosyl myoglobin Mb(NO) (Eq. (48)) might superficially appear similar it is much slower and follows a different rate law (107). Possible mechanisms will be discussed below.

$$NO + Mb(O_2) \longrightarrow metMb + NO_3^- \qquad (47)$$

$$O_2 + Mb(NO) \longrightarrow metMb + NO_3^- \qquad (48)$$

The possible use of metal nitrosyls to activate the 4-electron oxidant O_2 has been of interest for some time and Scheme 7 illustrates some hypothetical pathways for accomplishing this (108,109).

Conceivably, autoxidation of a metal nitrosyl complex might involve NO dissociation followed by uncatalyzed reaction of free NO with O_2 to give NO_2, and rebinding to the metal center to give a nitro or nitrito complex (pathway (C) in Scheme 7). However, since the NO autoxidation rate is second-order in [NO] in aqueous or aprotic media (99a), accumulation of sufficient free NO to make such a sequence viable seems unlikely. Alternatively, NO dissociation might be followed by reaction of the denitrosylated metal center with O_2 to give a metal superoxide species (110) of the type $L_nM–OO$ known to react rapidly with free NO to form an O-bound peroxynitrite complex $L_nM–\underline{O}ONO$ (106,111). The latter species may undergo unimolecular isomerization to the nitrato complex (pathway (D) in Scheme 7) or O–O bond fragmentation to $L_nM = O + NO_2$ followed by recombination to give a nitrato complex (pathway (E)) or reaction of the putative oxo complex $L_nM = O$ with NO to give the nitrito analog (pathway (F)). The latter reaction has precedence for being quite rapid. A second-order rate constant of $3.1 \times 10^6 \, M^{-1} s^{-1}$ has been obtained for reaction of a $Cr^{IV} = O$ species to produce a Cr(III) nitrito complex (112), while a value of $1.8 \times 10^7 \, M^{-1} s^{-1}$ was determined for the reaction of NO with $MbFe^{IV} = O$, the "ferryl" form of myoglobin generated by H_2O_2 oxidation of myoglobin (113). It might also be noted that the reaction of NO with oxo complexes such as $Cr^{IV} = O$ to give a nitrite ligand finds analogy in a recent report that NO reacts with metal nitride L_nMN complexes to yield the respective $\underline{N}NO$ complexes that are relatively labile (114).

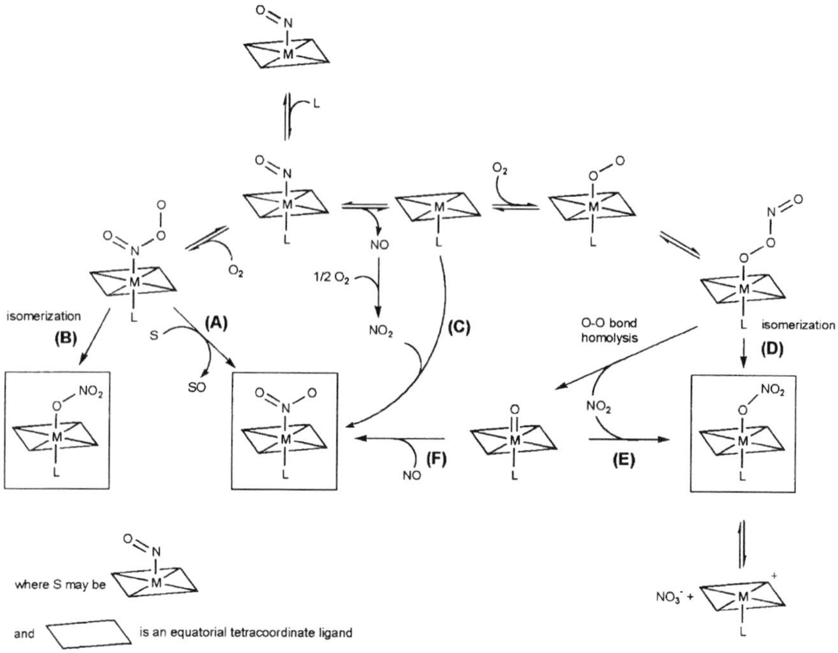

Among the important sinks for endogenously generated NO are the very fast reactions with oxyhemoglobin to form the nitrate ion plus methemoglobin with a second-order rate constant of $8.9 \times 10^7 \, M^{-1} s^{-1}$ (Eq. (49), pH 7.0) (111c). The analogous reaction of NO with oxymyoglobin (Eq. (47)), is also quite fast with a second-order rate constant $4 \times 10^7 \, M^{-1} s^{-1}$ (pH 7.0), (106a,111c). Herold and coworkers (111b,c) have examined the time resolved spectroscopy of the reactions depicted in Eqs. (47) and (49) and have concluded that NO reacts with the $Fe^{II}(O_2)$ species to give peroxynitrito intermediates; $Fe^{III}(\underline{O}ONO)$. Under neutral or acidic conditions, the latter rapidly decays to the "met", i.e. iron(III), form of the proteins with the quantitative formation of nitrate. Thus, the metal mediates the isomerization of peroxynitrite to nitrate (*111b,c*).

$$Hb(O_2) + NO \longrightarrow NO_3^- + metHb \qquad (49)$$

Reaction of the analogous nitrosyl myoglobin complex with dioxygen, a reaction of very great importance regarding the stability of cured meats, is much slower. The kinetics of Eq. (48) were studied by Skibsted *et al.* (*107*), who reported that, even at low dioxygen concentrations, the rate displayed limiting first-order behavior with a k_{obs} of $2.3 \times 10^{-4}\,s^{-1}$ in aqueous solution at 298 K, with $\Delta H^{\ddagger} = 110\,kJ\,mol^{-1}$ and $\Delta V^{\ddagger} = +8\,cm^3\,mol^{-1}$. These authors proposed that the reaction proceeded via prior formation of an O_2 complex with nitrosyl myoglobin, for example an N bonded peroxynitrite (the analog of pathway (A) in Scheme 7). However, the similarity of the limiting rate constant to that for NO dissociation $(2 \times 10^{-4}\,s^{-1})$ from Mb(NO) (*115*) tempts one to think in terms of a mechanism such as pathway (C) in Scheme 7 where NO dissociates from the iron center to allow formation of the dioxygen species which can react according to Eq. (48) above (*116*). Regardless, formation of NO_3^- as the nitrogen product indicates that the metal must be involved in the eventual oxidation step, since uncatalyzed NO autoxidation in aqueous media yields nitrite ion (Eq. (45)).

IV. Examples from the Chemical Biology of Metal Nitrosyl Complexes

The principal targets for NO under bioregulatory conditions are metal centers, primarily iron proteins (*117*). The best characterized example is the ferro-heme enzyme, soluble guanylyl cyclase (sGC) which is activated by formation of a nitrosyl complex with the iron(II) center. Other reports describe NO as an inhibitor for metalloenzymes such as cytochrome P450 (*118*), cytochrome oxidase (*119*), nitrile hydratase (*120*) and catalase (*121*), as a substrate for mammalian peroxidases (*122*), and as a contributor to the vasodilator properties of a salivary ferri-heme protein of blood sucking insects (*123*). Heme centers are also involved in the in vivo generation of NO by oxidation of arginine catalyzed by nitric oxide synthase (NOS) enzymes (*124*). There have been thousands of research articles published on the physiology and pharmacology of nitric oxide. Here we present selected examples of biological NO activity from the recent literature.

NO concentrations generated for bioregulatory purposes are low, and sub-micromolar values have been reported in endothelium cells for

blood pressure control (*125*). In contrast, the NO concentrations produced during immune response to pathogen invasion are much higher,
and under these conditions, reactive nitrogen species such as the peroxynitrite anion ($OONO^-$) and N_2O_3 may have physiological importance.
The biological relevance of the "on" and "off" reactions (Eq. (9)) discussed
above is emphasized by noting that activation of sGC involves such an
"on" reaction where the acceptor site of sGC is a $Fe^{II}(PPIX)$ moiety (*126*).
Other biological functions of NO such as inhibition of cytochrome
oxidase or catalase also apparently involve coordination at a heme iron,
so delineation of the dynamics and mechanisms of the "on" reaction is
essential to understanding the biochemistry of NO. Additional biological
processes such as sGC deactivation must involve labilization of M–NO
bonds, so the "off" reaction mechanism is equally important. For
example, the release of NO by the ferric-heme nitrophorin proteins is
the mechanism by which certain blood-sucking insects increase blood
flow to the site of the bite.

A. ACTIVATION OF HEME CONTAINING ENZYMES

Nitric oxide helps to regulate blood pressure in mammalian systems by
activation of the enzyme soluble guanylyl cyclase (sGC). This ferroheme
enzyme catalyzes the conversion of guanosine-5′-triphosphate (GTP) to
cyclic-3′,5′-monophosphate (cGMP) and inorganic pyrophosphate (PP_i).
cGMP promotes the activation of protein kinase G leading to phosphory-
lation reactions and smooth muscle relaxation (*127*). sGC is a heterodi-
mer with two identified isoforms $\alpha_1\beta_1$ and $\alpha_2\beta_2$ with molecular masses of
73 and 70 kDa for the α and β subunits, respectively (*128a*). The imidazole
group of the H105 residue in the β subunit has been identified as the
ligand that coordinates to the heme iron. While no crystal structure of
sGC has yet been obtained, the heme binding region is located in a
region of the β subunit that is likely to be responsible for dimerization
(*128*).

In the inactive state, sGC is linked to the ferroheme group via coordination to the imidazole nitrogen of a histidine side chain. Resonance
Raman and magnetic circular dichroism (MCD) spectra of sGC indicate
that the iron center is a mixture of low spin six-coordinate and high
spin five-coordinate species. It was postulated that the photolabile
distal ligand was an additional histidine imidazole based on the
similarity of the Raman spectrum of sGC and the model complex
$Fe(PPIX)(ImH_2)$ (*129*). Other workers report that sGC in the inactivated
state is exclusively high spin five-coordinate under aerobic conditions
(*130*). Regardless of the coordination environment of the heme in the

inactive state, the distal ligand must at most be weakly coordinated, given the very facile binding of NO to the sGC heme. The kinetics studies of the "on" and "off" rates of water soluble iron porphyrin model compounds and metMb with NO, discussed above, indicate that rapid reaction with NO involves dissociation of a labile ligand or a vacant coordination site. Based on these model systems a likely reaction mechanism is given in Scheme 8 below. Zhao *et al.* have measured the rate constant for the "on" reaction of NO with sGC as $>1.4 \times 10^8 \, M^{-1} \, s^{-1}$ at $4\,^{\circ}C$ using stopped-flow methods *(131)*. The k_{on} value is significantly larger than for other ferroheme proteins such as Hb and Mb which are already five-coordinate (see Table I). The considerable magnitude of the rate constant for the k_{on} step is not surprising given the fact that NO is generated at sub-micromolar concentrations in vivo for bioregulation *(125)*. Thus, sGC necessarily must have a high rate of reaction to compete effectively with other chemical and physical processes leading to NO depletion. The distal histidine must either be very labile or the first equilibrium must lie significantly to the right toward the resting state of the protein.

Measurements of the proximal histidine-iron stretching frequency by Resonance Raman spectroscopy revealed that this bond is very weak in relation to other heme protein systems ($\nu_{Fe\text{-}His} = 204 \, cm^{-1}$) *(130)*. Formation of the sGC–NO complex labilizes this ligand resulting in the formation of a 5-coordinate high spin iron(II) complex, and the conformational change responsible for the several hundred-fold increase in catalytic activity *(126,129,130)*.

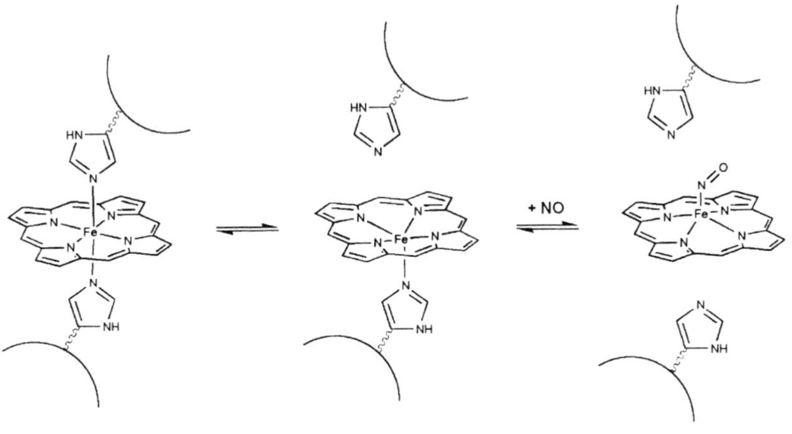

SCHEME 8.

The large equilibrium constants for Fe^{II}–NO bond formation and small k_{off} values for the ferroheme proteins are of biological interest with regard to deactivation of heme proteins. How, for example, does an enzyme such as soluble guanylyl cyclase, once activated by forming an NO complex, undergo deactivation? Kharitonov *et al.* (*27b*) have used stopped-flow kinetic methods to determine the first-order loss of NO from sGC–NO and determined a rate constant of $\sim 7 \times 10^{-4}\,s^{-1}$ in pH 7.4 buffered solution at 298 K. This rate is comparable to those for various ferroheme proteins listed in Table I, but is much less than required for reversible deactivation of the enzyme. However, in the presence of excess substrate GTP (5 mM) and the Mg^{2+} cofactor (3 mM) the rate was about 70-fold faster ($k_{off} \sim 5 \times 10^{-2}\,s^{-1}$ at 293 K), although the rate acceleration with GTP alone was only ~ 10-fold. A recent in vivo study (*132*) suggests that the actual rate of sGC deactivation is several orders of magnitude higher ($3.7\,s^{-1}$ at 310 K). Such differences illustrate potential complexities in comparing in vitro kinetics of purified proteins to analogous reactions in vivo.

Bohle and co-workers (*133*) have demonstrated that varying the electronic and stereochemical properties of porphyrin substituents can strongly influence the rates of NO labilization (Eq. (11)). For example, the displacement of NO from Fe(TPP)(NO) by pyridine is many orders of magnitude slower than from Fe(OBTPP)(NO) (OBTPP = octabromo-tetraphenylporphyrin). An analysis of the kinetics of the latter reaction indicated saturation in [L], and the mechanism was suggested to involve reversible formation of Fe(OBTPP)(L)(NO) followed by NO dissociation (Eq. (50)). Clearly changes in porphyrin properties can lead to enhanced reactivity toward NO loss.

$$Fe(Por)(NO) \xrightleftharpoons{+L} Fe(Por)(NO)(L) \xrightleftharpoons{+L} Fe(Por)(L_2) + NO \qquad (50)$$

A similar type of assisted NO loss in sGC is suggested by the observation of shifts in the NO and Fe–NO stretching frequencies of sGC–NO in the presence of cGMP or GTP by Resonance Raman spectroscopy (*134*). However, the Fe–Im stretching frequency was unchanged in the presence of cGMP or GTP indicating that the interaction with the heme center occurs on the distal side. A mechanism for deactivation of activated sGC can be suggested if indeed cGMP assists in labilization of NO from the active enzyme. A negative feedback mechanism would occur as the concentration of cGMP builds up and the enzyme is turned off by loss of NO. A mechanism of this sort would explain the differences in measured values of k_{off} and provide a reasonable pathway for enzyme deactivation in vivo. Additional experiments are required in order to explore this possibility further.

Myeloperoxidase (MPO) is a heme containing protein that catalyzes the two electron oxidation of halides (Cl^-, Br^- and I^-) and pseudohalide (SCN^-) to the corresponding hypohalous acid (Eq. (51)) in a process that is dependent upon $[X^-]$, $[H_2O_2]$ and $[H^+]$ (135). MPO is found in high concentrations in neutrophiles and plays important roles in immune response and inflammation (136).

$$Cl^- + H_2O_2 + H^+ \longrightarrow HOCl + H_2O \qquad (51)$$

Recently, NO has been shown to modulate the activity of MPO in two ways depending upon the relative concentrations of NO and H_2O_2 (Scheme 9) (29,137). When H_2O_2 is present, NO at low concentrations ($<2.5\,\mu M$) serves as a one-electron substrate for compounds I and II in the classic peroxidase cycle (Scheme 9, pathway C). NO accelerates the formation and decay of compound II by 20 and 44 times, respectively, effectively increasing the overall catalytic rate by over three orders of magnitude (29). Rapid hydrolysis of the NO oxidation product NO^+ generates nitrite ion which is also a one electron substrate for compounds I and II. The reactive nitrogen species NO_2Cl and NO_2 are formed by the reaction of nitrite with HOCl and by one electron oxidation of NO_2^-

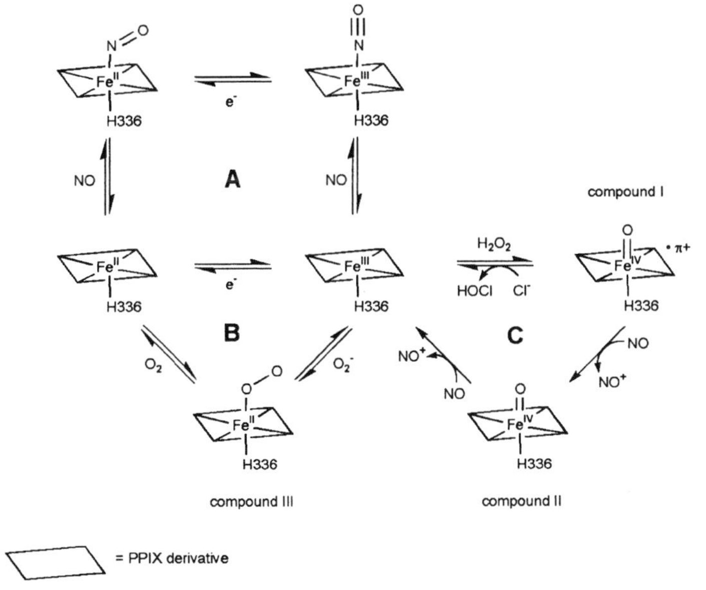

SCHEME 9.

with compound I, respectively (*138*). Thus peroxidases may serve as an important sink in vivo for NO giving rise to strong oxidizing species during inflammatory response.

At higher NO concentrations, MPO activity is inhibited through formation of an inactive ferric nitrosyl complex MPO(NO); the rate constant k_{on} is $1.07 \times 10^6 \, M^{-1} s^{-1}$ and the dissociation rate constant, k_{off}, is $10.8 \, s^{-1}$ (pH 7.0 phosphate buffer at $10\,°C$) (Scheme 9, pathway A). However, the inhibitory effects of NO are reduced in the presence of plasma levels of Cl^- (100 mM) where k_{on} and k_{off} rate constants were determined to be $1.5 \times 10^5 \, M^{-1} s^{-1}$ and $22.8 \, s^{-1}$, respectively. The modulating effects of NO on MPO activity parallel that of O_2^- which accelerates activity by serving as a substrate for compound II and inhibits activity by acting as a ligand for MPO (Scheme 9, pathway B) (*29*).

Nitric oxide and nitrite react with other peroxidase enzymes such as horseradish peroxidase (HRP) (*138a,139*), lactoperoxidase (*138a*) and eosinophil peroxidase (*140*) similarly. The rate constants for reaction of NO with compounds I and II in HRP were found to be $7.0 \times 10^5 \, M^{-1} s^{-1}$ and $1.3 \times 10^6 \, M^{-1} s^{-1}$, respectively (*139*). Catalytic consumption of NO as measured by an NO sensitive electrode in the presence of HRP compounds I and II is shown in Fig. 5 where accelerated consumption of NO is achieved even in deoxygenated solutions (*140*).

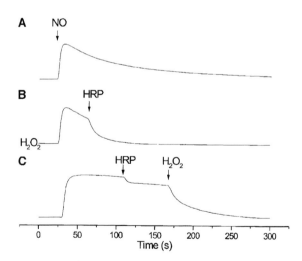

Fig. 5. NO consumption by the HRP/H_2O_2 system as recorded by an NO sensitive electrode. (A) autoxidation of 5 mM NO in air saturated phosphate buffer (pH 7.4). (B) Addition of 5 mM HRP to a solution containing 5 mM NO and 5 mM H_2O_2. (C) Addition of 0.5 mM HRP followed by addition of 5 mM H_2O_2 in anaerobic conditions (*139*). (Reprinted with permission from Ref. (*139*). Copyright 1999 Academic Press.)

A detailed mechanistic study of the interactions of NO and NO_2^- with $MbFe^{IV} = O$ (an analog of compound II above) was recently conducted by Herold and Rehmann (141). These authors demonstrated that $MbFe^{IV} = O$ reacts rapidly with NO to form an intermediate possessing a characteristics spectrum, consistent with the formation of the ferric nitrito complex $MbFe^{III}(ONO)$. This reaction has analogy with the reactions of NO with the $Cr^{IV} = O$ complex described above (112).

B. Formation of SNO-Hb

Given the many chemical and physiological pathways by which NO may be consumed one may question how NO can serve as an effective physiological signaling agent. It has been suggested that the physiological activity of NO may be preserved by sequestering NO in the form of thionitrosyl compounds, specifically S-nitrosylated cysteine-β93 in hemoglobin (SNO-Hb) (142,143). In an experiment by Gow and Stamler, a bolus injection of NO is made into aerated solutions of Hb to give a final ratio of NO : Hb of 1 : 100 (142). At sub-stoichiometric concentrations, NO binds to a portion of the available iron(II) heme sites in the T-state (tensed) or the unsaturated R-state (relaxed) Hb. SNO-Hb was reported to be formed when solutions of Hb(NO) are rapidly exposed to air and the NO : Hb ratio is in the physiological range of approximately 1 : 100. The allosteric transition from the T to R state is induced by an increase in oxygen tension. It was claimed that 75–85% of the added NO resulted in formation of Cys-β93-SNO through intramolecular NO transfer and reduction of O_2 to O_2^- (143,144).

A proposed physiological NO cycle is summarized in Scheme 10. In this scheme NO binds to Hb in the T-state in veinous (hypoxic) blood. Oxygenation in the lungs is accompanied by an allosteric transition and an autonitrosylation reaction to form SNO-Hb(O_2). Further transport through the circulatory system to regions of lower oxygen tension induces an allosteric transition back to the T-state with subsequent release of NO and O_2. While in the R-state NO may also be released from SNO-Hb to other biological thiols such as glutathione (GSH) through transnitrosylation reactions (142). The presence of SNO-Hb was measured in-vivo, in veinous and arterial blood in rats revealing an increased concentration of SNO-Hb in arterial (oxygenated) blood (145). Subsequent to these reports, an X-ray crystal structure of SNO-Hb(NO) has been obtained at 1.8 Å resolution (PDB ascension code 1BUW) (146).

A recent investigation into the reaction of NO with red blood cells (RBC) and with hemoglobin provides additional insight into the formation of Hb(NO) and SNO-Hb in vitro (147). In this work it is suggested

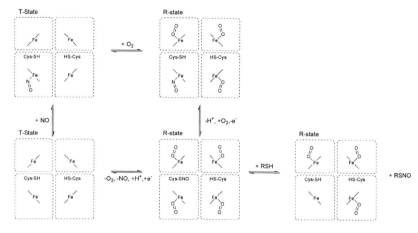

SCHEME 10.

that the heterogeneous conditions present during bolus addition of NO
to solutions of either RBC or Hb(O$_2$) result in the formation of Hb(NO),
not through the expected direct nitrosylation of Hb(O$_2$), but through a
three-step mechanism outlined in Eqs. (52)–(54), where a key step is the
reductive nitrosylation (Eq. (53)).

$$NO + Hb(O_2) \longrightarrow metHb + NO_3^- \qquad (52)$$

$$metHb + NO \longrightarrow Hb + \text{"}NO^+\text{"} \qquad (53)$$

$$NO + Hb \longrightarrow Hb(NO) \qquad (54)$$

Reaction (52) occurs at the gradient interface of the bolus addition until
local Hb(O$_2$) concentrations have been reduced, at which point additional NO reduces the iron(III) to iron(II) which can further react with
free NO to form Hb(NO). The validity of this mechanism was verified by
the observation that addition of CN$^-$ ion, which binds irreversibly to
metHb to form metHb(CN), significantly attenuated the formation of
Hb(NO) in both cell-free Hb and RBC. Mathematical models used to
simulate bolus addition of NO to cell-free Hb and RBC were compatible
with the experimental results (147). In the above experiments, SNO-Hb
was a minor reaction product and was formed even in the presence of
10 mM CN$^-$, suggesting that RSNO formation does not occur as a result
of (hydrolyzed) NO$^+$ formation during metHb reduction. However, formation of SNO-Hb was not detectable when NO was added as a bolus injection to RBC or through thermal decomposition of DEA/NO in cell free
Hb (DEA/NO = 2-(N,N-diethylamino)diazenolate). SNO-Hb was observed

when bolus NO was added to cell-free Hb in aerobic conditions. Furthermore, the SNO-concentration increased with each bolus addition suggesting that contamination may be occurring through the needle used to make the NO injections.

Zhang and Hogg have recently reported similar artifacts when NO is added as a bolus injection to solutions of $Mb(O_2)$ (148). These authors observed an oxygen concentration dependence when NO was added as a bolus injection, and no oxygen concentration dependence when NO was added homogenously with DEA/NO. As mentioned above the reaction of NO with $Mb(O_2)$ (Eq. (47)) is used as a quantitative test for NO (105). When NO was added as a bolus injection to aerated $Mb(O_2)$, the measured [NO] was half that found in similar experiments with partially deaerated $Mb(O_2)$. Analogous experiments were performed in the presence of GSH to test for the formation of GSNO. No S-nitrosylated product was observed when NO was added to a solution of $Mb(O_2)$ and GSH using an NO donor compound (DEA/NO). However, when NO was added as a bolus injection to a similar solution of $Mb(O_2)$ and GSH, formation of GSNO was observed (148). These observations suggest that while the reaction of NO with O_2 is third-order overall (Eq. (46)), and slow at physiological concentrations of NO and O_2 (99,100), the locally high concentrations of NO that exist during heterogeneous mixing from bolus injection of NO to aerated solutions can generate nitrosating species such as N_2O_3. As noted above, nitrite ion, another ubiquitous impurity in aqueous NO solutions, can catalyze reductive nitrosylation, quite possibly via N_2O_3 formation (69). Such considerations place increasing emphasis on the importance of evaluating the role of the NO_x impurities common to nitric oxide solutions on the potential artifacts which may obscure the true reactivity of NO under physiological conditions (90b).

V. Overview and Summary

This Chapter has reviewed certain substitution and redox mechanism studies involving the interaction of nitric oxide with transition metal complexes with a view toward illustrating how these may be relevant to the known biological functions of NO in mammalian biology (1,2). The large volume of relatively recent information regarding the chemistry, biochemistry and pathobiology of NO is daunting, but one may expect certain features of NO chemistry to dominate. Clearly, NO as a stable free radical participates very readily in one electron events such as coupling to other free radicals. For example, the NO reaction with O_2^-

to form $OONO^-$ proceeds at rates near the diffusion limit (*97*). However, whether the trapping of superoxide in this manner is a particularly dangerous contributor to oxidative and nitrosative stress, owing to formation of the peroxynitrite ion, or is protective to the organism, owing to deactivation of the even more deleterious superoxide ion, remains hotly debated. Such rapid reactions may also be expected with other free radicals and very likely participate in the mechanism for NO sensitization of radiation damage to cells (*149*). Such chemistry may offer therapeutic potential for NO donors in radiation treatment of tumors.

NO is also very reactive with redox active metal centers, especially if these are ligand substitution labile. Substitution reactions of NO on metal centers as well as reactions with radical species generally display kinetic rate laws that show a first-order dependence on [NO]. This contrasts to reactions where the substrate undergoes two electron changes, for example, the oxidation of Ph_3P (*150*). The latter requires two equivalents of NO in a third-order process unless promoted or catalyzed by another reagent. Of particular importance is direct autoxidation, the kinetics of which are third order, second-order in [NO]. Thus NO autoxidation and related third order processes are relatively slow under bioregulatory conditions, i.e. low [NO]. Autoxidation and the formation of highly reactive oxidizing and nitrosative species such as N_2O_3 may be important in the locale of immune response to pathogen infection, where higher [NO] is the norm (*99*). Although the short term function is protection of the host organism, the resulting oxidative and nitrosative stress from the generation of N_2O_3 and perhaps other species such as peroxynitrite may have a long term deleterious effects on the host.

With respect to the bioregulatory roles of NO in blood pressure control and neurological function, the principal action centers on the reaction with a metal center to form a nitrosyl complex, namely the activation of sGC by reaction with the iron(II) site of that ferroheme enzyme (*130,131*). Given the low NO concentrations generated for such functions, the "on" reaction must be very fast in order to provide the appropriate response to stimuli. Such substitution reactions are generally facile when the metal center has a vacant coordination site or is very labile, but are likely to be quite slow for non-labile coordinatively saturated metals even when the net reaction is very favorable. Accelerating ligand lability by an associative mechanism is certainly plausible, but the only convincing example of this mechanism involves NO displacement of NH_3 on the Ru^{III} complex $Ru(NH_3)_6^{3+}$ (Eq. (7)), and the second-order rate constant for this reaction is slow (*17*). The "off" reaction of metal nitrosyls may be equally important given, for example, that this is a likely mechanism for deactivation of sGC. Alternatively, since NO has been demonstrated to inhibit certain metalloenzymes, the "off" reaction serves to reactivate such systems. Clearly understanding the mechanisms of such metal nitrosyl reactions as functions of the media, conditions and the ligand field is crucial to interpreting the role of NO in activation as well as inhibition of metalloproteins.

There is also considerable biological interest in reactions of NO with ligands coordinated to a redox active metal. For example, the facile second-order trapping of NO by $Mb(O_2)$ or $Hb(O_2)$ is very fast and is mechanistically very distinct (*106,111*) from the third order autoxidation of NO (*99*). The result is oxidation of Fe(II) to Fe(III) with simultaneous NO oxidation to NO_3^-. In contrast, the facile reaction of $M^{IV} = O$ species (M = Fe or Cr) with NO to give $M^{III}(ONO)$ (*112,113*) leads to reduction of the metal along with oxidation of NO. The reaction with $Hb(O_2)$ is generally believed to be an important sink for NO in the cardiovascular system, while trapping of ferryl intermediates (or other strong oxidants) by NO may play a role in reducing oxidative stress. Oxidative stress may also be reduced by NO coordination to metal centers that are catalysts for Fenton chemistry (the generation of strongly oxidizing intermediates from H_2O_2) (*151*). It might also be noted that, when oxidative degradation of meats is inhibited by curing with nitrite ions, stable metal nitrosyls are formed (*152*). On the other hand, the ambiguous nature of possible NO function in oxidative stress is illustrated by its reaction with catalase (*121*), thereby inhibiting the protective function of this enzyme which destroys endogenous H_2O_2.

Forming a metal complex may serve to activate NO toward either nucleophilic or electrophilic attack depending on the nature of the metal complex and its oxidation state. Of particular interest biologically are the reactions with nucleophiles since this may well be a mechanism for thionitrosyl formation (e.g. Eq. (55)) as well as for reductively labilizing metals in insoluble matrices like ferritin.

$$L_nM^{III}(NO) + RSH \longrightarrow RSNO + L_nM^{II} + H^+ \qquad (55)$$

The metal center may also promote NO reactivity toward disproportionation or substrate oxidation ($2NO + S \rightarrow SO + N_2O$) by serving as a template where several NO molecules are gathered in association with a substrate molecule. These reactions are less likely to be important biologically given the relatively low [NO] generated with the possible exceptions of localized higher concentrations generated during immune response.

248 P.C. FORD et al.

Lastly, as an editorial comment, it should be noted that there is a legitimate concern that a significant percentage of the seemingly countless reports of the potential roles played by nitric oxide in biomedical issues may be compromised by the presence of NO_x impurities such as N_2O_3 and NO_2^-. Such impurities may have an excessive impact in laboratory studies involving bolus addition of NO owing to the third-order autoxidation mechanism as well as the tendency for NO_x species to be much more soluble in water and other solvents than NO itself. While these concerns should not discredit such reports, they add to the importance that any experimental demonstration of chemical or biological mechanism be supported by careful control studies to assess the impacts of NO_x impurities.

ACKNOWLEDGEMENTS

Studies related to the mechanisms of nitric oxide reactions with transition metal complexes in this laboratory were supported by grants from the U.S. National Science Foundation, by a Collaborative UC/Los Alamos National Laboratory Research grant, by a grant from the U.S. Japan Cooperative Research Program (Photoconversion/ Photosynthesis) (NSF INT 9116346), and by a grant from the ACS Petroleum Research Fund. We thank the students and postdoctoral fellows at UC Santa Barbara who participated in this research and acknowledge collaborative studies with Dr. David Wink (National Cancer Institute, Bethesda MD, USA), Dr. Mikio Hoshino (RIKEN, Wako-shi, Japan) and Dr. Jon Schoonover (Los Alamos National Laboratory).

VI. List of Abbreviations

Ar	p-$CH_3OC_6H_4^-$
bpy	2,2′-bipyridine
Cat	Catalase
cGMP	cyclic guanylyl monophosphate
Cyt	cytochrome-c
DEA/NO	2-(N,N-diethylamino)diazenolate
dmp	2,9-dimethyl-1,10-phenanthroline
DMSO	Dimethylsulfoxide
edta	ethylendiaminetetraacetic acid
ESR	electron spin resonance
FTIR	Fourier transform infrared
GSH	Glutathione
GTP	guanylyl triphosphate
Hb	ferro-hemoglobin

Hedtra	hydroxyethylenediamintriaceteic acid
HRP	horseradish peroxidase
ImH$_2$	Imidazole
IR	Infrared
Mb	ferro-myoglobin
metMb	ferri-myoglobin or met-myoglobin
MCPH	monochelated protoheme
metHb	ferri-hemoglobin or met-hemoglobin
MLCT	metal to ligand charge transfer
MPO	Myeloperoxidase
NHE	normal hydrogen electrode
NOS	nitric oxide synthase
NP	nitroprusside ion
Nuc	a general nucleophile
Nta	nitrilotriacetic acid
OBTPP	Octabromotetraphenylporphine
OEP	Octaethylporphine
Pc	Phthalocyaninato
Phen	1,10-phenanthroline
Por	a general porphyrinato ligand
PPIX	protoporphyrin IX
RBC	red blood cell
RSNO	a general thionitrosyl
Salen	N,N'-bis(salicylidenato)ethylenediamine
salophen	N,N'-1,2-phenylenediamine-bis (salicylidenimato)
SCE	standard calomel electrode
SGC	soluble guanylyl cyclase
SNO-Hb	S-nitrosylated Cys 93 Hemoglobin
Sol	a solvent molecule
^{t}Bu$_4$-salen	N,N'-ethylenebis(3,5-di-t-butyl-salicylideneiminato) dianion
^{t}Bu$_4$-salophen	N,N'-1.2-phenylenediaminebis (3-t-butyl-salicylideneiminato) dianion
TC-5,5	tropocoronand ligand
TMPS	tetra(sulfonatomesityl)porphyrin
TmTP	tetra-m-tolylporphyrin
Tp^{R2}	tris(3-(R^2)-5-methylpyrazol-1-yl)hydroborate
tpa	tris[(2-pyridyl)methyl]amine
TPP	meso-tetraphenylporphyrin
TPPS	tetra(4-sulfonatophenyl)porphine

REFERENCES

1. (a) Moncada, S.; Palmer, R. M. J.; Higgs, E. A. *Pharmacol. Rev.* **1991**, *43*, 109; (b) Feldman, P. L.; Griffith, O. W.; Stuehr, D. J. *Chem. Eng. News* **1993**, *71*, 10, 26; (c) Butler, A. R.; Williams, D. L. *Chem. Soc. Rev.* **1993**, 233; (d) *Methods in Nitric Oxide Research*; Feelisch M.; Stamler, J. S. Eds., John Wiley and Sons, Chichester, England, 1996, and references therein; (e) Wink, D. A.; Hanbauer, I.; Grisham, M. B.; Laval, F.; Nims, R. W.; Laval, J.; Cook, J.; Pacelli, R.; Liebmann, J.; Krishna, M.; Ford, P. C.; Mitchell, J. B. *Curr. Top. Cell. Regul.* **1996**, *34*, 159.
2. *"Nitric Oxide: Biology and Pathobiology"*; Eds. Ignarro, L. J.2000. Academic Press: San Diego, 2000.
3. Schwartz, S. E.; White, W. H. *"Trace Atmospheric Constitutents. Properties, Transformation and Fates"*; J. Wiley and Sons: New York, 1983.
4. *"Nitric Oxide and Infection"*; Eds. Fang, F. C.; Kluwer Academic/Plenum Publishers: New York, 1999.
5. (a) Eisenberg, R.; Meyer, C. D. *Acc. Chem. Res.* **1975**, *8*, 26; (b) McCleverty, J. A. *Chem. Rev.* **1979**, *79*, 53; (c) Bottomley, F. *React. Coord. Ligands* **1989**, *2*, 115; (d) Fanning, J. C. *Coord. Chem. Rev.* **1991**, *110*, 235; (e) Richter-Addo, G. B.; Legzdins, P. *"Metal Nitrosyls"*; Oxford Univ. Press: NY, 1992; (f) Westcott, B. L.; Enemark, J. H. Ed. Solomon, E. I., Lever, A. B. P., Eds.; In *Inorganic Electronic Structure and Spectroscopy*, 1999; Vol. 2; John Wiley and Sons, Inc.: New York, p. 403; (g) Cheng, L.; Richter-Addo, G. B. Chap. 33 in *The Porphyrin Handbook*, Vol. 4, Kadish, K. M.; Smith, K. M.; Guilard, R. Eds., Academic Press: New York, 2000; (h) Greenwood, N. N.; Earnshaw, A. *Chemistry of the Elements*; Pergamon Press: Oxford, 1993, p. 508; (i) Mingos, D. M. P.; Sherman, D. J. *Adv. Inorg. Chem.* **1989**, *34*, 293; (j) Bonner, F. T.; Stedman, G. Chapter 1 in Ref. (*1d*); (k) Ford, P. C.; Lorkovic, I. M. *Chem. Rev.* **2002**, *102*, 993.
6. (a) Dean J. A. *Lange's Handbook of Chemistry*, 13th edn., McGraw-Hill Book Company: New York, 1985; (b) Armor, J. N. *J. Chem. Eng. Data* **1974**, *19*, 84; (c) *IUPAC Solubility Series: Oxides of Nitrogen*; Ed. Young, C.L.; Pergamon Press: Oxford, UK, 1983.
7. (a) Gally, J. A.; Montague, P. R.; Reeke, G. N.; Edelman, G. M. *Proc. Natl. Acad. Sci. USA* **1990**, *87*, 3547; (b) Lancaster, J. R., Jr. *Nitric Oxide, Biology and Chemistry* **1997**, *1*, 18.
8. (a) Enemark, J. H.; Feltham, R. D. *J. Am. Chem. Soc.* **1974**, *96*, 5002; (b) Feltham, R. D.; Enemark, J. H. *Topics in Stereochemistry* **1981**, *12*, 155.
9. (a) Fomitchev, D. V.; Furlani, T. R.; Coppens, P. *Inorg. Chem.* **1998**, *37*, 1519; (b) Fomitchev, D. V.; Novozhilova, I.; Coppens, P. *Tetrahedron* **2000**, *56*, 6813; (c) Cheng, L.; Novozhilova, I.; Kim, C.; Kovalevsky, A.; Bagley, K. A.; Coppens, P.; Richter-Addo, G. B. *J. Am. Chem. Soc.* **2000**, *122*, 7142; (d) Fomitchev, D. V.; Coppens, P.; Li, T. S.; Bagley, K. A.; Chen, L.; Richter-Addo, G., *Chem. Commun.* **1999**, 2013.
10. (a) Carducci, M. D.; Pressprich, M. R.; Coppens, P. *J. Am. Chem. Soc.* **1997**, *119*, 2669; (b) Kim, C.; Novozhilova, I.; Goodman, M. S.; Bagley, K. A.; Coppens, P. *Inorg. Chem.* **2000**, *39*, 5791; (c) Coppens, P.; Novozhilova, I.; Kavalevky, A. *Chem. Rev.* **2002**, *102*, 861.
11. Norton, J. R.; Collman, J. P.; Dolcetti, G.; Robinson, W. T. *Inorg. Chem.* **1972**, *11*, 382.
12. Lee, K. Y.; Kuchynka, D. J.; Kochi, J. K. *Inorg. Chem.* **1990**, *29*, 4196.
13. (a) Shafirovich, V.; Lymar, S. V. *Proc. Natl. Acad. Sci. USA* **2002**, *99*, 7340; (b) Bartberger, M. D.; Liu, W.; Ford, E.; Miranda, K. M.; Switzer, C.; Fukuto, J. M.; Farmer, P. J.; Wink, D. A.; Houtz, K. N. *Proc. Natl. Acad. Sci. USA* **2002**, *99*, 10958.
14. *Handbook of Chemistry and Physics*; Ed. Lide, D. R. CRC Press: Boca Raton, 1990.
15. Stanbury, D. M. *Adv. Inorg. Chem.* **1989**, *33*, 69.

16. Examples are: (a) Fukuto, J. M.; Hobbs, A. J.; Ignarro, L. J. *Biochem. Biophys. Res. Commun.* **1993**, *196*, 707; (b) Schmidt, H. H. H. W.; Hofmann, H.; Schindler, U.; Shutenko, Z. S.; Cunningham, D. D.; Feelisch, M. *Proc. Natl. Acad. Sci. USA* **1996**, *93*, 14 492; (c) DeMastter, E. G.; Redfern, B.; Nagasawa, H. T. *Biochem. Pharmacol.* **1998**, *55*, 20 007; (d) Hughes, M. N. *Biochim. Biophys. Acta-Bioenergetics* **1999**, *1411*, 263; (e) Chazotte-Aubert, L.; Oikawa, S.; Gilibert, I.; Bianchini, F.; Kawanishi, S.; Ohshima, H. *J. Biol. Chem.* **1999**, *274*, 20 909; (f) Ellis, A.; Li, C.; Rand. M. J. *Br. J. Pharm.* **2000**, *129*, 315; (g) Xu, Y. P.; Alavanja, M. M.; Johnson, V. L.; Yasaki, G.; King, S. B. *Tet. Lett.* **2000**, *41*, 4265; (h) Miranda, K. M.; Espey, M. G.; Yamada, K.; Krishna, M.; Ludwick, N.; Kim, S.; Jourd'heuil, D.; Grisham, M. B.; Feelisch, M.; Fukuto, J. M.; Wink, D. A. *J. Biol. Chem.* **2001**, *276*, 1720; (i) Bartberger, M. D.; Fukuto, J. M.; Houk, K. N. *Proc. Natl. Acad. Sci. USA* **2001**, *98*, 2194; (j) Fukuto, J. M. Personal communication.

17. (a) Armor, J. N.; Scheidegger, H. A.; Taube, H. *J. Am. Chem. Soc.* **1968**, *90*, 5928; (b) Armor, J. N.; Pell, S. D. *J. Am. Chem. Soc.* **1973**, *95*, 7625.

18. Czap, A.; van Eldik, R. *Dalton Trans.* **2003**, 665.

19. (a) Works, C. F.; Ford, P. C. *J. Am. Chem. Soc.* **2000**, *122*, 7592; (b) Works, C. F. Ph.D. Dissertation, University of California, Santa Barbara, 2001.

20. E.g., (a) Moore, E.G.; Gibson, Q. H. *J. Biol. Chem.* **1976**, *251*, 2788; (b) Tamura, M.; Kobayashi, K.; Hayashi, K. *FEBS Lett.* **1978**, *88*, 124. (c) Rose, E. J.; Hoffman, B. M. *J. Am. Chem. Soc.* **1983**, *105*, 2866; (d) Hoshino, M.; Laverman, L.; Ford, P.C. *Coord. Chem. Rev.* **1999**, *187*, 75 and references therein.

21. (a) Laverman, L. E.; Ford, P. C. *Chem. Commun.* **1999**, 1843; (b) Laverman, L. E.; Hoshino, M.; Ford, P. C. *J. Am. Chem. Soc.* **1997**, *119*, 12 663; (c) Laverman, L. E.; Ford, P. C. *J. Am. Chem. Soc.* **2001**, *123*, 11 614.

22. Hoshino, M.; Ozawa, K.; Seki, H.; Ford, P. C. *J. Am. Chem. Soc.* **1993**, *115*, 9568.

23. Laverman, L. E.; Wanat, A.; Oszajca, J.; Stochel, G.; Ford, P. C.; van Eldik, R. *J. Am. Chem. Soc.* **2001**, *123*, 285.

24. (a) Abu-Soud, H. M.; Ichimori, K.; Presta, A.; Stuehr, D. J. *J. Biol. Chem.* **2000**, *275*, 17 349; (b) Scheele, J. S.; Bruner, E.; Kharitonov, V. G.; Martásek, P.; Roman, L. J.; Masters, B. S.; Sharma, V. S.; Magde, D. *J. Biol. Chem.* **1999**, *274*, 13 105.

25. Andersen, J. F.; Ding, X. D.; Balfour, C.; Shokhireva, T. K.; Champagne, D. E.; Walker, F. A.; Montfort, W. R. *Biochemistry* **2000**, *39*, 10 118.

26. (a) Olson, J. S.; Phillips, G. N. *J. Biol. Chem.* **1996**, *271*, 17 593; (b) Ikeda-Saito, M.; Dou, Y.; Yonetani, T.; Olson, J. S.; Li, T.; Regan, R.; Gibson, Q. H. *J. Biol. Chem.* **1993**, *268*, 6855.

27. (a) Sharma, V. S.; Magde, D. *Methods – Comp. to Meth. Enzym.* **1999**, *19*, 494; (b) Kharitonov, V. G.; Russwurm, M.; Madge, D.; Sharma, V. S.; Koesling, D. *Biochem. Biophys. Res. Comm.* **1997**, *239*, 284.

28. Takano, T.; Kallai, O. B.; Swanson, R.; Dickerson, R. E. *J. Biol. Chem.* **1973**, *248*, 5244.

29. Abu-Soud, H. M.; Hazen, S. L. *J. Biol. Chem.* **2000**, *275*, 37 524.

30. Wanat, A.; Schneppensieper, T.; Stochel, G.; van Eldik, R.; Eckhard, B.; Wieghardt, K. *Inorg. Chem.* **2002**, *41*, 4.

31. (a) Schneppensieper, T.; Wanat, A.; Stochel, G.; Goldstein, S.; Meyerstein, D.; van Eldik, R. *Euro. J. Inorg. Chem.* **2001**, 2317; (b) Schneppensieper, T.; Wanat, A.; Stochel, G; van Eldik, R. *Inorg. Chem.* **2002**, *41*, 2565.

32. Ostrich, I. J.; Gordon, L.; Dodgen, H. W.; Hunt, J. P. *Inorg. Chem.* **1980**, *19*, 619.

33. Schneppensieper, T.; Zahl, A.; van Eldik, R. *Angew. Chem., Int. Ed. Eng.* **2001**, *40*, 1678.

34. (a) Frauenfelder, H.; McMahon, B. H.; Austin, R. H.; Chu, K.; Groves, J. T. *Proc. Natl. Acad. Sci. USA* **2001**, *98*, 2370; (b) Cao, W.; Christian, J. F.; Champion, P. M.; Rosca, F.; Sage, J. T. *Biochemistry* **2001**, *40*, 5728.

35. Hoshino, M.; Kogure, M. *J. Phys. Chem.* **1989**, *93*, 5478.
36. Hoshino, M.; Nagashima, Y.; Seki, H.; De Leo, M.; Ford, P. C. *Inorg. Chem.* **1998**, *37*, 2464.
37. Traylor, T. G.; Luo, J.; Simon, J. A.; Ford, P. C. *J. Am. Chem. Soc.* **1992**, *114*, 4340.
38. Turley, W. D.; Offen, H. W. *J. Phys. Chem.* **1984**, *88*, 3605.
39. Caldin, E. F. *"Fast Reactions in Solution"*; Blackwell: Oxford, 1964.
40. Glasstone, S.; Laidler, K. J.; Eyring, H. *"The Theory of Rate Processes"*; McGraw-Hill: New York, 1941.
41. Barley, M. H.; Meyer, T. J. *J. Am. Chem. Soc.* **1986**, *108*, 5876.
42. (a) Tran, D.; Shelton, B. W.; White, A. H.; Laverman, L. E.; Ford, P. C. *Inorg. Chem.* **1998**, *37*, 2505; (b) Eberson, L.; Radner, F. *Acta Chem. Scand.* **1984**, *B38*, 861.
43. (a) Greenberg, S. S.; Xie, J.; Zatariarn, J. M.; Kapusta, D. R.; Miller, M. J. S. *J. Pharmacol. Exp. Ther.* **1995**, *273*, 257; (b) Brouwer, M.; Chamulitrat, W.; Ferruzzi, G.; Sauls, D. L.; Weinberg, J. B. *Blood* **1996**, *88*, 1857; (c) Kruszyna, H.; Magyar, J. S.; Rochelle, L. G.; Russell, M. A.; Smith, R. P.; Wilcox, D. E. *J. Pharmacol. Exp. Ther.* **1998**, *285*, 665.
44. (a) Wolak, M.; Stochel, G.; Hamza, M.; van Eldik, R. *Inorg. Chem.* **2000**, *39*, 2018; (b) Zheng, D.; Birke, R. L. *J. Am. Chem. Soc.* **2001**, *123*, 4637; (c) Wolak, M.; Zahl, A.; Schneppensieper, T.; Stochel, G.; van Eldik, R. *J. Am. Chem. Soc.* **2001**, *123*, 9780.
45. (a) Gorren, A. C. F.; de Boer, E.; Wever, R. *Biochim. Biophys. Acta* **1987**, *916*, 38; (b) Torres, J.; Wilson, M. T. *Biochim. Biophys. Acta* **1999**, *1411*, 310.
46. (a) Karlin, K. D.; Tyeklár, Z. Eds., *Bioinorganic Chemistry of Copper*; Chapman & Hall, Inc.: New York, 1993; (b) Averill, B. A. *Chem. Rev.* **1996**, *96*, 2951; (c) Halfen, J. A.; Mahapatra, S.; Wilkinson, E. C.; Gengenbach, A. J.; Young, V. G., Jr.; Que, L., Jr.; Tolman, W. B. *J. Am. Chem. Soc.* **1996**, *118*, 763.
47. Firth, R. A.; Hill, H. A. O.; Pratt, J. M.; Williams, R. J. P. *J. Chem. Soc. A* **1969**, 381.
48. Lei, Y.; Anson, F. C. *Inorg. Chem.* **1994**, *33*, 5003.
49. Wolfe, S. K.; Swinehart, J. H. *Inorg. Chem.* **1994**, *33*, 5003.
50. Marcus, R. A. *J. Phys. Chem.* **1968**, *72*, 891.
51. (a) Swinehart, J. H. *Coord. Chem. Rev.* **1967**, *2*, 385; (b) Swinehart, J. H.; Rock, P. A. *Inorg. Chem.* **1966**, *5*, 573; (c) Olabe, J. A.; Gentil, L. A.; Rigotti, G.; Navaza, A. *Inorg. Chem.* **1984**, *23*, 4297; (d) Baraldo, L. M.; Bessega, M. A.; Rigotti, G. E.; Olabe, J. A. *Inorg. Chem.* **1994**, *33*, 5890.
52. (a) Chen, Y.; Lin, F.-T.; Shepherd, R. E. *Inorg. Chem.* **1999**, *38*, 973, and references therein; (b) Reed, C. A.; Roper, W. R. *J. Chem. Soc., Dalton Trans.* **1972**, 1243; (c) Maciejowska, I.; Stasicka, Z.; Stochel, G.; van Eldik, R. *J. Chem. Soc., Dalton Trans.* **1999**, 3643; (d) Douglas, P. G.; Feltham, R. D. *J. Am. Chem. Soc.* **1972**, *94*, 5254; (e) Toma, H. E.; Silva, D. de O.; Saika, J. J. *J. Chem. Res.* **1996**, 456.
53. (a) Butler, A. R.; Calsy-Harrison, A. M.; Glidewell, C.; Sorensen, P. E. *Polyhedron* **1988**, *7*, 1197; (b) Szacilowski, K.; Stochel, G.; Stasicka, Z.; Kisch, H. *New J. Chem.* **1997**, *21*, 893.
54. Baran, E. J.; Muller, A *Angew. Chem. Int. Ed. Engl.* **1969**, *8*, 890.
55. (a) Kathò, Á.; Bodi, Z.; Dozsa, L.; Beck, M. T. *Inorg. Chim. Acta* **1984**, *83*, 145; (b) Dozsa, L.; Kormos, V.; Beck, M. T. *Inorg. Chim. Acta* **1984**, *82*, 69.
56. Chevalier, A. A.; Gentil, L. A.; Amorebieta, V. T.; Gutiérrez, M. M.; Olabe, J. A. *J. Am. Chem. Soc.* **2000**, *122*, 11 238.
57. Bowden, W. L.; Little, W. F.; Meyer, T. J. *J. Am. Chem. Soc.* **1977**, *99*, 4340.
58. (a) Ford, P. C.; Rokicki, A. *Adv. Organometallic Chem.* **1988**, *28*, 139; (b) Angelici, R. J. *Acc. Chem. Res.* **1972**, *5*, 335.
59. Kruszyna, H.; Kruszyna, R.; Rochelle, L. G.; Smith, R. P.; Wilcox, D. E. *Biochem. Pharmacol.* **1993**, *46*, 95.

60. Katz, N. E.; Blesa, M. A.; Olabe, J. A.; Aymonino, P. J. *J. Inorg. Nucl. Chem.* **1980**, *42*, 581.
61. Kathò, Á; Beck, M. T. *Inorg. Chim. Acta* **1988**, *154*, 99.
62. March, J. "*Advanced Organic Chemistry*"; John Wiley and Sons: New York, 1992.
63. Beck, W.; Smedal, H. S. *Angew. Chem. Int. Ed. Engl.* **1966**, *5*, 253.
64. Mu, X. H.; Kadish, K. M. *Inorg. Chem.* **1988**, *27*, 4720.
65. (a) Keilin, D.; Hartree, E. F. *Nature* **1937**, *139*, 548; (b) Ehrenberg, A.; Szczepkowski, T. W. *Acta Chem. Scand.* **1960**, *14*, 1684; (c) Gwost, D.; Caulton, K. G. *J. Chem. Soc. Chem. Comm.* **1973**, 64; *Inorg. Chem.* **1973**, *12*, 2095; (d) Yoshimura, T.; Suzuki, S.; Nakahara, A.; Iwasaki, H.; Masuko, M.; Matsubara, T. *Biochemistry* **1986**, *25*, 2436.
66. (a) Chien, J. C. W. *J. Am. Chem. Soc.* **1969**, *91*, 2166; (b) Addison, A. W.; Stephanos, J. J. *Biochemistry* **1986**, *25*, 4104; (c) Rogers, K. R.; Lukat-Rogers, G. S. *J. Biol. Inorg. Chem.* **2000**, *5*, 642.
67. Hoshino, M.; Maeda, M.; Konishi, R.; Seki, H.; Ford, P. C. *J. Am. Chem. Soc.* **1996**, *118*, 5702.
68. Reichenbach, G.; Sabatini, S.; Palombari, R.; Palmerini, C. A. *Nitric Oxide* **2001**, *5*, 395.
69. Fernandez, B. O.; Lorkovic, I. M.; Ford, P. C. *Inorg. Chem.* **2003**, *42*, 2.
70. (a) *Standard Reduction Potentials in Aqueous Solution*; Eds. Bard, A. J.; Parsons, R.; Jordan, J.; Marcel Dekker, Inc.: New York, 1985, p. 127; (b) Pearson, R. G. *J. Am. Chem. Soc.* **1986**, *108*, 6109.
71. Fukuto, J. M.; Cho, J. Y.; Switzer, C. H. "*Nitric Oxide: Biology and Pathobiology*"; Ed. Ignarro, L. J.; Academic Press: San Diego, 2000, Chapter 2.
72. (a) Enemark, J. H.; Feltham, R. D.; Rikker-Nappier, J.; Bizot, K. F. *Inorg. Chem.* **1975**, *14*, 624; (b) Stevens, R. E.; Guettler, R. D.; Gladfelter, W. L. *Inorg. Chem.* **1990**, *29*, 451; (c) Legzdins, P.; Nurse, C. R.; Rettig, S. J. *J. Am. Chem. Soc.* **1983**, *105*, 3727; Sharp, W. B.; Legzdins, P.; Patric, B. O. *J. Am. Chem. Soc.* **2001**, *123*, 8144.
73. (a) Lançon, D.; Kadish, K. M. *J. Am. Chem. Soc.* **1982**, *104*, 2042; (b) Liu, Y. M.; Desilva, C.; Ryan, M. D. *Inorg. Chim. Acta* **1997**, *258*, 247; (c) Liu, Y.; Ryan, M. D. *J. Electroanal. Chem.* **1994**, *368*, 201.
74. (a) Bayachou, M.; Lin, R.; Cho, W.; Farmer, P. J. *J. Am. Chem. Soc.* **1998**, *102*, 9888; (b) Lin, R.; Farmer, P. J. *J. Am. Chem. Soc.* **2000**, *122*, 2393.
75. Tovrog, B. S.; Diamond, S. E.; Mares, F.; Szalkiewicz, A. *J. Am. Chem. Soc.* **1981**, *103*, 3522.
76. (a) Berthelot, P.-E. M. *Compt. Rend.* **1891**, *112*, 1243; (b) Griffith, W. P.; Lewis, J.; Wilkinson, G. *J. Chem. Soc.* **1959**, 1775; (c) Booth, G.; Chatt, J. *J. Chem. Soc.* **1962**, 2099; (d) Feltham, R. D.; Carriel, J. T. *Inorg. Chem.* **1964**, *3*, 121.
77. (a) Hughes, W. B. *J. Chem. Soc., Chem. Comm.* **1969**, 1126; (b) Gans, P. *J. Chem. Soc. A.* **1967**, 943; (c) Rossi, M.; Sacco, A. *J. Chem. Soc., Chem. Comm.* **1971**, 694; (d) Gargano, M.; Giannoccaro, P.; Rossi, M.; Sacco, A.; Vasapollo, G. *Gazz. Chim. Ital.* **1975**, *105*, 1279; (e) Gwost, D.; Caulton, K. G. *Inorg. Chem.* **1974**, *13*, 414.
78. (a) Ruggiero, C. E.; Carrier, S. M.; Tolman, W. B. *Angew. Chem., Int. Ed. Engl.* **1994**, *33*, 895; (b) Schneider, J. L.; Carrier, S. M.; Ruggiero, C. E.; Young, V. G., Jr.; Tolman, W. B. *J. Am. Chem. Soc.* **1998**, *120*, 11 408.
79. Casewit, C. J.; Rappe, A. K. *J. Catal.* **1984**, *89*, 250.
80. Mahapatra, S. M.; Halfen, J. A.; Tolman, W. B. *J. Chem. Soc., Chem Comm.* **1994**, 1625.
81. (a) Komeda, N.; Nagao, H.; Kushi, Y.; Adachi, G.; Suzuki, M.; Uehara, A.; Tanaka, K. *Bull. Chem. Soc. Jpn.* **1995**, *68*, 581; (b) Tanaka, K.; Komeda, N.; Matsui, T. *Inorg. Chem.* **1991**, *30*, 3282.
82. (a) Franz, K. J.; Lippard, S. J. *J. Am. Chem. Soc.* **1998**, *120*, 9034; (b) Franz, K. J.; Lippard, S. J. *J. Am. Chem. Soc.* **1999**, *121*, 10 504.
83. Miranda, K. M.; Bu, X.; Lorkovic, I. M.; Ford, P. C. *Inorg. Chem.* **1997**, *36*, 4838.

84. (a) Kadish, K. M.; Adamian, V. A.; Caemelbecke, E. V.; Tan, Z.; Tagliatesta, P.; Bianco, P.; Boschi, T.; Yi, G.-B.; Khan, M. A.; Richter-Addo, G. B. *Inorg. Chem.* **1996**, *35*, 1343; (b) Bohle, D. S.; Goodson, P. A.; Smith, B. D. *Polyhedron* **1996**, *15*, 3147; (c) Yi, G.-B.; Khan, M. A.; Richter-Addo, G. B. *Chem. Commun.* **1996**, 2045; (d) Yi, G.-B.; Khan, M. A.; Richter-Addo, G. B. *Inorg. Chem.* **1996**, *35*, 3453; (f) Hodge, S. J.; Wang, L.-S.; Khan, M. A.; Young, V. G.; Richter-Addo, G. B. *Chem. Commun.* **1996**, 2283; (g) Bohle, D. S.; Hung, C.-H.; Smith, B. D. *Inorg. Chem.* **1998**, *37*, 5798.

85. Lorkovic, I. M.; Ford, P. C. *Inorg. Chem.* **1999**, *38*, 1467.

86. Lorkovic, I. M.; Ford, P. C. *Chem. Commun.* **1998**, 1225.

87. Lorkovic, I. M.; Miranda, K. M.; Lee, B.; Bernhard, S.; Schoonover, J. R.; Ford, P. C. *J. Am. Chem. Soc.* **1998**, *120*, 11 674.

88. Leal, F.; Lorkovic, I.; Ford, P. C. Unpublished observations. Presented at the National Meeting of the American Chemical Society, San Diego, April 2001.

89. (a) Yoshimura, T. *Inorg. Chem. Acta* **1984**, *83*, 17; (b) Settin, M. F.; Fanning, J. C. *Inorg. Chem.* **1988**, 27, 1431; (c) Ellison, M.; Schulz, C. E.; Scheidt, W. R. *Inorg. Chem.* **1999**, *38*, 100; (d) Lin, R.; Farmer, P. J. *J. Am. Chem. Soc.* **2001**, *123*, 1143.

90. (a) Lorkovic, I. M.; Ford, P. C. *Inorg. Chem.* **2000**, *39*, 632; (b) Lim, M. D.; Lorkovic, I. M.; Wedeking, K.; Zanella, A. A.; Works, C. F.; Massick, S. M.; Ford, P. C. *J. Am. Chem. Soc.* **2002**, *124*, 9737; (c) Kurtikyan, T. S.; Martrosyan, G. G.; Lorkovic, I. M.; Ford, P. C. *J. Am. Chem. Soc.* **2002**, *124*, 10 124.

91. Wayland, B. B.; Olsen, L. W. *J. Am. Chem. Soc.* **1974**, *96*, 6037.

92. Lorkovic, I. M.; Ford, P. C. *J. Am. Chem. Soc.* **2000**, *122*, 6516.

93. Olson, L. W.; Schaeper, D.; Lançon, D.; Kadish, K. M. *J. Am. Chem. Soc.* **1982**, *104*, 2042.

94. Göldner, M; Geniffke, B.; Franken, A.; Murray, K. S.; Homborg, H. *Z. Anorg. Allg. Chem.* **2001**, *627*, 935.

95. (a) Wink, D. A.; Mitchell, J. B. *Free Radicals Biol. Med.* **1998**, *25*, 434; (b) Wink, D. A.; Hanbauer, I.; Krishna, M. C.; DeGraff, W.; Gamson, J.; Mitchell, J. B. *Proc. Natl. Acad. Sci. USA* **1993**, *90*, 9813; (c) Kagan, V. E.; Kozlov, A. V.; Tyurina, Y. Y.; Shvedova, A. A.; Yalowich, J. C. *Antioxid. Redox Signaling* **2001**, *3*, 189.

96. (a) Betteridge, D. *J. Metabolism: Clinical and Experimental* **2000**, *49*, 3; (b) Sies, H. *Dev. Cardiovasc. Med.* **2000**, *224*, 1. *Peroxidases in Chemistry and Biology*; Eds. Everse, J.; Everse, K. E.; Grisham, M. B.; CRC Press: Boston, 1991.

97. Several values have been reported, all close to $10^{10}\,M^{-1}\,s^{-1}$. (a) Huie, R. E.; Padmaja, S. *Free Radicals Res. Commun.* **1993**, *18*, 195; (b) Goldstein, S.; Czapski, G. *Free Radicals Biol. Med.* **1995**, *34*, 4041; (c) Kissner, R.; Nauser, T.; Bougnon, P.; Lye, P. G.; Koppenol, W. H. *Chem. Res. Toxicol.* **1997**, *10*, 1285.

98. (a) Wink, D. A.; Darbyshire, J. F.; Nims, R. W.; Saveedra, J. E.; Ford, P. C. *Chem. Res. Toxicol.* **1993**, *6*, 23; (b) Awad, H. H.; Stanbury, D. M. *Int. J. Chem. Kinet.* **1993**, *25*, 375.

99. (a) Ford, P. C.; Wink, D. A.; Stanbury, D. M. *FEBS Lett.* **1993**, *326*, 1; (b) Thomas, D. D.; Liu, X.; Kantrow, S. P.; Lancaster, J. R., Jr. *Proc. Natl. Acad. Sci. USA* **2001**, *98*, 355.

100. (a) Miranda, K. M.; Espey, M. G.; Jourd'heuil, D.; Grisham, M. B.; Fukuto, J. M.; Feelisch, M.; Wink, D. A. Chapter 3 in Ref. (*2*) and references therein; (b) Wink, D. A.; Kasprzak, K. S.; Maragos, C. M.; Elsepuru, R. K.; Misra, M.; Dunams, T. M.; Cebula, T. A.; Koch, W. H.; Andrews, A. W.; Allen, J. S.; Keefer, L. K. *Science* **1991**, *254*, 1001; (c) Nguyen, T.; Brunson, D.; Crespi, C. L.; Penman, B. W.; Wishnok, J. S.; Tannenbaum, S. R. *Proc. Natl. Acad. Sci. USA* **1992**, *89*, 3030; (d) deRojas-Walker, T.; Tamir, S. J. H.; Wishnok, J. S.; Tannenbaum, S. R. *Chem. Res. Toxicol.* **1995**, *8*, 473.

101. (a) Lewis, R. S.; Deen, W. M. *Chem. Res. Toxicol.* **1994**, *7*, 568; (b) Pires, M.; Ross, D. S.; Rossi, M. J. *Int. J. Chem. Kinet.* **1994**, *26*, 1207; (c) Kharitonov, V. G.; Sundquist,

A. R.; Sharma, V. S., *J. Biol. Chem.* **1994**, *269*, 5881; (d) Goldstein, S.; Czapski, G. *J. Am. Chem. Soc.* **1995**, *117*, 12078.

102. (a) Radi, R.; Denicola, A.; Ferrer-Sueta, G.; Rubbo, H. Chapter 4 in Ref. (*2*) and refs therein; (b) Beckman, J. S.; Beckman T. W.; Chen, J.; Marshall, P. A.; Freeman, B. A. *Proc. Natl. Acad. Sci. USA* **1990**, *87*, 1620.

103. Nottingham, W. C.; Sutter, J. R. *Int. J. Chem. Kinet.* **1989**, *18*, 1289.

104. Liu, X.; Miller, M. J. S.; Joshi, M. S.; Thomas, D. D.; Lancaster, J. R., Jr. *Proc. Natl. Acad. Sci. USA* **1998**, *95*, 2175.

105. (a) Feelisch, M.; Noack, E. A. *Eur. J. Pharmacol.* **1987**, *139*, 19; (b) Kelm, M.; Dahmann, R.; Wink, D.; Feelisch, M. *J. Biol. Chem.* **1997**, *272*, 9922.

106. (a) Doyle, M. P.; Hoekstra, J. W. *J. Inorg. Biochem.* **1981**, *14*, 351; (b) Feelisch, M. *Cardiovas. Pharmacol.* **1991**, *17*, S25.

107. (a) Andersen, H. J.; Skibsted, L. H. *J. Agric. Food Chem.* **1992**, *40*, 1741; (b) Bruun-Jensen, L.; Skibsted, L. H. *Meat Sci.* **1996**, *44*, 145.

108. See, for example: (a) Clarkson, S. G.; Basolo, F. *Inorg. Chem.* **1973**, *12*, 1528; (b) Trogler, W. C.; Marzilli, L. G. *Inorg. Chem.* **1974**, *13*, 1008; (c) Kubota, M.; Phillips, D. A. *J. Am. Chem. Soc.* **1975**, *97*, 5637; (d) Tovrog, B. S.; Diamond, S. E.; Mares, F. *J. Am. Chem. Soc.* **1979**, *101*, 270; (e) Ugo, R.; Bhaduri, S.; Johnson, B. F. G.; Khair, A.; Pickard, A.; Benn-Taarit, Y. *J. Chem. Soc., Chem. Comm.* **1976**, 694; (f) Chen, L.; Powell, D. R.; Khan, M. A.; Richter-Addo, G. B. *Chem. Commun.* **2000**, 2301; (g) Ercolani, C.; Paoletti, A. M.; Pennisi, G.; Rossi, G. *J. Chem. Soc., Dalton Trans.* **1991**, 1317; (h) Andrews, M. A.; Chang, T. C.-T.; Cheng, C.-W.; Kelly, K. P. *Organometallics* **1984**, *3*, 1777; (i) Andrews, M. A.; Chang, T. C.-T.; Cheng, C.-W. *Organometallics* **1985**, *4*, 268; (j) Mares, F.; Diamond, S. E.; Regina, F. J.; Solar, J. P. *J. Am. Chem. Soc.* **1985**, *107*, 3545; (k) Muccigrosso, D. A.; Mares, F.; Diamond, S. E.; Solar, J. P. *Inorg. Chem.* **1983**, *22*, 960.

109. Goodwin, J.; Baily, R.; Pennington, W.; Rasberry, R.; Green, T.; Shasho, S.; Yongsavanh, M.; Echevarria, V.; Tiedeken, J.; Brown, C.; Fromm, G.; Lyerly, S.; Watson, N.; Long, A.; De Nitto, N. *Inorg. Chem.* **2001**, *40*, 4217.

110. See, for example:Carter, M. J.; Rillema, D. P.; Basolo, F. *J. Am. Chem. Soc.* **1974**, *96*, 392.

111. (a) Doyle, M. P.; Pickering, R. A.; Dykstra, R. L.; Cook, B. R. *J. Am. Chem. Soc.* **1982**, *104*, 3392; (b) Herold, S.; Matsui, T.; Watanabe, Y. *J. Am. Chem. Soc.* **2001**, *123*, 4085; (c) Herold, S.; Exner, M.; Nauser, T. *Biochemistry* **2001**, *40*, 3385.

112. DeLeo, M. A.; Ford, P. C. *J. Am. Chem. Soc.* **1999**, *121*, 1980.

113. (a) Herold, S. *FEBS Lett.* **1998**, *439*, 85; (b) Herold, S.; Rehmann, F.-J. K. *J. Biol. Inorg. Chem.* **2001**, *6*, 543. Private communication from S. Herold.

114. McCarthy, M. R.; Crevier, T. J.; Bennett, B.; Dehestani, A.; Mayer, J. M. *J. Am. Chem. Soc.* **2000**, *122*, 12391.

115. Moore, E. G.; Gibson, Q. H. *J. Biol. Chem.* **1976**, *251*, 2788.

116. Møller, J.K.S; Skibsted, L. H. *Chem. Rev.* **2002**, *102*, 1167.

117. (a) Traylor, T. G.; Sharma, V. S. *Biochemistry* **1992**, *31*, 2847; (b) Radi, R. *Chem. Res. Toxicol.* **1996**, *9*, 828.

118. Minamiyama, Y; Takemura, S.; Imaoka, S.; Funae, Y.; Tanimoto, Y.; Inoue, M. *J. Pharmacol. Exp. Ther.* **1997**, *283*, 1479.

119. Cleeter, M. W. J.; Cooper, J. M.; Darley-Usmar, V. M.; Moncada, S.; Scapira, A. H. V. *FEBS Lett.* **1994**, *345*, 50.

120. (a) Noguchi, T.; Hoshino, M.; Tsujimura, M.; Odaka, M.; Inoue, Y.; Endo, I. *Biochemistry* **1996**, *35*, 16777; (b) Odaka, M.; Fujii, K.; Hoshino, M.; Noguchi, T.; Tsujimura, M.; Nagashima, S.; Yohda, M.; Nagamune, T.; Inoue, Y.; Endo, I. *J. Am.*

Chem. Soc. **1997**, *119*, 3785; (c) Tsujimura, M.; Dohmae, N.; Odaka, M.; Chijimatsu, M.; Takio, K.; Yohda, M.; Hoshino, M.; Nagashima, S.; Endo, I. *J. Biol. Chem.* **1997**, *272*, 29 454.

121. Brown, G. C. *Eur. J. Biochem.* **1995**, *232*, 188.

122. Abu-Soud, H. M.; Khassawneh, M. Y.; Sohn, J.-T.; Murray, P.; Haxhiu, M. A.; Hazen, S. L. *Biochemistry* **2001**, *40*, 11 866.

123. (a) Ribiero, J. M. C.; Hazzard, J. M. H.; Nussenzveig, R. H.; Champagne, D. E.; Walker, F. A. *Science* **1993**, *260*, 539; (b) Walker, F. A.; Ribiero, J. M. C.; Monfort, W. R. *Metal Ions in Biological Systems*; Eds. Sigel, H.; Sigel, A.; **1998**, 36, 621.

124. (a) Tayeh, M. A.; Marletta, M. A. *J. Biol. Chem.* **1989**, *264*, 19 654; (b) Abu-Soud, H. M.; Ichimori, K.; Nakazawa, H.; Stuehr, D. J. *Biochemistry* **2001**, *40*, 6876; (c) Stuehr, D. J.; Pou, S.; Rosen, G. M. *J. Biol. Chem.* **2001**, *276*, 14 533 and references therein.

125. Malinski, T.; Czuchajowski, C. Chapter 22 in Ref. (*1d*).

126. Kim, S.; Deinum, G.; Gardner, M. T.; Marletta, M. A.; Babcock, G. T. *J. Am. Chem. Soc.* **1996**, *118*, 8769, and references therein.

127. Ignarro, L. J. "*Nitric Oxide: Biology and Pathobiology*"; Eds. Ignarro, L. J.; Academic Press: San Diego, 2000, Chapter 1.

128. (a) Koesling, D.; Friebe, A. "*Nitric Oxide: Biology and Pathobiology*"; Ed. Ignarro, L. J.; Academic Press: San Diego, 2000, Chapter 24; (b) Koesling, D. *Methods* **1999**, *19*, 485.

129. (a) Yu, A. E.; Hu, S.; Spiro, T. G.; Burstyn, J. N. *J. Am. Chem. Soc.* **1994**, *116*, 4117; (b) Burstyn, J. N.; Yu, A. E.; Dierks, E. A.; Hawkins, B. K.; Dawson, J. H. *Biochemistry* **1995**, *34*, 5896.

130. (a) Deinum, G.; Stone, J. R.; Babcock, G. T.; Marletta, M. A. *Biochemistry* **1996**, *35*, 1540; (b) Tomita, T.; Ogura, T.; Tsuyama, S.; Imai, Y.; Kitagawa, T. *Biochemistry* **1997**, *36*, 10 155.

131. Zhao, Y.; Brandish, P. E.; Ballou, D. P.; Marletta, M. A. *Proc. Natl. Acad. Sci. USA* **1999**, *96*, 14 753.

132. Bellamy, T. C.; Garthwaite, J. *J. Biol. Chem.* **2001**, *276*, 4287.

133. (a) Bohle, D. S.; Hung, C.-H. *J. Am. Chem. Soc.* **1995**, *117*, 9584; (b) Bohle, D. S.; Debrunner, P.; Fitzgerald, J. P.; Hansert, B.; Hung, C.-H.; Thomson, A. J. *Chem. Commun.* **1997**, 91.

134. Tomita, T.; Ogura, T.; Tsuyama, S.; Imai, Y.; Kitagawa, T. *Biochemistry* **1997**, *36*, 10 155.

135. (a) Furtmüller, P. G.; Burner, U.; Obinger, C. *Biochemistry* **1998**, 37, 17 923; (b) Kettle, A. J.; Winterbourn, C. C. *Biochemistry* **2001**, *40*, 10 204.

136. Blair-Johnson, M.; Fiedler, T.; Fenna, R. *Biochemistry* **2001**, *40*, 13 990.

137. Abu-Soud, H. M.; Hazen, S. L. *J. Biol. Chem.* **2000**, *275*, 5425.

138. (a) van der Vliet, A.; Eiserich, J. P.; Halliwell, B.; Cross, C. *J. Biol. Chem.* **1997**, *272*, 7617; (b) Byun, J.; Henderson, J. P.; Mueller, D. M.; Heincke, J. W. *Biochemistry* **1999**, *38*, 2590; (c) Eiserich, J. P.; Hristova, M.; Cross, C. E.; Jones, A. D.; Freeman, B. A.; Halliwell, B.; van der Vliet, A. *Nature* **1998**, *391*, 393.

139. (a) Glover, R. E.; Koshkin, V.; Dunford, H. B.; Mason, R. P. *Nitric Oxide: Biol. Chem.* **1999**, *3*, 439; (b) Ischiropoulos, H.; Nelson, J.; Duran, D.; Al-Mehdi, A. *Free Rad. Biol. Med.* **1996**, *20*, 373.

140. Wu, W.; Chen, Y.; Hazen, S. L. *J. Biol. Chem.* **1999**, 25 933.

141. Herold, S.; Rehmann, F. K. *J. Biol. Inorg. Chem.* **2001**, *6*, 543.

142. McMahon, T. J., Gow, A., Stamler, J. A. *Nitric Oxide: Biology and Pathobiology*, Chapter 15; Ed. Ignarro, L.J.; Academic Press: San Diego, 2000, and references therein.

143. Gow, A. J.; Stamler, J. S. *Nature* **1998**, *391*, 169.

144. Gow, A. J.; Buerk, D. D.; Ischiropoulos, H. *J. Biol. Chem.* **1997**, *272*, 2841.
145. Stamler, J. S.; Jia, L.; Eu, J. P.; McMahon, T. J.; Demchenko, I. T.; Bonaventura, J.; Gernert, K.; Piantadosi, C. A. *Science* **1997**, *276*, 2034.
146. Chan, N. L.; Rogers, P. H.; Arnone, A. *Biochemistry* **1998**, *37*, 16 459.
147. Han, T. H.; Hyduke, D. R.; Vaughn, M. W.; Fukuto, J. M.; Liao, J. C. *Proc. Natl. Acad. Sci. USA* **2002**, *99*, 7763.
148. Zhang, Y.; Hogg, N. *Free Rad. Biol. Med.* **2002**, *32*, 1212.
149. (a) Howard-Flanders, P. *Nature (London)* **1957**, *180*, 1191; (b) Mitchell, J. B.; Wink, D. A.; DeGraff, W.; Gamson, J. *Cancer Res.* **1993**, *53*, 5845; (c) Bourassa, J.; DeGraff, W.; Kudo, S.; Wink, D. A.; Mitchell, J. B.; Ford, P. C. *J. Am. Chem. Soc.* **1997**, *119*, 2853.
150. Lim, M. D.; Lorkovic, I. M.; Ford, P. C. *Inorg. Chem.* **2002**, *41*, 1026.
151. Wink, D. A.; Vodovotz, Y.; DeGraff, W.; Cook, J. A.; Pacelli, R.; Krishna, M.; Mitchell, J. B. *"Trace Atmospheric Constituents. Properties, Transformations and Fates"*; John Wiley and Sons: New York, 1983, Chapter 9.
152. Kroger-Ohlsen, M. V.; Skibsted, L. H. *Food Chem.* **2000**, *70*, 209.

HOMOGENEOUS HYDROCARBON C–H BOND ACTIVATION AND FUNCTIONALIZATION WITH PLATINUM

ULRICH FEKL and KAREN I. GOLDBERG

Department of Chemistry, Box 351700, University of Washington, Seattle, WA 98195-1700, USA

ADVANCES IN INORGANIC CHEMISTRY
VOLUME 54 ISSN 0898-8838

I. Introduction

The selective functionalization of hydrocarbon C–H bonds under mild conditions is one of the most difficult and most promising challenges facing chemists today (1–5). It has even been referred to as a "holy grail" in chemistry (1). By functionalization, it is meant that the reaction results in the exchange of hydrogen for a non-hydrogen atom or group. As shown in Scheme 1(**A**), R–H is functionalized to form R–X. Reactions of this type offer immense technological promise. Hydrocarbons are abundant and inexpensive feedstocks and such a functionalization step leads directly to value-added products. To be commercially viable, however, the reaction would have to be selective, involve an affordable and environmentally benign oxidant and avoid harsh and expensive reaction conditions.

The functionalization reaction as shown in Scheme 1(**A**) clearly requires the breaking of a C–H bond at some point in the reaction sequence. This step is most difficult to achieve for R = alkyl as both the heterolytic and homolytic C–H bond dissociation energies are high. For example, the pK_a of methane is estimated to be ca. 48 (6,7). Bond heterolysis, thus, hardly appears feasible. C–H bond homolysis also appears difficult, since the C–H bonds of alkanes are among the strongest single bonds in nature. This is particularly true for primary carbons and for methane, where the radicals which would result from homolysis are not stabilized. The bond energy (homolytic dissociation enthalpy at 25 °C) of methane is 105 kcal/mol (8).

Thus, a highly reactive species is needed to make this type of bond activation reaction feasible under mild conditions. In addition, to be useful, the C–H bond activation must occur with both high chemo- and regioselectivity. Over the past several decades, it has been shown that transition metal complexes are able to carry out alkane activation reactions (1–5). Many of these metal-mediated reactions operate under mild to moderate conditions and exhibit the desirable chemoselectivity and regioselectivity. Thus, using transition metal complexes, alkane activation can be preferred over product activation, and the terminal positions of alkanes, which actually contain the stronger C–H bonds, can be selectively activated. The fact that a hydrocarbon C–H bond has been broken in a

A)   R-H ·············▶ R-X

B)   R-H ·············▶ R-D

SCHEME 1.

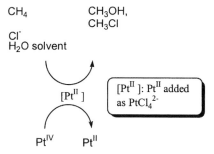

SCHEME 2.

system under investigation is often inferred from the fact that hydrogen is exchanged, which can be seen, for example, by deuterium labeling as shown in Scheme 1(**B**). Breaking of the C–H bond is called C–H bond activation, and is necessary but not sufficient for hydrocarbon functionalization. For functionalization, the C–H bond must actually be replaced by another functionality.

Among the very few catalytic systems that allow not only C–H bond activation but also functionalization are those based on platinum(II) catalysts. Soon after the discovery that platinum salts in aqueous solution catalyze H/D exchange in hydrocarbons (*9,10*), a hydrocarbon functionalization cycle was developed on the basis of this system (*11*). This cycle is depicted in Scheme 2.

This cycle, often referred to as the "Shilov-cycle", converts methane into methanol and chloromethane in homogeneous aqueous solution at mild temperatures of 100–120 °C (*11*). However, while Pt(II) (added to the reaction as $PtCl_4^{2-}$) serves as the catalyst, the system also requires Pt(IV) (in the form of $PtCl_6^{2-}$) as a stoichiometric oxidant. Clearly, this system impressively demonstrates functionalization of methane under mild homogeneous conditions, but is impractical due to the high cost of the stoichiometric oxidant used. A recent development by Catalytica Advanced Technology Inc., often referred to as the "Catalytica system" used platinum(II) complexes as catalysts to convert methane into methylbisulfate (*12*). The stoichiometric oxidant in this case is SO_3, dissolved in concentrated H_2SO_4 solvent. This cycle is depicted in Scheme 3.

Sulfur trioxide is not only much cheaper than hexachloroplatinate, but the sulfur dioxide formed can in principle be converted back to sulfur trioxide by oxidation with air. Methylbisulfate can be hydrolyzed to methanol, and so a complete cycle that converts CH_4 and O_2 into CH_3OH can be envisioned. Overoxidation of methane to CO_2 – a major side reaction of many procedures – was relatively limited in the

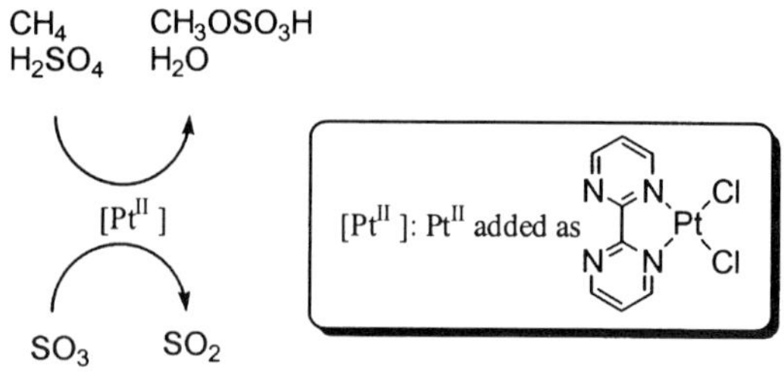

SCHEME 3.

Catalytica system. The methanol produced accounted for 81% of the oxidized methane. This was determined after 90% of the methane initially present was consumed in a single run. The Catalytica system is, in terms of methanol yield and selectivity, an unmatched achievement in homogeneous low-temperature methane functionalization (13). Although this cycle produces methanol in high yield and selectivity, it is still not economically competitive with established high-temperature (and generally heterogeneously catalyzed) procedures. This is primarily due to the high cost of product separation, the handling of H_2SO_4 as a solvent, and the fact that the system is inhibited by the water produced in the reaction.

Thus, an efficient, selective and economically competitive process that directly oxidizes methane to methanol or other alkanes to their desirable terminal alcohols has yet to be developed. However, there is considerable evidence to support that platinum(II) catalyst systems are some of the most promising leads in this effort. In the past several years, a number of research groups around the world have systematically studied the reactivity of model Pt(II) and Pt(IV) complexes. Valuable insight into the mechanisms of the individual reaction steps needed to accomplish the desired transformations using platinum complexes and suitable oxidants has been gained in these studies. A 1998 review article by Stahl, Labinger and Bercaw is highly recommended to readers as a detailed source of information concerning the mechanism of Shilov's platinum-catalyzed alkane oxidations (14). The systems based on platinum chlorides have been extensively reviewed in 2000 in a book by Shilov and Shul'pin (15). We attempt in the current review to present more recent developments in the field, highlighting particularly those studies which have provided significant mechanistic insight into the individual reaction steps involved in alkane oxidation by Pt(II) catalysts. With detailed knowledge of the underlying reaction mechanisms, the rational design of C–H activation and functionalization systems can become a realistic goal instead of a "holy grail".

The latest developments in the mechanistic understanding of platinum catalyzed alkane oxidation have involved Pt complexes with chelating ligands having mainly – but not exclusively – nitrogen donors. This review will focus on recent studies of these chelated Pt systems. The selection of the literature is somewhat subjective and will not be comprehensive. However, it should provide a true flavor of the current state of progress in the field.

II. Classic Division of the Hydrocarbon Functionalization Cycle into Three Parts

Relatively soon after the discovery that aqueous solutions containing $PtCl_4^{2-}$ and $PtCl_6^{2-}$ can functionalize methane to form chloromethane and methanol, a mechanistic scheme for this conversion was proposed (*16,17*). As shown in Scheme 4, a methylplatinum(II) intermediate is formed (step **I**), and this intermediate is oxidized to a methylplatinum(IV) complex (step **II**). Either reductive elimination involving the Pt(IV) methyl group and coordinated water or chloride or, alternatively, nucleophilic attack at the carbon by an external nucleophile (H_2O or Cl^-) was proposed to generate the functionalized product and reduce the Pt center back to Pt(II) (step **III**) (*17*). This general mechanism has received convincing support over the last two decades (comprehensive reviews can be found in Refs. (*2,14,15*)). Carbon-heteroatom bond formation from Pt(IV) (step **III**) has been shown to occur via nucleophilic attack at a Pt-bonded methyl, as discussed in detail below (Section V.A).

The cycle shown in Scheme 4 is also thought to form the basis of other Pt based C–H functionalization systems (*12,18,19*). However, the cycle as pictured in Scheme 4 is quite general and the intimate mechanisms of the individual steps (C–H activation (**I**), oxidation (**II**), and functionalization (**III**)) may differ somewhat in each particular system.

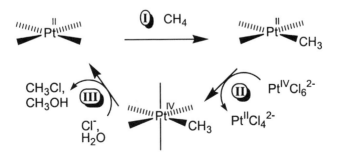

SCHEME 4.

The geometries and energies of the intermediates and transition states for each of these steps have been a topic of active research in the past several years and it is these recent developments that will be highlighted in this review.

III. Mechanisms for the C–H Activation Sequence: Formation of a Pt(II) Hydrocarbyl Complex from Pt(II) and Hydrocarbon

For the C–H activation sequence, the different possibilities to be considered are shown in Scheme 5: (a) direct oxidative addition to square-planar Pt(II) to form a six-coordinate Pt(IV) intermediate and (b, c) mechanisms involving a Pt(II) alkane complex intermediate. In (b) the alkane complex is deprotonated (which is referred to as the electrophilic mechanism) while in (c) oxidative addition occurs to form a five-coordinate Pt(IV) species which is subsequently deprotonated to form the Pt(II) alkyl product.

A. ON THE DIRECT OXIDATIVE ADDITION OF R–H TO SQUARE-PLANAR Pt(II)

Direct oxidative addition of C–H bonds to square-planar Pt(II) to form octahedral Pt(IV) alkyl hydrides (Scheme 5, a) appears to be extremely uncommon. In fact, there are no unequivocal examples of this reaction pathway. However, a couple of examples have been reported wherein such a direct mechanism may provide the best explanation for the available data.

Studies of intramolecular C–H bond activation reactions at Pt(II) (and Pd(II)) leading to cyclometalated products have typically found that these reactions are facilitated by ancillary ligands that can easily dissociate (20,21). In the absence of any weakly bonded ligand, unusually

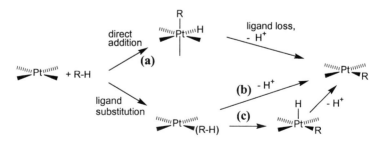

SCHEME 5.

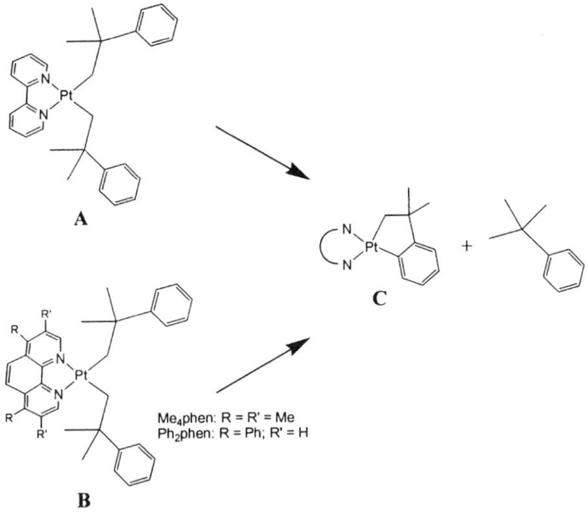

Me$_4$phen: R = R' = Me
Ph$_2$phen: R = Ph; R' = H

SCHEME 6.

slow reactions are observed, and a direct oxidative addition pathway has been proposed. For example, the complexes shown as reactants in Scheme 6(**A** and **B**) undergo thermal C–H bond activation reactions to generate cyclometalated products of the type **C** and *t*-butylbenzene (*22*). Regardless of whether tetramethylphenanthroline or diphenylphenanthroline was the ligand in complex **B**, the cyclometalation reaction of complex **B** was an order of magnitude slower than the reaction of the bipyridine complex (**A**). In addition, the kinetic deuterium isotope effect for **B** ($k_H/k_D = 3.3$ for the Ph$_2$phen example) was significantly larger than the corresponding effect measured for **A** ($k_H/k_D = 1.3$). These results were interpreted in terms of a dechelation mechanism (κ^1-bipyridine) for complex **A** to yield a highly reactive three-coordinate Pt(II) intermediate prior to C–H activation. In contrast, a direct oxidative addition mechanism was deemed a satisfactory explanation for **B** (*22*).

For *inter*molecular hydrocarbon activation, we are not aware of any example where compelling evidence exists for C–H activation directly by square-planar four-coordinate Pt(II), without preceding (or concomitant) ligand loss. In one example, such a direct reaction may take place but the alternative explanation involving ligand loss is also consistent with the data. The compound (dmpe)PtMe(O$_2$CCF$_3$) (dmpe = bis(dimethylphosphino)ethane) activates benzene C–H bonds at elevated temperature (125 °C) (*23*). This reactivity contrasts with that of (dmpe)PtMeCl which is inactive at the same temperature or even at

150 °C (*23*). While these results would be consistent with a mechanism whereby a labile ligand has to be replaced by benzene, no inhibition of the reaction was observed with added trifluoroacetate. Instead a slight enhancement occurred. Since the data do not exclude either mechanism, the authors explicitly left open the question of whether this reaction occurs by replacement of trifluoroacetate by benzene or via the direct pathway.

Ultimately, direct oxidative addition of C–H bonds to Pt(II) is not an intrinsically forbidden reaction (*24,25*). However, the paucity of verifiable examples of such a pathway and the significantly larger and increasing pool of examples in which ligand loss occurs prior to C–H activation appears to point to a conclusion that C–H activation will be a more facile reaction if at least one ancillary ligand on the metal center is labile. As will be discussed below, this labile ligand can be replaced by hydrocarbon such that the hydrocarbon binds within the square-planar coordination environment of the Pt(II) center. This binding of the hydrocarbon to the metal dramatically increases the facility of the C–H bond cleavage reaction.

B. ACTIVATION OF R–H BY Pt(II) COMPLEXES POSSESSING A LABILE LIGAND

Facile C–H bond activation by Pt(II) metal centers seems to require at least one labile ligand in the coordination sphere of platinum. One of the earliest intermolecular examples of this is the activation of C–D bonds in benzene-d_6 by *trans*-$(PMe_3)_2Pt(neopentyl)(OTf)$ at 133 °C, where trifluoromethanesulfonate (triflate, OTf) provides the labile group (Scheme 7, **A**) (*26*).

Consistent with the proposal that triflate is replaced in the coordination sphere by hydrocarbon, inhibition by added triflate was observed. The labile group pentafluoropyridine in the cation $[(tmeda)PtMe(NC_6F_5)]^+$ (tmeda = N,N,N',N'-tetramethylethylenediamine) serves a similar role, and activation of methane, as indicated by incorporation of ^{13}C from $^{13}CH_4$ into the methyl group of the complex, was observed at 85 °C (Scheme 7, **B**) (*27*). Water acts as the labile ligand in recently developed diimine ligated Pt(II) systems which activate C–H bonds in trifluoroethanol solvent (*28–33*). An example is shown in Scheme 7(**C**). Similarly, the tetrahydrofuran (THF) complex shown in Scheme 7(**D**) activates the benzene solvent under mild conditions. The reaction is slowed down in the presence of added THF (*34*). In Shilov's system based on $PtCl_4^{2-}$, substitution of at least one chloride by water is assumed to be necessary to create the active species having a more labile ligand. Inhibition of activity is observed if excess chloride is present (*2,15*).

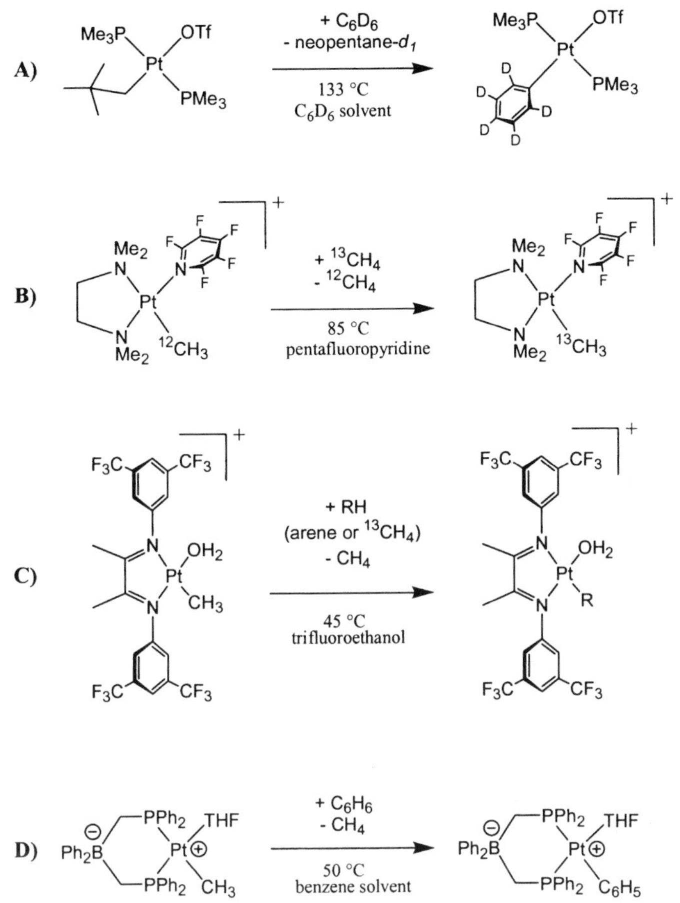

SCHEME 7.

A related base-promoted C–H activation of benzene by Pt(II) was recently reported. With the tridentate monoanionic amido 'pincer' ligand 'N$_3$', the triflate complex 'N$_3$'Pt(OTf), depicted in Scheme 8 was shown to activate benzene in the presence of base (*35*). It was noted that the chloro complex 'N$_3$'PtCl was not reactive under these conditions. The activity of the triflate complex again appears to result from the higher lability of triflate which can allow for coordination of the hydrocarbon.

All these examples support the concept that hydrocarbon may need to bind directly to the Pt(II) center as shown in Scheme 9 in order to facilitate the C–H activation reaction. The product of the reaction

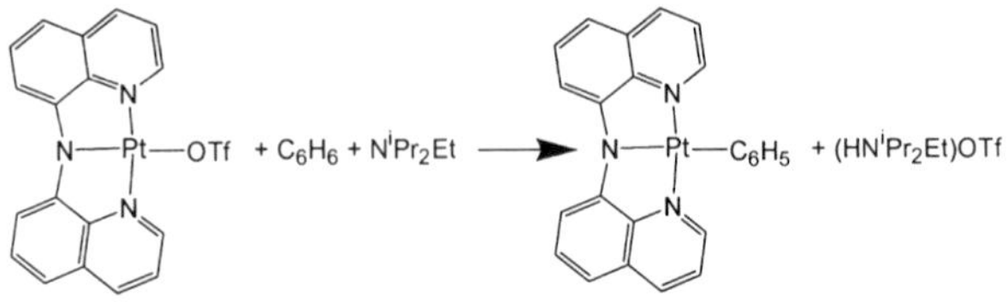

SCHEME 8.

SCHEME 9.

shown in Scheme 9 is referred to as a σ-complex of the hydrocarbon. For the alkane example, this would be a σ-alkane complex.

What purpose does the alkane binding to the Pt(II) center serve? For the electrophilic pathway (Scheme 5, **b**), this is immediately apparent. σ-Alkane complexes should be considerably more acidic than free alkanes, such that deprotonation may become a viable C–H activation pathway. While the acidic character of alkane complexes has not been directly observed, it can be inferred from the measured acidity of analogous agostic complexes (*36*) and from the acidity of the σ-complexes of dihydrogen (*37*), both of which can be regarded models for alkane complexes (see Section III.E).

For the oxidative addition pathway, however, it is not obvious why the C–H bond cleavage reaction should be more facile if the hydrocarbon first binds in the coordination sphere of the metal (Scheme 5, **c**). One argument could be that the equilibrium between the Pt(II) alkane complex and the five-coordinate Pt(IV) alkyl hydride has an intrinsically low activation barrier. Insight into this question together with detailed information about the mechanisms of these Pt(II) σ-complex/Pt(IV) alkyl hydride interconversions has been gained via detailed studies of reductive elimination reactions from Pt(IV), as discussed below.

C. FIVE-COORDINATE Pt(IV) INTERMEDIATES AND Pt(II) σ-COMPLEXES IN REDUCTIVE ELIMINATION FROM Pt(IV) AND OXIDATIVE ADDITION TO Pt(II)

1. Reductive elimination to form carbon–carbon bonds

While our focus in this section of the review is mainly on reactions which make and break carbon–hydrogen bonds, similar reactions involving carbon–carbon bonds can serve as excellent models. Both hydride ligands and metal-bonded alkyls are σ-donors and bind similarly to the metal. Neither has any ability to act as a π-donor or -acceptor. An in-depth discussion of C–C bond cleavage (*38–43*) and formation (*44,45*) at transition metals is beyond the scope of this review. However, mechanistic studies of sp^3-C–sp^3-C reductive elimination from Pt(IV) are of significant value as this reaction serves as a model for the more facile sp^3-C–H coupling reaction.

C–C reductive elimination from Pt(IV) has been well studied over the past 30 years. One of the earliest mechanistic investigations of this reactivity concentrated on the formation of ethane from the Pt(IV) complexes L$_2$PtMe$_3$X (L = PMe$_2$Ph, PMePh$_2$, PMe$_3$, X = halide) (*46*). The rate of reductive elimination was inhibited by added phosphine, and this result, among others, provided strong evidence for a mechanism involving phosphine dissociation to form a five-coordinate intermediate prior to the C–C coupling reaction (Scheme 10).

A similar mechanism was proposed, based on analogous kinetic behavior, for Pt(IV) tetramethyl complexes with isonitrile ligands (*47*). Isonitrile dissociation preceded the release of ethane.

Chelating ligands have a much lower propensity for dissociation. Yet detailed kinetic studies on similar systems with bidentate phosphine ligands, *fac*-(L$_2$)PtMe$_3$X (L$_2$ = dppe (bis(diphenylphosphino)ethane), dppbz (bis(diphenylphosphinobenzene); X = I, OAc, OPh) also showed that ligand dissociation was required prior to any C–C coupling (*48–51*). In this case, however, the X$^-$ group rather than the phosphine was lost to form a five-coordinate intermediate, as shown in Scheme 11. A competitive C–X reductive elimination also occurs from these complexes and involves the same five-coordinate cation (Section V.A).

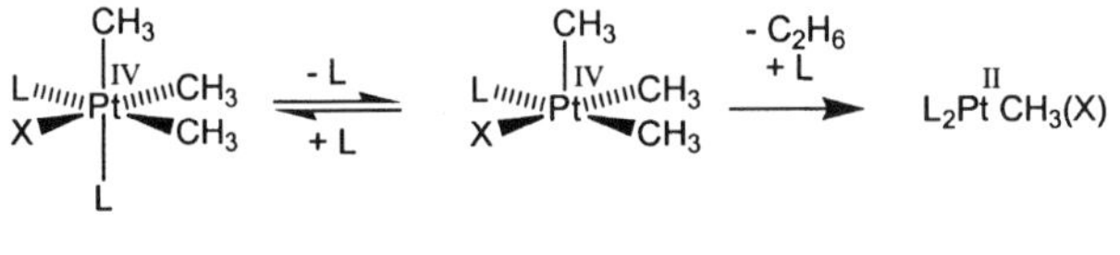

SCHEME 10.

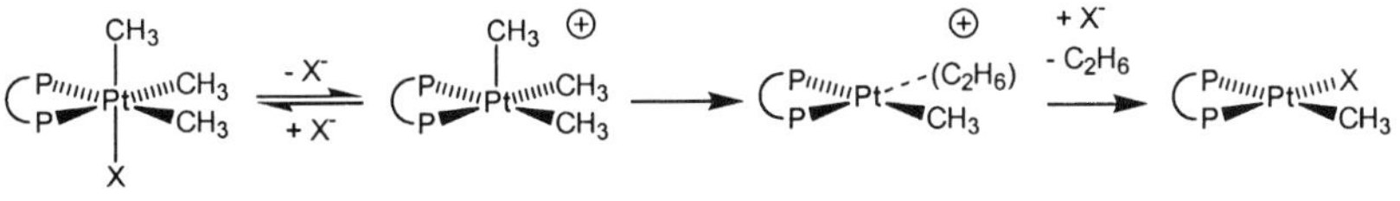

SCHEME 11.

It might be suggested that the pathway involving a five-coordinate Pt(IV) intermediate could be shut down for a C–C reductive elimination reaction from a *tetra*methylplatinum(IV) complex of a chelating bidentate ligand, L_2PtMe_4. In this case, no ligand which could be expected to easily dissociate would be present. Could this situation promote the elusive direct C–C reductive elimination? A recent detailed study on (dppe)PtMe₄ complexes (dppe = bis(diphenylphosphino)ethane) addressed this issue and found that even in this case, a five-coordinate intermediate was involved (52). Dissociation of one end of the chelating phosphine ligand dppe (Scheme 12) provided access to the reactive unsaturated species.

Thus to date, virtually all studies of C–C reductive elimination to form alkanes from Pt(IV) have found that these reactions proceed via five-coordinate intermediates. Only very recently have stable examples of Pt(IV) alkyl hydrides been synthesized (53–69). Detailed studies of C–H reductive elimination to form alkanes from these related complexes have identified similar five-coordinate intermediates on the reaction pathway (see following section).

2. Reductive Elimination to form C–H bonds from Pt(IV) and Oxidative Addition to Cleave C–H bonds at Pt(II)

There are now a number of quite stable Pt(IV) alkyl hydride complexes known and the synthesis and characterization of many of these complexes were covered in a 2001 review on platinum(IV) hydride chemistry (69). These six-coordinate Pt(IV) complexes have one feature in common: a ligand set wherein none of the ligands can easily dissociate from the metal. Thus it would appear that prevention of access to a five-coordinate Pt(IV) species contributes to the stability of Pt(IV) alkyl hydrides. The availability of Pt(IV) alkyl hydrides has recently allowed detailed studies of C–H reductive elimination from Pt(IV) to be carried out. These studies, as described below, also provide important insight into the mechanism of oxidative addition of C–H bonds to Pt(II).

a. Methyl(hydrido)platinum(IV) Complexes of Moderate to High Stability The cationic complex (bpma)PtMe₂H⁺ (bpma = bis(pyridyl-methyl)amine) is an alkyl(hydrido)platinum(IV) complex which exhibits moderate stability in solution at room temperature (*62*). Soon after its discovery, it was found to be an ideal system for studying the mechanism of methane reductive elimination (*64*). As shown in Scheme 13, the tridentate nitrogen donor ligand contributes to the octahedral environment stabilizing the Pt(IV) alkylhydride. The $(\kappa^3\text{-bpma})Pt^{IV}Me_2H^+$ complex is obtained by protonation of the platinum(II) precursor $(\kappa^2\text{-bpma})Pt^{II}Me_2$. When the Pt(II) precursor is protonated with deuterated acid, the Pt-deuteride complex $(\text{bpma})PtMe_2D^+$ is obtained (Scheme 13). Scrambling of deuterium occurs selectively into the methyl group *trans* to amine (Scheme 13), whereas it can be demonstrated that the methyl group that is eliminated is the group *cis* to amine (italics in Scheme 13). That methane is eliminated at a moderate rate at room temperature seems to be due to the fact that the bpma ligand can coordinate in the square plane of the reductive elimination product $(\text{bpma})PtMe^+$. A comparison of the (very similar) scrambling rate and the elimination rate, along with the activation parameters for reductive elimination, provided support for the proposed mechanism shown in Scheme 13, which involves σ-methane complexes and five-coordinate Pt(IV) as common intermediates for both scrambling and methane reductive elimination (*64*). To achieve the methane elimination, the "dangling" pyridyl residue replaces the methane from a methane σ-complex. The different reactivities (scrambling vs. elimination) of the two non-equivalent methyl groups (*trans* to amine and *trans* to pyridine) were explained in a straightforward manner in terms of the geometric arrangement of the "dangling" pyridyl

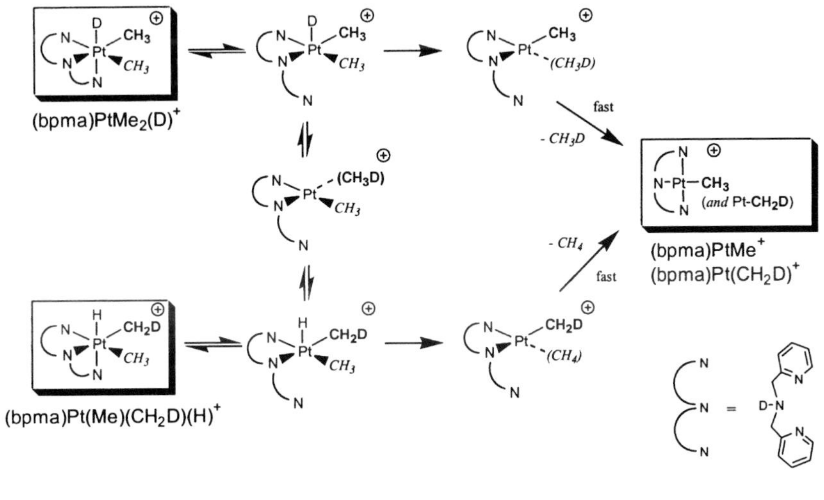

SCHEME 13.

residue with respect to the corresponding two types of σ-bonded methane: the methane *trans* to amine cannot be replaced by the uncoordinated pyridyl, whereas the methane *trans* to coordinated pyridine is within reach of the "dangling" pyridyl residue and can be substituted and thus eliminated (*64*).

Trispyrazolylborate ligands have proven very useful in the stabilization of Pt(IV) alkyl hydrides. Trispyrazolylborate ligands (*70*) can coordinate with all three nitrogen donors by capping the face of an octahedron (facial coordination). However, unlike the bpma ligand described above, this tridentate ligand is constrained geometrically such that it cannot coordinate in a meridional arrangement. Trispyrazolylborate ligands are abbreviated in this review according to the Trofimenko notation (*70*): trispyrazolylborate = Tp, and the related tris(3,5-dimethylpyrazolyl)borate = Tp^{Me_2} (Scheme 14). The use of the notation "Tp^{R_2}" in this review implies either of these very similar ligands.

Virtually the entire series of Pt(IV) complexes of the type $Tp^{R_2}PtX_3$ (X = Me, H). are now known. The trimethyl complexes $TpPtMe_3$ (*71*) and $Tp^{Me_2}PtMe_3$ (*72*) were reported in 1974 and 1990, respectively. Structural characterization by X-ray crystallography was published in 2000 for $Tp^{Me_2}PtMe_3$ (*73*). The same paper also contains a structural and spectroscopic comparison of $Tp^{Me_2}PtMe_3$ with the fluorinated analog, $Tp^{(CF_3)_2}PtMe_3$, along with a comprehensive overview of the *trans*-influence of ligands coordinated to the trimethylplatinum(IV) unit. The trimethylplatinum(IV) complexes of Tp^{R_2} are very stable even at elevated temperature, and ethane reductive elimination to generate a platinum(II) product from these compounds has never been reported. Very recently, the corresponding trihydride complex $Tp^{Me_2}PtH_3$ was synthesized and characterized (*74*). Whether it undergoes reductive elimination of H_2 was not noted.

In contrast, C–H coupling has been observed from the mixed methyl (hydrido) compounds $Tp^{R_2}PtMeH_2$ and $Tp^{R_2}PtMe_2H$. This reactivity and

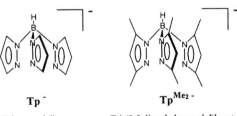

Tp⁻

Trispyrazolylborate

Tp^{Me₂⁻}

Tris(3,5-dimethylpyrazolyl)borate

SCHEME 14.

the mechanistic insight into C–H reductive elimination from Pt(IV) gained from studies of these complexes are described below.

The complex $TpPtMeH_2$ was synthesized by reacting $TpPtMe(CO)$ with water (66). While it is stable towards reductive elimination of methane at 55 °C, deuterium incorporation from methanol-d_4 solvent occurs rapidly into the hydride positions and subsequently, more slowly, into the methyl position (Scheme 15). The scrambling into the methyl position has been attributed to reversible formation of a methane complex which does not lose methane under the reaction conditions (75,76). Similar scrambling reactions have been observed for other metal alkyl hydrides at temperatures below those where alkane reductive elimination becomes dominant (77–84). This includes examples of scrambling without methane loss at elevated temperature (78).

A computational (DFT) study on the $TpPtMeH_2$ system has given activation parameters for methane loss versus scrambling of hydride protons into methane positions, and, consistent with the experimental data, shows that the barrier for the loss of methane from the σ-complex (Scheme 16, **A**) to yield free methane and the $TpPt^{II}H$ species **B** or **C** is higher than the barrier for the hydrogen/deuterium scrambling reaction (85). Scrambling was computed to be more favorable than methane loss by a $\Delta\Delta G_{298}^{\ddagger}$ of ca. 5 kcal/mol. It is interesting that the "free" pyrazolyl residue, which due to the tripodal geometry is not

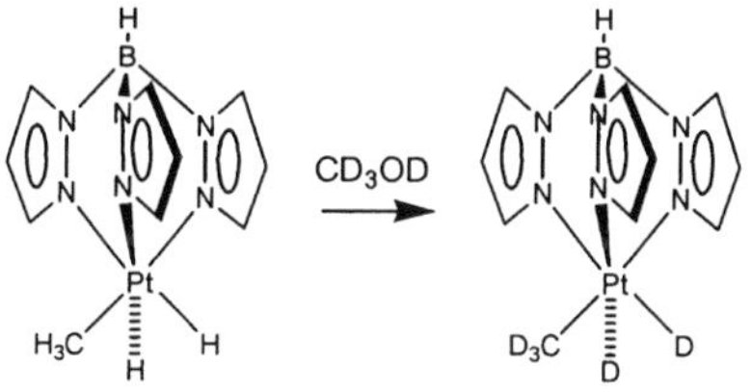

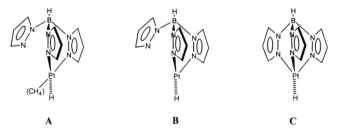

SCHEME 15.

SCHEME 16.

capable of coordinating in the square plane of Pt(II), can still coordinate (in computations) to the axial position. The energy change associated with this coordination, however, is close to zero: $\Delta H = -0.7$ kcal/mol, $\Delta G_{298} = +0.6$ kcal/mol for the conversion of **B** to **C** (Scheme 16).

The reactivity of the closely related system $\text{Tp}^{\text{Me}_2}\text{PtMeH}_2$ toward electrophiles in arene solvents has also been reported recently (68). The boron-based Lewis acid $\text{B}(\text{C}_6\text{F}_5)_3$ induced elimination of methane and formation of an aryl(dihydrido) platinum(IV) complex via arene C–H activation (Scheme 17, **A** → **C**). The active acid may be either $\text{B}(\text{C}_6\text{F}_5)_3$ or alternatively a proton generated from $\text{B}(\text{C}_6\text{F}_5)_3$ and trace water. It was proposed that the acid coordinates to a pyrazole nitrogen (shown in Scheme 17, **B**) forming an intermediate five-coordinate platinum(IV) complex, which readily eliminates methane.

Protonation reactions of the related dimethyl(hydrido)platinum(IV) complex $\text{Tp}^{\text{Me}_2}\text{PtMe}_2\text{H}$ (58) leading to rapid methane reductive elimination have also been reported (86). This protonation was shown to occur exclusively at the pyrazole nitrogen, presumably forming a five-coordinate Pt(IV) intermediate. This species should undergo C–H coupling, and while a Pt(II) methane complex is not observed, trapping with

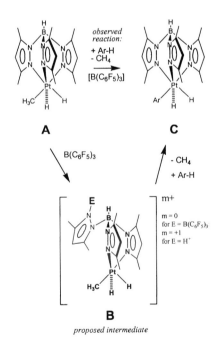

A

observed reaction:
+ Ar-H
- CH$_4$
$\xrightarrow{\;\;[\text{B}(\text{C}_6\text{F}_5)_3]\;\;}$

C

$\text{B}(\text{C}_6\text{F}_5)_3$

- CH$_4$
+ Ar-H

B

$m+$

$m = 0$
for $E = \text{B}(\text{C}_6\text{F}_5)_3$
$m = +1$
for $E = \text{H}^+$

proposed intermediate

SCHEME 17.

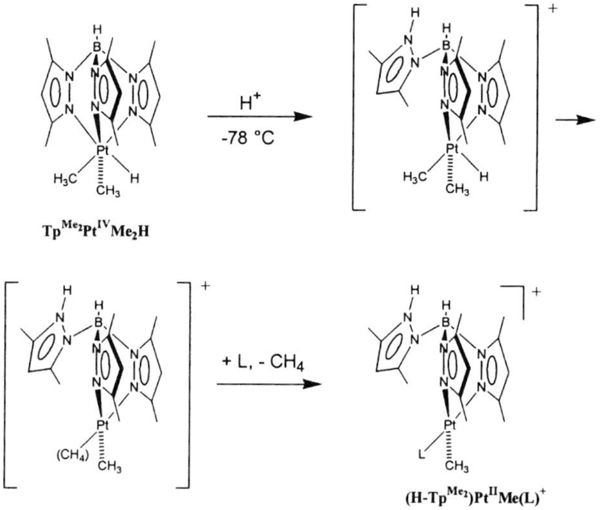

SCHEME 18.

a variety of ligands leads to the platinum(II) complexes shown in Scheme 18 which contain a κ^2-Tp^{Me_2} ligand having one protonated pyrazole group (86).

At elevated temperatures, the compound $Tp^{Me_2}PtMe_2H$ undergoes reductive elimination of methane without added acid. This thermolysis was recently studied in detail (87). In particular, the behavior of the deuterated species $Tp^{Me_2}PtMe_2D$ was instructive since two processes were observed: scrambling of deuterium into the methyl positions at 46–66 °C and reductive elimination of methane accompanied by activation of C_6D_6 solvent at 90–130 °C. The kinetic data provided evidence that methane loss is dissociative in this case (see Section III.H), and the species $Tp^{Me_2}Pt^{II}Me$ is involved as an intermediate. This species is trapped by C_6D_6 as the Pt(IV) product, and the oxidative addition product $Tp^{Me_2}Pt(Me)(C_6D_5)D$ is observed. From the difference in activation enthalpies for the H/D scrambling reaction and the C–H reductive elimination reaction, a lower limit for the binding enthalpy of methane (in the σ-methane complex) of 9 kcal/mol was estimated (87).

The oxidative addition of alkane C–H bonds to Pt(II) has also been observed in these Tp^{R_2}-based platinum systems. As shown in Scheme 19, methide abstraction from the anionic Pt(II) complex (κ^2-Tp^{Me_2})PtMe$_2^-$ by the Lewis acid $B(C_6F_5)_3$ resulted in C–H oxidative addition of the hydrocarbon solvent (88). When this was done in pentane solution, the pentyl(hydrido)platinum(IV) complex **E** (R = pentyl) was observed as a

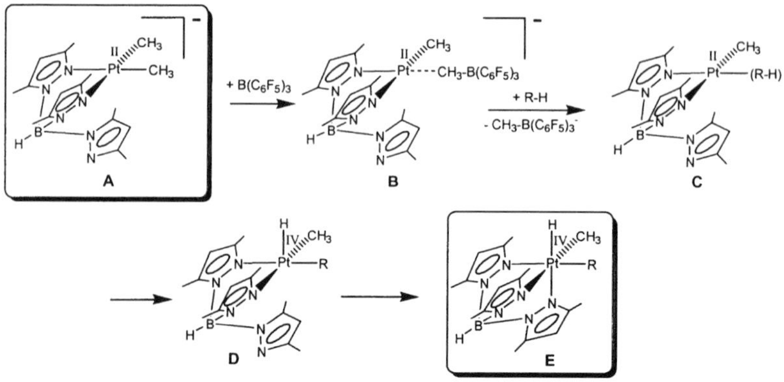

SCHEME 19.

stable product. Similarly, using benzene as solvent, the phenyl(hydrido) platinum(IV) complex was obtained, or using cyclohexane, the corresponding cyclohexyl(hydrido) complex. The intermediates shown in Scheme 19 (**B**, **C**, and **D**) were not observed.

b. Platinum(IV) Alkyl Hydrides as Intermediate Species in the Protonation of Platinum(II) Alkyl Complexes Several of the alkyl(hydrido) platinum(IV) complexes discussed above, for example (bpma)PtMe$_2$H$^+$ and Tp^{Me_2}PtMe$_2$H (*58,62,64*), can be synthesized by protonation of platinum(II) precursors. In the precursors to these platinum(IV) alkyl hydrides, a pendant nitrogen donor is present. This group coordinates to Pt in the platinum(IV) product, as for example depicted in Scheme 20. The protonation of platinum(II) alkyl complexes which do not have a pendant donor group can also lead to observable alkyl(hydrido) platinum(IV) complexes provided that an additional ligand which can occupy the sixth site is present. In most cases, such reactions have to be carried out at low temperature (*53–56*).

The observation of stable Pt(IV) alkyl hydrides upon protonation of Pt(II) alkyls has provided support for the idea that the methane which had been observed in earlier studies (*89–92*) of protonation of Pt(II) methyls could be produced via a reductive elimination reaction from Pt(IV). An extensive study of protonation of Pt(II) methyl complexes was carried out in 1996 (*56*) and an excellent summary of these results appeared in a recent review article (*14*). Strong evidence was presented to support the involvement of both Pt(IV) methyl hydrides and Pt(II) σ-methane complexes as intermediates in the rapid protonolysis reactions of Pt(II) methyls to generate methane. The principle of microscopic

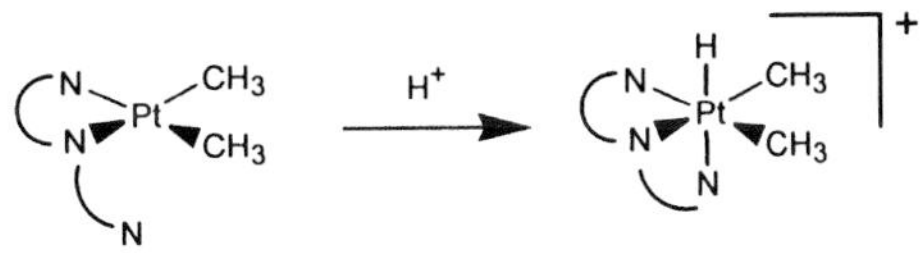

SCHEME 20.

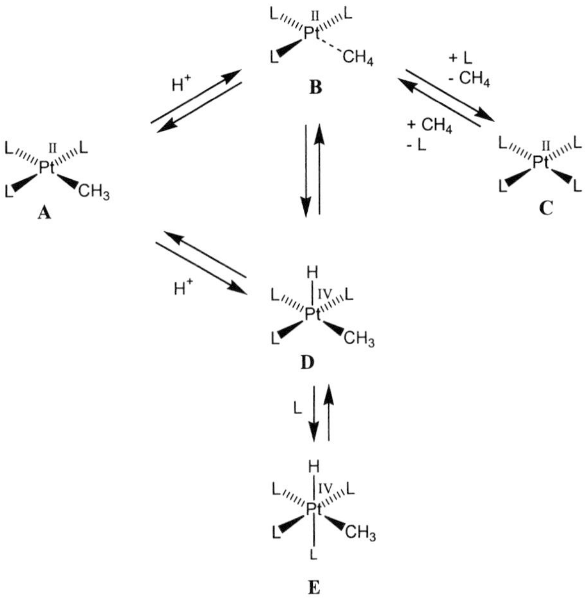

FIG. 1. Unified scheme (similar to that presented in Ref. (56)) for protonation of platinum(II) methyl compounds and for methane activation. "L" is a general two-electron donor ligand. The ligands L on Pt need not be identical, and charges are not shown.

reversibility provides that these species would also be involved in C–H activation reactions at Pt(II). Thus, if we consider Fig. 1, which is similar to the scheme presented in Ref. (56), there are two reasonable possibilities for the protonation of a Pt(II) methyl group to form methane: protonation directly at the methyl group to form in one step a σ-methane complex (**B**) or protonation at platinum to form a methyl(hydrido)platinum(IV) complex (**D**). If the reverse reaction is considered, activation of methane by Pt(II) (**C** in Fig. 1), it can easily be seen that the reaction will proceed via either direct deprotonation of the methane complex **B** (electrophilic pathway) or via oxidative addition to form **D**.

The question of which pathway is preferred was very recently addressed for several diimine-chelated platinum complexes (93). It was convincingly shown for dimethyl complexes chelated by a variety of diimines that the metal is the kinetic site of protonation. In the system under investigation, acetonitrile was used as the trapping ligand L (see Fig. 1) which reacted with the methane complex **B** to form the elimination product **C** and also reacted with the five-coordinate alkyl hydride species **D** to form the stable six-coordinate complex **E** (93). An increase in the concentration of acetonitrile led to increased yields of the methyl (hydrido)platinum(IV) complex **E** relative to the platinum(II) product **C**. It was concluded that the equilibration between the species **D** and **B** and the irreversible and associative[1] reactions of these species with acetonitrile occur at comparable rates such that the kinetic product of the protonation is more efficiently trapped at higher acetonitrile concentrations. Thus, in these systems protonation occurs preferentially at platinum and, by the principle of microscopic reversibility, this indicates that C–H activation with these systems occurs preferentially via oxidative addition (93).

D. STABLE FIVE-COORDINATE Pt(IV) COMPLEXES

The five-coordinate alkylhydridoplatinum(IV) complexes, shown as intermediates in the discussion above, have never been directly observed. However, recently five-coordinate Pt(IV) models for such species have been synthesized and even crystallographically characterized (94,95). A perspectives article describing the road to finding stable examples of five-coordinate platinum(IV) species and highlighting these recent discoveries has also appeared (96). The complexes and their syntheses are depicted in Scheme 21. **A** is a trimethylplatinum(IV) complex of a 'nacnac'-type β-diiminate ligand (94). It was synthesized by reacting the potassium salt of the ligand with tetrameric trimethylplatinum triflate. **C** is a complex of the protonated Tp^{Me_2} ligand and can be synthesized by protonation of the six-coordinate precursor **B** (95). Related derivatives of the five-coordinate dihydrido(silyl)platinum(IV) complex depicted in Scheme 21(**C**) were also prepared having SiPh$_3$ or SiPh$_2$H groups instead of a SiEt$_3$ group (95).

As might be expected given the discussion above concerning the role of five-coordinate Pt(IV) in reductive elimination, complex **A** reductively eliminates ethane at elevated temperature (97). This is in contrast to

---

[1]See Section III.H.

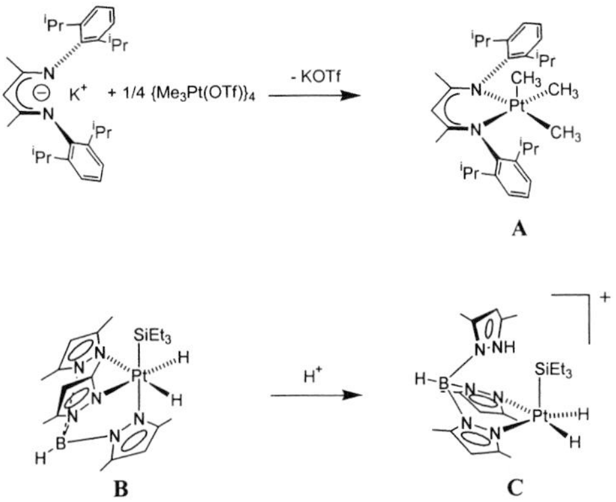

SCHEME 21.

six-coordinate Pt(IV) alkyl complexes with nitrogen ligands which are typically stable to reductive elimination (*98–104*). The unusually facile carbon–carbon reductive elimination from **A** is likely due to its penta-coordinate nature. Also remarkable is the highly fluxional nature of complex **A** in solution, such that the two equatorial and the single apical methyl group exchange on the ^{1}H NMR timescale (750 MHz) even at $-50\,^\circ$C. This observation provides support for the proposal that five-coordinate Pt(IV) intermediates may be involved in isomerization reactions of octahedral six-coordinate Pt(IV) compounds (*52,105–107*). Another interesting feature of **A** is the photolability of the complex. One methyl group is, upon irradiation, easily transferred to the central carbon of the ligand, as shown in Scheme 22 (*94*).

What specific properties of these complexes have allowed isolation of five-coordinate Pt(IV), in the form of the trimethyl complex and the dihydridosilyl complexes? These two types of complexes are significantly different, and their stability is apparently due to different factors. Comparing the trimethyl complex in Scheme 21(**A**) with the related but six-coordinate complexes of a similarly bulky α-diimine ligand (*98*), shown in Scheme 23, is instructive. In Scheme 23A, triflate is clearly coordinated, exhibiting an O–Pt distance of 2.276(3) Å (*98*), which is typical for Pt-coordinated triflate (*108*). This triflate complex **A** in Scheme 23 was obtained from dry tetrahydrofuran. The aqua complex cation **B**, also structurally characterized, was obtained from acetone containing trace water. An equilibrium between coordinated triflate and coordinated water, very likely via a common five-coordinate intermediate, was indicated by NMR spectroscopy (*98*).

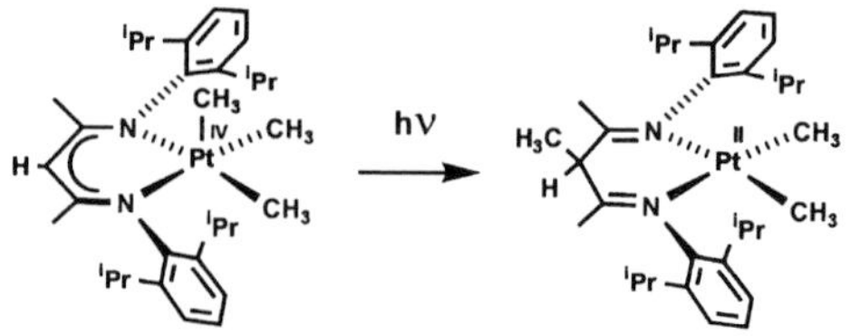

SCHEME 22.

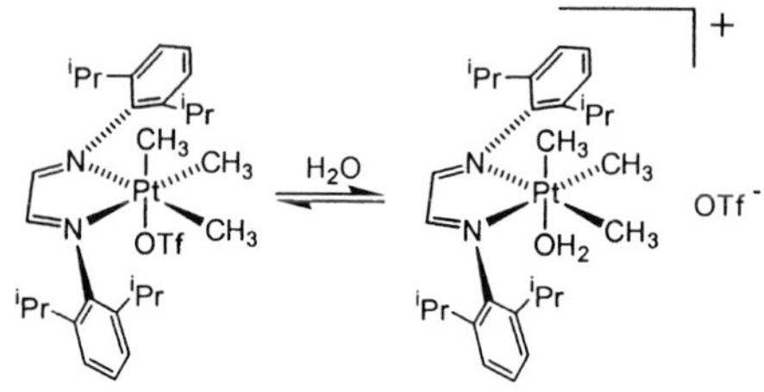

SCHEME 23.

It appears that the advantage of the anionic nacnac-ligand in stabiliz-ing five-coordinate species like **A** in Scheme 21 mainly lies in its charge. Since the resulting five-coordinate complex is not cationic but neutral, anion coordination does not occur. The uncharged nature of the com-pound, and presumably also the presence of the isopropyl groups, allow solubility in alkane solvent, such that the use of potentially coordinating polar solvents is not necessary. In addition, the isopropyl groups of **A** (Scheme 21), although not agostic, effectively shield the open site. It is worth noting here that the anionic 'nacnac'-type β-diketiminate ligands (109,110) have proven to be very useful chelating ligands in recent years (111–114). A comprehensive review on metal complexes of β-diketiminate ligands was published in 2002 (115).

Complex **C** (Scheme 21) seems to shows stable penta-coordination for apparently very different reasons. The compound is cationic, but the counterion is a non-coordinating tetraarylborate. Interestingly, neither the dichloromethane solvent nor the diethylether present in the reaction mixture seem to coordinate to the open site in solution. The compound was crystallized from a tetrahydrofuran/pentane mixture as the tetrahy-drofuran (THF) solvate, but in the crystal structure, the THF is remote from the open site at platinum. The open site is shielded somewhat by the methyl groups of the protonated Tp^{Me_2} ligand, but it does not appear completely inaccessible. A reasonable explanation for the observed five-coordinate structure is that silicon electronically stabilizes high formal oxidation states of transition metals without inducing a pronounced positive charge on the metal. Silicon is relatively electropositive, and it is only in a formal sense that silyl ligands are considered as R_3Si^- anions. This idea is perhaps supported by the fact that a variety of transition metal complexes have been prepared in unusually high oxidation states using silicon-donors (*116–120*). If it is true that the electron-density in the silicon–metal bond is more localized at the metal, the position *trans* to the silicon should display reduced electrophilicity. This was already indicated by the crystal structure of the six-coordinate platinum(IV) complex $Tp^{Me_2}Pt(SiEt_3)H_2$ (compound **B** in Scheme 21), where the Pt–N bond *trans* to silicon was unusually long (2.30 Å) (*74*).

It is interesting to compare an alternative synthesis of the five-coordinate Pt(IV) dihydridosilyl complex, by silane activation, with the synthesis of a somewhat related Pt(IV) silyl hydride complex containing a functionalized diimine ligand. Both Pt(IV) species can be formed by silane oxidative addition as shown in Scheme 24. Note that in the formation of the five-coordinate Pt(IV) complex (Scheme 24, **A**), the precursor has a labile ligand (solvent, most likely diethylether) (*95*). This labile ligand is lost in the reaction to allow silane coordination within the square-planar geometry of Pt(II). Notably, in the related silane σ-bond activation using the diimine ligand with two pendant ether arms (Scheme 24, **B**), the precursor has one ether residue coordinated (*121*). This ether group probably functions as a labile ligand to allow silane coordination to Pt(II). However, because the ether is covalently linked and the ligand has sufficient flexibility, this group would be able

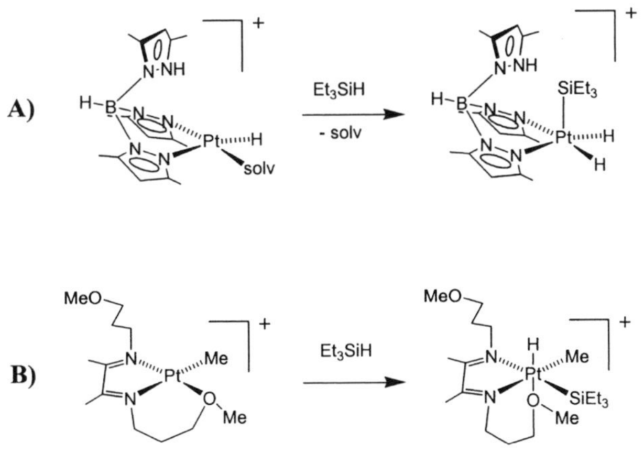

SCHEME 24.

to coordinate to the Pt(IV) center, and a six-coordinate structure for the product has been proposed.

E. STABLE σ-COMPLEXES AND RELATED SPECIES

1. Dihydrogen Complexes of Pt(II)

Both σ-silane and σ-dihydrogen complexes are useful and in some cases isolable models for the rarely observed σ-alkane complexes. Unfortunately, silane complexes of Pt(II) are generally dinuclear (37,122) rendering them less desirable models for Pt(II) σ-alkane complexes. In contrast, the closely related dihydrogen complexes of Pt(II) are mostly monomeric (123–126).

In a particularly interesting study of Pt(II) dihydrogen complexes, Stahl, Labinger and Bercaw reported the reaction of the complexes trans-Pt(PCy$_3$)$_2$(X)H (Cy = cyclohexyl, X = SiH$_3$, H, CH$_3$, Ph, Cl, Br, I, CN, F$_3$CSO$_3$) with acids (125). Dihydrogen complexes were only observed for those compounds which have a strong σ-donor ligand trans to the hydride (X = H, CH$_3$ or Ph). In contrast, when halides were the trans ligands (reactants of the type trans-Pt(PCy$_3$)$_2$(X)H), protonation occurred at the platinum center. However, only six-coordinate Pt(IV) dihydride complexes were observed. The sixth ligand was halide, either from protonation with HCl or formed by halide abstraction from starting material. No direct evidence for five-coordinate Pt(IV) dihydride species was found.

The authors point out that the dependence of the site of electrophilic attack on the ligand trans to the hydride in the model systems may be important with respect to alkane activation. If the information is transferable to Pt-alkyls, protonation at the metal rather than the alkyl should be favored with weak (and "hard") σ-donor ligands like Cl$^-$ and H$_2$O. These are the ligands involved in Shilov chemistry and so by the principle of microscopic reversibility, C–H oxidative addition may be favored over electrophilic activation for these related complexes.

2. η^2-Benzene Complexes of Pt(II)

The discussion has focused so far on activation of alkanes, where formation of the σ-complex seems to precede oxidative addition. For arenes, formation of the analogous σ$_{(CH)}$-arene complex is thought to occur before oxidative addition to form an aryl hydride. These σ-complexes have never been observed, presumably because they are unstable with respect to the π-complexes. Both types of arene complexes are, for the case of benzene, shown in Scheme 25: the σ$_{(CH)}$-arene complex as **A** and

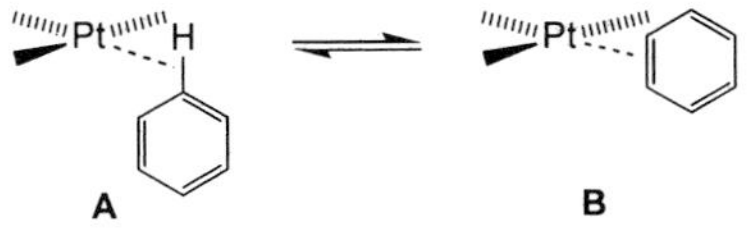

SCHEME 25.

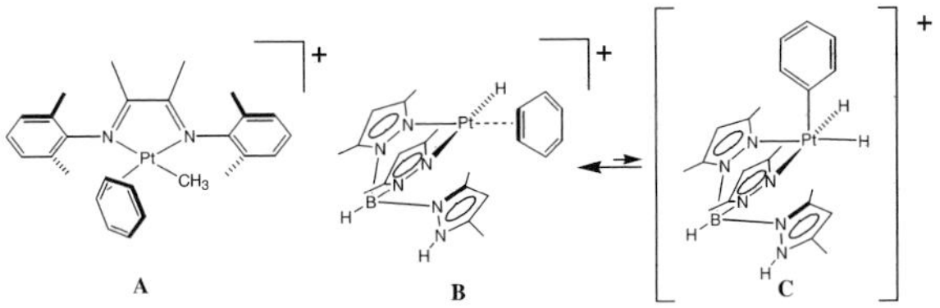

SCHEME 26.

the π-complex as **B**. While Pt(II) σ-complexes of arenes as shown in Scheme 25(**A**) have not been observed, Pt(II) π-complexes depicted in Scheme 25(**B**) have been recently observed, and – for one example – even crystallographically characterized. These species are shown in Scheme 26.

Complex **A** (Scheme 26) has been observed by ^{1}H NMR spectroscopy at $-69\,^\circ$C. At $-23\,^\circ$C, benzene and methane are lost to form new species within 1 h (*29*). The related complex **B**, however, is relatively stable, and loses benzene only slowly at room temperature (*127*). Complex **B** was stable enough to be characterized by X-ray crystallography in addition to NMR studies in solution. Using NMR line-broadening and spin-saturation studies, it was shown that the hydrogens on the coordinated arene exchange with the metal hydride. This was interpreted in terms of C–H activation by oxidative addition, involving the five-coordinate structure **C**, shown in Scheme 26, and a barrier $\Delta G^{\ddagger}$ of 12.7 kcal/mol was obtained at $-21\,^\circ$C (*127,128*). Elimination of benzene from complex **B** in Scheme 26 leads to a dihydrido-bridged dimer (*129*).

F. FIVE-COORDINATE INTERMEDIATES AND σ-COMPLEXES: COMPUTATIONAL APPROACHES

The experimental data available to date consistently indicate that ligand dissociation precedes reductive elimination from six-coordinate platinum(IV). In the reverse direction (oxidative addition), it seems necessary that the hydrocarbon molecule coordinates in the square plane of platinum(II). C–H bond cleavage then forms a five-coordinate Pt(IV) species consistent with the principle of microscopic reversibility.

The important question then arises of whether this scenario involving five-coordinate and σ-complex intermediates must occur for every system. This question becomes more intriguing when one considers that the related C–H reductive elimination from platinum(II) complexes to yield platinum(0) products does not require prior dissociation of a ligand. For example, strong mechanistic evidence has been presented to support the direct elimination of methane from $(PPh_3)_2Pt^{II}CH_3(H)$ (130). This difference in reaction mechanisms of C–H reductive elimination from Pt(IV) and Pt(II) complexes has been examined using density functional theory (DFT). DFT has been used to study reductive elimination from a number of platinum and palladium systems (131). The compounds $(PR_3)_2Pt^{II}CH_3(H)$ (R = Me, H) were used as models for $(PPh_3)_2$ $Pt^{II}CH_3(H)$ (132,133). For both Pt(II) systems, it was found that direct reductive elimination to form the Pt(0) complex of the type $(PR_3)_2Pt$ was favored over a mechanism involving prior ligand dissociation. The energetics of relevant compounds and transition states (ΔH_{298}, kcal/mol) are shown in Fig. 2. Note that although the barrier to reductive elimination is lower from the three-coordinate Pt(II) species, the enthalpic cost of losing the ligand is significant and this results in the direct reductive elimination pathway being the lower energy route.

It is also interesting to consider that by the principle of microscopic reversibility, the oxidative addition of methane to Pt(0) should then also follow a direct pathway of addition to a two-coordinate Pt(0) starting complex. It can be noted as well that the transition state for methane addition shows a 'product-like' P–Pt–P angle: 109.2° for the bis-PH_3 system and 114.1° for the bis-PMe_3 system. These data are consistent with many computational findings on related systems (134) and explain the experimental observation that Pt(0) complexes of chelating phosphines are more active in C–H activation than corresponding linear complexes of non-chelating phosphines (135–137). That a smaller bond angle in the Pt(0) reactant leads to more facile C–H activation has been rationalized using extended Hückel calculations (138).[2]

The two different routes, direct reductive elimination and a preliminary ligand dissociation pathway, were similarly investigated for the Pt(IV) complexes, $(PR_3)_2Cl_2PtCH_3(H)$ (R = Me, H) (132,133). A part of this study (dealing with one possible isomer) is summarized in Fig. 3.

---

[2]The classic platinum(0) approach to C–H activation, yielding platinum(II) alkyl hydrides as the oxidative addition products, contributed significantly to our understanding of C–H activation. However, the platinum(II)/(IV) approach has proven capable of achieving oxidative *functionalization* of hydrocarbons, and so this review focuses on the higher oxidation state.

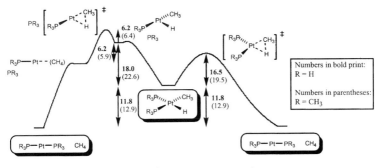

FIG. 2. Enthalpy diagram (ΔH_{298}, kcal/mol; B3LYP level of DFT) for reductive elimination of methane from cis-$(R_3P)_2PtCH_3(H)$, $PR_3=P(CH_3)_3$ or PH_3. The diagram was drawn using the data from Refs. (132,133).

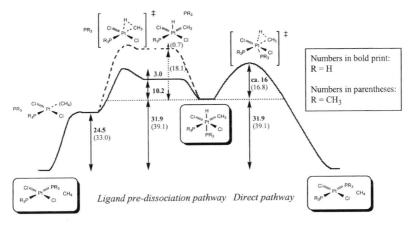

FIG. 3. Enthalpy diagram (ΔH_{298}, kcal/mol; B3LYP level of DFT) for reductive elimination of methane from one isomer of $(R_3P)_2Cl_2PtCH_3(H)$, $PR_3=P(CH_3)_3$ or PH_3. The dotted line refers to the $P(CH_3)_3$ system, where the relative order of barrier heights changes in comparison to the PH_3 system. The diagram was drawn using the data from Refs. (132,133).

A transition state for the direct methane elimination from the Pt(IV) complex having two PH_3 ligands was not observed. Phosphine loss occurred concomitantly with the reductive elimination. However, the authors were able to estimate an activation barrier of ca. 16 kcal/mol for direct elimination from this Pt(IV) complex $(PH_3)_2Cl_2PtCH_3(H)$ using artificial restraints for the geometry optimization. This value is very close to the 16.5 kcal barrier obtained for reductive elimination from the Pt(II) analog $(PH_3)_2PtCH_3(H)$. In contrast however, the ligand loss pathway is the favored one for Pt(IV). While the barrier for reductive elimination from the five-coordinate Pt(IV) intermediate is somewhat lower than the barrier for reductive elimination from the related three-coordinate Pt(II)intermediate (3.0 vs. 6.2 kcal/mol), the primary reason that the ligand dissociation pathway is favored for $(PH_3)_2Cl_2PtCH_3(H)$ is that the enthalpic cost of ligand dissociation is much lower in the platinum(IV) system (10.2 kcal/mol instead of 18.0 kcal/mol for the Pt(II) analog). Conversely, oxidative addition of methane is, according to these calculations, enthalpically favored to occur via a methane complex. Thus one ligand has to be lost from Pt(II) to form a σ-complex intermediate.

In contrast, when the more electron-rich phosphine PMe_3 is employed, the situation looks somewhat different (Fig. 3). The direct elimination pathway was calculated to have a barrier of 16.8 kcal/mol which is very close to the barrier estimated for PH_3 ligands. However, the increased binding enthalpy of PMe_3 versus PH_3 is significant enough to cause the ligand dissociation pathway to be enthalpically less favorable by 2 kcal/mol relative to the direct pathway. Entropic contributions will account for this small energy difference and, at least for monodentate phosphines, a ligand dissociation pathway will likely dominate. On the other hand, a Pt(IV) center with an electron-rich phosphine ligand which is *chelating* may undergo C–H reductive elimination directly.

Note that the oxidative addition of methane to $(PR_3)_2PtCl_2$ is calculated to be uphill by ca. 32 kcal/mol for the PH_3 system and by ca. 39 kcal/mol for the PMe_3 system (Fig. 3). Interestingly, this result indicates that ligation by the weaker ligand PH_3 leads to a thermodynamically more favorable oxidative addition of hydrocarbons. A related ligand effect was reported by Hill and Puddephatt (*139*) in a DFT study, where the equilibrium between the cationic methane complexes $L_2PtMe(CH_4)^+$ and the corresponding five-coordinate oxidative addition products $L_2PtMe_2H^+$ (L = NH$_3$, PH$_3$) was investigated. In accord with the idea that PH_3 should be a better σ-donor than NH_3, the electron density at platinum was calculated to be greater with L = PH$_3$ than with L = NH$_3$, both in the Pt(II) σ-methane complex and in the corresponding five-coordinate Pt(IV) cation. Yet when the thermodynamics of oxidative addition were investigated, it was found that the less donating spectator ligand results in oxidative addition of methane being more favorable: Using L = NH$_3$, the five-coordinate oxidative addition product is uphill from the methane complex by only 9 kcal/mol, as opposed to 15 kcal/mol for L = PH$_3$. These computational results are in agreement with the general experimental observation that nitrogen donor ligands stabilize the higher oxidation state of Pt and Pd (IV vs. II) more than phosphine derived ligands (*98–100,140,141*).

Finally, computational methods have also been used to point out that ligand loss to generate unsaturated species is not the only way C–H oxidative addition at Pt(II) can be facilitated. Just as bent bisphosphine-ligated Pt(0) complexes significantly favor oxidative addition of C–H bonds relative to the linear Pt(PR$_3$)$_2$ species, Pt(II) complexes which are constrained to a see-saw type geometry rather than a square-planar arrangement might be predicted to similarly favor hydrocarbon activation. This effect has been examined computationally (*142*). Summarized in Scheme 27 are the ways in which C–H oxidative addition can be favored at Pt(0) and Pt(II) via the geometrical arrangement of ligands. Perhaps not surprisingly, the fragments which can easily insert into C–H bonds are isolobal to CH$_2$ (*143*).

As seen from the above discussion concerning geometrical constraints of ligand environments and the effects of ligand donor atoms, computational work may prove useful to the experimental chemist in the search for promising new ligands. For example, several tridentate donor systems where ligand constraints stabilize Pt(IV) over Pt(II) have been investigated computationally (*144*). Two of the ligands predicted to stabilize Pt(IV) methyl hydrides are 1,4,7-triazacyclonane-2,5,8-triene and cyclohexane-1,3,5-triamine (**A** and **B**, respectively in Scheme 28). Both are currently underutilized experimentally in Pt(IV) chemistry. **A** is predicted to stabilize d^6 systems to such an extent that even a *palladium*(IV) methyl hydride was predicted to be stable (*144*). Ligand **B** has been used in experimental work with platinum (*145,146*) and does efficiently

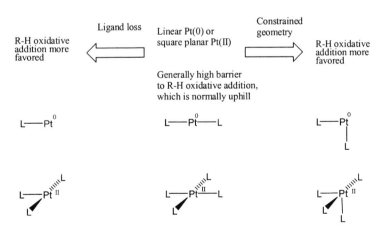

SCHEME 27.

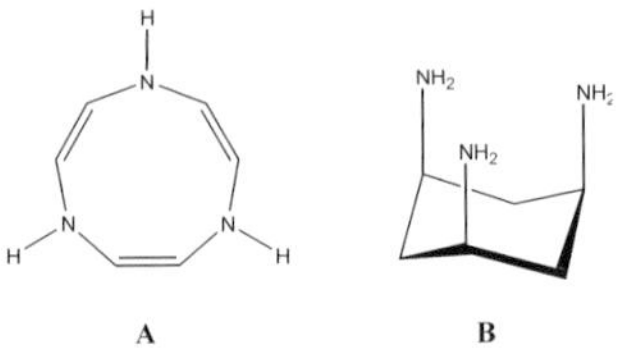

SCHEME 28.

stabilize Pt(IV) (see Section IV.C below). However, hydrocarbon activation has not yet been pursued with this ligand on platinum.

G. THE ELECTROPHILIC PATHWAY: COMPUTATIONAL APPROACHES

It should be pointed out that the studies described above (both experimental and theoretical) do not provide an answer to the question of whether C–H activation at Pt(II) occurs primarily via an oxidative addition pathway or an electrophilic pathway. The studies do indicate that σ-complexes are competent intermediates and that C–H oxidative addition can occur to generate a five-coordinate Pt(IV) alkyl hydride. Thus, the oxidative addition pathway must be considered, but need not be operating in all Pt(II)-based C–H activation systems. Modifications of the coordination sphere of the platinum center may lead to a change in the electronic situation such that oxidative addition is favored for one system, but the electrophilic pathway is favored for a different system. Which mechanism is operating in a particular system wherein a true Pt(IV) alkyl hydride species is not observed, is difficult to address experimentally. However, this question has been treated computationally by several researchers working with different Pt(II) species. The differences in activation energies for these alternative mechanisms can be quite small. This was demonstrated in 1996 in a computational (DFT) study on methane activation by $PtCl_2(OH_2)_2$, clearly relevant to the Shilov system, where the conclusion was drawn that the oxidative addition pathway and the σ-bond metathesis pathway are too close in energy to allow a prediction as to which one would be preferred (*147*). Notably, in this work, an attempt was made to model solvent effects on the reaction. One or two water molecules of the second coordination sphere were included in addition to Pt-coordinated water molecules.

A similar conclusion, that the alternative pathways sometimes differ only slightly in the energy barrier, was reached in several more recent theoretical papers which investigated methane activation in ("N_2")$PtCl_2$ systems relevant to the Catalytica process. It has been demonstrated experimentally that compared with (bipyrimidine)$PtCl_2$,

$(NH_3)_2PtCl_2$ exhibits higher catalytic activity toward methane function-alization (12). However, under acidic conditions, (bipyrimidine)$PtCl_2$ is much more stable than $(NH_3)_2PtCl_2$ towards dissociation of the nitrogen donor ligand, and this makes the bipyrimidine system superior (12,148). Using DFT calculations and employing a dielectric continuum model to correct for solvation, the activation of methane by $(NH_3)_2PtCl_2$ was stud-ied (149). It was assumed in this work that methane replaces an ammonia ligand from the first coordination sphere to form a σ-methane complex. Both pathways, one in which oxidative addition then occurs to form a five-coordinate methyl hydride or deprotonation of the coordinated methane occurs to form the Pt(II) methyl were investigated. The Pt-bonded chloride ligand acts as a base for the deprotonation of the methane. It was found that the activation barriers are similar for both processes using cis-$(NH_3)_2PtCl_2$. Solvation-corrected $\Delta G_{298}^{\ddagger}$ values of 4.2 and 11.0 kcal/mol (gas phase $\Delta E^{\ddagger}$: 7.9 and 20.2 kcal/mol) were found for C–H bond breaking by oxidative addition and by the electrophilic path-way, respectively. However, for the case of the trans-complex, it was predicted that oxidative addition would be strongly favored, the solva-tion-corrected $\Delta G_{298}^{\ddagger}$ being lower by 20.5 kcal/mol (10.3 vs. 30.8 kcal/mol; gas-phase $\Delta E^{\ddagger}$: 11.7 vs. 32.9 kcal/mol) (149).

Computational results were also obtained, in a different study, for the possibility that the chloride rather than the ammonia in cis-$(NH_3)_2PtCl_2$ was substituted by methane (148). In the same contribution, an analogous study examined the reactivity of the (bipyrimidine)$PtCl_2$ complex (Fig. 4).

Whether the resulting ion pair-type methane complex [("N_2") $PtCl(CH_4)]^+Cl^-$ undergoes the C–H activation reaction by oxidative addition or by the electrophilic pathway was answered differently for cis-$(NH_3)_2PtCl_2$ and for the bipyrimidine-based catalyst. For cis-$(NH_3)_2PtCl_2$, the oxidative addition pathway was favored by 10 kcal/mol, whereas for the bipyrimidine system the electrophilic pathway was calculated to be more favorable by 9 kcal/mol (148). In an earlier DFT com-putational study by another group (150), the cation (bipyrimidine) $PtCl(CH_4)^+$ was investigated instead of the tight ion pair [(bipyrimi-dine)$PtCl(CH_4)]^+Cl^-$, and the opposite conclusion was reached. Oxidative addition was favored over the σ-bond metathesis pathway. The reason for this discrepancy appears to be inclusion or omission of the counterion. If the chloride counterion is not included, the coordi-nated chloride has to deprotonate the alkane complex. This may have a higher barrier than deprotonation by non-bonded chloride in the ion pair. Since ion pairing and solvation effects are large in these systems, and calculations of these effects still contain considerable potential

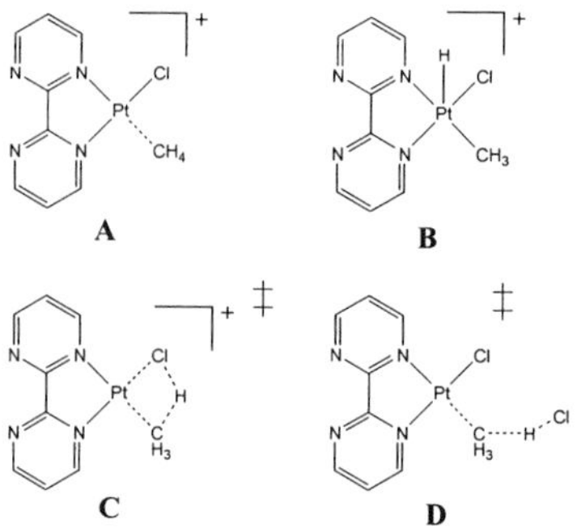

FIG. 4. Relevant structures for the discussion of methane activation by (bipyrimidine)PtCl$_2$: Methane complex of Pt(II) (**A**); methyl(hydrido)platinum(IV) complex, the product of the oxidative addition (**B**); transition state for intramolecular deprotonation of the methane complex ("σ-bond metathesis", sometimes also called "electrophilic", **C**); intermolecular deprotonation of the methane complex ("electrophilic pathway", **D**).

errors, it is not yet clear whether calculated small preferences in favor of one or the other mechanism represent the physical reality. They do, however, impressively demonstrate that activation energies for these different mechanisms can be very close in energy.

In a study of the methane complex [(diimine)Pt(CH$_3$)(CH$_4$)]$^+$ (diimine = HN=C(H)–C(H)=NH), relevant to the diimine system experimentally investigated by Tilset *et al.* (*28*), theoretical calculations indicate preference for the oxidative addition pathway (*30*). When one water molecule was included in these calculations, the preference for oxidative addition increased due to the stabilization of Pt(IV) by coordinated water (*30*). The same preference for oxidative addition was previously calculated for the ethylenediamine (en) system [(en)Pt(CH$_3$)(CH$_4$)]$^+$ (*151*). This model is relevant for the experimentally investigated tmeda system [(tmeda)Pt(CH$_3$)(solv)]$^+$ discussed above (Scheme 7, **B**) (*27,152*). For the bis-formate complex Pt(O$_2$CH)$_2$, a σ-bond metathesis was assumed and the energies of intermediates and transition states were calculated (*153*). Unfortunately the potential alternative of an oxidative addition pathway was not investigated in this work.

Phosphine complexes are generally regarded more electron-rich than the corresponding ammine complexes, and which pathway is preferred under these electronic conditions has also been investigated. For *trans*-PtCl$_2$(PH$_3$)$_2$, oxidative addition has been calculated to be much more favorable than the electrophilic pathway for the activation of methane (*154*). While the lowest-energy pathway involved, again, ligand dissociation (phosphine) followed by coordination of methane to form the methane complex $(CH_4)PtCl_2(PH_3)$, the lowest-energy pathway for the actual C–H activation step was the oxidative addition of a C–H bond of coordinated methane in the complex $(CH_4)PtCl_2(PH_3)$ to form $PtCH_3(H)Cl_2(PH_3)$. For this reaction, a reaction energy of +11.4 kcal/mol was calculated, and the activation energy was 16.2 kcal/mol. This is in sharp contrast to the electrophilic pathway, where the coordinated chloride in $trans$-$(CH_4)PtCl_2(PH_3)$ intramolecularly deprotonates the coordinated methane to yield $PtCH_3(Cl–H)Cl(PH_3)$. For this electrophilic pathway, a reaction energy of 34.2 kcal/mol and an activation energy of 37.3 kcal/mol were calculated. Thus, both kinetically and thermodynamically the oxidative addition pathway in this system is more favorable by more than 20 kcal/mol. These gas-phase calculations may have some relevance to the reactivity in such non-polar media as hydrocarbons (154). The situation may be quite different for polar media, but it is interesting to note that a ca. 20 kcal/mol preference for oxidative addition versus electrophilic pathway was found for the analogous $trans$-$(NH_3)_2PtCl_2$ system discussed above, and that this barrier did not change significantly upon correction for bulk solvation (149). Thus, it seems to be the case that oxidative addition is preferred over the electrophilic pathway for the $trans$-diammine and the $trans$-diphosphine complexes when the electrophilic pathway involves deprotonation by $coordinated$ chloride. However, this may not be true if the deprotonation occurs by external base. For the cis-complexes where the energy differences are much smaller, the ambiguity between the mechanisms appears to be greater. Our impression is that the small preferences recently calculated for one or the other mechanism do not necessarily represent the chemical reality. More extensive efforts to include explicitly treated solvent molecules and counterions in addition to corrections for bulk solvent are probably needed.

In the end, while computations on these alternative pathways, oxidative addition and σ-bond metathesis, have provided some insight, the question by which way does a particular reaction proceed cannot yet be answered definitively. It is very interesting, however, that both mechanisms involve the same Pt(II) intermediate in which the hydrocarbon binds in the square-planar coordination sphere of the metal.

H. Associative or Dissociative Substitution at Pt(II) by Hydrocarbon?

Although there is now considerable evidence to support the proposal that a labile group must be replaced by the hydrocarbon prior to a facile C–H activation step, studies to answer the question of how such a substitution to form a σ-complex of the hydrocarbon proceeds have only recently been undertaken. It is conceivable that the labile group dissociates first to create a highly reactive, three-coordinate 14-electron species (dissociative mechanism, generally referred to as "**D**"). However, substitution mechanisms are also possible which do not involve a three-coordinate species. The substitution may either involve a five-coordinate intermediate (associative mechanism, "**A**") or might be a concerted interchange ("**I**"). Interchange processes are classified associative or dissociative interchange, ("I_a" or "I_d") if the bond-making or -breaking, respectively, is more advanced in the transition state (*155,156*). These possibilities are depicted in Scheme 29.

Most substitutions at Pt(II) have been found to proceed via an associative mechanism (*157,158*). Very few exceptions are known, and it is interesting to note that all involve complexes wherein a very strong σ-donor (like a Pt-bonded carbon) is present *trans* to the group which is being substituted (*159–162*). This electronic situation does not always lead to dissociative substitution at Pt(II) (*162–164*) but has been documented to do so in some well-studied cases. The majority of platinum(II) complexes utilized for C–H activation do not have exceptionally strong *trans*-directing ligands opposite to the labile group, so an associative mechanism might be anticipated. However, the classic studies on substitution reactions at Pt(II) have traditionally employed relatively strong nucleophiles, and never a hydrocarbon C–H bond, as an incoming ligand.

The idea that a hydrocarbon, an extremely weak donor, might associatively attack a square-planar platinum center and thus assist in displacing the leaving group seems at first difficult to accept. However, two recent mechanistic studies have found convincing evidence to support associative pathways for these reactions. One study investigated the microscopic reverse reaction, methane loss following protonation of a (diimine)PtMe$_2$ complex (Scheme 30) (*165*).

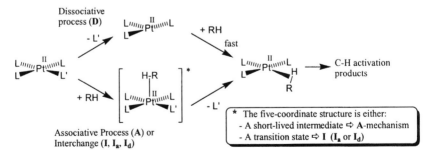

SCHEME 29.

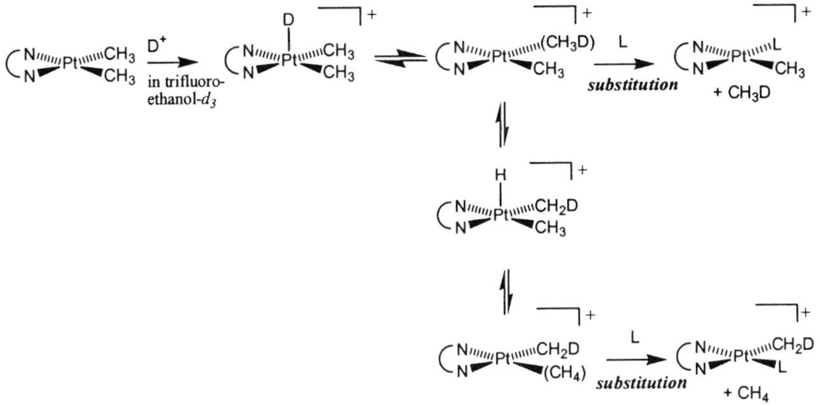

Scheme 30.

Scrambling of deuterium into the methyl position of the product was observed, and the extent of scrambling depended strongly on the concentration of the added ligand L (acetonitrile). The amount of deuterium scrambling decreased significantly with increasing ligand concentration. This is exactly what would be expected for associative substitution of methane by the incoming ligand L. In contrast, for a dissociative mechanism, no dependence of the extent of scrambling on the concentration of L would be expected. Thus, the conclusion was drawn that methane is associatively replaced by the ligand acetonitrile. Based on the principle of microscopic reversibility, the substitution of acetonitrile by methane, which has not yet been observed experimentally, would in this same system involve an associative mechanism.

In the second study, the actual hydrocarbon activation mechanism was studied and the mechanistic conclusions were drawn on the basis of the activation parameters, including the activation volume (33). Since the activation volume ($\Delta V^{\ddagger}$) represents the volume change on going from the reactants to the transition state, this parameter gives direct information about the associative or dissociative nature of a process, which is particularly useful for the elucidation of reaction mechanisms (166,167). The Pt(II) system that was investigated is shown in Scheme 31 (33). The "N$_2$" ligand is again an α-diimine. Kinetic studies were carried out at various pressures using high-pressure NMR spectroscopy (168). Since the observed apparent rate constant included the pre-equilibrium constant K_1, substitution of water by more weakly bonded trifluoroethanol (TFE) solvent to give the more active trifluoroethanol complex, the $\Delta H°$, $\Delta S°$, and $\Delta V°$ parameters for the pre-equilibrium were also

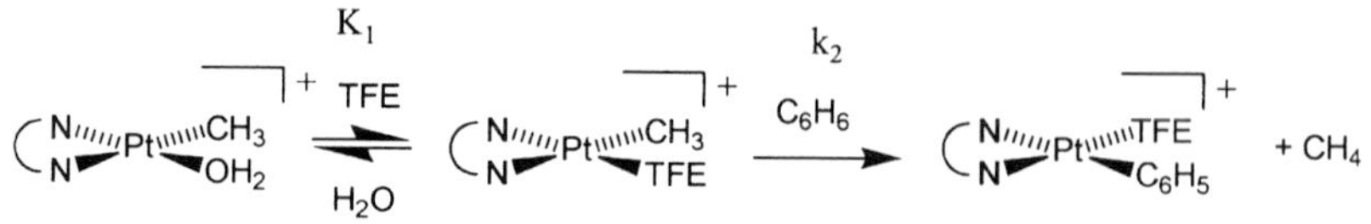

SCHEME 31.

determined. This allowed, together with the apparent activation parameters (which contained the K_1-equilibrium *and* the k_2-step), the determination of the true activation parameters for the k_2-step, the C–H activation of benzene. For this crucial step, an activation volume of $-9.5 \pm 1.3 \, cm^3/mol$ was obtained (*33*). This value is very typical for associative substitution at platinum(II) (*169,170*) and supports the argument that the (diimine)PtMe(solvento)$^+$ cation, reactive in hydrocarbon activation, undergoes associative substitution reactions, even when hydrocarbon is involved. Since α-diimines are moderate σ-donors, somewhat similar to chloride, this may suggest that in Shilov's system the hydrocarbon also binds via an associative step. For systems, however, which are electronically or even sterically very different from the diimine systems, the mechanistic pathway may differ and a dissociative mechanism remains a possibility. For example, kinetic data for the solution thermolysis of the complex $Tp^{Me_2}PtMe_2H$ (Section III.C.2) indicate that the σ-methane intermediate $Tp^{Me_2}Pt^{II}Me(CH_4)$ dissociatively loses methane prior to oxidative addition of benzene-d_6 to form $Tp^{Me_2}PtMe(C_6D_5)D$ (*87*). Further mechanistic investigations can be expected, and should be of considerable value for understanding hydrocarbon activation.

The intimate mechanism of ligand substitution has been suggested to play an important role in determining the product selectivity of arene C–H bond activation reactions. In a recent study on the regioselectivity of hydrocarbon C–H activation with two α-diimine-chelated Pt(II) complexes possessing a labile solvent ligand (H$_2$O/trifluoroethanol, Fig. 5), surprising selectivities were found. For some substrates, it was observed that the selectivities depended strongly on the substituents of the chelate ligand (*31*).

The *meta*-selectivity for toluene activation, observed for both systems, is very unusual (Fig. 5). Also remarkable is the switch in selectivity from aryl C–H activation to benzylic activation in *p*-xylene, just by changing the chelate ligand from the diimine equipped with trifluoromethyl substitutents in the *meta*-positions of the phenyl residue to the diimine bearing methyl substituents in the *ortho*-positions (Fig. 5). The authors suggested that the C–H bond activation is reversible and the isomeric σ-methane complexes are in equilibrium prior to the substitution of

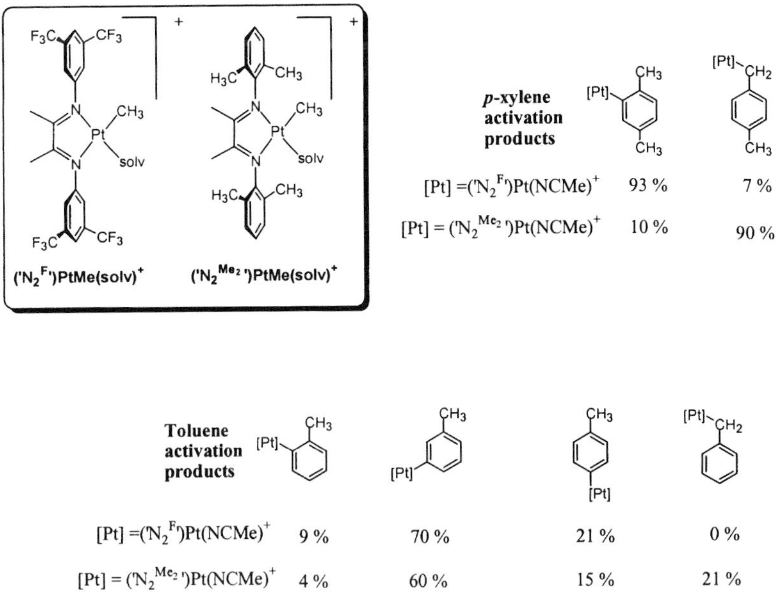

FIG. 5. Product selectivities for two diimine-chelated C–H activation systems (*31*). The coordinated solvent ('solv') in the starting complex is H$_2$O, but the C–H activating species is thought to have trifluoroethanol (the solvent employed) coordinated. The product distribution was determined after the reaction mixture was quenched with acetonitrile.

methane by solvent. For example, Scheme 32 can be used to explain the *ortho*-, *meta*-, and *para*-selectivities observed for toluene activation. The product solvent complexes do not isomerize under the conditions employed. The different barriers (arising both from steric and electronic factors) for the formation of the solvent complexes by means of an associative substitution were proposed to determine the selectivities (*31*).

I. HOW ELECTROPHILIC ARE "ELECTROPHILIC" Pt(II) COMPLEXES?

Given the discussion above detailing that in many systems C–H activation at Pt(II) occurs via an oxidative addition pathway, it is perhaps not surprising that Pt(II) species capable of C–H activation need not be very electrophilic. On the contrary, it might be intuitively expected that electron-rich Pt(II) centers are more suitable to oxidatively add a hydrocarbon molecule. Theoretical analyses, in contrast, seem to suggest rather that electron-poor Pt(II) centers add hydrocarbon molecules more favorably (see above, Section III.F). A recent experimental study

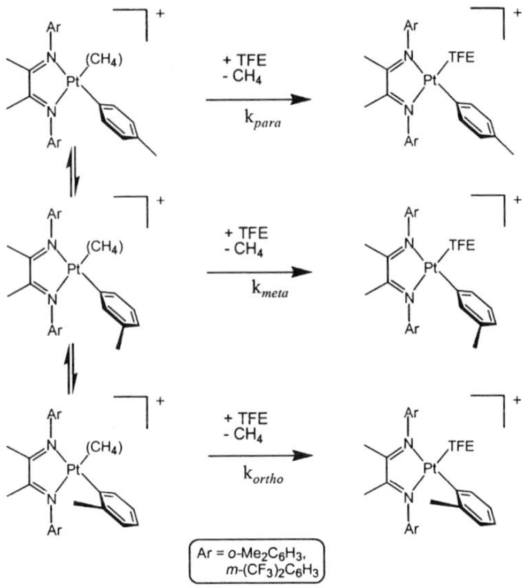

Scheme 32.

has shown that the situation is indeed very complex (32). The C–H activation of the cationic species $(L)PtMe(solv)^+$ (L = an α-diimine ligand, solv = trifluoroethanol/H_2O) was investigated in great detail as a function of various substituents on the arene moiety of the ligand (Scheme 33). The influence of the substituents on the electron density at the platinum center was determined using the CO stretching frequencies of the corresponding methyl carbonyl cations (similar to the compounds shown in the left half of Scheme 33 but with CO in place of "solv"). The observed CO stretching frequencies correlated well with the known σ-values, but one important exception was noted: complexes having methyl groups in the 2,6-positions (R_2, R_6) showed lower electron density at platinum (higher stretching frequencies) compared to complexes having hydrogens in these positions. This was rationalized in terms of the dihedral angle between arene planes and the diimine plane, which is probably different for the two series. The more bulky methyl substituents were suggested to lock the dihedral angle to ca. 90°, whereas sterically less demanding hydrogen substituents may allow for greater orbital overlap and electron-donation from the arene to the platinum via the diimine. The observation was made that the more electron-rich complexes were more reactive with respect to hydrocarbon activation (32). Remarkably, the two series, the "ortho-dimethyl" and the "non-bulky" series followed the same trend, but apparently *for different reasons*.

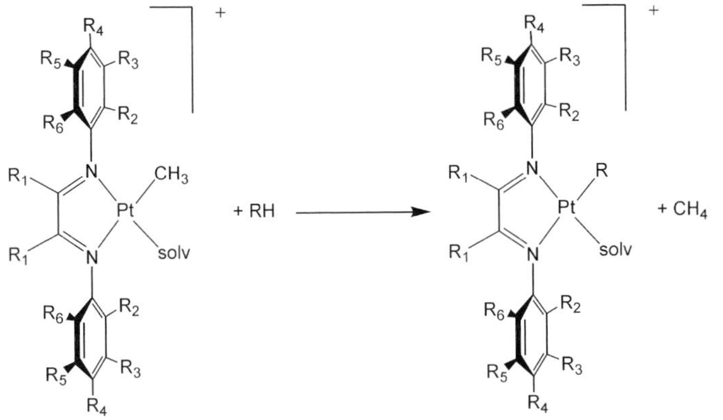

SCHEME 33.

For the "non-bulky" series it was concluded from isotope effects that the C–H activation (by oxidative addition) is rate-determining. The rate of this oxidative addition was proposed to increase with more electron-rich metal centers. In the case of the "ortho-dimethyl" series, it was concluded that substitution of trifluoroethanol by the hydrocarbon is rate-determining. Since these C–H activation rates also increased with more electron-rich metal centers, it was suggested that the substitution of solvent by benzene might be faster for electron-rich metal centers. It is worth commenting here that there exists experimental support for this hypothesis. Electron-rich Pt(II) centers undergo faster associative (or interchange) substitution reactions than their electron-poor analogs, provided that the increased electron density comes from σ-donation (155,163,164,171). On the other hand, the opposite is true concerning π-effects, since substitution rates generally increase with increasing π-acidity of the spectator ligands (158,172–175). The σ-effects are normally explained in terms of ground-state labilization of the reactant complex, and in the work discussed here, ground state labilization was also invoked (32). The authors noted that a complete interpretation is still lacking for some data. For example, the complexes having hydrogen instead of methyl in the backbone (R_1-position in Scheme 33) react much slower than expected (32). A greater distinction between σ- and π-effects might help in resolving the remaining ambiguities.

An important lesson from this study (32) is, that even within the same ligand series, both electronic and steric effects change the relative energies of reactive intermediates and transition states such that different steps can become rate-determining. Only after it has been understood which step is rate-determining, might a prediction be possible which electronic situation renders the platinum center more active in this C–H activation. A very similar insight was gained from a study on iridium-based hydrocarbon activation, where electronic effects governed the overall rate via their influence on substitution rates rather than on the actual C–H activation process (176).

From the discussion above, it also follows that platinum complexes capable of C–H activation need not be cationic, as might wrongly be suggested by the fact that most systems that have been studied are cations. However, this cationic charge is, in general, the simple consequence of the fact that most chelating ligands used are neutrally charged. The chemistry of anionic chelate ligands is now rapidly developing, and a few platinum complexes which are not cationic are already known to be capable of C–H activation. One example is the active species in the $B(C_6F_5)_3$-induced C–H activation by $Tp^{Me_2}PtMe_2^-$ (see Section III.C.2; Scheme 19). The active species is either the uncharged three-coordinate complex $Tp^{Me_2}PtMe$ (88) or the Lewis acid adduct $[Tp^{Me_2}Pt(Me)-Me-B(C_6F_5)_3]^-$, which is even anionic! Another example of a neutrally charged Pt(II) complex capable of hydrocarbon activation was shown above in Section III.B (Scheme 7, **D**). In this case, the complex contains a novel anionic chelating diphosphine (34). Benzene activation was observed, but alkane activation by this diphosphine complex has not been reported.

A very different neutrally charged complex for alkane activation has been reported recently and is shown in Scheme 34(**A**). The compound is a hydridoplatinum(II) complex bearing an anionic ligand based on the familiar nacnac-type, but with a pendant olefin moiety (97). This complex is extremely soluble in arenes and alkanes and activates C–H bonds in both types of hydrocarbons. This is indicated by deuterium incorporation from deuterated hydrocarbon into the substituents on the arene of the ligand and into the Pt hydride position ($\mathbf{A} \rightarrow \mathbf{A}$-$d_{27}$, Scheme 34). The open site needed for hydrocarbon coordination at Pt(II) is created by olefin insertion instead of anion or solvent substitution (97).

It is now becoming increasingly evident that platinum(II) complexes capable of C–H activation do not have to bear a positive charge. It has already been proposed for Shilov's system that the active species is an uncharged aquated chloro complex, namely $Pt(Cl)_2(OH_2)_2$ (2,16,147). However, even aquated chloro complexes are insoluble in higher hydrocarbons. The synthesis of platinum(II) complexes which are soluble in hydrocarbons due to the specific properties of the spectator ligands is an interesting approach. A different concept, solubilization of partially aquated tetrachloroplatinate in higher alkanes using surfactants to generate inverted micelles, was partly encouraging, but appeared to be

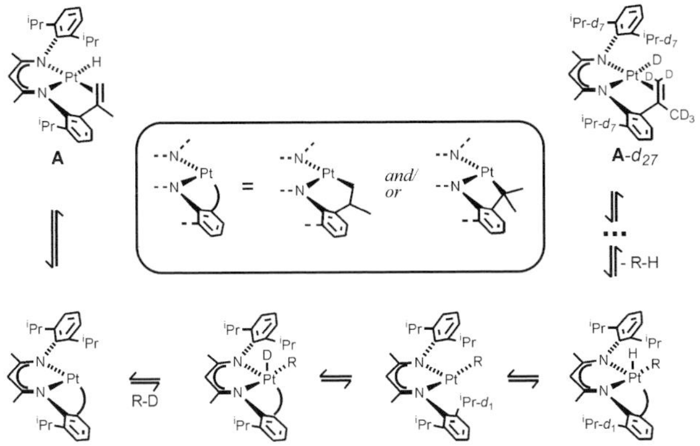

SCHEME 34.

limited due to decomposition of the surfactant used (177). It seems now that many hydrocarbon–soluble platinum(II) complexes will be designed and applied to the problem of alkane functionalization.

IV. Mechanisms for the Oxidation Step: From Pt(II)-Hydrocarbyl to Pt(IV)-Hydrocarbyl

A. OXIDATIVE ADDITION VERSUS ELECTRON TRANSFER

The oxidation of Pt(II) complexes is thought to proceed in most cases via addition of an electrophile to the Pt(II) center (141,178–182). This process does not involve outer-sphere electron transfer. The initial product of the electrophile addition is possibly a five-coordinate Pt(IV) species, however, the observable Pt(IV) product is six-coordinate. Coordination of a sixth ligand, e.g. solvent, occurs (183). This may proceed concertedly with the addition of the electrophile, in particular if the sixth ligand is solvent, or alternatively after the addition of the electrophile, as shown in Scheme 35.

The oxidant in the original Shilov-system was hexachloroplatinate. Convincing evidence has been presented (see review Ref. (14)) that the mechanism of oxidation of the Pt(II) alkyl $PtCl_3R^{2-}$ by $PtCl_6^{2-}$ is very similar to the mechanism shown in Scheme 35. Formally, Cl^+ is transferred in this inner-sphere process (184–186), depicted in Scheme 36. In particular, from a ^{195}Pt-labeling study on both a stable model system (185) and the highly reactive species $PtCl_3Me^{2-}$ itself (184), it was found that the alkyl group stays on the Pt center that is oxidized. This demonstration that alkyl transfer from Pt(II) to Pt(IV) is not

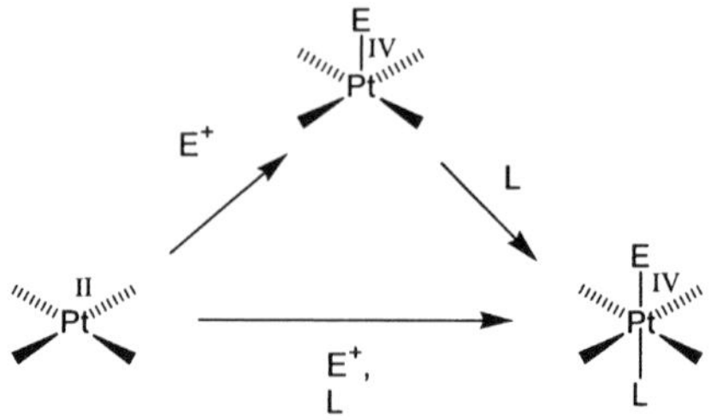

SCHEME 35.

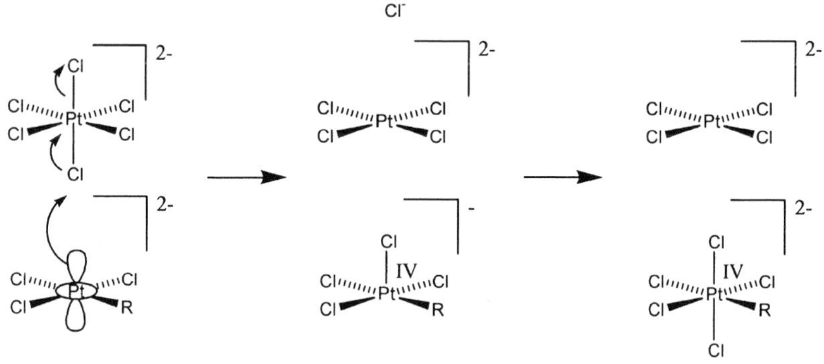

SCHEME 36.

involved is important since it implies that the oxidation of the Pt(II) alkyl should not be specific to Pt(IV) as an oxidant. Thus, a variety of other oxidants may be viable for modified Shilov-type systems.

In the Catalytica system, SO_3 acts as the oxidant to convert Pt(II) to Pt(IV). In this system, the oxidation step was found to be rate-determining (12). Some other oxidants have also been employed, e.g. elemental chlorine (18) and hydrogen peroxide (19). To date, the oxidation reaction has received less attention than the C–H activation step, and the general factors governing the oxidation of platinum(II) to platinum(IV) are incompletely understood. The difficulty in establishing a general theory that allows one to predict the thermodynamics as well as the kinetics for this oxidation reaction lies in the fact that every system is unique due to the complication that the oxidant forms bonds of varying strength to the metal center in the oxidized product. A different oxidant brings a different set of ligands into the coordination sphere of platinum. Thus, a strong oxidant (based on tabulated redox potentials) can be a weak oxidant towards platinum(II) if it forms weak bonds to platinum in the Pt(IV) product. Furthermore, the *two-electron* redox potentials for

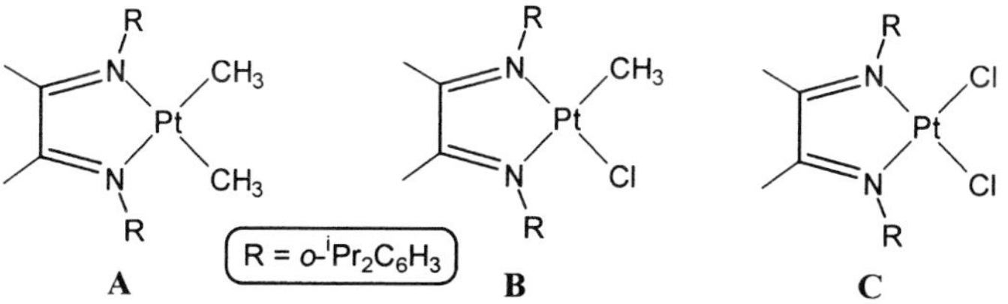

Scheme 37.

oxidation of platinum(II) complexes to Pt(IV) are not easily measured, since single-electron transfer to form Pt(III) normally occurs upon oxidation of Pt(II). This step is often followed by decomposition or disproportionation (*187*). Thus, the ease of oxidation of platinum(II) to platinum(IV) cannot be accurately predicted from electrochemical measurements of one-electron redox potentials. However, an effort to rationalize the ease of oxidation of platinum(II) in a qualitative fashion, based on one-electron redox potentials, has been made (*188*). The diimine-chelated platinum(II) complexes shown in Scheme 37 were used. The highly irreversible one-electron oxidation reactions showed that the methyl chloro complex **B** is oxidized at a potential that is ca. 500 mV higher than that of the corresponding dimethyl complex **A**. Oxidation of the corresponding dichloro complex **C** is more difficult by another ca. 400 mV. Oxidation by inner-sphere oxidants was also investigated, which allows semi-quantitative determinations of the ease of oxidation by the oxidative addition mechanism (*188*). For example, as shown in Scheme 38, a variety of diimines appear to stabilize the oxidation state IV better than the more basic diamine ligand tmeda (tetramethylethylenediamine). This observation appears to support the theory (discussed above in Section III.F) that less donating ligands may stabilize Pt(IV) better.

B. Oxidation by SO_3 and Redox Mediators

Concerning the oxidation of Pt(II) to Pt(IV) by SO_3 in the Catalytica system (*12*), it seems reasonable to assume that SO_3 adds to Pt(II) in an oxidative fashion to generate Pt(IV) and a probably protonated form of SO_3^{2-}. Protonation of the sulfite formed (to yield H_2SO_3 and finally the experimentally observed products SO_2 and H_2O) would be expected under the strongly acidic reaction conditions. If the protonation of the sulfur species occurs during the oxidation of Pt(II), this might lower the reaction barrier for this oxidation reaction. In addition, further stabilization might be expected if a coordinating ligand is available *trans* to

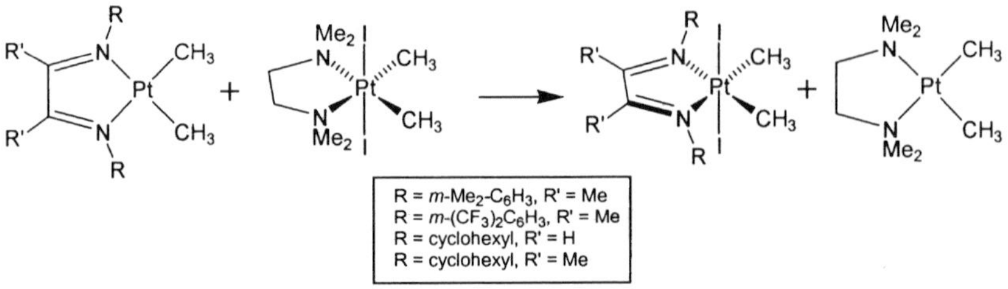

R = m-Me₂-C₆H₃, R' = Me
R = m-(CF₃)₂C₆H₃, R' = Me
R = cyclohexyl, R' = H
R = cyclohexyl, R' = Me

Scheme 38.

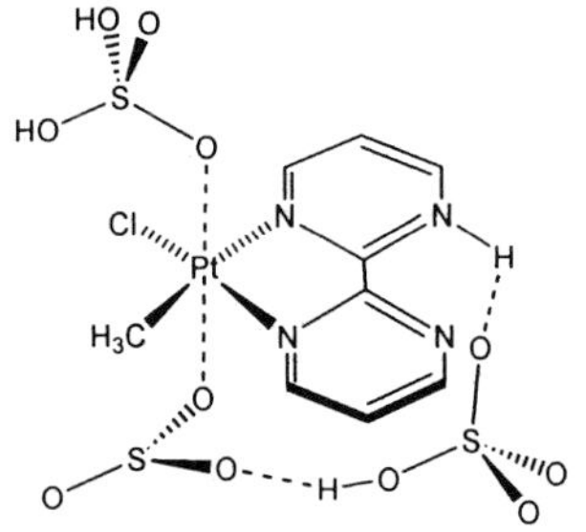

Scheme 39.

the SO$_3$ to stabilize the Pt(IV) state. Based on these arguments, a cyclic structure involving hydrogen bonding to the SO$_3$ has been proposed (148), depicted in Scheme 39. While this ground-state structure – which might precede the transition state – has been computationally optimized, calculations on the corresponding transition state have not been reported yet. In fact, the system is complex enough to offer a variety of possibilities in the order in which the oxidation of the platinum and the protonation of the reduced sulfur species might occur.

Since SO$_3$/H$_2$SO$_4$ is clearly not the most desirable system for industrial applications, a formidable challenge is to find an oxidant that oxidizes Pt(II) much faster than SO$_3$ does, operates in an environmentally friendly solvent, and can be (like S^{VI}/S^{IV}) reoxidized by oxygen from air. Ideally, the reduced oxidant would get reoxidized in a continuous process, such that the oxidant acts as a redox mediator. In addition, the redox behavior has to be tuned such that the platinum(II) alkyl intermediate would be oxidized but the platinum(II) catalyst would not be completely oxidized. Such a system that efficiently transfers oxidation equivalents from oxygen to Pt(II) would be highly desirable. A redox mediator system based on heteropolyacids has been reported for the Pt-catalyzed oxidation of C–H bonds by O$_2$, using Na$_8$HPMo$_6$V$_6$O$_{40}$

(*189*). However, not more than three turnovers were observed. Heteropolyacids were also active as redox mediators in an Pt-based electrocatalytic C–H bond functionalization system, but the selectivity was lost after six turnovers (*190*). Significant practical progress has been made very recently using copper chloride as a co-catalyst for the functionalization of C–H bonds using O_2 as the terminal oxidant (*191*). This system involves K_2PtCl_4 in H_2O/sulfuric acid in the presence of O_2. Higher turnover numbers (ca. 50) were achieved in the presence of the co-catalyst $CuCl_2$. The observed selectivity for the terminal position in the oxidation of ethanesulfonic acid led to the proposal that the same mechanism for the C–H activation step as in the original Shilov system is operating. However, exactly how copper is involved in this cycle is mechanistically unknown. It is reasonable to assume that copper(II), easily regenerated using air, oxidizes the platinum(II) alkyl intermediate to Pt(IV). Several hypotheses have been discussed to explain the oxidation of platinum(II) to platinum(IV) by copper halides, and the involvement of polynuclear aggregates having at least two copper centers has been proposed (*192–194*). However, detailed mechanistic studies are needed to completely understand the mechanism of Pt(II) oxidation by Cu(II). Indeed, further mechanistic studies in general, concerning the role of redox mediators in platinum systems utilizing O_2, will assist in the rational design of such systems.

C. OXIDATION OF Pt(II) BY DIOXYGEN

It is conceivable that oxygen, the most readily available and environmentally benign oxidant, could be directly incorporated into a platinum-catalyzed alkane functionalization cycle. Thus, redox mediators, which were discussed above, might not be needed, since O_2 itself might be able to oxidize suitably ligated Pt(II) alkyls. The oxidation of platinum(II) by dioxygen has been an active area of recent investigation. Several studies, all employing Pt(II) complexes chelated by nitrogen ligands, have provided significant insight into these oxidations by O_2. However, a full understanding is still lacking. One of the earliest examples where dioxygen from air appeared to be responsible for the oxidation of Pt(II) to Pt(IV) was the demonstration that bipyridine-chelated Pt(II) in the presence of the ligand cyclohexane-1,3,5-triamine ("tach") can be oxidized to Pt(IV) in aqueous solution (*146*). The product was crystallographically characterized, and the structure shown as **A** in Scheme 40 was assigned (*146*). Later, using NMR spectroscopy, the solution structure at neutral pH was also determined (Scheme 40, **B**) (*145*). It has been suggested that the tendency of the tach ligand to coordinate

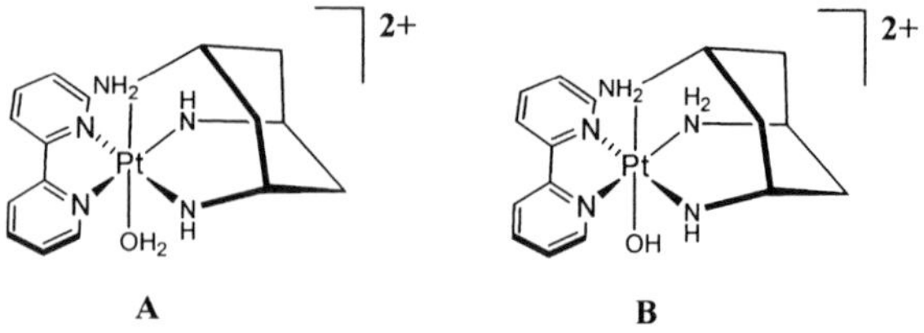

SCHEME 40.

in a *fac*-tridentate fashion stabilizes Pt(IV), and it was also suggested (although not conclusively shown) that O_2 from air is responsible for the oxidation of Pt(II) to Pt(IV) in this case (*145,146*). It is notable in this context that the facially coordinating ligands tris(pyridin-2-yl)methanol (*195,196*) and tris(pyrazolyl)borate (*60,197*) facilitate the oxidation of platinum(II) dialkyls or diaryls such that the corresponding Pt(IV) hydroxo complexes are formed in the presence of water. In these cases, however, the authors attribute the oxidation to water (and not O_2) (*60,195–197*). In contrast, the involvement of O_2 as the oxidant was suggested by the authors in the reaction of tetrachloroplatinate and 1-thia-4,7-diazacyclonane in aqueous solution to yield a Pt(IV) product (*198*). A system for which it was clearly shown that oxidation of Pt(II) to Pt(IV) in aqueous solution requires the presence of O_2 is the oxidation of (κ^2-tacn)$_2$Pt^{2+} to yield (κ^3-tacn)$_2$Pt^{4+} (tacn = 1,4,7-triazacyclononane) (*199*). Similarly, O_2 is required for the oxidation of (tacn)PtMe$_2$ to give the hydroxo complex (tacn)PtMe$_2$(OH)$^+$ (*65*).

Some evidence to suggest that peroxo complexes can be intermediates in the oxidation of Pt(II) by O_2 has been presented. As shown in Scheme 41, a Pt(IV) peroxo complex was obtained by reacting *cis*-PtCl$_2$(DMSO)$_2$ and 1,4,7-triazacyclononane (tacn) in ethanol in the presence of air (*200*). An alkylperoxoplatinum(IV) complex is obtained in the reaction of (phen)PtMe$_2$ (phen = 1,10-phenanthroline) with dioxygen and isopropyl-iodide. Under conditions that favor radical formation (light or radical initiators), an isopropylperoxoplatinum(IV) compound was obtained (*201,202*), depicted in Scheme 42.

A well-defined system that has recently been thoroughly investigated involves the oxidation of platinum(II) dimethyl complexes chelated by nitrogen donors (*203,204*). Oxidation by oxygen occurs in methanol solvent, and a methoxo(hydroxo)platinum(IV) complex is ultimately formed, shown in Scheme 43 (overall reaction, $\mathbf{A} \rightarrow \mathbf{C}$) (*203*). This reaction occurs within some hours or days (depending on the nitrogen ligand and the concentrations employed) at room temperature or slightly elevated temperature (20–50 °C). Oxidation does not occur if oxygen is

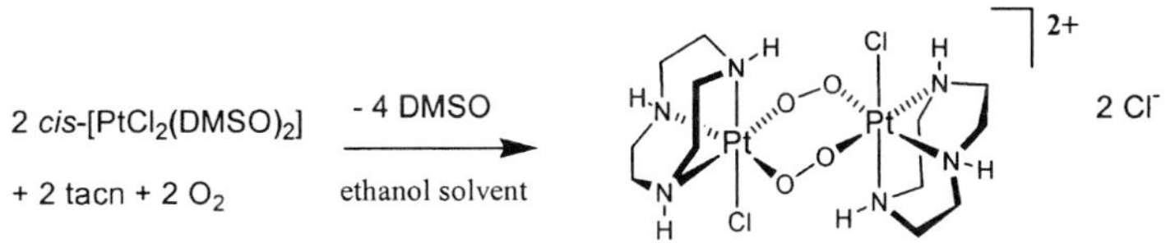

SCHEME 41.

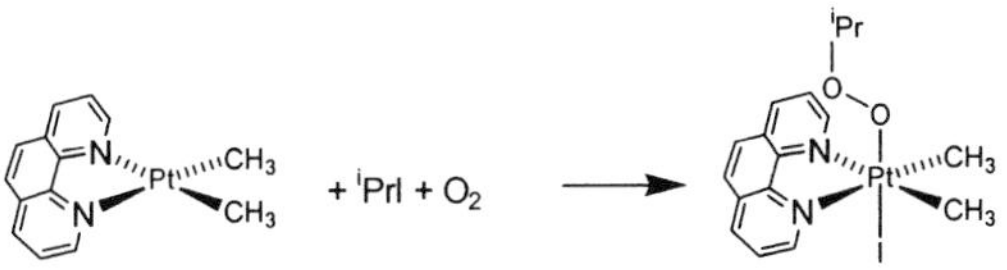

SCHEME 42.

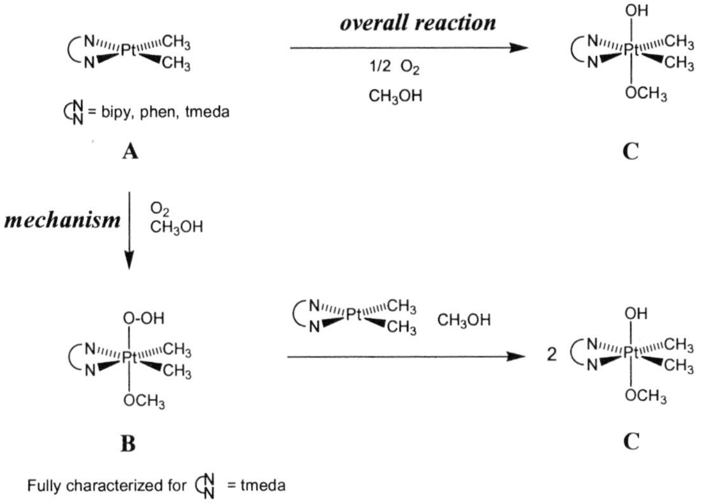

SCHEME 43.

strictly excluded, which disproved an earlier proposal (205) that alcohols or water act as the oxidant in this reaction. A hydroperoxo complex of Pt(IV), shown in Scheme 43 (as **B**), was identified as an intermediate in this reaction (204). The hydroperoxo complex was generated preferentially if the concentration of Pt(II) dimethyl complex **A** was low and the oxygen concentration very high. This inter-mediate **B**, with the ligand tmeda (*N,N,N′,N′*-tetramethylethylenedi-amine), was fully characterized. When reacted with one equivalent of

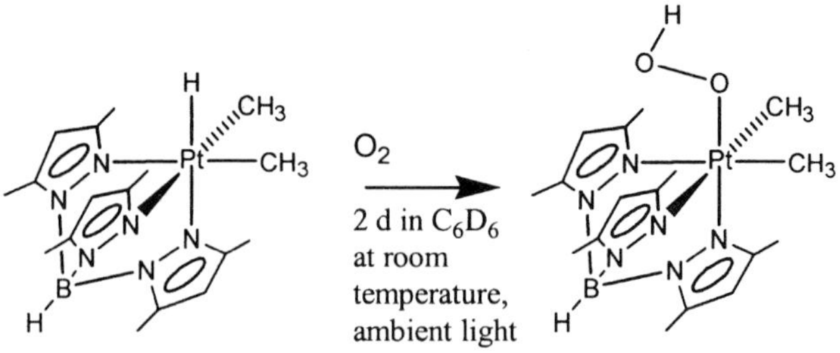

SCHEME 44.

the platinum(II) dimethyl complex **A**, two equivalents of the final product **C** were generated. Using kinetic simulations of product distribution under various conditions, it could be convincingly shown that **B** is truly an intermediate in the formation of **C** from **A** (Scheme 43) (*204*).

A Pt(IV) hydroperoxo complex has also been observed upon reaction of a Pt(IV) hydride with dioxygen. As shown in Scheme 44, $Tp^{Me_2}PtMe_2H$ (Tp^{Me_2} = tris(3,5-dimethylpyrazolyl)borate) reacts with O_2 to form $Tp^{Me_2}PtMe_2OOH$ which has been fully characterized (*206*). Based on experiments with radical initiators and inhibitors, a radical pathway has been proposed for this formal insertion of O_2 into the Pt(IV)–hydride bond. Note that this reaction is quite different from the Pt(II) oxidations discussed above, since the complex that reacts with dioxygen is already in the oxidation state IV. Thus the platinum center itself is not oxidized by O_2, but the hydride ligand is oxidized instead. However, this reaction is of particular interest to the selective oxidation of alkanes due to the fact that the reactant $Tp^{Me_2}PtMe_2H$ is very similar to $Tp^{Me_2}PtMe(R)H$ which can be formed by R–H oxidative addition to Pt(II) (Scheme 19) (*88*). This type of O_2 insertion into a metal–hydride bond may also be relevant to routes of hydrocarbon functionalization using other metals and dioxygen.

V. Mechanisms for the Functionalization Sequence: Carbon-Heteroatom Coupling to Release the Product

In the preceding discussion, we have covered recent discoveries which shed light on the first two steps of Pt(II) catalyzed alkane functionalization (Section III; Scheme 4): alkane activation by Pt(II) and oxidation of the resulting platinum(II) alkyl. The final step is the release of the

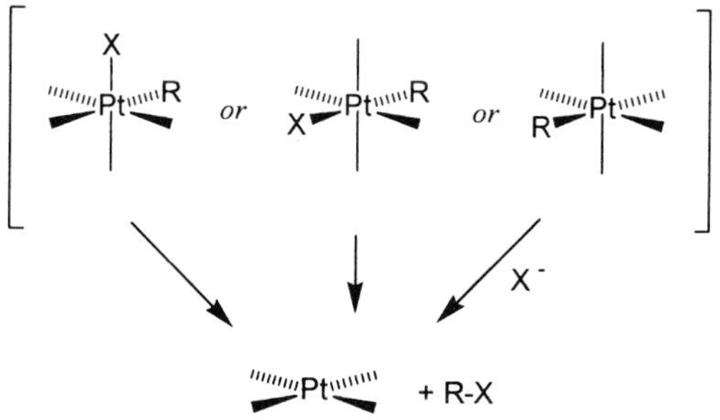

Scheme 45

functionalized organic product RX and regeneration of a Pt(II) species. The X^- group may either be already attached to the Pt(IV) center or may react by an intermolecular mechanism with the the Pt(IV) alkyl as shown in Scheme 45. Convincing evidence for an intermolecular mechanism is discussed below. A variety of C–X bond forming reactions from Pt(IV) have all been found to involve multiple elementary steps. Convincing evidence has been presented for the dissociation of a ligand from the Pt(IV) center prior to the actual reductive coupling. Thus once again we see the proposed involvement of five-coordinate Pt(IV) intermediates.

A. FIVE-COORDINATE INTERMEDIATES INVOLVED IN CARBON-HETEROATOM COUPLING REACTIONS FROM Pt(IV)

The involvement of carbon-heteroatom coupling from Pt(IV) as the product release step in Shilov's Pt(II) catalyzed alkane functionalization scheme has prompted several investigations into the mechanistic details of this type of reaction. A significant portion of these studies has already been covered in a 1998 review on platinum catalyzed alkane oxidation (14). As discussed in this review, the mechanism shown in Scheme 46 is widely accepted and most consistent with the large array of available data. The platinum(IV) alkyl complexes of the type $PtRCl_5^{2-}$, including the actual "Shilov-intermediate" $PtCH_3Cl_5^{2-}$, undergo competitive R–OH and R–Cl reductive elimination in the presence of chloride in aqueous solution, a process which has received attention from detailed mechanistic studies (185,207). The reductive elimination step involves S_N2 attack of the nucleophile at the alkyl group, demonstrated by showing that inversion occurred when Cl–CHD–CHD–OH was released upon reaction of the stereospecifically deuterium-labeled model system $[Pt(-CHD–CHD–OH)Cl_5]^{2-}$ with chloride (185,207). Kinetic data for

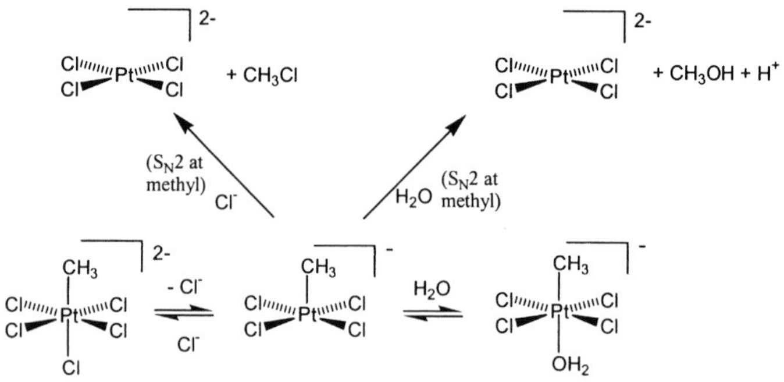

SCHEME 46.

competitive C–Cl and C–O reductive elimination for the reaction shown
in Scheme 46 were inconsistent with nucleophilic attack at the penta-
chloro(alkyl) complex directly and consistent only with nucleophilic
attack at either the five-coordinate species or the aqua complex.
Although nucleophilic attack at the aqua complex could not be ruled
out on kinetic grounds, the authors regarded it more likely that the
five-coordinate species is involved, in light of the many studies pointing
towards five-coordinate Pt(IV) intermediates in reductive elimination
and oxidative addition (see references in Ref. (14)). The recent
discovery of stable five-coordinate Pt(IV) complexes lends additional
support to the concept of five-coordinate Pt(IV) species as intermediates
(94,95,97).

Further insight into the carbon–oxygen reductive elimination from
Pt(IV) and the involvement of five-coordinate Pt(IV) intermediates has
been provided recently. The first direct observation of high-yield C–O
reductive elimination from Pt(IV) was described and studied in detail
(50,51). Carbon–oxygen coupling to form methyl carboxylates and methyl
aryl ethers was observed upon thermolysis of the Pt(IV) complexes
("P$_2$")PtMe$_3$(OR) ("P$_2$"=bis(diphenylphosphino)ethane or o-bis(diphenylphosphino)benzene; OR=carboxylate, aryl oxide). As shown in Scheme 47,
competitive C–C reductive elimination to form ethane was also observed.

Strong support for the mechanism shown in Scheme 48 was presented.
Dissociation of OR$^-$ to form a five-coordinate cation occurs prior to
reductive coupling. C–C reductive elimination occurs in an intramolecular fashion from this cation. In contrast, C–O reductive elimination
occurs by nucleophilic attack of the OR$^-$ on the Pt(IV)-methyl group of
the five-coordinate cation.

Although the rate-determining step for C–O bond formation involves
nucleophilic attack of OR$^-$ at the platinum bound methyl group (k_2),

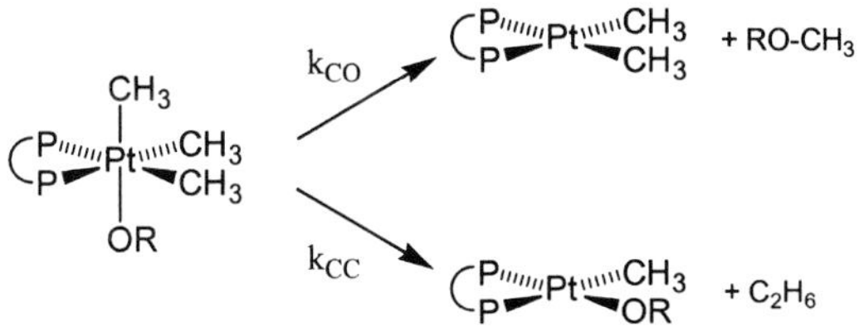

SCHEME 47.

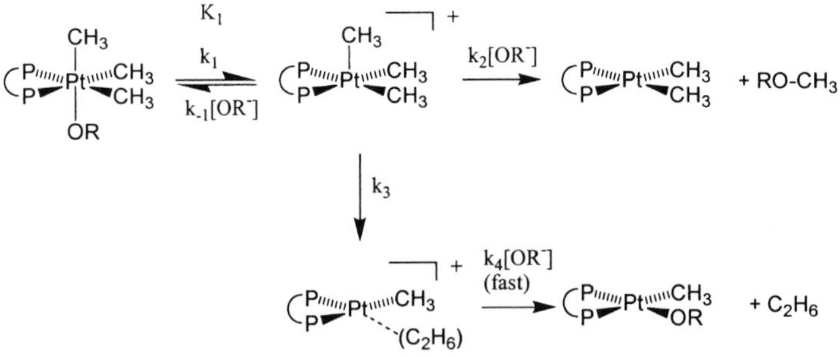

SCHEME 48.

the interesting observation was made that the use of less nucleophilic OR^- groups actually led to increased rates of C–O coupling. For example, a Hammett plot using the rates of C–O coupling from an analogous series of complexes with substituted aryloxide groups yielded a positive ρ-value of 1.44. While this result may initially seem counterintuitive, it is a direct consequence of the required OR^- dissociation in the first step (K_1); this equilibrium process involving formation of the five-coordinate cation will be favored by more electron-withdrawing OR^- groups.

Thus, it is a significant finding that the five-coordinate Pt(IV) complex is the species which undergoes nucleophilic attack. It implies that to facilitate this last step of functionalization in platinum catalyzed systems, the most useful coupling groups may be sulfonates, trifluoroacetate, or water, rather than halides, hydroxides, or alkoxides, despite the initially attractive nucleophilicities of the latter species. It is interesting in this context that the Shilov catalyst system forms methanol by nucleophilic attack of water on a platinum bound methyl group (2,14,15) and the Catalytica system couples a methyl group with bisulfate (12). Electron-withdrawing OR^- groups like sulfonates, trifluoroacetate, and bisulfate also appear to protect the C–O coupled alcohol-derivative product from overoxidation (3,12,208,209).

B. AN ALTERNATIVE TO ALKANE OXYGENATION: ALKANE DEHYDROGENATION

As discussed in preceding sections of this review, five-coordinate platinum(IV) species have been identified as key intermediates in both C–H oxidative addition reactions to Pt(II) and in the functionalization step, carbon-heteroatom bond formation from Pt(IV). These 16-electron-d^6 intermediates contain an open site in the coordination sphere of the metal. This should make β-hydrogen elimination from an alkyl group extremely facile (*210*). Yet it has been observed using platinum chloride complexes as catalysts, that functionalization (with reasonable selectivity for the terminal position) is much preferred over dehydrogenation for higher alkanes (*15*). In a recent short "opinion" article an explanation for this result has been offered (*211*). Relevant structures are shown in Scheme 49. While there is a highly electrophilic open site *cis* to the alkyl in the five-coordinate Pt(IV) species A that could participate in β-H elimination, it was argued that the deprotonation in aqueous solution to form the complex $[RPtCl_3]^{2-}$ (**C**) is much faster than β-hydrogen elimination. For **B**, the *trans*-directing influence of alkyl makes positioning of the open site *cis* to alkyl (as required for β-hydride elimination) thermodynamically unfavorable. The same argument applies to chloride dissociation from **C**, where chloride dissociation would be favored *trans* to alkyl and not in the *cis* position. This situation was contrasted (*211*) with the alkane dehydrogenation system (CF_3CO_2)-$Ir(PR_3)_2(H)_2$ (*212*), where an open site *cis* to the alkyl is favored in the five-coordinate d^6 intermediate, $(CF_3CO_2)Ir(PR_3)_2(R)(H)$, again by the *trans*-influence specific to this ligand set.

If alkyl groups having β-hydrogens are present on platinum *cis* to an open site, β-H-elimination will indeed occur, reversibly sometimes, and it can occur both from Pt(II) and Pt(IV) (*52,97,213–219*). Catalytic dehydrogenation of an alkane using a soluble platinum complex has been reported in an early study on acceptorless thermal dehydrogenation. At 151 °C, cyclooctane was catalytically dehydrogenated (up to 10 turnovers)

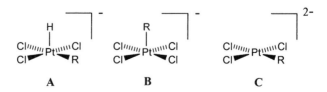

SCHEME 49.

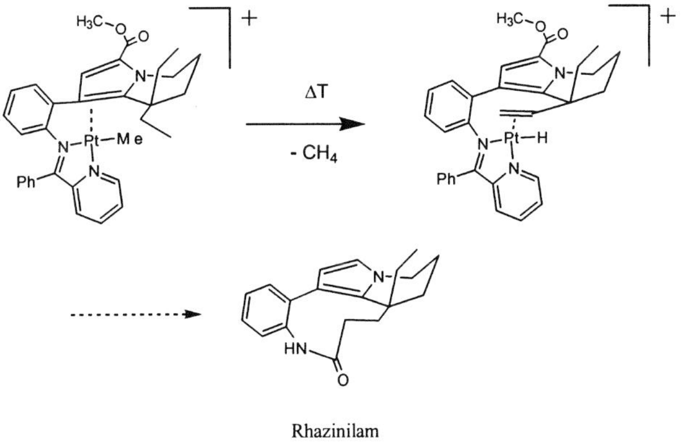

Rhazinilam

SCHEME 50.

to yield cyclooctene and H_2 with the Pt(II) catalyst precursor $Pt(SnCl_3)_2(P(OR)_3)_3$ (R = Me, Et) (*220*).

A stoichiometric dehydrogenation reaction has been utilized for the synthesis of the natural product Rhazinilam in which the key step was a platinum-mediated C–H activation, followed by β-elimination (*213,214*). This step is shown in Scheme 50.

This elegant organic transformation is, along with the recent development of a Pt(II)-based derivatization procedure for amino acids (*221*), an impressive example that platinum-mediated C–H activation offers great potential for not only the functionalization of simple small molecules but also the synthesis of complicated organic targets.

VI. Summary and Concluding Comments

The last several years have seen impressive achievements in elucidating the mechanism of hydrocarbon activation and functionalization with platinum(II)/(IV) systems. In particular, the actual hydrocarbon activation step has been studied in tremendous detail. Kinetic studies, isotope labeling, modeling of intermediates, and computational methods have all been used to determine the key factors necessary for achieving facile C–H activation by Pt(II). A general understanding that the alkane must coordinate to a site within the square plane of the Pt(II) center to form a σ-complex has been reached. Information about exactly how this arrangement can be obtained (substitution, olefin insertion, etc.) is rapidly accumulating and being exploited. Oxidative addition of a C–H bond of the alkane in the σ-complex to generate a five-coordinate Pt(IV) alkyl hydride appears to be a viable mechanism for C–H activation and in many cases favored over an electrophilic pathway.

Two additional steps are, however, needed to achieve functionalization of the alkane: oxidation of the intermediate platinum(II) hydrocarbyl intermediate, and reductive coupling to release the product. Recent studies have contributed significantly to our understanding of the carbon-heteroatom bond formation at Pt(IV). A classic study demonstrating stereochemical inversion supported nucleophilic attack of the heteroatom at the Pt(IV)-bonded carbon as the product forming step. Evidence has also been presented in support of the nucleophilic attack occurring at a five-coordinate Pt(IV) center. From a model study using oxygen donors to form carbon–oxygen bonds, it was concluded that less nucleophilic oxygen donors enhance carbon–oxygen reductive elimination from Pt(IV), presumably by favoring the five-coordinate species. This last concept is promising, since weakly coordinating ligands are also needed in the C–H activation step, such that synergistic effects can be expected from the use of relatively weak donors. In addition, the C–X coupled products with electron withdrawing X groups appear to be more resistant to overoxidation.

Progress has also been made in understanding the factors involved in the oxidation of the intermediate platinum(II) hydrocarbyl complexes. It is encouraging that organometallic platinum(II) methyl complexes appear to be more easily oxidized than the corresponding inorganic coordination compounds. Fine-tuning of an oxidant to achieve selective oxidation of the platinum(II) hydrocarbyl intermediate and not the platinum(II) catalyst appears to be a reasonable goal. Interesting approaches to incorporate dioxygen as the oxidant have been reported. Both oxidation of Pt(II) centers by dioxygen and the reaction of a Pt(IV) hydride with dioxygen show promise in this regard. The challenge remains to find a system which quickly and efficiently incorporates oxygen from air to form a platinum(IV) alkyl. This species, however, should still be labile enough that fast reductive elimination of the product is possible. This will be a key point. A subtle balance must be found between stabilizing Pt(II) too strongly and stabilizing Pt(IV) too strongly. Extensive work will be needed to find the ideal compromise.

In summary, the advances of the past few years have well demonstrated both the challenges and the promise of the Pt(II)/Pt(IV) redox couple for alkane functionalization. It should also be mentioned that the emerging conceptual understanding of the elementary steps involved in this process has also contributed to the development of methods to activate and functionalize alkyl groups in complex organic molecules.

This methodology clearly enriches the tool box of the synthetic organic chemist. Other spin-offs from the studies described above, such as the incorporation of dioxygen as an oxidant and the use of alkyl carbon-heteroatom coupling as a product release step for other metal-mediated organic transformations, may also emerge over the next several years.

The platinum(II)/(IV) approach to alkane functionalization has long been viewed as one of the most promising. The rapid progress in the understanding of this system that has occurred over the past few years, combined with the increasing numbers of scientists working in this area and the active current efforts into the development of viable commercial Pt systems suggest that selective alkane functionalization should no longer be viewed as a "holy grail". Instead, it should be viewed as an achievable goal.

VII. Note Added in Proof

Several relevant papers appeared while this manuscript was in press, listed here with their relevance to particular sections of this review:

Section III.B: "C–H Activation and C–C Coupling Arenes by Cationic Pt(II) Complexes" (222) were reported. These complexes have diethylether as a labile ligand (see Scheme 7). The C–C coupling is unusual.

Section III.C: A "Hydrido(methyl)carbene Complex of Platinum(IV)" (223) and "Methyl(hydrido)platinum(IV) Complexes with Flexible Tridentate Nitrogen-Donor Ligands" (224) are structurally related to the system shown in Scheme 13 and give additional information on how steric and electronic factors influence the stability of platinum(IV) methyl hydrides.

"Isotope Effects in C–H Bond Activation Reactions by Transition Metals" (225) were reviewed, and some pitfalls in interpreting kinetic isotope were pointed out. The interpretation of the kinetic isotope effects offered by the authors of the original reports (75,76,85) on the system shown in Schemes 15, 16 was criticized.

"Control of H–C(sp^3) Bond Cleavage Stoichiometry: Clean Reversible Alkyl Ligand Exchange with Alkane in [LPt(Alk)(H)$_2$]$^+$ (L=[2.1.1]-(2,6)-Pyridinophane)" (226): this complex activates hydrocarbons RH to yield LPtRH$_2^+$. This is similar to the C–H bond activation shown in Scheme 17 but occurs without added acid.

Section III.E: "Barriers for Arene C–H Bond Activation in Platinum(II) η^2-Arene Intermediates" (227) were measured for the complex shown in Scheme 26 (**B**) and similar derivatives.

ACKNOWLEDGEMENTS

For support, we are grateful to the National Science Foundation. U. F. is additionally grateful to DAAD (Deutscher Akademischer Austauschdienst) for various fellowships. We would like to thank A. M. Brackett, J. L. Look, A.V. Pawlikowski, and N. A. Smythe for valuable comments and for proof-reading of the manuscript.

REFERENCES

1. Arndtsen, B. A.; Bergman, R. G.; Mobley, T. A.; Peterson, T. H. *Acc. Chem. Res.* **1995**, *28*, 154.
2. Shilov, A. E.; Shul'pin, G. B. *Chem. Rev.* **1997**, *97*, 2879.
3. Sen, A. *Acc. Chem. Res.* **1998**, *31*, 550.
4. Crabtree, R. H. *J. Chem. Soc., Dalton Trans.* **2001**, 2437.
5. Labinger, J. A.; Bercaw, J. E. *Nature* **2002**, *417*, 507.
6. Lowry, T. H.; Richardson, K. S. *"Mechanism and Theory in Organic Chemistry"*; Harper & Row: New York, 1976.
7. Streitwieser, A., Jun; Taylor, D. R. *J. Chem. Soc., Chem. Comm.* **1970**, 1248.
8. Berkowitz, J.; Ellison, G. B.; Gutman, D. *J. Phys. Chem.* **1994**, *98*, 2744.
9. Garnett, J. L.; Hodges, R. J. *J. Am. Chem. Soc.* **1967**, *89*, 4546.
10. Gol'dshleger, N. F.; Tyabin, M. B.; Shilov, A. E.; Shteinman, A. A. *Zh. Fiz. Khim.* **1969**, *43*, 2174.
11. Gol'dshleger, N. F.; Es'kova, V. V.; Shilov, A. E.; Shteinman, A. A. *Russ. J. Phys. Chem.* **1972**, *46*, 785.
12. Periana, R. A.; Taube, D. J.; Gamble, S.; Taube, H.; Satoh, T.; Fuji, H. *Science* **1998**, *280*, 560.
13. Wolf, D. *Angew. Chem. Int. Ed.* **1998**, *37*, 3351.
14. Stahl, S. S.; Labinger, J. A.; Bercaw, J. E. *Angew. Chem. Int. Ed.* **1998**, *37*, 2181.
15. Shilov, A. E.; Shul'pin, G. B. *"Activation and Catalytic Reactions of Saturated Hydrocarbons in the Presence of Metal Complexes"*; Kluwer: Boston, 2000.
16. Shilov, A. E.; Shteinman, A. A. *Coord. Chem. Rev.* **1977**, *24*, 97.
17. Kushch, L. A.; Lavrushko, V. V.; Misharin, Y. S.; Moravsky, A. P.; Shilov, A. E. *Nouv. J. Chim.* **1983**, *7*, 729.
18. Horváth, I. T.; Cook, R. A.; Millar, J. M.; Kiss, G. *Organometallics* **1993**, *12*, 8.
19. DeVries, N.; Roe, D. C.; Thorn, D. L. *J. Mol. Catal. A* **2002**, *189*, 17.
20. Ryabov, A. D. *Chem. Rev.* **1990**, *90*, 403.
21. Wong-Foy, A. G.; Henling, L. M.; Day, M.; Labinger, J. A.; Bercaw, J. E. *J. Mol. Catal. A* **2002**, *189*, 3.
22. Griffiths, D. C.; Young, G. B. *Organometallics* **1989**, *8*, 875.
23. Peters, R. G.; White, S.; Roddick, D. M. *Organometallics* **1998**, *17*, 4493.
24. Tatsumi, K.; Hoffmann, R.; Yamamoto, A.; Stille, J. K. *Bull. Chem. Soc. Jpn.* **1981**, *54*, 1857.
25. Saillard, J.-Y.; Hoffmann, R. *J. Am. Chem. Soc.* **1984**, *106*, 2006.
26. Brainard, R. L.; Nutt, W. R.; Lee, T. R.; Whitesides, G. M. *Organometallics* **1988**, *7*, 2379.
27. Holtcamp, M. W.; Labinger, J. A.; Bercaw, J. E. *J. Am. Chem. Soc.* **1997**, *119*, 848.
28. Johansson, L.; Ryan, O. B.; Tilset, M. *J. Am. Chem. Soc.* **1999**, *121*, 1974.
29. Johansson, L.; Tilset, M.; Labinger, J. A.; Bercaw, J. E. *J. Am. Chem. Soc.* **2000**, *122*, 10846.

30. Heiberg, H.; Johansson, L.; Gropen, O.; Ryan, O. B.; Swang, O.; Tilset, M. *J. Am. Chem. Soc.* **2000**, *122*, 10 831.
31. Johansson, L.; Ryan, O. B.; Rømming, C.; Tilset, M. *J. Am. Chem. Soc.* **2001**, *123*, 6579.
32. Zhong, H. A.; Labinger, J. A.; Bercaw, J. E. *J. Am. Chem. Soc.* **2002**, *124*, 1378.
33. Procelewska, J.; Zahl, A.; van Eldik, R.; Zhong, H. A.; Labinger, J. A.; Bercaw, J. E. *Inorg. Chem.* **2002**, *41*, 2808.
34. Thomas, J. C.; Peters, J. C. *J. Am. Chem. Soc.* **2001**, *123*, 5100.
35. Harkins, S. B.; Peters, J. C. *Organometallics* **2002**, *21*, 1753.
36. Brookhart, M.; Green, M. L. H.; Wong, L. L. *Prog. Inorg. Chem.* **1988**, *36*, 1.
37. Kubas, G. J. *"Metal Dihydrogen and σ-Bond Complexes"*; Kluwer Academic/Plenum: New York, 2001.
38. Taw, F. L.; White, P. S.; Bergman, R. G.; Brookhart, M. *J. Am. Chem. Soc.* **2002**, *124*, 4192.
39. Rybtchinski, B.; Oevers, S.; Montag, M.; Vigalok, A.; Rozenberg, H.; Martin, J. M. L.; Milstein, D. *J. Am. Chem. Soc.* **2001**, *123*, 9064.
40. Rybtchinski, B.; Milstein, D. *Angew. Chem. Int. Ed.* **1999**, *38*, 870.
41. Albrecht, M.; Gossage, R. A.; Spek, A. L.; van Koten, G. *J. Am. Chem. Soc.* **1999**, *121*, 11 898.
42. Wick, D. D.; Northcutt, T. O.; Lachicotte, R. J.; Jones, W. D. *Organometallics* **1998**, *17*, 4484.
43. Murakami, M.; Amii, H.; Shigeto, K.; Ito, Y. *J. Am. Chem. Soc.* **1996**, *118*, 8285.
44. Guari, Y.; Sabo-Etienne, S.; Chaudret, B. *Eur. J. Inorg. Chem.* **1999**, 1047.
45. Yamamoto, T.; Murakami, Y.; Abla, M. *Chem. Lett.* **1999**, 419.
46. Brown, M. P.; Puddephatt, R. J.; Upton, C. E. E. *J. Chem. Soc., Dalton Trans.* **1974**, 2457.
47. Roy, S.; Puddephatt, R. J.; Scott, J. D. *J. Chem. Soc., Dalton Trans.* **1989**, 2121.
48. Goldberg, K. I.; Yan, J.; Winter, E. L. *J. Am. Chem. Soc.* **1994**, *116*, 1573.
49. Goldberg, K. I.; Yan, J.; Breitung, E. M. *J. Am. Chem. Soc.* **1995**, *117*, 6889.
50. Williams, B. S.; Holland, A. W.; Goldberg, K. I. *J. Am. Chem. Soc.* **1999**, *121*, 252.
51. Williams, B. S.; Goldberg, K. I. *J. Am. Chem. Soc.* **2001**, *123*, 2576.
52. Crumpton, D. M.; Goldberg, K. I. *J. Am. Chem. Soc.* **2000**, *122*, 962.
53. Hill, G. S.; Rendina, L. M.; Puddephatt, R. J. *Organometallics* **1995**, *14*, 4966.
54. Stahl, S. S.; Labinger, J. A.; Bercaw, J. E. *J. Am. Chem. Soc.* **1995**, *117*, 9371.
55. De Felice, V.; De Renzi, A.; Panunzi, A.; Tesauro, D. J. *J. Organomet. Chem.* **1995**, *488*, C13.
56. Stahl, S. S.; Labinger, J. A.; Bercaw, J. E. *J. Am. Chem. Soc.* **1996**, *118*, 5961.
57. Hill, G. S.; Puddephatt, R. J. *J. Am. Chem. Soc.* **1996**, *118*, 8745.
58. O'Reilly, S. A.; White, P. S.; Templeton, J. L. *J. Am. Chem. Soc.* **1996**, *118*, 5684.
59. Canty, A. J.; Dedieu, A.; Jin, H.; Milet, A.; Richmond, M. K. *Organometallics* **1996**, *15*, 2845.
60. Canty, A. J.; Fritsche, S. D.; Jin, H.; Patel, J.; Skelton, B. W.; White, A. H. *Organometallics* **1997**, *16*, 2175.
61. Hill, G. S.; Vittal, J. J.; Puddephatt, R. J. *Organometallics* **1997**, *16*, 1209.
62. Jenkins, H. A.; Yap, G. P. A.; Puddephatt, R. J. *Organometallics* **1997**, *16*, 1946.
63. Holtcamp, M. W.; Labinger, J. A.; Bercaw, J. E. *Inorg. Chim. Acta* **1997**, *265*, 117.
64. Fekl, U.; Zahl, A.; van Eldik, R. *Organometallics* **1999**, *18*, 4156.
65. Prokopchuk, E. M.; Jenkins, H. A.; Puddephatt, R. J. *Organometallics* **1999**, *18*, 2861.
66. Haskel, A.; Keinan, E. *Organometallics* **1999**, *18*, 4677.
67. Hinman, J. G.; Baar, C. R.; Jennings, M. C.; Puddephatt, R. J. *Organometallics* **2000**, *19*, 563.

68. Reinartz, S.; White, P. S.; Brookhart, M.; Templeton, J. L. *Organometallics* **2001**, *20*, 1709.
69. Puddephatt, R. J. *Coord. Chem. Rev.* **2001**, *219–221*, 157.
70. Trofimenko, S. *"Scorpionates. "The Chemistry of Polypyrazolylborate Ligands"*; Imperial College Press: London, 1999.
71. King, R. B.; Bond, A. *J. Am. Chem. Soc.* **1974**, *96*, 1338.
72. Roth, S.; Ramamoorthy, V.; Sharp, P. R. *Inorg. Chem.* **1990**, *29*, 3345.
73. Fekl, U.; van Eldik, R.; Lovell, S.; Goldberg, K. I. *Organometalllics* **2000**, *19*, 3535.
74. Reinartz, S.; White, P. S.; Brookhart, M.; Templeton, J. L. *Organometallics* **2000**, *19*, 3748.
75. Lo, H. C.; Haskel, A.; Kapon, M.; Keinan, E. *J. Am. Chem. Soc.* **2002**, *124*, 3226.
76. Lo, H. C.; Haskel, A.; Kapon, M.; Keinan, E. *J. Am. Chem. Soc.* **2002**, *124*, 12 626.
77. Gould, G. L.; Heinekey, D. M. *J. Am. Chem. Soc.* **1989**, *111*, 5502.
78. Chernega, A.; Cook, J.; Green, M. L. H.; Labella, L.; Simpson, S. J.; Souter, J.; Stephens, A. H. H. *J. Chem. Soc., Dalton Trans.* **1997**, 3225.
79. Gross, C. L.; Girolami, G. S. *J. Am. Chem. Soc.* **1998**, *120*, 6605.
80. Flood, T. C.; Janak, K. E.; Iimura, M.; Zhen, H. *J. Am. Chem. Soc.* **2000**, *122*, 6783.
81. Zhou, R.; Wang, C.; Hu, Y.; Flood, T. C. *Organometallics* **1997**, *16*, 434.
82. Wang, C.; Ziller, J. W.; Flood, T. C. *J. Am. Chem. Soc.* **1995**, *117*, 1647.
83. Wick, D. D.; Reynolds, K. A.; Jones, W. D. *J. Am. Chem. Soc.* **1999**, *121*, 3974.
84. Northcutt, T. O.; Wick, D. D.; Vetter, A. J.; Jones, W. D. *J. Am. Chem. Soc.* **2001**, *123*, 7257.
85. Iron, M. A.; Lo, H. C.; Martin, J. M. L.; Keinan, E. *J. Am. Chem. Soc.* **2002**, *124*, 7041.
86. Reinartz, S.; White, P. S.; Brookhart, M.; Templeton, J. L. *Organometallics* **2000**, *19*, 3854.
87. Jensen, M. P.; Wick, D. D.; Reinartz, S.; White, P. S.; Templeton, J. L.; Goldberg, K. I. Manuscript submitted.
88. Wick, D. D.; Goldberg, K. I. *J. Am. Chem. Soc.* **1997**, *119*, 10 235.
89. Belluco, U.; Giustiniani, M.; Graziani, M. *J. Am. Chem. Soc.* **1967**, *89*, 6494.
90. Jawad, J. K.; Puddephatt, R. J.; Stalteri, M. A. *Inorg. Chem.* **1982**, *21*, 332.
91. Alibrandi, G.; Minniti, D.; Romeo, R.; Uguagliati, P.; Calligaro, L.; Belluco, U.; Crociani, B. *Inorg. Chim. Acta* **1985**, *100*, 107.
92. Alibrandi, G.; Minniti, D.; Romeo, R.; Uguagliati, P.; Calligaro, L.; Belluco, U. *Inorg. Chim. Acta* **1986**, *112*, L15.
93. Wik, B. J.; Lersch, M.; Tilset, M. *J. Am. Chem. Soc.* **2002**, *124*, 12 116.
94. Fekl, U.; Kaminsky, W.; Goldberg, K. I. *J. Am. Chem. Soc.* **2001**, *123*, 6423.
95. Reinartz, S.; White, P. S.; Brookhart, M.; Templeton, J. L. *J. Am. Chem. Soc.* **2001**, *123*, 6425.
96. Puddephatt, R. J. *Angew. Chem. Int. Ed.* **2002**, *41*, 261.
97. Fekl, U.; Goldberg, K. I. *J. Am. Chem. Soc.* **2002**, *124*, 6804.
98. Hill, G. S.; Yap, G. P. A.; Puddephatt, R. J. *Organometallics* **1999**, *18*, 1408.
99. van Asselt, R.; Rijnberg, E.; Elsevier, C. J. *Organometallics* **1994**, *13*, 706.
100. Jain, V. K.; Rao, G. S.; Jain, L. *Adv. Organomet. Chem.* **1987**, *27*, 113.
101. Clark, H. C.; Ferguson, G.; Jain, V. K.; Parvez, M. *J. Organomet. Chem.* **1984**, *270*, 365.
102. Clark, H. C.; Ferguson, G.; Jain, V. K.; Parvez, M. *Organometallics* **1983**, *2*, 806.
103. Usón, R.; Forniés, J.; Espinet, P.; Garín, J. *J. Organomet. Chem.* **1976**, *105*, C25.
104. Kharchevnikov, V. M. *Tr. Khim. Khim. Tekhnol.* **1972**, 36.
105. Crespo, M.; Puddephatt, R. J. *Organometallics* **1987**, *6*, 2548.
106. Abel, E. W.; Khan, A. R.; Kite, K.; Orrell, K. G.; Sik, V. *J. Chem. Soc., Dalton Trans.* **1980**, 2208.

107. Clark, H. C.; Manzer, L. E. *Inorg. Chem.* **1973**, *12*, 362.
108. Schlecht, S.; Magull, J.; Fenske, D.; Dehnicke, K. *Angew. Chem. Int. Ed.* **1997**, *36*, 1994.
109. Parks, J. E.; Holm, R. H. *Inorg. Chem.* **1968**, *7*, 1408.
110. McGeachin, S. G. *Can. J. Chem.* **1968**, *46*, 1903.
111. MacAdams, L. A.; Kim, W.-K.; Liable-Sands, L. M.; Guzei, I. A.; Rheingold, A. L.; Theopold, K. H. *Organometallics* **2002**, *21*, 952.
112. Spencer, D. J. E.; Aboelella, N. W.; Reynolds, A. M.; Holland, P. L.; Tolman, W. B. *J. Am. Chem. Soc.* **2002**, *124*, 2108.
113. Budzelaar, P. H. M.; de Gelder, R.; Gal, A. W. *Organometallics* **1998**, *17*, 4121.
114. Cheng, M.; Moore, D. R.; Reczek, J. J.; Chamberlain, B. M.; Lobkovsky, E. B.; Coates, G. W. *J. Am. Chem. Soc.* **2001**, *123*, 8738.
115. Bourget-Merle, L.; Lappert, M. F.; Severn, J. R. *Chem. Rev.* **2002**, *102*, 3031.
116. Brookhart, M.; Grant, B. E.; Lenges, C. P.; Prosenc, M. H.; White, P. S. *Angew. Chem. Int. Ed.* **2000**, *39*, 1676.
117. Klei, S. R.; Tilley, T. D.; Bergman, R. G. *J. Am. Chem. Soc.* **2000**, *122*, 1816.
118. Corey, J. Y.; Braddock-Wilking, J. *Chem. Rev.* **1999**, *99*, 175.
119. Suginome, M.; Ito, Y. *J. Chem. Soc., Dalton Trans.* **1998**, 1925.
120. Levy, C. J.; Puddephatt, R. J. *Organometallics* **1997**, *16*, 4115.
121. Fang, X.; Scott, B. L.; Watkin, J. G.; Kubas, G. J. *Organometallics* **2000**, *19*, 4193.
122. Choi, S.-H.; Lin, Z. *J. Organomet. Chem.* **2000**, *608*, 42.
123. Gusev, D. G.; Notheis, J. U.; Rambo, J. R.; Hauger, B. E.; Eisenstein, O.; Caulton, K. G. *J. Am. Chem. Soc.* **1994**, *116*, 7409.
124. Butts, M. D.; Scott, B. L.; Kubas, G. J. *J. Am. Chem. Soc.* **1996**, *118*, 11831.
125. Stahl, S. S.; Labinger, J. A.; Bercaw, J. E. *Inorg. Chem.* **1998**, *37*, 2422.
126. Kimmich, B. F. M.; Bullock, R. M. *Organometallics* **2002**, *21*, 1504.
127. Reinartz, S.; White, P. S.; Brookhart, M.; Templeton, J. L. *J. Am. Chem. Soc.* **2001**, *123*, 12724.
128. Reinartz, S.; White, P. S.; Brookhart, M.; Templeton, J. L. *J. Am. Chem. Soc.* **2002**, *124*, 7249.
129. Reinartz, S.; Baik, M.-H.; White, P. S.; Brookhart, M.; Templeton, J. L. *Inorg. Chem.* **2001**, *40*, 4726.
130. Abis, L.; Sen, A.; Halpern, J. *J. Am. Chem. Soc.* **1978**, *100*, 2915.
131. Dedieu, A. *Chem. Rev.* **2000**, *100*, 543.
132. Bartlett, K. L.; Goldberg, K. I.; Borden, W. T. *J. Am. Chem. Soc.* **2000**, *122*, 1456.
133. Bartlett, K. L.; Goldberg, K. I.; Borden, W. T. *Organometallics* **2001**, *20*, 2669.
134. Niu, S.; Hall, M. B. *Chem. Rev.* **2000**, *100*, 353.
135. Hackett, M.; Ibers, J. A.; Whitesides, G. M. *J. Am. Chem. Soc.* **1988**, *110*, 1436.
136. Hackett, M.; Whitesides, G. M. *J. Am. Chem. Soc.* **1988**, *110*, 1449.
137. Hofmann, P.; Heiss, H.; Neiteler, P.; Müller, G.; Lachmann, J. *Angew. Chem. Int. Ed.* **1990**, *29*, 880.
138. Hofmann, P.; Heiss, H.; Müller, G. *Z. Naturforsch. B* **1987**, *42*, 395.
139. Hill, G. S.; Puddephatt, R. J. *Organometallics* **1998**, *17*, 1478.
140. Bayler, A.; Canty, A. J.; Skelton, B. W.; White, A. H. *J. Organomet. Chem.* **2000**, *595*, 296.
141. Rendina, L. M.; Puddephatt, R. J. *Chem. Rev.* **1997**, *97*, 1735.
142. Vedernikov, A. N.; Shamov, G. A.; Solomonov, B. N. *Russ. J. Gen. Chem.* **1997**, *67*, 1159.
143. Hoffmann, R. *Angew. Chem. Int. Ed.* **1982**, *21*, 711.
144. Vedernikov, A. N.; Shamov, G. A.; Solomonov, B. N. *Russ. J. Gen. Chem.* **1999**, *69*, 1102.
145. Erickson, L. E.; Cook, D. J.; Evans, G. D.; Sarneski, J. E.; Okarma, P. J.; Sabatelli, A. D. *Inorg. Chem.* **1990**, *29*, 1958.

146. Sarneski, J. E.; McPhail, A. T.; Onan, K. D.; Erickson, L. E.; Reilley, C. N. *J. Am. Chem. Soc.* **1977**, *99*, 7376.
147. Siegbahn, P. E. M.; Crabtree, R. H. *J. Am. Chem. Soc.* **1996**, *118*, 4442.
148. Kua, J.; Xu, X.; Periana, R. A.; Goddard, W. A., III *Organometallics* **2002**, *21*, 511.
149. Mylvaganam, K.; Bacskay, G. B.; Hush, N. S. *J. Am. Chem. Soc.* **2000**, *122*, 2041.
150. Gilbert, T. M.; Hristov, I.; Ziegler, T. *Organometallics* **2001**, *20*, 1183.
151. Heiberg, H.; Swang, O.; Ryan, O. B.; Gropen, O. *J. Phys. Chem. A* **1999**, *103*, 10 004.
152. Holtcamp, M. W.; Henling, L. M.; Day, M. W.; Labinger, J. A.; Bercaw, J. E. *Inorg. Chim. Acta* **1998**, *270*, 467.
153. Biswas, B.; Sugimoto, M.; Sakaki, S. *Organometallics* **2000**, *19*, 3895.
154. Vedernikov, A. N.; Shamov, G. A.; Solomonov, B. N. *Russ. J. Gen. Chem.* **2000**, *70*, 1184.
155. Langford, C. H.; Gray, H. B. *"Ligand Substitution Processes"*; Benjamin: New York, 1965.
156. Lincoln, S. F.; Merbach, A. E. *Adv. Inorg. Chem.* **1995**, *42*, 1.
157. Basolo, F.; Gray, H. B.; Pearson, R. G. *J. Am. Chem. Soc.* **1960**, *82*, 4200.
158. Basolo, F.; Pearson, R. G. *"Mechanisms of Inorganic Reactions"*, 2nd edn.; John Wiley and Sons, Inc.: New York, 1967.
159. Minniti, D.; Alibrandi, G.; Tobe, M. L.; Romeo, R. *Inorg. Chem.* **1987**, *26*, 3956.
160. Alibrandi, G.; Bruno, G.; Lanza, S.; Minniti, D.; Romeo, R.; Tobe, M. L. *Inorg. Chem.* **1987**, *26*, 185.
161. Frey, U.; Helm, L.; Merbach, A. E.; Romeo, R. *J. Am. Chem. Soc.* **1989**, *111*, 8161.
162. Romeo, R.; Grassi, A.; Scolaro, L. M. *Inorg. Chem.* **1992**, *31*, 4383.
163. Schmülling, M.; Grove, D. M.; van Koten, G.; van Eldik, R.; Veldman, N.; Spek, A. L. *Organometallics* **1996**, *15*, 1384.
164. Frey, U.; Grove, D. M.; van Koten, G. *Inorg. Chim. Acta* **1998**, *269*, 322.
165. Johansson, L.; Tilset, M. *J. Am. Chem. Soc.* **2001**, *123*, 739.
166. van Eldik, R., Ed. *Inorganic High Pressure Chemistry: Kinetics and Mechanisms*; Studies in Inorganic Chemistry, vol. 7; Elsevier: Amsterdam, 1986.
167. van Eldik, R.; Dücker-Benfer, C.; Thaler, F. *Adv. Inorg. Chem.* **2000**, *49*, 1.
168. Zahl, A.; Neubrand, A.; Aygen, S.; van Eldik, R. *Rev. Sci. Instrum.* **1994**, *65*, 882.
169. van Eldik, R.; Asano, T.; le Noble, W. J. *Chem. Rev.* **1989**, *89*, 549.
170. Drljaca, A.; Hubbard, C. D.; van Eldik, R.; Asano, T.; Basilevsky, M. V.; le Noble, W. J. *Chem. Rev.* **1998**, *98*, 2167.
171. Wendt, O. F.; Elding, L. I. *Inorg. Chem.* **1997**, *36*, 6028.
172. Jaganyi, D.; Hofmann, A.; van Eldik, R. *Angew. Chem. Int. Ed.* **2001**, *40*, 1680.
173. Fekl, U.; van Eldik, R. *Eur. J. Inorg. Chem.* **1998**, 389.
174. Romeo, R.; Scolaro, L. M.; Nastasi, N.; Arena, G. *Inorg. Chem.* **1996**, *35*, 5087.
175. Haake, P.; Cronin, P. A. *Inorg. Chem.* **1963**, *2*, 879.
176. Tellers, D. M.; Yung, C. M.; Arndtsen, B. A.; Adamson, D. R.; Bergman, R. G. *J. Am. Chem. Soc.* **2002**, *124*, 1400.
177. Gaemers, S.; Keune, K.; Kluwer, A. M.; Elsevier, C. J. *Eur. J. Inorg. Chem.* **2000**, 1139.
178. Jones, M. M.; Morgan, K. A. *J. Inorg. Nucl. Chem.* **1972**, *34*, 259.
179. Morgan, K. A.; Jones, M. M. *J. Inorg. Nucl. Chem.* **1972**, *34*, 275.
180. Elding, L. I.; Gustafson, L. *Inorg. Chim. Acta* **1976**, *19*, 165.
181. Wilkins, R. G. *"Kinetics and Mechanism of Reactions of Transition Metal Complexes"*, 2nd edn.; VCH: Weinheim, 1991.
182. Hindmarsh, K.; House, D. A.; van Eldik, R. *Inorg. Chim. Acta* **1998**, *278*, 32.
183. Puddephatt, R. J.; Scott, J. D. *Organometallics* **1985**, *4*, 1221.
184. Wang, L.; Stahl, S. S.; Labinger, J. A.; Bercaw, J. E. *J. Mol. Catal. A* **1997**, *116*, 269.

185. Luinstra, G. A.; Wang, L.; Stahl, S. S.; Labinger, J. A.; Bercaw, J. E. *J. Organomet. Chem.* **1995**, *504*, 75.
186. Luinstra, G. A.; Wang, L.; Stahl, S. S.; Labinger, J. A.; Bercaw, J. E. *Organometallics* **1994**, *13*, 755.
187. Johansson, L.; Ryan, O. B.; Rømming, C.; Tilset, M. *Organometallics* **1998**, *17*, 3957.
188. Scollard, J. D.; Day, M.; Labinger, J. A.; Bercaw, J. E. *Helv. Chim. Acta* **2001**, *84*, 3247.
189. Geletii, Y. V.; Shilov, A. E. *Kinet. Catal.* **1983**, *24*, 413.
190. Freund, M. S.; Labinger, J. A.; Lewis, N. S.; Bercaw, J. E. *J. Mol. Catal.* **1994**, *87*, L11.
191. Lin, M.; Shen, C.; Garcia-Zayas, E. A.; Sen, A. *J. Am. Chem. Soc.* **2001**, *123*, 1000.
192. van Koten, G. *Pure Appl. Chem.* **1990**, *62*, 1155.
193. van Beek, J. A. M.; van Koten, G.; Wehman-Ooyevaar, I. C. M.; Smeets, W. J. J.; van der Sluis, P.; Spek, A. L. *J. Chem. Soc., Dalton Trans.* **1991**, 883.
194. Terheijden, J.; van Koten, G.; de Booys, J. L.; Ubbels, H. J. C.; Stam, C. H. *Organometallics* **1983**, *2*, 1882.
195. Canty, A. J.; Honeyman, R. T.; Roberts, A. S.; Traill, P. R.; Colton, R.; Skelton, B. W.; White, A. H. *J. Organomet. Chem.* **1994**, *471*, C8.
196. Canty, A. J.; Fritsche, S. D.; Jin, H.; Honeyman, R. T.; Skelton, B. W.; White, A. H. *J. Organomet. Chem.* **1996**, *510*, 281.
197. Canty, A. J.; Fritsche, S. D.; Jin, H.; Skelton, B. W.; White, A. H. *J. Organomet. Chem.* **1995**, *490*, C18.
198. Heinzel, U.; Mattes, R. *Polyhedron* **1991**, *10*, 19.
199. Wieghardt, K.; Köppen, M.; Swiridoff, W.; Weiss, J. *J. Chem. Soc., Dalton Trans.* **1983**, 1869.
200. Davies, M. S.; Hambley, T. W. *Inorg. Chem.* **1998**, *37*, 5408.
201. Ferguson, G.; Parvez, M.; Monaghan, P. K.; Puddephatt, R. J. *J. Chem. Soc., Chem. Comm.* **1983**, 267.
202. Ferguson, G.; Monaghan, P. K.; Parvez, M.; Puddephatt, R. J. *Organometallics* **1985**, *4*, 1669.
203. Rostovtsev, V. V.; Labinger, J. A.; Bercaw, J. E.; Lasseter, T. L.; Goldberg, K. I. *Organometallics* **1998**, *17*, 4530.
204. Rostovtsev, V. V.; Henling, L. M.; Labinger, J. A.; Bercaw, J. E. *Inorg. Chem.* **2002**, *41*, 3608.
205. Monaghan, P. K.; Puddephatt, R. J. *Organometallics* **1984**, *3*, 444.
206. Wick, D. D.; Goldberg, K. I. *J. Am. Chem. Soc.* **1999**, *121*, 11 900.
207. Luinstra, G. A.; Labinger, J. A.; Bercaw, J. E. *J. Am. Chem. Soc.* **1993**, *115*, 3004.
208. Sen, A. *Acc. Chem. Res.* **1988**, *21*, 421.
209. Muehlhofer, M.; Strassner, T.; Herrmann, W. A. *Angew. Chem. Int. Ed.* **2002**, *41*, 1745.
210. Collman, J. P.; Hegedus, L. S.; Norton, J. R.; Finke, R. G. "*Principles and Applications of Organotransition Metal Chemistry*"; University Science Books: Mill Valley, CA, 1987.
211. Eisenstein, O.; Crabtree, R. H. *New J. Chem.* **2001**, *25*, 665.
212. Burk, M. J.; Crabtree, R. H. *J. Am. Chem. Soc.* **1987**, *109*, 8025.
213. Johnson, J. A.; Li, N.; Sames, D. *J. Am. Chem. Soc.* **2002**, *124*, 6900.
214. Johnson, J. A.; Sames, D. *J. Am. Chem. Soc.* **2000**, *122*, 6321.
215. Spencer, J. L.; Mhinzi, G. S. *J. Chem. Soc., Dalton Trans.* **1995**, 3819.
216. Carr, N.; Mole, L.; Orpen, A. G.; Spencer, J. L. *J. Chem. Soc., Dalton Trans.* **1992**, 2653.
217. Carr, N.; Dunne, B. J.; Orpen, A. G.; Spencer, J. L. *J. Chem. Soc., Chem. Comm.* **1988**, 926.
218. Azam, K. A.; Brown, M. P.; Cooper, S. J.; Puddephatt, R. J. *Organometallics* **1982**, *1*, 1183.
219. Chaudhury, N.; Puddephatt, R. J. *J. Chem. Soc., Dalton Trans.* **1976**, 915.
220. Yamakawa, T.; Fujita, T.; Shinoda, S. *Chem. Lett.* **1992**, 905.

221. Dangel, B. D.; Johnson, J. A.; Sames, D. *J. Am. Chem. Soc.* **2001**, *123*, 8149.
222. Konze, W. V.; Scott, B. L.; Kubas, G. J. *J. Am. Chem. Soc.* **2002**, *124*, 12 550.
223. Prokopchuk, E. M.; Puddephatt, R. J. *Organometallics* **2003**, *22*, 563.
224. Prokopchuk, E. M.; Puddephatt, R. J. *Organometallics* **2003**, *22*, 787.
225. Jones, W. D. *Acc. Chem. Res.* **2003**, *36*, 140.
226. Vedernikov, A. N.; Caulton, K. G. *Angew. Chem. Int. Ed.* **2002**, *41*, 4102.
227. Norris, C. M.; Reinartz, S.; White, P. S.; Templeton, J. L. *Organometallics* **2002**, *21*, 5649.

DENSITY FUNCTIONAL STUDIES OF IRIDIUM CATALYZED ALKANE DEHYDROGENATION

MICHAEL B. HALL[a] and HUA-JUN FAN[b]

[a]Department of Chemistry, Texas A&M University, 3255 TAMU, College Station, TX 77843-3255, USA
[b]Department of Chemistry, Prairie View A&M University, PO Box 4107, Prairie View, TX 77446-4107, USA

I. Introduction

Conversions of alkanes to alkenes are important reactions because saturated alkanes are the world's most abundant organic resource but are not as easily functionalized as alkenes (*1*). The high strength of the C–H bond relative to the π bond in the alkene, makes direct alkane dehydrogenation highly endothermic with an extremely high barrier (*2*). Fortunately, transition metal complexes are able to catalyze the reaction by opening lower activation-energy pathways. In the past 20 years, most studies in this area have been focused on searching for

ADVANCES IN INORGANIC CHEMISTRY
VOLUME 54 ISSN 0898-8838

efficient catalysts among the late transition metals, such as Ru (*3*), Os (*4*), Rh (*5*), Ir (*6*), Pd (*7*) and Pt (*8*) complexes. Studies have shown that these endothermic dehydrogenation reactions can be driven thermodynamically through either the transfer dehydrogenation reaction (*9*) (transferring hydrogen to a sacrificial hydrogen acceptor) or the acceptorless dehydrogenation reaction (photochemically (*10*) or thermally (*11*)).

The first examples of transfer dehydrogenation catalysts were reported in 1979 by Crabtree (*12a*) and in 1980 by Felkin (*12b*). However, because of the harsh reaction conditions needed, the catalysts decomposed and produced low turnovers. Because slow reaction rates and catalyst decomposition were problems with most early catalysts, attention was focused on thermally stable homogeneous catalysts. The higher reaction temperature available with these thermally stable catalysts provides more thermodynamically and kinetically favorable reaction paths. This allows use of less reactive alkanes, and provides easier educt and product separation (*13*). Among the catalysts developed, pincer iridium complexes such as $(\eta^3\text{-}C_6H_3(CH_2PR_2)_2\text{-}1,3)IrH_2$ complex, first discovered by Moulton and Shaw (*14*), showed promising properties such as thermal stability up to $200\,^{\circ}\text{C}$ (*11k*), and reactivity with both aliphatic and cycloaliphatic C–H bonds (*15*). The chelating ligand of this catalyst is abbreviated as ^{R}PCP because the available chelating atoms of the pincer ligand are the two P and C from the arene ring (*16*). Other pincer rhodium (*17*) and osmium (*4*) and ruthenium (*18*) complexes and pincer-like ligands such as PCN (*19*), NCN (*20*), are also under investigation.

The pincer complexes (^{R}PCP)IrH$_2$ were the first homogeneous systems to catalyze alkane dehydrogenation thermally (*11k*). Although the first reaction was carried out with ($^{i\text{-}Bu}$PCP)IrH$_2$, the ($^{i\text{-}Pr}$PCP)IrH$_2$ complex has shown even higher turnover rates (*1,11*). Haenel *et al.* recently reported an even more effective, higher temperature ($250\,^{\circ}\text{C}$) acceptorless catalyst, in which the pincer ligand is replaced by anthracene-1,8-diphosphine (anthraphos, $\eta^3\text{-}C_{14}H_7(PR_2)_2\text{-}3,13$, abbreviated as ^{R}AP where in initial studies R = ^{t}Bu) (*13a*). These experimental studies have provided an excellent opportunity for computational chemists to examine the various mechanisms for these reactions and to seek the factors that play the key roles in the catalytic properties of these two systems (*13a*). As a result, the mechanisms for the transfer and acceptorless reactions have been thoroughly examined. There are three mechanisms proposed for the acceptorless reaction: the dissociative pathway (D), associative pathway (A) and interchange pathway (I) (*13a,21*), whose importance may depend on the temperature, concentration of reactive species (alkane, alkene, and dihydrogen), and phosphine substituents (*22*).

In this chapter, we will study the elementary reaction steps of these mechanisms focusing primarily on the anthraphos systems. This chapter begins with a description of the impact of different methods (coupled cluster, configuration interaction and various DFT functionals), different basis sets, and phosphine substituents on the oxidative addition of methane to a related Ir system, $[CpIr(III)(PH_3)Me]^+$. Then, it compares the elementary reaction steps, including the effect of reaction conditions such as temperature, hydrogen pressure, alkane and alkene concentration, phosphine substituents and alternative metals (Rh). Finally, it considers how these elementary steps constitute the reaction mechanisms. Additional computational details are provided at the end of the chapter.

II. Cyclopentadienyl Iridium Complex

A. Background

Bergman and coworkers reported that the compound $[Cp^*Ir(PMe_3)(CH_3)(Solv)]^+$ ($Cp^* = C_5(CH_3)_5$, Solv = solvent (CH_2Cl_2)) can break alkane C–H bonds at, or below, room temperature (23). Among the mechanisms proposed (23–25), Hall and coworkers (21a,25a–d), showed that the oxidative-addition / reductive-elimination (OA/RE) pathway through an Ir(V) intermediate is the lowest energy path. No transition state could be located for any route involving a four-center intermolecular σ-bond metathesis pathway, where all hypothetical routes were found to be higher in energy (see Fig. 1). In the OA/RE pathway the R–H bond is broken and two new bonds to the Ir are made. Thus, the Ir increases its formal oxidation state by two. Then, this Ir intermediate reductively eliminates methane. In contrast, the σ-bond metathesis

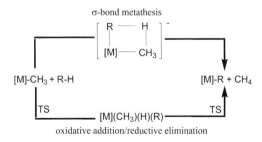

FIG. 1. The proposed mechanisms for $[Cp^*Ir(PMe_3)(CH_3)]^+$.

would have proceeded through a single transition state (TS) without changing the formal oxidation state of Ir. When this system was examined in the gas-phase and in the absence of alkane, Chen and coworkers found that the Ir attacks one of the CH_3 groups on its phosphine ligand and eliminates methane (*24b,c*). Calculations by Niu and Hall have confirmed this behavior and have shown that this intramolecular reaction is also an OA/RE process with somewhat higher activation energy than the intermolecular one (*25a–d*).

We choose this reaction system as a model to test the basis set and methodology because the system is similar to the Ir dehydrogenation system but is smaller and relatively well understood. In the following, we will discuss the impact of methods, basis set and phosphine substituents.

B. Methodology

In this section, we will examine how the theoretical methodology with basis set **1** (BS1, double-zeta with polarization functions on all atoms except the metal atom) influences the energy of the $[CpIr(III)(PH_3)Me]^+$ reaction. In addition to Hartree–Fock (*26*) (HF) and Møller–Plesset perturbation theory (*27*) (MPx), we report configuration interaction (CI), including singles and doubles (CISD) (*28*) (with and without Davidson's size-consistency correction), coupled cluster singles and doubles (CCSD) (*29*), perturbative corrections for triples (CCSD(T)) (*30*), and DFT functional (*31*) (B3PW91, MPW1PW91, BP86, B3P86, and B3LYP) calculations for $[CpIr(III)(PH_3)Me \cdots CH_4]^+$ (**1**), TS (**2**) and $[CpIr(V)(PH_3)(Me)(H)(Me)]^+$ (**3**) based on the optimized geometries from B3LYP/BS1. The results are summarized in Table I.

For the σ-bond (agostic) species (**1**), we calculated the basis set superposition error (BSSE) between $CpIr(III)(PH_3)Me$ and CH_4 fragments. The largest BSSE error is found for the CCSD method (3.3 kcal/mol), while HF (0.6 kcal/mol) and DFT, B3LYP (1.1 kcal/mol), BP86 (0.4 kcal/mol) are smaller. These BSSEs are comparable to the calculated binding energies of this σ-bond species (**1**), which are 3.24 (CCSD), 0.55 (HF), 0.55 (B3LYP), 0.71 kcal/mol (BP86) before the BSSE correction. Therefore, this σ-bond species (**1**) is barely bound and would have low concentration at higher temperature although larger van der Waals interactions for larger substrates could provide stronger attraction.

The next step, the oxidative addition of alkane to form the Ir(V) intermediate is quite sensitive to the methods and presents a challenging problem to determine which method gives the most accurate answer. The calculated activation ($\Delta E^{\neq}$) and reaction (ΔE°) energies show

TABLE I

The Calculated Energies (kcal/mol) for the Reaction Based on the Geometries Optimized at B3LYP/BS1

	$\Delta E^{\neq}$	ΔE°
HF	21.74	19.68
MP2	−5.17	−16.59
MP3	6.96	−0.37
MP4(SDQ)	1.93	−6.52
CCSD	5.63	−1.96
CCSD(T)	3.63	−4.39
CISD	11.32	6.05
CISD [a]	4.6	−3.16
B3PW91	4.91	−0.44
MPW1PW91	4.12	−0.47
BP86	4.25	−1.78
B3P86	4.20	−1.54
B3LYP	9.95	4.02

[a] With Davidson's correction.

substantial variations. For example, HF predicts a very endoenergetic reaction ($\sim$ +20 kcal/mol), while MP2 predicts a very exoenergetic one ($\sim$ −16 kcal/mol). These two methods represent the two extremes of the results. HF, of course, has no electron correlation, while MP2's perturbative correction often overcorrects the energy, especially for transition metal systems. The other methods predict values between these two extremes and range from $\sim$ −7 kcal/mol for MP4(SDQ), which may still be overestimating the correlation effect, to $\sim$ +6 kcal/mol for CISD, which is expected to underestimate the correlation effect. For the most accurate methods, CCSD and CCSD(T), the energetics are $\sim$ −2.0 and $\sim$ −4.4 kcal/mol. The CISD with Davidson's size-consistency correction produces a similar value of $\sim$ −3.2 kcal/mol.

Although the B3LYP method generally produces excellent results, this study shows that both BP86 and B3P86 methods predict a more exothermic reaction in closer agreement with the most accurate ab initio techniques. Because these latter methods are expected to yield the most accurate results (32), one may conclude that B3LYP underestimates the stability of the higher oxidation states for Ir(III)↔Ir(V). Further comparison to other DFT methods suggests that, of the five DFT functionals examined, the BP86 functional produces an energy difference closer to the most accurate CCSD, CCSD(T) and CISD (with Davidson's correction) methods.

C. Basis set Impact

Four basis sets were examined: BS1 and BS3 are based on the Couty-Hall modification of the Hay and Wadt ECP, and BS2 and BS4 are based on the Stuttgart ECP. Two basis sets, BS1 and BS2, are used to optimize the geometries of species in the OA reaction, $[CpIr(PH_3)(CH_3)]^+ + CH_4 \leftrightarrow [CpIr(PH_3)(H)(CH_3)_2]^+$, at the B3LYP level, while the other basis sets, BS3 and BS4, are used only to calculate energies at the previously optimized B3LYP/BS1 geometries. BS1 is double-zeta with polarization functions on every atom except the metal atom. BS2 is triple-zeta with polarization on metal and double-zeta correlation consistent basis set (with polarization functions) on other atoms. BS3 is similar to BS1 but is triple-zeta with polarization on the metal. BS4 is similar to BS2 but is triple-zeta with polarization on the C and H that are involved in the reaction. The basis set details are described in the Computational Details section at the end of this chapter.

The optimized geometries for the σ-bond species **1**, the transition state **2**, and the intermediate product **3** from BS1 at B3LYP level are shown in Fig. 2. The calculated geometries from BS2 are similar to those from BS1 and the calculated bond lengths are listed in parenthesis in

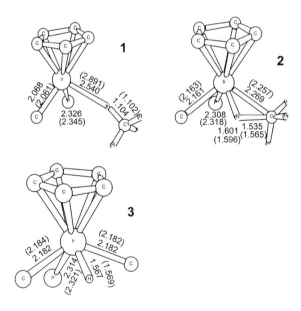

Fig. 2. The bond lengths of the optimized geometries from BS1 and BS2 (in parenthesis) at B3LYP level.

the figure. The largest difference between two calculated geometries is in the Ir–C(CH$_4$) distance of **1**. Other bond length differences in the σ-bond species **1** occur for the Ir–C(Cp) and Ir–P distances, which are 0.01–0.03 Å shorter in BS2 and 0.02 Å longer in BS2, respectively. The differences at the TS and intermediate product are even smaller (about 0.01 Å). Examining the thermodynamic parameters of the reaction listed in Table II, one can see that BS2 destabilizes both the TS and the intermediate product by approximately the same amount of energy. However, the free-energy differences are somewhat larger because of the change (noted above) in the binding of CH$_4$ to Ir in **1**.

Next, we compare the HF, MP2 and B3LYP energies for basis sets: BS1, BS2, BS3 and BS4, at the B3LYP/BS1 geometry. The results are summarized in Table III. Although MPx perturbation methods have been

TABLE II

THE CALCULATED THERMODYNAMIC PARAMETERS (kcal/mol) OF THE REACTION WITH B3LYP IN TWO BASIS SETS (BS1 AND BS2)

		$\Delta E^{\neq}$	$E_{\mathrm{ZPE}}^{\neq}$	$\Delta H^{\neq}$	$\Delta G^{\neq}$
$1 \to 2$	BS1	9.95	10.39	9.81	12.19
	BS2	11.63	12.50	10.56	18.34
		ΔE°	E_{ZPE}°	ΔH°	ΔG°
$1 \to 3$	BS1	4.02	6.24	5.51	8.49
	BS2	6.72	9.16	7.18	14.78

TABLE III

THE CALCULATED SINGLE POINT ENERGIES (WITHOUT ZERO-POINT AND ENTROPY CORRECTIONS) OF BS1 TO BS4 BASED ON B3LYP / BS1 GEOMETRY (IN kcal/mol)

	BS1	BS2	BS3	BS4
$\Delta E^{\neq}$				
HF	21.74	22.37	24.39	24.52
MP2	−5.17	−6.93	−8.76	−17.36
B3LYP	9.95	10.29	11.24	11.96
ΔE°				
HF	19.68	19.91	23.85	23.89
MP2	−16.59	−17.92	−19.97	−29.33
B3LYP	4.02	4.26	6.33	7.06

shown to give reliable results for many third row transition metal systems, they show energy oscillations in Tables I and III. MP2 seems to overestimate the stability of the B3LYP TS with respect to the reactant (negative activation energy). Much of this problem may come from the use of B3LYP geometries, where the Ir–CH$_4$ interaction, predicted by DFT to be weak, would be stronger by MP2 because of the improved treatment of dispersion interactions. The energy variation from BS1 to BS4 found at the HF and B3LYP level was +4 and +3 kcal/mol, respectively, while that for MP2 is −12 kcal/mol.

Since improved basis sets appear to make the Ir(V) intermediate (3) slightly less stable (B3LYP) with respect to the Ir(III) reactant (1), BS1 at the BP86 level may be even closer to large basis set CCSD(T) than to the smaller basis set CCSD(T) results. Thus, BP86 will assume the role of the most accurate energy method for the system to be studied here.

D. LIGAND REPLACEMENT

Conventionally, PH$_3$ has been used to substitute for PR$_3$ in quantum chemical calculations in order to save computer time and improve efficiency. Hall and coworkers have investigated the effectiveness and suitability of this replacement on several systems (33). Generally, these comparisons focus on structural and energetic changes. Usually, the optimized geometries using PH$_3$ are similar to those using PMe$_3$ or PPh$_3$. However, since PMe$_3$ is a better σ-donor than PPh$_3$ and PPh$_3$ is fairly close in σ-donor strength to PH$_3$ (34), PMe$_3$ would be expected to stabilize the higher oxidation-state pathway by donating more electron density to the metal. Consequently, model calculations with PH$_3$ would be expected to underestimate the stability of the higher oxidation-state species for alkyl phosphine such as that in [Cp*Ir(PMe$_3$)(CH$_3$)(Solv)]$^+$.

In addition to the electronic difference between PR$_3$ and PH$_3$, bulkier ligands on the phosphine can change the reaction through their steric effect. Using the R = ^{t}Bu on the anthraphos system, Haenel et al. calculated the available molecular surface (AMS) around the metal center as a measure of the space available to the alkane (13b). They correlated the AMS to the relative reactivities of the catalysts and the results show that two bulky tert-butyl groups on each P certainly limit the access to the metal center, and thus, may reduce the reactivity. Other theoretical studies on the pincer complexes showed that this steric contribution/limitation plays a less important role than the activation barriers introduced by the catalyst itself (22), where the increase in energy barrier induced by the bulky ^{t}Bu is smaller than the original barriers calculated using the PMe_3 model. Here, we will compare $R = H$ and $R = {}^tBu$ as substituents on the anthraphos system.

Because the $[Cp^*Ir(III)(PR_3)Me(Solv)]^+$ system has only one phosphine ligand and is less crowded than the pincer or anthraphos systems, we used PH_3 and methane as models in this section. Niu and Hall also calculated the reaction with a basis sets similar to BS1 but with $R = methyl$ in the model (25c). They calculated the transition state at 13.0 kcal/mol and the intermediate at 5.2 kcal/mol higher than the reactant. These results are similar to the ones shown in Table II. Thus, the selection of $[CpIr(PH_3)(CH_3)]^+$ as a model will not be a major concern because for this reaction pathway the increased steric bulk of PMe_3 (disfavoring the Ir(V) intermediate) is balanced by its stronger σ-donor strength (favoring the Ir(V) intermediate).

III. Fundamental Steps

Alkane dehydrogenation with the pincer complexes $(({}^RPCP)IrH_2)$ has been extensively investigated both experimentally (11) and theoretically (21,22). Recently, a more effective and higher reaction temperature (250 °C) acceptorless catalyst, the anthraphos iridium(III) complex $({}^RAP)IrH_2$, was reported (13). Previous studies show that both complexes could catalyze alkane dehydrogenation by either transfer or thermal (acceptorless) reactions, where the former reaction occurs at lower temperature (100 °C less) than the latter. Even though the transfer reaction utilizes a sacrificial olefin and the thermal reaction uses only a high reaction temperature for the alkane dehydrogenation, both reactions share many identical fundamental forward (and reverse) reaction steps. These fundamental reaction steps generally include the following seven elementary steps from (I) to (VII) (L = pincer or anthraphos).

$$(L)IrH_2 \ (\mathbf{4}) + RH \equiv (L)IrH_3R \ (\mathbf{5}) \tag{I}$$

$$(L)IrH_3R \ (\mathbf{5}) \equiv (L)IrHR \ (\mathbf{6}) + H_2 \tag{II}$$

$$(L)IrHR \ (\mathbf{6}) \equiv (L)IrH_2(olefin) \ (\mathbf{7}) \tag{III}$$

$$(L)IrH_2(olefin) \ (\mathbf{7}) \equiv (L)IrH_2 \ (\mathbf{4}) + (olefin) \tag{IV}$$

$$(L)IrH_2 \ (\mathbf{4}) \equiv (L)Ir \ (\mathbf{8}) + H_2 \tag{V}$$

$$(L)Ir \ (\mathbf{8}) + RH \equiv (L)IrHR \ (\mathbf{6}) \tag{VI}$$

$$(L)IrH_2 \ (4) + RH \equiv (L)IrHR \ (6) + H_2 \qquad \text{(VII)}$$

The intermediates **4** to **8** with the pincer complex as an example are shown in Fig. 3. There are two conformations for all of these intermediates except **8**. Our early studies demonstrated that these two conformations represent two reaction pathways beginning with **4a** and **4b** (L = pincer). The major difference between **4a** and **4b** is the relative position of two hydrides. Species **4a** has a small H–Ir–H angle (62.7°) and species **4b** has a large one (176.6°). Due to the unfavorability of two *trans* hydrides in **4b**, it is about ∼9 kcal/mol less stable than **4a**. Furthermore, the Ir–H bond lengths in **4b** are 0.08 Å longer than those in **4a**, and the Ir–C(^{H}PCP) bond length is 0.09 Å shorter than that in **4a**. The structure of the TS shows an unsymmetrical structure, in which one of the H–Ir–C(^{H}PCP) angles is 79.9° and the other is 112.5°. Thus, the conversion between them involves an asymmetric motion, which

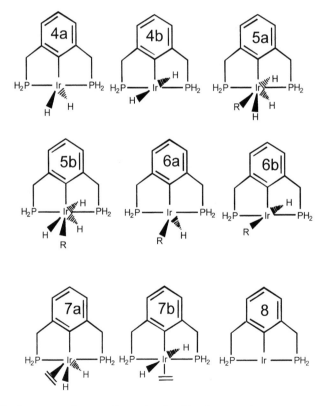

FIG. 3. The important intermediates involved in the transfer and acceptorless reactions of the pincer complex (anthraphos species are similar).

occurs when both H atoms initially move the same direction (clockwise or counter clockwise) rather than toward (or away from) each other. This result is attributed to an orbital crossing between occupied and empty orbitals in the symmetric transformation from **4a** to **4b** (35). Except for the 9.0 kcal/mol energy difference between **4a** and **4b** (21a), the energy profiles for intermediates and TS states along routes beginning with **4a** and **4b** are quite similar (see Fig. 5). Therefore, in the following discussion, we focus on reactions starting with species **4a** and related species without trans-H atoms, and the notations **a** and **b** are dropped for simplicity. The calculated thermodynamic parameters ($\Delta H^{\neq}$, ΔH°, $\Delta G^{\neq}$, and ΔG°) for gas-phase species at standard thermodynamic conditions (RTP: 298.15 K and 1.0 atm pressure) with B3LYP// B3LYP and BP86//B3LYP methods are shown in Tables IV and V.

TABLE IV

THE CALCULATED THERMODYNAMIC PARAMETERS (ΔH AND ΔG) FOR THESE SEVEN FUNDAMENTAL REACTION STEPS AT STP AND WITH THE B3LYP METHOD

Reaction step	$\Delta H^{\neq}$	ΔH°	$\Delta G^{\neq}$	ΔG°
(I)	15.69	9.37	29.08	22.88
(II)	14.14	14.14	14.14 [a]	3.05
(III)	3.70	−10.05	6.72	−7.74
(IV)	20.82	20.82	20.82 [a]	7.68
(V)	27.54	27.54	27.54 [a]	18.95
(VI)	1.09	−4.03	13.09	6.98
(VII)	(11.31) [b]	(23.51) [b]	(24.42) [b]	(25.93) [b]

[a] Estimated by assuming no $T\Delta S$ contribution at transition state.
[b] There is a dihydrogen complex formed before H_2 loss.

TABLE V

THE CALCULATED THERMODYNAMIC PARAMETERS (ΔH AND ΔG) FOR THESE SIX FUNDAMENTAL REACTION STEPS AT STP AND WITH THE BP86 ENERGIES

Reaction step	$\Delta H^{\neq}$	ΔH°	$\Delta G^{\neq}$	ΔG°
(I)	9.71	3.51	23.10	17.02
(II)	19.41	19.41	19.41 [a]	8.32
(III)	−0.25	−13.54	2.87	−11.23
(IV)	25.78	25.78	25.78 [a]	12.64
(V)	30.69	30.69	30.69 [a]	22.10
(VI)	−3.98	−7.77	8.02	3.24

[a] Estimated by assuming no $T\Delta S$ contribution at transition state.

The calculated reaction parameters with BP86//B3LYP methods at two higher temperatures (150 and 250 °C) are shown in Tables VI and VII, respectively. Further corrections for low H_2 and olefin pressure/concentration and high alkane pressure/concentration (36,37) on BP86//B3LYP values are shown in Tables VIII and IX.

The elementary reaction step I is the oxidative addition of the alkane (RH) to the Ir(III) species, (L)IrH$_2$ (4), to form an Ir(V) species, (L)IrH$_3$R (5). At STP conditions, the B3LYP method shows an enthalpy barrier of 15.69 kcal/mol at the TS and an endothermicity of 9.37 kcal/mol (see Table IV). The calculated Gibbs free-energy barrier for this reaction is almost double, $\Delta G^{\neq} = 29.08$ kcal/mol and $\Delta G° = 22.88$ kcal/mol; this free-energy increase comes from the loss of entropy in this associative step. Under the same conditions and with B3LYP geometries, the BP86 energies are similar but generally about

TABLE VI

THE CALCULATED THERMODYNAMIC PARAMETERS (ΔH AND ΔG) FOR THESE SIX FUNDAMENTAL REACTION STEPS AT 150 °C AND WITH BP86 ENERGIES

Reaction step	$\Delta H^{\neq}$	ΔH	$\Delta G^{\neq}$	ΔG
(I)	9.83	3.79	28.70	22.64
(II)	19.93	19.93	19.93 [a]	3.64
(III)	−0.37	−13.58	4.18	−10.25
(IV)	25.64	25.64	25.64 [a]	7.14
(V)	31.01	31.01	31.01 [a]	18.50
(VI)	−3.67	−7.29	13.00	7.78

[a] Estimated by assuming no $T\Delta S$ contribution at transition state.

TABLE VII

THE CALCULATED THERMODYNAMIC PARAMETERS (ΔH AND ΔG) FOR THESE SIX FUNDAMENTAL REACTION STEPS AT 250 °C AND WITH BP86 ENERGIES

Reaction step	$\Delta H^{\neq}$	ΔH	$\Delta G^{\neq}$	ΔG
(I)	10.02	4.12	33.13	27.06
(II)	20.15	20.15	20.15 [a]	−0.24
(III)	−0.50	−13.54	5.28	−9.46
(IV)	25.42	25.42	25.42 [a]	2.79
(V)	31.11	31.11	31.11 [a]	15.53
(VI)	−3.38	−6.84	16.91	11.29

[a] Estimated by assuming no $T\Delta S$ contribution at transition state.

TABLE VIII

THE CALCULATED THERMODYNAMIC PARAMETERS (ΔH AND ΔG) FOR THESE SIX
FUNDAMENTAL REACTION STEPS AT 150 °C, WITH BP86 ENERGIES, AND WITH
CORRECTION FOR HIGH ALKANE PRESSURE AND LOW H_2 AND OLEFIN PRESSURE

Reaction step	$\Delta H^{\neq}$	ΔH	$\Delta G^{\neq}$	ΔG
(I)	9.83	3.79	22.27	16.21
(II)	19.93	19.93	19.93 [a]	−5.49
(III)	−0.37	−13.58	4.18	−10.25
(IV)	25.64	25.64	25.64 [a]	−1.99
(V)	31.01	31.01	31.01 [a]	9.37
(VI)	−3.67	−7.29	6.57	1.35

[a] Estimated by assuming no $T\Delta S$ contribution at transition state.

TABLE IX

THE CALCULATED THERMODYNAMIC PARAMETERS (ΔH AND ΔG) FOR THESE SIX
FUNDAMENTAL REACTION STEPS AT 250 °C, WITH BP86 ENERGIES AND WITH
CORRECTION FOR HIGH ALKANE PRESSURE AND LOW H_2 AND OLEFIN PRESSURE

Reaction step	$\Delta H^{\neq}$	ΔH	$\Delta G^{\neq}$	ΔG
(I)	10.02	4.12	24.97	18.90
(II)	20.15	20.15	20.15 [a]	−7.16
(III)	−0.50	−13.54	5.28	−9.46
(IV)	25.42	25.42	25.42 [a]	−4.13
(V)	31.11	31.11	31.11 [a]	8.61
(VI)	−3.38	−6.84	8.75	3.13

[a] Estimated by assuming no $T\Delta S$ contribution at transition state.

5 kcal/mol lower in energy for both the barrier and the endothermicity
(see Table V). From our results on the related $CpIr(PR_3)Me+CH_4$ system,
we expect that the BP86 energies will be closer to the most accurate
CCSD values.

Although the transfer reaction can occur at 100 °C, the anthraphos
iridium complex does not begin to show catalytic activity until 150 °C
and continues to be stable to 250 °C. Therefore, we have made
temperature corrections to 150 °C (423 K) in Table VI and to 250 °C
(523 K) in Table VII. Compared to STP values, the free-energy barriers
for this reaction increase by 5.6 kcal/mol for 423 K and 10.0 kcal/mol
for 523 K. As expected, the enthalpies ($\Delta H^{\neq}$ and ΔH) hardly change
(< 0.5 kcal/mol). One can also make corrections for the fact that the

concentrations of the various species involved are not at standard conditions (1 atm, 1 M). The corrections for high alkane concentration (effective "pressure" = 244.66 atm) are 6.43 (423 K) and 8.16 kcal/mol (523 K) (36,37). As shown in Tables VIII and IX for 423 and 523 K, respectively, the pressure corrections lower the free-energy barriers and ΔG values.

Reaction step II is the reductive elimination of dihydrogen from **5** to form the Ir(III) alkyl-hydride species (L)IrHR (**6**). This is a very endothermic elimination reaction (14.14 and 19.41 kcal/mol at STP with the B3LYP and BP86 methods, respectively). No conventional transition state is found for this elimination; in other words, the recombination reaction of (**6**) and H_2 has no enthalpic barrier. This type of recombination reaction is often regarded as having a significant entropic barrier, whose magnitude depends on the degree of entropy lost in forming the "variational" transition state. This lack of a barrier has been reported in previous studies on the pincer complex (21,22). In order to estimate the barrier for the forward reaction (loss of H_2), we have assumed that all of the endothermicity contributes to the free-energy barrier, i.e. the entropy contribution arising from separation of the H_2 molecule does not contribute to the free-energy until it begins to decrease (38). The free-energy of the reaction, shown in Tables VIII and IX, has also been corrected for low hydrogen pressure, 9.13 (423 K) and 6.92 kcal/mol (523 K) (36,37). Since H_2 is a product, this correction only affects the ΔG values. After both high temperature and low-pressure corrections, this reaction step becomes exoenergetic, but retains a substantial free-energy barrier.

Reaction step III involves rearrangement to put the β-H close to the metal then β-H transfer to make the olefin complex (L)IrH$_2$(olefin) (**7**) from the alkyl-hydride complex (L)IrHR (**6**). This exothermic reaction has a relatively small barrier ($\Delta H^{\neq} = 3.7$ and $\Delta H^{\circ} = -10.05$ kcal/mol (B3LYP, STP)). Since this reaction step only involves the intramolecular rearrangement, there is little entropic contribution and the free energies are similar to the enthalpies (see Table IV). Although Table V shows a negative $\Delta H^{\neq}$ barrier at the BP86 level, this "abnormality" arises because there are two rearrangement steps which proceed the β-H transfer. One is to rotate about the M–C$_\alpha$ bond to bring the C$_\beta$ close to the metal, and the other is to rotate about the C$_\alpha$–C$_\beta$ bond to bring the H–C$_\beta$ to an eclipsed position. The TS states and intermediates of both steps involve small energy changes, less than 1.5 kcal/mol. Thus, the "actual" intermediate just before the TS for the β-H transfer step (C$_\beta$ close to metal, eclipsed conformation) is 0.55 kcal/mol more stable than the TS listed in the table. Like B3LYP, BP86 shows a very exothermic reaction (-13.54 kcal/mol, STP, Table V) for olefin complex formation. The calculated free-energy surface gives a barrier of 2.87 kcal/mol and ΔG° value of -11.23 kcal/mol (BP86, STP, Table V). Because this reaction step is a rearrangement with an equal number of reactants and products, little change of the free-energy occurs when corrections from STP to reaction conditions are made (from Table VI to Table IX).

Reaction step IV is the final step in the dehydrogenation; olefin is lost from $(L)IrH_2(olefin)$ (7) to form the starting species $(L)IrH_2$ (4). Like the loss of H_2 in step II, this step is endothermic but has no additional enthalpic barrier. At STP, the B3LYP method gives 20.82 kcal/mol free-energy barrier (again estimated as $\Delta G^{\neq} = \Delta H^{\circ}$) and the calculated ΔG° is 7.68 kcal/mol (see Table IV). When corrected for high temperature and low olefin pressure, this reaction step also becomes exoenergetic.

Reaction step V also results in H_2 loss; however, unlike reaction step II, step V starts with the reactant $(L)IrH_2$ (4), an Ir(III) species and forms the Ir(I) species $(L)Ir$ (8). At STP conditions, B3LYP method gives a 27.54 kcal/mol barrier, (the endothermicity assuming no entropy gain) and a calculated ΔG° of 18.95 kcal/mol (Table IV), which reflects the entropy contribution due to the loss of H_2. As anticipated, the BP86 method gives values more positive by ~ 3 kcal/mol (Table V). Higher reaction temperature and low H_2 pressure corrections reduce the endoenergicity significantly but not the barrier (Tables VIII and IX).

Reaction step VI is the oxidative addition of alkane to the Ir(I) species $(L)Ir$ (8) to form the alkyl-hydride $(L)IrHR$ (6). This step has a relatively small enthalpic barrier of 1.09 kcal/mol (B3LYP, STP) and is exothermic by -4.03 kcal/mol. Because entropy is lost in this associative step, the calculated $\Delta G^{\neq}$ is 13.09 kcal/mol and ΔG° is 6.98 kcal/mol (Table IV). Although Table V shows that the BP86 method apparently gives a negative barrier (-3.98 kcal/mol), these systems form σ-bond alkane complexes and the barrier from this "agostic" complex is positive (3.10 kcal/mol). Regardless of this small barrier, the entropic contribution remains the main contribution to the free-energy barrier: 8.02 (BP86, STP), 6.57 (BP86, 150 °C, pressure corrected) and 8.75 kcal/mol (BP86, 250 °C, pressure corrected).

Reaction step VII involves the concerted oxidative addition of the incoming alkane and reductive elimination of the dihydrogen. Generally, interchange steps are often classified as either associative interchange or dissociative interchange. A preliminary identification of an interchange transition state has been made, but it appears that this TS leads to a dihydrogen intermediate, a σ-bond complex, which will subsequently need to eliminate H_2 to complete the reaction. Thus, with the transition states determined to date, one cannot confirm that this route will provide any advantage. Values listed in Table IV for reaction VII are in parenthesis to indicate their tentative nature.

IV. Transfer Reaction

The transfer reaction utilizes a sacrificial alkene to remove the dihydrogen from the pincer or anthraphos complex first, before the oxidative addition of the target alkane. The elementary reaction steps are slightly different from the thermal reaction, which is discussed in the next section, both in their order and their direction. For simplicity, we describe the symmetric reaction where the sacrificial alkene is ethylene and the reactant is ethane ($21b$). The elementary reaction steps for the mechanism of this transfer reaction involve IV^R, III^R, VI^R, VI, III and IV, where the superscript "R" stands for the reverse of the elementary steps listed in Section III. These reverse steps (IV^R, III^R, and VI^R) involve the sacrificial alkene extracting dihydride from the metal to create the Ir(I) species 8, while steps VI, III and IV involve oxidative addition of target alkane, β-H transfer and olefin loss.

The calculated thermodynamic parameters in Table X are for 150 °C, a value close to typical operating temperatures, and without pressure corrections, there is no H_2 and both alkane and alkene could be at unit concentration. The bottom part of the table is the same as Table VI. The free-energy profile of the transfer reaction mechanism is shown in Fig. 4. In the first step of the mechanism (IV^R), sacrificial alkene (in this case ethylene) was bound to the (L)Ir(H)$_2$ complex 4 to form an olefin complex 7. As mentioned above, the forward reaction of this step has no

TABLE X

THE CALCULATED THERMODYNAMIC PARAMETERS (ΔH AND ΔG)
FOR THE REACTION STEPS IN THE TRANSFER REACTION AT 150 °C,
WITH BP86 ENERGIES AND NO PRESSURE CORRECTIONS WERE MADE
FOR THE TRANSFER REACTION

Reaction Step	$\Delta H^{\neq}$	ΔH	$\Delta G^{\neq}$	ΔG
(IV^R)	0.00	−25.64	18.50	−7.14
(III^R)	13.21	13.58	14.43	10.25
(VI^R)	3.62	7.29	5.22	−7.78
(VI)	−3.67	−7.29	13.00	7.78
(III)	−0.37	−13.58	4.18	−10.25
(IV)	25.64	25.64	25.64	7.14

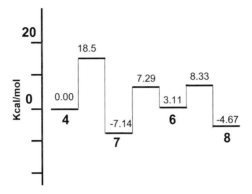

FIG. 4. The free-energy (G) profile of the transfer-reaction mechanism at 423 K with the BP86 method and without any pressure corrections. Reaction going from left to right uses up the sacrificial olefin and from right to left generates the target olefin.

enthalpic barrier and $\Delta G^{\neq}$ is assumed to be equal to ΔH°, which is 25.64 kcal/mol. This number now becomes the exothermicity of the reaction and the free-energy barrier comes from the entropy lost on forming the olefin complex. The next step (III^R) involves reaction of the bound olefin with one of the two hydride ligands to make the alkyl-hydride complex **6**. This reaction step has a 14.43 kcal/mol free-energy barrier and the intermediate product (**6**) is 10.25 kcal/mol less stable than the olefin species **7**. In the next step (VI^R) the sacrificial alkane is reductively eliminated to produce the (L)Ir species **8**. This exothermic step completes the transfer of dihydride from the metal to the sacrificial alkene over a small barrier. Upon formation of the Ir(I) species **8**, the reaction is ready for the conversion of the target alkane. The next three steps are symmetric with respect to the three reverse steps discussed above and can be followed in the figure by starting at the right side and moving back to the left. Thus, step VI oxidatively adds the target alkane, step III transfers a β hydrogen, and step IV eliminates the olefin. This last step provides the highest free-energy barrier in the transfer reaction.

V. Acceptorless Reaction

A recent study shows that the (^{R}AP)IrH$_2$ complex can catalytically convert alkane to alkene at temperatures ranging from 150 to 250 °C through an acceptorless, thermal reaction. Three suggested reaction mechanisms, associative (A), dissociative (D) and interchange (I) will be compared in the following sections. All of these mechanisms lead to the same key intermediate, the alkyl-hydride (L)Ir(R)(H), **6**. From this key intermediate all mechanisms, including the transfer reaction, follow the same sequence of elementary steps (III then IV). Thus, the alkyl-hydride proceeds through a series of low energy intermediates (not discussed in detail here) to align a β hydrogen for transfer to the iridium and formation of the olefin–dihydride complex **7**. Olefin elimination then regenerates the starting catalyst, the dihydride complex **4**. Although this last step has one of the highest enthalpic barriers at STP, both the entropic and pressure corrections favor the loss of olefin.

A. ASSOCIATIVE MECHANISM (A)

The first reaction step of the associative mechanism is the association step itself, i.e. the oxidative addition of alkane to form alkyl-trihydride Ir(V) species **5** (reaction step I). This mechanism then proceeds through reaction steps II, III and IV. Figure 5 shows the free-energy profile for this sequence of reactions at 150 °C with pressure corrections (also see Table VIII). Reaction step I (formation of the Ir(V) species **5**) has the largest single barrier; this step is followed by step II with another fairly large barrier for H_2 loss to form an alkyl-hydride Ir(III) species **6**. Species **6** is ~ 6 kcal/mol more stable than species **5**, and only ~ 11 kcal/mol less stable than species **4** plus ethane. Although the results reported here are for the anthraphos complex, similar barriers are found for the pincer complex (*21,22*).

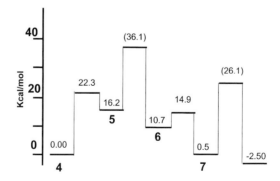

FIG. 5. The free-energy (G) profile of the associative mechanism at 423 K with the BP86 method and with pressure correction. Numbers in parenthesis are the highest estimated barrier for the H_2/olefin loss.

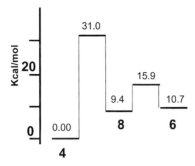

FIG. 6. The free-energy (G) profile of the dissociative mechanism at 423 K with the BP86 method and with pressure corrections. From **6** the reaction continues as in Fig. 5.

B. DISSOCIATIVE MECHANISM (D)

The first reaction step of the dissociative mechanism is the reductive elimination of H_2 from species **4** (step V). This step is followed by oxidative addition of alkane (step VI) to form the common intermediate **6**. Then, reaction steps III and IV complete the cycle. Figure 6 shows the free-energy profile at $150\,^\circ C$ with pressure corrections for this dissociative mechanism (also see Table VIII). Like H_2 loss from **5** (above), H_2 loss from **4** has no additional barrier beyond its endothermicity. So the free-energy barrier of reaction step V equals the endothermicity of this step, $\sim 31\,kcal/mol$. This barrier is $\sim 11\,kcal/mol$ higher than that in reaction step II, in which H_2 is lost from the Ir(V) species, (L)IrH$_3$R (**5**). The correction for low dihydrogen pressure reduces ΔG for step V by $\sim 9\,kcal/mol$, but the step retains a high kinetic barrier. The next reaction step is oxidative addition of alkane, in which the oxidation state of Ir changes from Ir(I) back to Ir(III). This reaction has a relatively small free-energy barrier before producing the common intermediate species **6**.

C. INTERCHANGE MECHANISM (I)

Comparing the free-energy profiles in Figs. 5 and 6 one finds similar high-energy points at $\sim 30\,kcal/mol$. An interchange mechanism would be an attractive alternative if it could take advantage of the favorable entropy for H_2 loss and avoid the steric problem from the high coordination number of **5** (*13*). In the interchange mechanism the reaction would proceed directly from **4** through a single transition state to **6**. However, all TS thus far determined lead to an intermediate with a fairly strongly bound H_2, which leads to a free-energy profile similar to that of Fig. 6.

VI. Geometric Factor

A particular geometric coordination at the metal can contribute to catalytic properties of these and other systems. Typically, multidentate ligands are used to restrict the allowed coordination geometry. For example, both pincer and anthraphos ligands impose a *mer* coordination geometry about the metal by linking a pair of *trans* phosphorus donors through a rigid arene ring. Such geometries turn out to be an important feature of these catalysts. Hall and coworkers (*21*) have shown that the *mer* geometry, as opposed to the *fac* geometry, stabilizes the 16 e⁻ Ir(III) complexes, lowers the barrier for reductive elimination from Ir(V), and weakens the metal–olefin interaction, all important aspects of these systems. These effects arise from the relative energy of the singlet and triplet state in the transition metal fragment; a *mer* geometry produces a lower energy singlet and a higher energy triplet than a *fac* geometry. Thus, the constrained geometry imposed by pincer and anthraphos ligands is particularly important to this catalytic reaction.

VII. Reaction Conditions

From the energy profiles of the associative and dissociative mechanisms (above), one can see that the critical energy barriers are oxidative addition of alkane, reductive elimination of dihydrogen, and olefin loss. At low temperature, the first associative step seems to be favored, while at elevated temperature, entropy contributions begin to favor the first dissociative step (*22*). Fig. 7, compares the free-energy profiles for both mechanisms at room temperature (298 K) and the reaction temperature of 523 K. At higher temperature, the associative pathway (Fig. 7a) shows increases in free energies (~ 10 kcal/mol) for the transition states and intermediate **5**, while intermediate **6** shows a slight increase. Interestingly, at the higher temperature, intermediate **8** in the dissociative mechanism has a significantly lower free-energy (~ 7 kcal/mol), while the free-energy at its transition state shows only small changes. From Fig. 8 one can see the effect of the pressure corrections (high alkane pressure and low olefin and dihydrogen pressure) on the

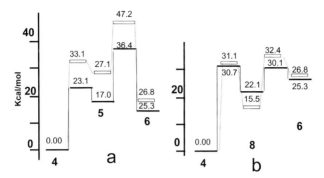

FIG. 7. The BP86 free-energy (G) profiles without pressure corrections from 298 to 523 K for associative (a) and dissociative (b) pathways (solid line is 298 K, open bar is 523 K).

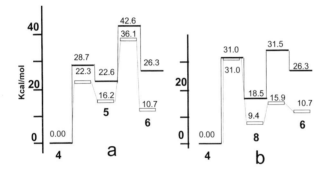

FIG. 8. The BP86 free-energy (G) profiles at 423 K with and without pressure corrections for associative (a) and dissociative (b) pathways (solid line is without pressure corrections, open bar is with pressure corrections).

associative and dissociative paths. While the barriers for the associative path are lowered significantly by these corrections, the key barrier on the dissociative path remains unchanged. From these changes it is clear that at a sufficiently high temperature the reaction will proceed by the dissociative mechanism. At lower temperatures and sufficiently high alkane pressure/concentration the reaction could proceed by the associative mechanism. However, the prediction of the most favorable path is still too dependent on the methodology to make a firm prediction. In fact the dynamics of the reaction may sample the entire range of these pathways and others such as the interchange pathway, mentioned above, or rebound pathways not discussed here.

TABLE XI

THE RELATIVE ENERGIES (ΔE) OF KEY SPECIES WITH PH_2
AND $P(^tBu)_2$ MODELS (BP86)

	4	5	6	8
PH_2 model	0.00	3.80	26.28	30.85
$P(^tBu)_2$ model	0.00	19.77	30.50	26.70

VIII. Model with *tert*-butyl Phosphine

Here again, we will focus on the steps between **4** and **6** as the remaining steps are identical for both mechanisms. Due to the large size of the system with ^tBu substituents, we carried out the geometry optimization with the key intermediates (**4**, **5**, **6**, and **8**) rather than the transition states.

The results in Table XI show that when ^tBu's replace H's, **5** is strongly destabilized, **6** is weakly destabilized, and **8** is weakly stabilized relative to **4** plus ethane. The calculated energy differences agree with the expectations of steric arguments. Thus, (L)IrRH$_3$, **5**, is much more crowded than (L)IrH$_2$, **4**, while (L)IrRH, **6**, and (L)Ir, **8**, are slightly more and less crowded, respectively. Similar results are reported in the other studies (*22*). Because the TS geometry of the associative step resembles **5**, and the TS geometry of the dissociative step resembles **8**, one expects similar corrections for the steric effects of phosphine susbstituents at these transition states. The largest correction (~ 16 kcal/mol) is for the associative Ir(V) intermediate, a result which suggests that the first barrier in the associative mechanism will rise significantly and could become the rate determining step for this mechanism. Of course, this correction adds further to the favorable aspects of the dissociative mechanism as the dominant route.

IX. Anthraphos Rhodium Complex

Table XII contains the enthalpy and free-energy differences of the critical intermediate species for the anthraphos Rh catalyst. Although no experimental results are available yet, our predicted energies show a much smaller dissociation energy for H_2 loss, the first step in the dissociative mechanism (**8** is more stable by ~ 13 kcal/mol relative to **4**). In contrast, the oxidative addition intermediate of the first step in the associative mechanism, the M(V) species **5**, is ~ 14 kcal/mol less stable

TABLE XII

THE BP86 RELATIVE ENERGIES OF CRITICAL INTERMEDIATE
SPECIES AT STP BETWEEN ANTHRAPHOS Rh AND Ir
COMPLEXES (kcal/mol)

		4	5	6	8
Ir	H	0.00	9.37	23.51	27.54
	G	0.00	22.88	25.93	18.95
Rh	H	0.00	25.04	34.54	14.25
	G	0.00	37.43	38.27	5.80

(relative to the respective starting complex) for Rh than for Ir. These
results are consistent with the known behavior of third vs. second
row transition metals. From these results one can predict that the Rh
catalyst would prefer the dissociative pathway to the associative one.
If the dehydrogenation reaction proceeds dissociatively, would the Rh
complex be a better catalyst?

X. Conclusions

The thermodynamic parameters for the alkane dehydrogenation
reaction are calculated for both the pincer and anthraphos iridium(III)
complexes. The mechanism of the transfer reaction, and the associative,
dissociative and interchange mechanisms for the acceptorless reactions
are discussed and compared. As these reactions typically occur at
conditions very different from STP, important corrections for high
temperature, high reactant (alkane) concentration and low product
(H_2, olefin) concentration are important.

Neglecting the small barriers in simple rearrangement steps and the
weakly bound σ-bond (agostic) complexes, these reactions can be
described by seven elementary reaction steps, I to VII, and the different
mechanisms can be described as different combinations of these seven
reaction steps. One key intermediate, the 16-electron (L)Ir(III)(H)
(alkyl) (6), arises in all of the acceptorless mechanisms and in the
transfer mechanism. This intermediate proceeds through a β-hydrogen
transfer (step III) to form an olefin complex (7), which then loses olefin
(step IV) to regenerate the starting complex (4). Although the barrier
for the olefin loss is rather high, entropy contributions reduce its free-
energy barrier at higher temperatures so that it is rate determining
only for the transfer reaction.

In the acceptorless (thermal) reaction, the choice of mechanism, associative vs. dissociative, is governed by the first two steps. With corrections for both entropy and concentration, one finds that the highest free-energy critical points are on the associative-mechanism profile. Thus, the calculations predict that the dissociative mechanism is favored. However, the difference is still rather small for calculations of this precision. Therefore, multiple or alternative reaction pathways may be available to a particular catalyst under specific conditions. The rigid *mer* geometries enforced by the pincer or anthraphos ligands are critical to the catalytic activity of these systems as they prevent the system from being trapped in too stable an intermediate. Although the PH_2 model is a reasonable choice for most of these calculations and usually provides qualitatively correct electronic results, the calculations with $P(^tBu)_2$ show that there is a substantial steric factor, further favoring the dissociative path.

The major difference between this work and similar calculations by Krogh-Jespersen *et al.* (22) is the choice of method for the final energy calculations, B3LYP vs. BP86. We choose the latter because it appears to give better results for a similar well studied reaction. The principal difference for the mechanisms that arises from our choice is that the steps for reductive elimination have higher barriers, while those for oxidative additions have lower barriers.

XI. Computational Details

All the intermediates and transition states (TS) are optimized using density functional theory with the B3LYP functional (39) implemented in a GAUSSIAN 98 program (40). All of the TS obtained in studies were optimized by a quasi-Newton method (41), and verified by separated frequency calculations, which showed only one imaginary frequency (42). Single point energies of these intermediates and TS in the anthraphos complex are also calculated at the BP86 level (39,43).

The model used in anthraphos complex is $(^HAp)IrH_2$, $(^HAp = \eta^3$-$C_{14}H_7(PH_2)_2$-3,13). The basis set for Ir was the modified LANL2DZ of Couty and Hall (44), whose outermost p orbital was replaced by (41) split of an optimized 6p function. The basis set for P was the standard LANL2DZ basis set augmented by a d-type polarization function (45). The effective core potential (ECP) was used for both Ir and P (44a–b). The carbon and hydrogen atoms bonded to Ir were described by the all electron Dunning-Huzinaga double-zeta basis set with polarization functions (D95**) (46), and the STO-3G basis set was used for the uncoordinated carbons and hydrogens in the ^{H}Ap ligand. We also optimized the geometry of starting complex 1 with the 3-21G basis set (47) for the uncoordinated carbons and hydrogens in the ^{H}Ap ligand, no geometrically significant difference was found compared to that with the STO-3G basis set. For the C and H of ethane, ethene, and H_2 ligands, a D95** basis set was used.

Because the calculations for the $(^{H}PCP)IrH_2$ model, where $^{H}PCP = \eta^3$-$C_6H_3(CH_2PH_2)_2$-1,3, were done at earlier times, the basis set used for the ligand was slightly different. The basis sets for Ir and P were the same as described above. In the study of the acceptorless reaction (21a), the Dunning-Huzinaga double-zeta basis set with polarization functions (D95**) (46) was used for the metal-coordinated carbons and hydrogens, and the STO-3G basis set was used for the uncoordinated carbons and hydrogens in the ^{H}PCP ligand. In the study of the transfer reaction (21b), the 6-31G(d, p) basis sets (48) were used for the metal-coordinated carbons and hydrogens, and the 3-21G basis set (47) was used for the uncoordinated carbons and hydrogens in ^{H}PCP ligand.

For the related $[CpIr(PH_3)(CH_3)]^+$ system, four basis sets were used. Basis set one (BS1) is the same as the ones described above for Ir and P, but the C and H are described as D95**. Basis set two (BS2) is the Stuttgart relativistic, small core ECP basis set (49) augmented with a polarization function for Ir, and Dunning's correlation consistent double-zeta basis set with polarization function (50) for P, C and H. Basis set three (BS3) is the same as BS1 except the d-orbital of Ir was described by further splitting into triple-zeta (111) from a previous double-zeta (21) description and augmented with a f-polarization function (51). Basis set four (BS4) is the same as BS2 for Ir, P, and most of the C and H, but the C and H atoms involved in the oxidative addition were described with Dunning's correlation consistent triple-zeta basis set with polarization.

REFERENCES

1. (a) Jia, C.; Kitamura, T.; Fujiwara, Y.; Acc. Chem. Res. 2001, 34, 633–639; (b) Slugovc, C.; Padilla-Martinez, I.; Sirol, S.; Carmona, E. Coord. Chem. Rev. 2001, 213, 129; (c) Kakiuchi, F.; Murai, S. "Activation of Unreactive Bonds and Organic Synthesis"; Ed. Murai, S.; Springer: Berlin, 1999, p. 47; (d) Shilov, A. E.; Shul'pin, G. B. Chem. Rev. 1997, 97, 2879; (e) Arndtsen, B. A.; Bergman, R. G.; Mobley, T. A.; Peterson, T. H. Acc. Chem. Res. 1995, 28, 154–162; (f) Hall, C. L. "Activation and Functionalization of Alkanes"; John Wiley & Sons: New York, 1989; (g) Crabtree, R. H. Chem. Rev. 1985, 85, 245–269.

2.   (a) For example, the ethane dehydrogenation requires a barrier of 123.9 kcal/mol and
     is endothermic by 33.0 kcal/mol with zero-point energy corrections; (b) Bromberg,
     S. E.; Yang, H.; Asplund, M. C.; Lian, T.; McNamara, B. K.; Kotz, K. T.; Yeston, J. S.;
     Wilkens, M.; Frei, H.; Bergman, R. G.; Harris, C. B. *Science* **1997**, *278*, 260–263.
3.   (a) Takaya, H.; Murahashi, S. I. *Synlett* **2001**, *1*, 991–994; (b) Busch, S.; Leitner, L. *Adv.
     Synth. Cat.* **2001**, *343*, 192–195.
4.   (a) Gusev, D. G.; Dolgushin, F. M.; Antipin, M. Y. *Organometallics* **2001**, *20*, 1001–1007;
     (b) Gauvin, R. M.; Rozenberg, H.; Shimon, L. J. W.; Milstein, D. *Organometallics*
     2001, *20*, 1719–1724; (c) Barrio, P.; Castarlenas, R.; Esteruelas, M. A.; Onate, E.
     Organometallics **2001**, *20*, 2635–2638; (d) Wen, T. B.; Cheung, Y. K.; Yao, J.; Wong,
     W. T.; Zhou, Z. Y.; Jia, G. *Organometallics* **2000**, *19*, 3803–3809.
5.   (a) Sundermann, A.; Uzan, O.; Martin, J. M. L. *Organometallics* **2001**, *20*, 1783–1791;
     (b) Grushin, V. V.; Marshall, W. J.; Thorn, D. L. *Adv. Synth. Cat.* **2001**, *343*, 161–
     165; (c) Woodmansee, D. H.; Bu, X. H.; Bazan, G. C. *Chem. Commun.* **2001**, 619–620;
     (d) Davies, H. M. L.; Antoulinakis, E. G. *J. Organomet. Chem.* **2001**, *617*, 47–55.
6.   (a) Golden, J. T.; Andersen, R. A.; Bergman, R. G. *J. Am. Chem. Soc.* **2001**, *123*, 5837–
     5838; (b) Peterson, T. H.; Golden, J. T.; Bergman, R. G. *J. Am. Chem. Soc.* **2001**, *123*,
     455–462; (c) Alaimo, P. J.; Arndtsen, B. A.; Bergman, R. G. *Organometallics* **2000**, *19*,
     2130–2143; (d) Tani, K.; Yamagata, T.; Kataoka, Y.; Mashima, K.; *J. Synth. Org.
     Chem. Japan* **1999**, *57*, 656–666; (e) Grigoryan, E. A. *Kinet. Cat.* **1999**, *40*, 350–363.
7.   (a) Martin-Matute, B.; Mateo, C.; Cardenas, D. J.; Echavarren, A. M. *Chem. Eur. J.*
     2001, *7*, 2341–2348; (b) Catellani, M.; Cugini, F.; Tiefenthaler, D.; *Can. J. Chem.* **2001**,
     79, 742–751; (c) Fang, X. G.; Scott, B. L.; Watkin, J. G.; Kubas, G. J. *Organometallics*
     2000, *19*, 4193–4195; (d) Catellani, M.; Motti, E.; Ghelli, S. *Chem. Commun.* **2000**,
     2003–2004; (e) Zhuravel, M. A.; Grewal, N. S.; Glueck, D. S.; Lam, K. C.; Rheingold,
     A. L. *Organometallics* **2000**, *19*, 2882–2890.
8.   (a) Johansson, L.; Ryan, O. B.; Romming, C.; Tilset, M. *J. Am. Chem. Soc.* **2001**, *123*,
     6579–6590; (b) Fekl, U.; Kaminsky, W.; Goldberg, K. I. *J. Am. Chem. Soc.* **2001**, *123*,
     6423–6424; (c) Reinartz, S.; White, P. S.; Brookhart, M.; Templeton, J. L. *J. Am.
     Chem. Soc.* **2001**, *123*, 6425–6426; (d) Thomas, J. C.; Peters, J. C. *J. Am. Chem. Soc.*
     2001, *123*, 5100–5101; (e) Reinartz, S.; White, P. S.; Brookhart, M.; Templeton, J. L.
     Organometallics **2001**, *20*, 1709–1712; (f) Falvello, L. R.; Fernandez, S.; Larraz, C.;
     Llusar, R.; Navarro, R.; Urriolabeitia, E. P. *Organometallics* **2001**, *20*, 1424–1436.
9.   (a) Crabtree, R. H.; Mihelcic, J. M.; Quirk, J. M. *J. Am. Chem. Soc.* **1979**, *101*, 7738-7740
     (b) Baudry, D.; Ephritikhine, M.; Felkin, H.; Holmes-Smith, R. *Chem. Comm.* **1983**,
     788–789; (c) Maguire, J. A.; Goldman, A. S. *J. Am. Chem. Soc.* **1991**, *113*, 6706–6708;
     (d) Fujii, T.; Higashino, Y.; Satio, Y. *J. Chem. Soc. Dalton Trans.* **1993**, 517–520;
     (e) Belli, J.; Jensen, C. M. *Organometallics* **1996**, *15*, 1532–1534.
10. (a) Burk, M. J.; Crabtree, R. H.; Parnell, C. P.; Uriarte, R. J. *Organometallics* **1984**,
     3, 816–817; (b) Burk, M. J.; Crabtree, R. H.; McGrath, D. V. *Chem. Comm.* **1985**,
     1829–1830; (c) Burk, M. J.; Crabtree, R. H. *J. Am. Chem. Soc.* **1987**, *109*, 8025–8032;
     (d) Nomura, K.; Saito, Y. *Chem. Comm.* **1988**, 161; (e) Nomura, K.; Saito, Y. *J. Mol.
     Catal.* **1988**, *5*, 57; (f) Sakakura, T.; Sodeyama, T.; Tanaka, M. *New. J. Chem.* **1989**, *13*,
     737; (g) Maguire, J. A.; Boese, W. T.; Goldman, A. S. *J. Am. Chem. Soc.* **1989**, *111*,
     7088; (h) Sakakura, T.; Sodeyama, T.; Abe, F.; Tanaka, M. *Chem. Lett.* **1991**, 297.
11. (a) Gupta, M.; Hagen, C.; Flesher, R. J.; Kaska, W. C.; Jensen, C. M. *Chem. Commun.*
     1996, 2083; (b) Gupta, M.; Hagen, C.; Kaska, W. C.; Cramer, R. E.; Jensen, C. M. *J.
     Am. Chem. Soc.* **1997**, *119*, 840–841; (c) Gupta, M.; Kaska, W. C.; Jensen, C. M. *Chem.
     Commun.* **1997**, 461; (d) Crabtree, R. H.; Mihelcic, J. M.; Quirk, J. M. *J. Am. Chem.
     Soc.* **1979**, *101*, 7738; (e) Baudry, D.; Ephritikhine, M.; Felkin, H.; Holmes-Smith, R.

Chem. Comm. **1983**, 788; (f) Maguire, J. A.; Goldman, A. S. *J. Am. Chem. Soc.* **1991**, *113*, 6706; (g) Fujii, T.; Higashino, Y.; Satio, Y. *J. Chem. Soc. Dalton Trans.* **1993**, 517; (h) Belli, J.; Jensen, C. M. *Organometallics* **1996**, *15*, 1532; (I) Lee, D. W.; Kaska, W. C.; Jensen, C. M. *Organometallics* **1998**, *17*, 1; (j) Jensen, C. M. *Chem. Commun.* **1999**, 2443; (k) Xu, W. W.; Rosini, G. P.; Gupta, M.; Jensen, C. M.; Kaska, W. C.; Krogh-Jespersen, K.; Goldman, A. S. *Chem. Commun.* **1997**, 2273; (l) Liu, F.; Goldman, A. S. *Chem. Commun.* **1999**, 655.

12. (a) Crabtree, R. H.; Mihelcic, J. M.; Qirk, J. M. *J. Am. Chem. Soc.* **1979**, *101*, 7738; (b) Baudry, D.; Ephrittikhine, M.; Felken, H. *J. Chem. Soc. Chem. Commun.* **1980**, *24*, 1243; (c) Bianchi, F.; Gallazzi, M. C.; Porri, L.; Diversi, P. *J. Organomet. Chem.* **1980**, *202*, 99.

13. (a) Oevers, S.; Haenel, M. W.; Kaska, W. C. 219th ACS National Meeting, Inorg. Paper #144, March 26–30, **2000**, San Francisco, California; (b) Haenel, M. W.; Oevers, S.; Angemund, K.; Kaska, W. C.; Fan, H. J.; Hall, M. B. *Angew. Chem. Int. Ed.* **2001**, *40*, 3596; (c) "Anthraphos" has previously been used to refer to the chiral ligand *trans*-11,12-bis(diphenylphosphino)-9,10-dihydro-9,10-ethanoanthracene; (d) Recent communication at 224th ACS National meeting in Boston, **2002** shows that anthraphos Ir(III) complex also shows catalytic property at 150 °C.

14. Moulton, C. J.; Shaw, B. L. *J. Chem. Soc. Dalton Trans.* **1976**, 1020.

15. Jensen, C. M. *Chem. Commun.* **1999**, 2443–2449 and references therein.

16. In following discussion, we are referring to the entire class of ligands when R is unspecified. The same notation is used for anthraphos ligand.

17. Sundermann, A.; Uzan, O.; Milstein, D.; Martin, J. M. L. *J. Am. Chem. Soc.* **2000**, *122*, 7095–7104.

18. Gusev, D. G.; Dolgushin, F. M.; Antipin, M. Y. *Organometallics* **2000**, *19*, 3429–3434.

19. Gandelman, M.; Vigalok, A.; Shimon, L. J. W.; Milstein, D. *Organometallics* **1997**, *16*, 3981.

20. Gerisch, M.; Krumper, J. R.; Bergman, R. G.; Tilley, T. D. *J. Am. Chem. Soc.* **2001**, *123*, 5818–5819

21. (a) Niu, S.; Hall, M. B. *J. Am. Chem. Soc.* **1999**, *121*, 3992–3999; (b) Li, S.; Hall, M. B. *Organometallics* **2001**, *20*, 2153, erreta: **2001**, *20*, 3210; (c) Fan, H.-J.; Hall, M. B. *J. Mol. Catal. A* **2002**, *189*, 111–118.

22. (a) Krogh-Jespersen, K.; Czerw, M.; Kanzelberger, M.; Goldman, A. S. *J. Chem. Infor. Comp. Sci.* **2001**, *41*, 56; (b) Krogh-Jespersen, K.; Czerw, M.; Summa, N.; Renkema, K. B.; Achord, P. D.; Goldman, A. S. *J. Am. Chem. Soc.* **2002**, *124*, 11 404–11 416; (c) Krogh-Jespersen, K.; Czerw, M.; Goldman, A. S. *J. Mol. Catal. A* **2002**, *189*, 95–110.

23. (a) Arndtsen, B. A.; Bergman, R. G. *Science* **1995**, *270*, 1970; (b) Burger, P.; Bergman, R. G. *J. Am. Chem. Soc.* **1993**, *115*, 10 462; (c) For a review see: Lohrenz, J. C. W.; Hacobsen, H. *Angew. Chem., Int. Ed. Engl.* **1996**, *35*, 1305.

24. (a) Lohrenz, J. C. W.; Hacobsen, H. *Angew. Chem. Int. Ed. Engl.* **1996**, *35*, 1305; (b) Hingerling, C.; Plattner, D. A.; Chen, P. *Angew. Chem. Int. Ed. Engl.* **1997**, *36*, 243; (c) Hinderling, C.; Feichtinger, D.; Plattner, D. A.; Chen, P. *J. Am. Chem. Soc.* **1997**, *119*, 10 793–10 804.

25. (a) Strout, D.; Zaric, S.; Niu, S.; Hall, M. B. *J. Am. Chem. Soc.* **1996**, *118*, 6068–6069; (b) Jimenez-Catano, R.; Niu, S.; Hall, M. B. *Organometallics* **1997**, *16*, 1962; (c) Niu, S.; Hall, M. B. *J. Am. Chem. Soc.* **1998**, *120*, 6169–6170; (d) Niu, S.; Zaric, S.; Bayse, C. A.; Strout, D. L.; Hall, M. B. *Organometallics* **1998**, *17*, 5139–5147; (e) Su, M. D.; Chu, S. Y. *J. Am. Chem. Soc.* **1997**, *119*, 5373; (f) Han, Y. Z.; Deng, L. Q.; Ziegler, T. *J. Am. Chem. Soc.* **1997**, *119*, 5939.

26. (a) Roothaan, C. C. J. *Rev. Mod. Phys.* **1951**, *23*, 69; (b) Pople, J. A.; Nesbet, R. K. *J. Chem. Phys.* **1954**, *22*, 571–574; (c) McWeeny, R.; Dierksen, G. *J. Chem. Phys.* **1968**, *49*, 4852.

27. (a) Møller, C.; Plesset, M. S. *Phys. Rev.* **1934**, *46*, 618; (b) Head-Gordon, M.; Pople, J. A.; Frisch, M. J. *Chem. Phys. Lett.* **1988**, *153*, 503; (c) Frisch, M. J.; Head-Gordon, M.; Pople, J. A. *Chem. Phys. Lett.* **1990**, *166*, 275; (d) Frisch, M. J.; Head-Gordon, M.; Pople, J. A. *Chem. Phys. Lett.* **1990**, *166*, 281; (e) Head-Gordon, M.; Head-Gordon, T. *Chem. Phys. Lett.* **1994**, *220*, 122 (1994); (f) Saebo, S.; Almlof, J. *Chem. Phys. Lett.* **1989**, *154*, 83; (g) Pople, J. A.; Binkley, J. S.; Seeger, R. *Int. J. Quant. Chem. Symp.* **1976**, *10*, 1; (h) Krishnan, R.; Pople, J. A. *Int. J. Quant. Chem.* **1978**, *14*, 91; (i) Raghavachari, K.; Pople, J. A.; Replogle, E. S.; Head-Gordon, M. *J. Phys. Chem.* **1990**, *94*, 5579.

28. (a) Pople, J. A.; Seeger, R.; Krishnan, R. *Int. J. Quant. Chem. Symp.* **1977**, *11*, 149; (b) Krishnan, R.; Schlegel, H. B; Pople, J. A. *J. Chem. Phys.* **1977**, *72*, 4654; (c) Raghavachari, K.; Pople, J. A. *Int. J. Quant. Chem.* **1981**, *20*, 167.

29. (a) Pople, J. A.; Krishnan, R.; Schlegel, H. B.; Binkley, J. S. *Int. J. Quant. Chem. XIV* **1978**, 545–560; (b) Bartlett, R. J.; Purvis, G. D. *Int. J. Quant. Chem.* **1978**, *14*, 516; (c) Cizek, J. *Adv. Chem. Phys.* **1969**, *14*, 35; (d) Purvis, G. D.; Bartlett, R. J. *J. Chem. Phys.* **1982**, *76*, 1910; (e) Scuseria, G. E.; Janssen, C. L.; Schaefer, H. F. III. *J. Chem. Phys.* **1988**, *89*, 7382; (f) Scuseria, G. E.; Schaefer, H. F. III. *J. Chem. Phys.* **1989**, *90*, 3700.

30. Pople, J. A.; Head-Gordon, M.; Raghavachari, K. *J. Chem. Phys.* **1987**, *87*, 5968.

31. Parr, R. G.; Yang, W. Density-Functional Theory of Atoms and Molecules; Oxford University Press: New York, 1989.

32. Niu, S.; Hall, M. B. *J. Phys. Chem. A* **1997**, *101*, 1360–1365.

33. (a) Lin, Z.; Hall, M. B. *J. Am. Chem. Soc.* **1992**, *114*, 2928–2932; (b) Song, J.; Hall, M. B. *J. Am. Chem. Soc.* **1993**, *115*, 327–336.

34. Streuli, C. A. *Anal. Chem.* **1960**, *32*, 985; (b) Tolman, C. A. *Chem. Rev.* **1977**, *77*, 313; (c) Rahman, M. M.; Liu, H. Y.; Prock, A.; Giering, W. P. *Organometallics* **1987**, *6*, 50; (d) Bush, R. C.; Angelici, R. J. *Inorg. Chem.* **1988**, *27*, 681; (e) Poe, A. J. *Pure Appl. Chem.* **1988**, *60*, 1209; (f) Bartlett, K. L.; Goldberg, K. I.; Borden, W. T. *Organometallics* **2001**, *20*, 2669–2670.

35. Riehl, J. F.; Jean, Y.; Eisenstein, O.; Pelissier, M. *Organometallics* **1992**, *11*, 729–737.

36. (a) Moore, W. J. "Physical Chemistry", 3rd edn.; Englewood Cliffs, N.J.: Prentice-Hall Inc., 1962, p. 191.

37. Since our calculations are for ethane and ethene, the corrections for pressure/concentration are based on the equilibrium $C_2H_{6(g)} \leftrightarrow C_2H_{4(g)} + H_{2(g)}$ and our calculated values are $\Delta G = 23.17$ and 20.15 kcal/mol (BP86) at 423 and 523 K, respectively. Neat alkane concentration is ~ 10 mol/L, which is equivalent to an ideal pressure of 244 atm. The equilibrium constants calculated from ΔG are 1.08×10^{-12} and 3.80×10^{-9}, and the olefin and dihydrogen pressure are 1.93×10^{-5} atm and 1.28×10^{-3} atm at 423 and 523 K, respectively. The calculated correction for ethane including the statistical factor is RTln(6xP) and the ethene and dihydrogen corrections are RTln(P). The final corrections are 6.43 and 8.16 kcal/mol for ethane and −9.13 and −6.92 kcal/mol for ethene and dihydrogen at 423 and 523 K, respectively.

38. Cooper, A. C.; Caulton, K. G. *Inorg. Chem.* **1998**, *37*, 5938–5940.

39. (a) Becke, A. D. *Phys. Rev.* **1988**, A38, 3098; (b) Becke, A. D. *J. Chem. Phys.* **1993**, *98*, 1372; (c) Becke, A. D. *J. Chem. Phys.* **1993**, *98,* 5648; (d) Lee, C.; Yang, W.; Parr, R. G. *Phys. Rev.* **1988**, *B37*, 785.

40. Frisch, M. J.; Trucks, G. W.; Schlegel, H. B.; Scuseria, G. E.; Robb, M. A.; Cheeseman, J. R.; Zakrzewski, V. G.; Montgomery, Jr., J. A.; Stratmann, R. E.; Burant, J. C.; Dapprich, S.; Millam, J. M.; Daniels, A. D.; Kudin, K. N.; Strain, M. C.; Farkas, O.;

Tomasi, J.; Barone, V.; Cossi, M.; Cammi, R.; Mennucci, B.; Pomelli, C.; Adamo, C.; Clifford, S.; Ochterski, J.; Petersson, G. A.; Ayala, P. Y.; Cui, Q.; Morokuma, K.; Malick, D. K.; Rabuck, A. D.; Raghavachari, K.; Foresman, J. B.; Cioslowski, J.; Ortiz, J. V.; Stefanov, B. B.; Liu, G.; Liashenko, A.; Piskorz, P.; Komaromi, I.; Gomperts, R.; Martin, R. L.; Fox, D. J.; Keith, T.; Al-Laham, M. A.; Peng, C. Y.; Nanayakkara, A.; Gonzalez, C.; Challacombe, M.; Gill, P. M. W.; Johnson, B.; Chen, W.; Wong, M. W.; Andres, J. L.; Gonzalez, C.; Head-Gordon, M.; Replogle, E. S.; Pople, J. A. Gaussian, Inc., Pittsburgh PA, 1998.

41. Schlegel, H. B. *Theor. Chim. Acta* **1984**, *66*, 33.

42. Foresman, J. B.; Frish, A. E. *"Exploring Chemistry with Electronic Structure Methods"*; Gaussian, Inc.: Pittsburgh, PA, 1993.

43. Perdew, J. P. *Phys. Rev.* **1986**, *B33*, 8822–8824.

44. (a) LANL2DZ: Dunning D95 basis sets on first row, Los Alamos ECP plus double-cheta basis set on Na–Bi; (b) Hay, P. J.; Wadt, W. R. *J. Chem. Phys.* **1985**, *82*, 299; (c) Couty, M.; Hall, M. B. *J. Comput. Chem.* **1996**, *17*, 1359.

45. Hollwarth, Bohme, M.; Dapprich, S.; Ehlers, A. W.; Gobbi, A.; Jonas, V.; Kohler, K. F.; Stegmann, R.; Veldkamp, A.; Frenking, G. *Chem. Phys. Lett.* **1993**, *208*, 237–240.

46. Dunning, T. H., Jr.; Hay, P. J. In: Schaefer, H. F., III, (Ed.), *Modern Theoret. Chem.*, vol. 3; Plenum: New York, 1976.

47. Hehre, W. J.; Radom, L.; Schleyer, P.v.R.; Pople, J. A. *"Ab initio Molecular Orbital Theory"*; Wiley: New York, 1986.

48. Harihara, P. C.; Pople, J. A. *Theor. Chim. Acta* **1973**, *28*, 213–222.

49. (a) Dolg, M.; Stoll, H.; Preuss, H.; Pitzer, R. M. *J. Phys. Chem.* **1993**, *97*, 5852; (b) These basis sets and ECPs correspond to Revision: Fri Jun 27 **1997** of the Stuttgart/Dresden groups; (c) Basis sets were obtained from the Extensible Computational Chemistry Environment Basis Set Database, Version 1/02/**2002**, as developed and distributed by the Molecular Science Computing Facility, Environmental and Molecular Sciences Laboratory which is part of the Pacific Northwest Laboratory, P.O. Box 999, Richland, Washington 99352, USA, and funded by the U.S. Department of Energy. The Pacific Northwest Laboratory is a multi-program laboratory operated by Battelle Memorial Institute for the U.S. Department of Energy under contract DE-AC06-76RLO 1830. Contact David Feller or Karen Schuchardt for further information.

50. Woon, D. E.; Dunning, T. H., Jr. *J. Chem. Phys.* **1993**, *98*, 1358.

51. Ehlers, A. W.; Bohme, M.; Dapprich, S.; Gobbi, A.; Hollwarth, A.; Jonas, V.; Kohler, K. F.; Stegmann, R.; Veldkamp, A.; Frenking, G. *Chem. Phys. Lett.* **1993**, *208*, 111–114.

RECENT ADVANCES IN ELECTRON-TRANSFER REACTIONS

DAVID M. STANBURY

Department of Chemistry, Auburn University, Auburn, AL 36849, USA

ADVANCES IN INORGANIC CHEMISTRY
VOLUME 54 ISSN 0898-8838

I. Introduction

The field of electron-transfer chemistry is presently in a stage of rapid growth and diversification. Evidence for this vigorous state of affairs is overwhelming from a survey of the presentations given at the 2002 Dalton Discussion meeting in Kloster Banz and the 2001 and 1999 Gordon Research Conferences on Inorganic Reaction Mechanisms in Ventura. The objective of this chapter is to sketch aspects of current areas of research activity, using the presentations at these three conferences as a guide to significant recent advances.

The diversity of topics to be discussed is immense and includes experimental studies of single-electron transfer at dinuclear transition-metal complexes, across electrode interfaces, between and within metalloproteins, to main-group radicals, between organic and inorganic molecules; in parallel with these experimental studies, theoretical studies are proceeding apace, with notable advances in the understanding of outer-sphere electron-transfer and two-electron transfer. Another notable trend has been the increasing technical sophistication in the field: molecules and mechanisms of ever-increasing complexity have been studied, techniques with increasingly improved time resolution, reaction conditions that are increasingly demanding of control, and calculations that require ever greater computational power are hallmarks of these studies.

While the following review can hardly be considered comprehensive, it is hoped that the reader finds it interesting, stimulating, and provocative.

II. Outer-Sphere Electron Transfer Reactions

A. PAF STUDIES OF VERY RAPID ELECTRON TRANSFER

A significant technical development is the pulsed-accelerated-flow (PAF) method, which is similar to the stopped-flow method but allows much more rapid reactions to be observed (1). Margerum's group has been the principal exponent of the method, and they have recently refined the technique to enable temperature-dependent studies. They have reported on the use of the method to obtain activation parameters for the outer-sphere electron transfer reaction between $[IrCl_6]^{2-}$ and $[W(CN)_8]^{4-}$. This reaction has a rate constant of $1 \times 10^8 \, M^{-1} s^{-1}$ at $25\,^{\circ}C$, which is too fast for conventional stopped-flow methods. Since the reaction has a large driving force it is also unsuitable for observation by rapid relaxation methods.

The overall design of the PAF apparatus has evolved extensively over the years and is now in its fifth iteration. This latest iteration has primarily involved modifications to permit accurate temperature control of the sample solution from 40 °C to less than 0 °C. The fundamental PAF operation concept is derived from classical continuous-flow instruments. Two stable solutions are flowed turbulently through a mixer, and the steady-state properties of the mixed reacting solution are measured. There are two major alterations to the basic continuous-flow design. One alteration is that the flow rate is accelerated during a single experiment, while the solution absorbance is monitored at a fixed position along the flow path; this accelerated flow method greatly facilitates the collection of data as a function of flow rate, or, equivalently, of time. The other alteration is that the absorbance of the solution is monitored coaxially with the flow path, which means that at any instant the absorbance is integrated over a continuous range of solution ages. This "integrated observation" method complicates the mathematical handling of the data, but it yields the benefit of allowing solutions to be observed immediately beyond the physical point of mixing, which effectively eliminates the dead time (typically 2 ms) associated with stopped-flow methods. A consequence of this experimental design is that solutions are not completely mixed prior to observation; it is thus necessary to deconvolute the effects of the "mixing time" characteristic of the instrument from the effects of chemical reaction. The mathematics associated with this mixing-time deconvolution and the integrated observation has been solved for a number of simple rate laws, most notably for first-order rate laws. Rapid computer codes are available to solve the ensuing equations and derive reliable first-order rate constants. Rate constants as large as $5 \times 10^5 \, \text{s}^{-1}$ have been determined in this way. A schematic design of the current PAF instrument is shown in Fig. 1. A single "push" with this instrument typically consumes 6 mL of reactant solution and is thus much more economical than a classical continuous-flow instrument.

Electron transfer to $[\text{IrCl}_6]^{2-}$ from $[\text{W(CN)}_8]^{4-}$ has long been used as a test reaction for rapid kinetics methods. In the cited study, Becker *et al.* conducted the reaction under pseudo-first-order conditions with 37 µM $[\text{IrCl}_6]^{2-}$ as the limiting reagent. The loss of $[\text{IrCl}_6]^{2-}$ was monitored at 487 nm at 0, 25, and 40 °C, yielding pseudo-first-order rate constants in the range of $2 \times 10^4 \, \text{s}^{-1}$ and activation parameters of $\Delta H^{\ddagger} = 10.0 \pm 0.8 \, \text{kJ mol}^{-1}$ and $\Delta S^{\ddagger} = -58 \pm 3 \, \text{J K}^{-1} \text{mol}^{-1}$. These are chemically reasonable results and demonstrate that the PAF method is one of the most powerful methods available in the study of fast aqueous reactions.

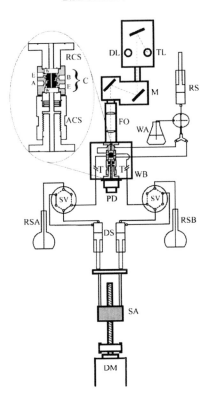

Fɪɢ. 1. Schematic of PAF–V. Key: DM, drive motor; SA, screw assembly; RSA, reactant solution A; RSB, reactant solution B; DS, drive syringes; SV, main switching valves; PD, photodetector; WB, water bath; WA, waste; FO, focusing optics; M, monochrometer; RS, receiving syringe; DL, deuterium lamp; TL, tungsten lamp; ACS, adjustable cell support; C, mixing/observation cell; W, quartz windows; A, reactant A entrance to cell; B, reactant B entrance to cell; E, product exit from cell; RCS, rigid cell support; T, a portion of the 4.6 m of coiled tubing not shown for clarity. Reproduced from Ref. (1) by permission of the Royal Society of Chemistry.

B. Outer-Sphere Electron Transfer Involving Mixed-Valent $[Cu_2]^{3+}$ Centers

Although there is a huge body of research on the kinetics of outer-sphere electron-transfer reactions of mononuclear transition-metal complexes, there are only a small number of papers on dinuclear systems. When the valences of the two metal centers are localized, current evidence indicates that the metal centers typically react essentially independently. On the other hand, for delocalized systems this can hardly be the case. Experimental study of electron transfer with such

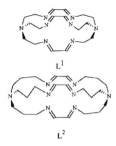

L^1

L^2

SCHEME 1. Adapted with permission from Ref. (2). Copyright 2001, Society of Biological Inorganic Chemistry.

delocalized dinuclear systems has been extremely limited, and, in the case of Cu_2 systems, nonexistent until the recent work from Jane Nelson's laboratories (2).

The subject molecules are obtained as dinuclear copper complexes with the octa-aza cryptate ligands L^1 and L^2 shown in Scheme 1.

X-ray crystallography shows that in the $[Cu_2L]^{3+}$ state the Cu–Cu distance is short (~ 2.4 Å), and ESR and electronic spectroscopy indicate that the systems are delocalized with both Cu centers in the 1.5 oxidation state. One-electron reduction of both $[Cu_2L^1]^{3+}$ and $[Cu_2L^2]^{3+}$ leads to the corresponding $[Cu_2L]^{2+}$ complexes without any drastic perturbation to the coordination environments at the metal centers. In the case of $[Cu_2L^1]^{3+}$, reduction causes the Cu–Cu distance to increase by 0.09 Å, but for $[Cu_2L^2]^{3+}$ the increase is much larger (0.51 Å).

The kinetics of five different electron-transfer reactions of these complexes were studied, all in water at 25 °C and $\mu = 0.1$ M ($NaNO_3$). The two self-exchange reactions were studied by ^1H NMR line broadening methods in the slow-exchange limit, leading to second-order self-exchange rate constants of 8.4×10^4 $M^{-1}s^{-1}$ for Cu_2L^1 and 2.2×10^5 $M^{-1}s^{-1}$ for Cu_2L^2. The cross electron-transfer reaction between $[Cu_2L^1]^{3+}$ and $[Cu_2L^2]^{2+}$ was studied by stopped-flow methods and found to have a rate constant of 1.7×10^5 $M^{-1}s^{-1}$, in excellent agreement with that predicted from the self-exchange rate constants and the Marcus cross relationship. The other two reactions were the oxidation of $[Cu_2L^1]^{2+}$ and $[Cu_2L^2]^{2+}$ by $[Co(ox)_3]^{3-}$; again, the observed rate constants are in excellent agreement with the predictions of Marcus theory. From these results it is evident that all of the reactions occur with rate-limiting electron transfer and are not complicated by conformational equilibria along the reaction coordinate.

The above results have great relevance to biochemistry, where it is found that dinuclear Cu_2 ("Cu_A") centers function as electron-transfer sites in cytochrome c oxidase and nitrous oxide reductase. In these enzymes the Cu_A site is required to transfer electrons very rapidly; it is gratifying to see that the first two examples of synthetic Cu_2 models also display high electron-transfer reactivity. However, this insight should be tempered by the astonishing lack of correlation between the self-exchange rate constants of the two model complexes and the corresponding changes in Cu–Cu distances. One may wonder how a 0.51 Å Cu–Cu distance change could be compatible with such a rapid self-exchange rate constant. Indeed, this would appear to contradict the central assumptions of the entatic hypothesis. A similar lack of correlation is found in the electron-transfer chemistry of mononuclear Cu polythiaether complexes (described below), where metal–ligand bond cleavage accompanies some very rapid self-exchange reactions. It may be that the Cu–Cu potential-energy surfaces are remarkably flat, which would allow a small reorganizational energy despite the large structural change, or else there is a two-step mechanism that is not exposed by the results obtained so far.

C. COMPUTATION OF SELF-EXCHANGE RATE CONSTANTS

The history of computing electron self-exchange rate constants for transition metal complexes goes back to the 1950s, but the increasing power of modern computers has allowed progressively more sophisticated models to be investigated. Rotzinger has recently introduced the idea of explicit inclusion of the second solvation sphere in quantum mechanical calculations. In a recent paper he has applied the method to the $[V(H_2O)_6]^{2+/3+}$, $[Ru(H_2O)_6]^{2+/3+}$, $[V(H_2O)_6]^{3+/4+}$, and $[Ru(H_2O)_6]^{3+/4+}$ aqueous systems, the latter two being hypothetical (3).

Considering the $[M(H_2O)_6]^{2+/3+}$ system as a typical example, Rotzinger has performed calculations on the species $[M(H_2O)_6 \cdot (H_2O)_{12}]^{n+}$, where the second coordination sphere is described by 12 water molecules that are hydrogen bound to the six water ligands, an example being shown in Fig. 2.

He calculates at the MCQDPT2 level the energies of the geometry-optimized molecules for $n = 2$ and 3, and then he calculates the energies of the two species with their charges reversed but with their geometries frozen. Taking appropriate sums and differences of these four energies yields the internal energy change for the vertical electron-transfer process, λ_{fs}. The solvent reorganizational energy, λ'_{ou}, is then calculated using the classical two-sphere dielectric continuum model, but starting with the third coordination sphere.

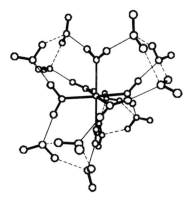

FIG. 2. Perspective view of the $V(OH_2)_6 \cdot (OH_2)_{12}^{3+}$ ion exhibiting D_2 symmetry. The dashed lines represent the hydrogen bonds within the cyclic water trimers in the second coordination sphere. Reproduced from Ref. (3) by permission of the Royal Society of Chemistry.

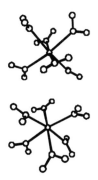

FIG. 3. Perspective view of the $[Ru(OH_2)_6]_2^{5+}$ dimer with S_6 symmetry (the Ru$\cdots$Ru distance is 6.80 Å). Reproduced from Ref. (3) by permission of the Royal Society of Chemistry.

Explicit calculation of the electronic coupling matrix element, H_{ab}, is performed by modeling the transition state (Fig. 3) as a supermolecule, $[M(H_2O)_6]_2^{5+}$, and optimizing its geometry under the constraint of having an inversion center of symmetry. The numerical value of H_{ab} is then obtained from the energy gap between the appropriate molecular orbitals of the supermolecule.

Nuclear frequency factors are calculated directly from the calculated molecular vibrational frequencies and the reorganizational energies, and these, in conjunction with the calculated H_{ab} values lead to values for the electronic transmission coefficient, κ_{el}.

Rotzinger then evaluated λ'_{ou} and H_{ab} as a function of the distance between the two reactant metal centers. He used the Fuoss equation to calculate the ion-pairing equilibrium constant to form the precursor complex at these internuclear distances. Assembly of these data then allowed the calculation of the self-exchange rate constants as a function of the internuclear distance in the transition state, the maximum rate being taken as the actual rate.

Several notable results emerge from these calculations. One is that all four of the reactions are predicted to have electronic transmission coefficients smaller than 5×10^{-3}, meaning that they are all significantly non-adiabatic. Another is that the magnitude of H_{ab} is quite sensitive to the symmetry of the orbitals involved, and hence there are ample grounds to expect the Marcus cross relationship to break down. Thirdly, comparison of the results with those obtained without explicit inclusion of the second coordination sphere shows that hydrogen bonding between the first and second coordination spheres can contribute significantly to the total reorganizational energy; this hydrogen-bonding contribution is calculated to be much stronger for the higher-charged 3+/4+ systems, although there are no experimental data to test this prediction. Finally, the calculated self-exchange rate constant for the $V^{2+/3+}$ system is ~ 10-fold faster than measured (4), which is taken as an indication that the measured rate constant is in error because of complications arising from the reaction of V^{2+} with ClO_4^- present in the solutions.

D. Ion-Pairing Inhibition of Electron-Transfer Kinetics

Ion pairing in aqueous solution often plays a minor role in electron-transfer kinetics, but in low-dielectric solvents such as dichloromethane the effects can be much stronger. Wherland's group has recently uncovered a fascinating case where ion pairing drastically inhibits the rates of outer-sphere electron transfer, even though one of the reactants is uncharged (5).

They have studied the effect in the oxidation of ferrocene by $[Co^{III}((dmg)_3(BF)_2)]^+$ in CH_2Cl_2. Here, $(dmg)_3(BF)_2^{2-}$ is a clathrochelate ligand composed of three dimethylglyoxime and two BF moieties, which forms a substitution-inert Co^{III} complex that is reversibly reduced to Co^{II}. The net electron transfer reaction is first-order in both $[Co^{III}]$ and $[Fe(cp)_2]$, and the apparent second-order rate constant, k, is found to be highly sensitive to identity and concentration of the background electrolyte. A typical sampling of data is presented in Fig. 4.

A two-term rate law is inferred, having one term (k_0) for reaction of the free Co^{III} cation with ferrocene and the other term (k_X) for reaction of

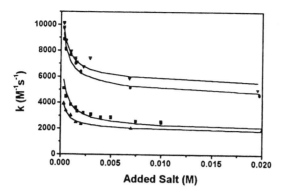

FIG. 4. The effect of added electrolyte (Bu_4N^+ salts) on the reduction of approximately 20 mM $[Co(dmg)_3(BF)_2]BF_4$ by ferrocene at $-20°C$ in CH_2Cl_2. The solid lines represent the parameters fit to all of the data, as described in the text. From top to bottom: Br^-, Cl^-, BF_4^-, NO_3^-. In all cases, the rate constant in the absence of ion pairing, from the fitting, is $2.8 \times 10^5 \, M^{-1} \, s^{-1}$. Reproduced from Ref. (5) by permission of Elsevier Science.

the neutral $Co^{III} \cdot X$ ion pair with ferrocene. Because of the extensive ion pairing in this solvent, even with no added electrolyte the Co^{III} complex exists primarily as an ion pair with its BF_4^- counter-ion. Thus, even with no added electrolyte the kinetic effects of ion pairing are dominant, and the rate constant for the free Co^{III} reactant can only be obtained by careful data fitting. The results at $-20 \, °C$ are $k_0 = 2.8 \times 10^5 \, M^{-1} s^{-1}$; k_X values are 1.9×10^3, 4.8×10^3, 4.1×10^3, and $1.7 \times 10^3 \, M^{-1} s^{-1}$ for the counter-ions BF_4^-, Br^-, Cl^-, and NO_3^-, respectively. We see that ion pairing can inhibit the rate constants by more than a factor of 100.

At first sight, these strong effects might not seem to be predictable, given that the ferrocene reactant is uncharged and thus the formation of the precursor complex should be unaffected by the charge of the other reactant. The reaction of the ion-paired species, however, is not a simple electron-transfer reaction, because transfer of the anion must also occur. A detailed understanding of the dynamics of the process remains to be developed.

E. ELECTRON TRANSFER COUPLED TO A CHANGE IN NUMBER OF LIGANDS

Cu(II/I) redox couples often present unusual complexities arising from the dual features of high lability in both oxidation states and a "preference" for highly different structures in the two oxidation states. The structural differences arising from reduction of Cu(II) to Cu(I) can appear either as a loss of coordination number (e.g. from 5 to 4) or as a change in the basic geometry at the metal center (e.g. from square planar to tetrahedral). Indeed, it is rather uncommon to find a copper(II/I) redox system that is free of these complications, and the more typical systems react with outer-sphere electron-transfer reagents through a sequence of steps including substitutional or conformational changes as well as redox steps. Disentangling these complexities in the electron-transfer kinetics of Cu(II/I) systems has been a major activity in the Rorabacher research group. They have recently applied their methods to the chemistry of copper ions in the presence of the 1,4,7-trithiacyclononane ligand ([9]aneS$_3$) (6). The work demonstrates that such complicated systems can be understood in depth, and it shows how electron transfer with major accompanying structural change can lead to the counterintuitive result of a large self-exchange rate constant for the net process.

In this comprehensive study of the Cu/[9]aneS$_4$ system, Kandegedara *et al.* have performed an unusually wide range of experiments, including electrospray mass spectrometry, various electrochemical methods, stopped-flow kinetics, conventional spectrophotometric equilibrium measurements, and in both aqueous and acetonitrile solutions (6). They used conventional spectrophotometry to determine the binding constants for formation of both the mono- and bis-[9]aneS$_3$ complexes of Cu(II) in water, which led to the unusual result that the second binding constant is about 100-fold larger than the first. Stopped-flow methods were used to obtain the rate constants for gain and loss of both ligands for aqueous Cu(II). Cyclic voltammetry was then used to determine the standard potentials for the various Cu(II/I) couples and the remaining stability constants for both Cu(II) and Cu(I) in water. In acetonitrile electrospray mass spectrometry was used to demonstrate that Cu(I) exists predominantly bound to a single [9]aneS$_3$ ligand, and square-wave voltammetry was used to determine the binding constant for the second ligand. Cyclic voltammmetry was then used to determine the relevant standard potentials and binding constants in acetonitrile. These studies showed that Cu(I) exists primarily as the mono-ligated complex in both solvents, which implies that a change in the number of coordinated ligands must accompany electron transfer. Finally, stopped-flow studies were conducted to determine the kinetics of oxidation and reduction in both solvents by a series of classical outer-sphere reagents.

Analysis of the above data led to the conclusion that all of the redox reactions proceed with electron transfer through the $[CuL_2]^{2+/+}$ redox couple, and that the change in number of ligands occurs in the Cu(I) oxidation state. This interpretation is given as pathway I in Fig. 5.

If self-exchange rate constants for the Cu(II/I) couple are calculated by applying the Marcus cross relationship to the observed second-order

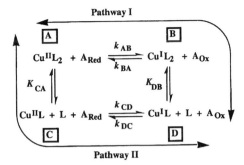

Pathway I

Pathway II

FIG. 5. Dual-pathway square scheme mechanism for electron transfer involving the $Cu^{II/I}([9]aneS_3)_n$ system. The vertical reactions involve ligand gain or loss while the horizontal reactions represent electron transfer. Reproduced from Ref. (6) by permission of the Royal Society of Chemistry.

rate constants and using standard potentials appropriate to the net reaction ($CuL_2^{2+} + e^- \rightleftharpoons CuL^+ + L$), the results span a range of about five orders of magnitude, with the largest being $\sim 10^6\,M^{-1}\,s^{-1}$. Note that this treatment makes the implicit assumption that electron transfer and ligand substitution are concerted. The wide range in calculated self-exchange rate constants is to be expected, given that an inappropriate driving force is used for the actual electron-transfer process. Nevertheless, the large calculated self-exchange rate constants show that a sequential mechanism can be a very effective means of performing electron transfer with accompanying major structural change.

When the Marcus analysis is corrected to use rate constants and driving forces characteristic of the CuL_2^{n+} species the derived self-exchange rate constants are much more self consistent. We caution, however, that the CuL_2^{2+} complex likely has all six thiaether atoms coordinated to the Cu^{II} center, while the CuL_2^+ complex is probably four coordinate. Since it is rather unlikely that electron transfer occurs in concert with this change in coordination number, a further correction will probably be required in order to obtain physically meaningful self-exchange rate constants.

III. Radical Electron-Transfer Reactions

Section II focuses on outer-sphere reactions of species that are stable in their adjacent oxidation states, which leads to a degree of confidence in the reaction mechanisms and the ability to define and measure the pertinent driving forces. In Section III we consider reactions involving unstable free radicals, which typically have more complex reaction mechanisms and less certainty about the one-electron standard potentials.

A. Oxidation of $Br_2^{\bullet-}$ by $ClO_2^{\bullet}$

In a conventional study of the aqueous oxidation of ClO_2^- by Br_2 it was found that a 10-step mechanism was required to explain the data, and the complex kinetics did not allow the precise resolution of several of the rate constants (7). The overall reaction is,

$$Br_2 + 2ClO_2^- \longrightarrow 2Br^- + 2ClO_2^{\bullet} \tag{1}$$

The first step of the proposed mechanism is the reversible electron transfer from ClO_2^- to Br_2 as in reaction (2):

$$Br_2 + ClO_2^- \rightleftharpoons Br_2^{\bullet-} + ClO_2^{\bullet} \qquad k_1, k_{-1} \tag{2}$$

The requisite value for k_{-1} was only approximately defined by the fitting procedures, and because of uncertainty in the standard potential for the $Br_2/Br_2^{\bullet-}$ redox couple it was likewise deemed unsuitable to use the value of k_1 and the principle of detailed balancing to derive the value of k_{-1}. Further reason to be doubtful of the derived value of k_{-1} was a major disagreement between it and the value predicted by the cross relationship of Marcus theory.

Tóth *et al.* then used laser flash photolysis as a means to determine the value of k_{-1} independently of the above study (8). They used 355 nm laser light to photolyze mixtures of $ClO_2^{\bullet}$ and Br_2/Br_3^-. Absorption of this light by Br_3^- led to the prompt formation of $Br_2^{\bullet-}$, and the subsequent loss of $Br_2^{\bullet-}$ was monitored by its absorbance at 360 nm. The loss of $Br_2^{\bullet-}$ occurred with mixed 2nd- and 1st-order kinetics due to the parallel 2nd-order self reaction of $Br_2^{\bullet-}$ and its pseudo-first-order reaction with $ClO_2^{\bullet}$. These experiments led to a value of $3.6 \times 10^9 \, M^{-1} s^{-1}$ for k_1, which is in good agreement with the approximate value ($1.1 \times 10^9 \, M^{-1} s^{-1}$) originally obtained.

There are three points of significance of this result. One is that it provides strong support for the 10-step mechanism originally proposed for reaction 1. Another is that it facilitates a more robust fitting of the mechanism to the kinetic data obtained for that reaction. Thirdly, it confirms that reaction 2 has a rate constant that is four orders of magnitude greater than predicted by Marcus theory. It is concluded that reaction 2 is poorly modeled as an outer-sphere process and is better described as having significant transition-state bonding between the two reactants (an inner-sphere mechanism).

B. DIVERSE ONE-ELECTRON PATHWAYS FOR REDUCTION OF AQUEOUS CHLORINE SPECIES

Hypochlorous acid (HOCl) is a very well-known oxidant, and there is good evidence that the chlorine atom undergoes hydrolysis as in

$$Cl^{\bullet} + H_2O \rightleftharpoons HOCl^{\bullet -} + H^+ \qquad pKa \sim 5.3 \qquad (3)$$

It is thus reasonable to anticipate that HOCl could behave as an outer-sphere one-electron oxidant. Indeed, the standard potential for the $HOCl/HOCl^{\bullet -}$ couple is estimated at 0.25 V (9). In prior reports where such a pathway might have been uncovered, alternative pathways generally have been found, such as inner-sphere mechanisms and reactions via Cl_2. Reaction via Cl_2 is often a viable pathway because of the presence of Cl^- either as a contaminant or reaction product and its reaction with HOCl as in Eq. (4).

$$Cl_2 + H_2O \rightleftharpoons HOCl + Cl^- + H^+ \qquad K_h \qquad (4)$$

An attempt to demonstrate the direct outer-sphere reduction of HOCl was recently published, in which the reductant was selected to be $[Ru(NH_3)_5isn]^{2+}$ because of its well-known behavior as an outer-sphere reductant and its relatively low standard potential (10). In this reagent the ligand isn is isonicotinamide. Further efforts to promote reaction via HOCl entailed the use of chloride-free preparations of HOCl.

Despite the careful selection of conditions and reactants, the reaction of $[Ru(NH_3)_5isn]^{2+}$ with HOCl proved to be unusually complex (10). The ruthenium(II) reactant is widely employed as a typical outer-sphere one-electron reductant, but a variety of Ru-containing products are generated in the reaction with an excess of aqueous HOCl. Rapid-scan stopped-flow experiments showed that the reaction occurs in at least two phases, and the formation and decay of an intermediate species was clearly demonstrated. Spectral comparisons indicated that the intermediate is $[Ru(NH_3)_5isn]^{3+}$, and direct experiments showed that solutions of authentic $[Ru(NH_3)_5isn]^{3+}$ react rapidly with aqueous HOCl.

Kinetic studies showed that the consumption of $[Ru(NH_3)_5isn]^{2+}$ occurs as rapidly as $[Ru(NH_3)_5isn]^{3+}$ is produced in the first phase of the reaction with HOCl, and hence there might be reasonable hope that the first phase of the reaction corresponds to the outer-sphere reduction of HOCl. Stopped-flow studies showed that the rate law for this first phase is,

$$-\frac{d[Ru(II)]}{dt} = 2(k_1 K_h[H^+] + k_3[H^+][Cl^-] + k_5 K_h[Cl^-]) \frac{[HOCl]_{tot}[Ru(II)]}{K_h + [H^+][Cl^-]}$$

$$(5)$$

Here, $[HOCl]_{tot}$ refers to the total reactive chlorine concentration, $[Cl_2]$ plus $[HOCl]$. The three terms in this rate law were attributed to the following three rate-limiting steps:

$$Ru(II) + HOCl + H^+ \longrightarrow Ru(III) + H_2O + Cl^\bullet \qquad k_1 \qquad (6)$$

$$Ru(II) + Cl_2 \longrightarrow Ru(III) + Cl_2^{\bullet -} \qquad k_3 \qquad (7)$$

$$Ru(II) + HOCl + Cl^- \longrightarrow Ru(III) + Cl_2^{\bullet -} + OH^- \qquad k_5 \qquad (8)$$

Thus, no evidence for the direct bimolecular one-electron reduction of HOCl could be obtained. Under conditions of relatively high pH and very low Cl^- concentrations, where the three pathways shown above should be minimized, the reaction is autocatalytic, presumably because of the production of Cl^- and its enhancement of reaction through reactions 7 and 8.

These results show that direct outer-sphere reduction of HOCl is not easily achieved, and they also show the remarkable diversity of pathways through which aqueous chlorine can react with outer-sphere reductants.

C. GENERATION OF $SO_3^{\bullet -}$ IN THE OXIDATION OF S(IV) BY Fe(III)

The Fe(III)/S(IV) reaction has long been of interest because of its importance in the catalytic autoxidation of S(IV). The latter reaction is known to have a complex chain mechanism, and the production of $SO_3^{\bullet -}$ radicals has been considered to be the essential chain-initiating step. It is also widely believed that the direct oxidation of S(IV) by Fe(III) is the source of $SO_3^{\bullet -}$. There is little agreement among the various papers published on the direct reaction of Fe(III) with S(IV) with regard to its mechanism, and much of this disagreement can be traced to the potential for Fe(III) to bind several S(IV) ligands under the typical conditions of excess S(IV).

Lente and Fábián have recently reported on the Fe(III)/S(IV) reaction with excess Fe(III), i.e. with the inverse of the typical concentration ratio (11). This procedure effectively limits the speciation to complexes having at most one S(IV) ligand. This simplification, however, comes at the cost of introducing dinuclear Fe(III) species into the reaction mixture. Nevertheless, the inverted concentration ratios lead to data that impose substantial constraints on proposed mechanisms, and they provide strong support for one specific pathway in the formation of $SO_3^{\bullet-}$. Also see Chapter 8.

When the reaction of Fe(III) with S(IV) is studied by stopped-flow methods with excess Fe(III) under rather acidic conditions (pH~1) a complex series of UV–vis absorbance changes is observed. As is shown in Fig. 6, three distinct phases occur, with variable results depending on the observation wavelength.

An important simplifying consequence of the use of inverted concentration ratios is that the reaction is independent of O_2 concentration, which means that unintended O_2 contamination should not distort the data. Because of the complexity of the reaction, the relatively new technique of Matrix Rank Analysis was used to sort out the speciation. This analysis led to the identification of two sulfur-containing intermediates: $[Fe_2(OH)SO_3]^{3+}$ and $[Fe(SO_3)]^+$. Other reactant species known to be present under these conditions include SO_2, HSO_3^-, Fe^{3+}, $Fe(OH)^{2+}$, and

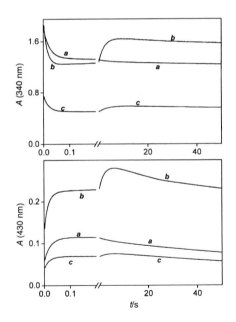

FIG. 6. Typical kinetic traces in the early phase of the iron(III)–sulfur(IV) reaction. pH = 1.40, [Fe(III)] = 25.0 mM, [S(IV)] = 0.50 mM (a); pH = 1.40, [Fe(III)] = 25.0 mM, [S(IV)] = 1.50 mM (b); pH = 1.70 [Fe(III)] = 7.5 mM, [S(IV)] = 0.50 mM (c); T = 10.0 °C; μ = 1.0 M (NaClO$_4$); optical pathlength 1 cm. Reproduced from Ref. (11) by permission of the Royal Society of Chemistry.

$Fe_2(OH)_2^{4+}$. A kinetic model was developed that consisted of 10 steps, and it was fit to the complete set of time-dependent traces by use of a combined least-squares optimization with numerical integration.

One key feature of the model is that the $[Fe_2(OH)SO_3]^{3+}$ species is a dead-end intermediate with respect to the overall redox process. The actual redox reaction consists of only two steps:

$$Fe^{3+} + HSO_3^- \rightleftharpoons FeSO_3^+ + H^+ \qquad K_H \qquad (9)$$

$$FeSO_3^+ \longrightarrow Fe^{2+} + SO_3^{\bullet -} \qquad k_9 \qquad (10)$$

The first step is the reversible formation of $FeSO_3^+$, which occurs in a few hundred ms. The second step is an intramolecular redox process, which has a rate constant of $0.2\,s^{-1}$ at $25\,°C$. Formation and dissociation of the dinuclear Fe(III) species are the main sources of the kinetic complexities.

By use of well-established standard potentials, the reported values for K_H and k_9, and the principle of detailed balancing, one can calculate that the reverse of reaction (10) has a rate constant (k_{-9}) of $2 \times 10^3\,M^{-1}\,s^{-1}$. Normal ligand substitution reactions at Fe^{2+} are much faster than this, which raises questions regarding the nature of the transition state for this reaction.

D. OUTER-SPHERE OXIDATION OF THIOLS

Organic thiols are widely encountered, and they are rather susceptible to oxidation. They have special significance in biochemistry, where cysteine undergoes oxidative coupling in the cross-linking of proteins and the oxidation of glutathione plays many physiological roles. The reactions of thiols with one-electron oxidants are ubiquitous, and there have been many reports on the kinetics of oxidation of thiols by classical outer-sphere oxidants. A survey of these reports reveals that copper catalysis is a widespread phenomenon in these reactions, and that outer-sphere oxidations of aliphatic thiols tend to be so sensitive to this effect that it is generally only the catalyzed pathway that can be detected. As a consequence, virtually nothing is known about the direct outer-sphere oxidation of aliphatic thiols.

This gap in our knowledge is now closed, as the first paper on the uncatalyzed outer-sphere oxidation of an aliphatic thiol was recently published (12). This work selected thioglycolic acid (TGA, mercaptoacetic acid, $HSCH_2CO_2H$) as a representative thiol because of its high water solubility, low vapor pressure, and simple structure. The oxidant was $[IrCl_6]^{2-}$, a well-characterized one-electron oxidant that frequently reacts through an outer-sphere mechanism. As is typical of such reactions, the oxidation of TGA by $[IrCl_6]^{2-}$ proved to be highly sensitive to trace copper catalysis. It was found, however, that this copper catalysis could be thoroughly suppressed through the addition of small amounts of bathophenanthrolinedisulfonate (bathophen). Presumably, bathophen functions by binding copper ions in solution and thus rendering them noncatalytic.

One fascinating outcome of this work is the finding that the stoichiometry of the redox reaction is altered by the addition of bathophen. In the absence of bathophen the reaction is,

$$2[IrCl_6]^{2-} + 2TGA \longrightarrow 2[IrCl_6]^{3-} + RSSR \qquad (11)$$

where RSSR represents the disulfide derived from TGA. In the presence of bathophen there is a significant yield of sulfoacetate ($^-O_3SCH_2CO_2H$), indicating that reaction (11) runs concurrently with

$$6[IrCl_6]^{2-} + TGA \longrightarrow 6[IrCl_6]^{3-} + {}^-O_3SCH_2CO_2H \qquad (12)$$

The conventional wisdom that one-electron oxidants react with thiols to yield disulfides is apparently derived from reactions in which trace copper catalysis dominated the chemistry.

The rate law for the noncatalyzed reaction is first-order in both [TGA] and [Ir(IV)], and it shows a complex pH dependence with the rates generally increasing with pH. It is inferred that only the thiolate forms of TGA are reactive and that the thiol forms are unreactive. Thus, at high pH the reaction has the following rate-limiting step:

$$^-SCH_2COO^- + [IrCl_6]^{2-} \longrightarrow {}^\bullet SCH_2COO^- + [IrCl_6]^{3-} \qquad k_d \qquad (13)$$

At lower pH the minor tautomeric form of the monoanion becomes the dominant reactant:

$$^-SCH_2COOH + [IrCl_6]^{2-} \longrightarrow {}^\bullet SCH_2COOH + [IrCl_6]^{3-} \qquad k_c \qquad (14)$$

A literature value for E° for the ${}^\bullet SCH_2COO^-/{}^-SCH_2COO^-$ redox couple (0.74 V) was then used in conjunction with the cross relationship of Marcus theory to derive a self-exchange rate constant of $1.5 \times 10^5\,M^{-1}\,s^{-1}$ for the ${}^\bullet SCH_2COO^-/{}^-SCH_2COO^-$ redox couple.

A high susceptibility of thiolates to outer-sphere oxidation is thus implied by the combined action of the relatively low standard potential and high self-exchange rate constant given above. However, the high pK_a of most thiols and the unreactivity of the thiol form suggests that thiols are resistant to oxidation through outer-sphere mechanisms at physiological pH or in acidic media. These factors may explain the sensitivity of these reactions to copper catalysis, but a full explanation can be formulated only after determination of the rate laws for the catalyzed pathways.

E. ELECTRON-TRANSFER REACTIONS OF PHENOXYL RADICALS

Phenoxyl radicals (PhO•), like the thyl radicals discussed above, are widespread reactive intermediates. The corresponding phenols (PhOH) typically have pK_a values around 10, so the properties of the PhO•/PhO⁻ redox couples become highly relevant above pH 10. Standard potentials for a few of these redox couples have been determined by use of pulse radiolysis to generate the unstable phenoxyl radicals in the presence of appropriate electron donors. These conditions lead to the rapid establishment of electron-transfer equilibria as in,

$$\text{PhO}^\bullet + \text{D}^- \rightleftharpoons \text{PhO}^- + \text{D}^\bullet \tag{15}$$

From the measured electron-transfer equilibrium constant and the known standard potential for the reference D•/D⁻ couple it has been possible to determine E° for the PhO•/PhO⁻ couple. The method, however, is non-trivial and does not lend itself to the rapid determination of standard potentials for a large series of related compounds.

An alternative electrochemical method has recently been used to obtain the standard potentials of a series of 31 PhO•/PhO⁻ redox couples (13). This method uses conventional cyclic voltammetry, and it is based on the CV's obtained on alkaline solutions of the phenols. The observed CV's are completely irreversible and simply show a wave corresponding to the one-electron oxidation of PhO⁻. The irreversibility is due to the rapid homogeneous decay of the PhO• radicals produced, such that no reverse wave can be detected. It is well known that PhO• radicals decay with second-order kinetics and rate constants close to the diffusion-controlled limit. If the mechanism of the electrochemical oxidation of PhO⁻ consists of diffusion-limited transfer of the electron from PhO⁻ to the electrode and the second-order decay of the PhO• radicals, the following equation describes the scan-rate dependence of the peak potential:

$$E_\text{p} = E_\text{red}^\circ + 0.902\,\frac{RT}{F} - \frac{RT}{3F} \ln\!\left(\frac{2kC^\circ 2RT}{v3F}\right) \tag{16}$$

Here, v is the scan rate, k is the radical self-reaction rate constant, and E_p is the CV wave peak potential. The standard potentials obtained ranged from 1.28 V (4-O₂NPhOH) to 0.17 V (4-HOPhOH). Good agreement with the literature values was obtained in those cases where the data were available.

An example where electron transfer from PhO^- is important comes from a related publication on the reaction of phenol with O_2 where $[Ru(bpy)_3]^{2+}$ is used as a photosensitizer (14). In acidic media the reaction involves generation of $^1O_2$ by quenching of excited $[Ru(bpy)_3]^{2+}$; reaction of $^1O_2$ with phenol leads to the production of benzoquinone. The quantum yields for benzoquinone production are highly pH dependent, showing a sharp peak at pH 8.4. This unusual pH dependence arises from the competition of several pathways, and one of the most important being the electron-transfer quenching of $[^*Ru(bpy)_3]^{2+}$ by PhO^-:

$$[^*Ru(bpy)_3]^{2+} + PhO^- \longrightarrow [Ru(bpy)_3]^+ + PhO^\bullet \qquad (17)$$

Examination of the relevant standard potentials indicates that this reaction is favorable.

IV. "Small-Molecule" Intramolecular Electron-Transfer Reactions

A. PICOSECOND EVENTS IN INTRAMOLECULAR ELECTRON TRANSFER

The molecule $[Re^I(MQ^+)(CO)_3(dmb)]^{2+}$ has been a model system for studying intramolecular electron transfer over the last two decades. Here, MQ^+ is the monodentate ligand N-methyl-4,4'-bipyridinium, dmb is the bidentate ligand 4,4'-dimethyl-2,2'-bipyridine, and the three CO ligands are facially coordinated. Irradiation of this complex at room temperature in solution with near-UV light leads to a sequence of intramolecular electron-transfer events as shown in Fig. 7.

In brief, irradiation leads to the sub-picosecond oxidation of the Re(I) center and the production of a mixture of two reduction sites: i.e. either the MQ^+ or dmb ligand can be reduced. On the time-scale of several picoseconds inter-ligand electron transfer occurs, converting the mixture of excited states to $[Re^{II}(MQ^\bullet)(CO)_3(dmb)]^{2+}$.

Three types of rapid-spectroscopic studies have recently added to our understanding of this system (15). In the first of these, 0.2 ps excitation is followed by picosecond time-resolved visible absorption spectroscopy. Spectral changes confirmed the 8.3 ps (in acetonitrile) and 14 ps (in ethylene glycol) interligand electron transfer process. The second method was based on picosecond-time-resolved resonance Raman spectroscopy (TR^3). The Raman probe wavelength was 600 nm, thus ensuring that the observed peaks corresponded to the $MQ^\bullet$ moiety. Line-narrowing and frequency shifts were observed on the ~ 10 ps timescale, indicating that the $[Re^{II}(MQ^\bullet)(CO)_3(dmb)]^{2+}$ state is initially produced vibrationally

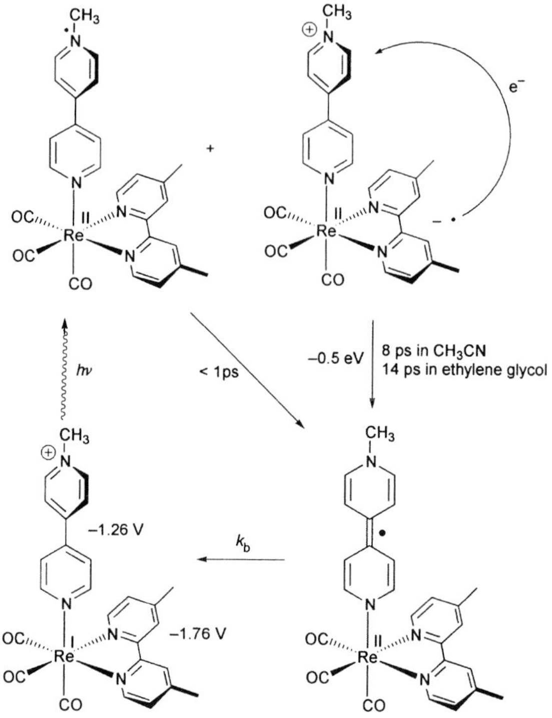

FIG. 7. Excited-state interligand electron transfer in fac-$[Re(MQ)(CO)_3(dmb)]^{2+}$. Reproduced from Ref. (15) by permission of the Royal Society of Chemistry.

hot and that this excess energy is not concentrated in any specific high frequency modes. The third method utilized picosecond time-resolved IR absorption spectroscopy. Features corresponding to the CO stretches were observed and taken in support of characterization of the excited state(s) as Re(II) species. Spectral evolution on the $\sim 10\,\mathrm{ps}$ timescale indicated that the two different MLCT excited states have significant structural differences in their $Re(CO)_3$ moieties. Data of this type should be of great value in developing theories to account for the rates of intramolecular electron transfer.

B. STRUCTURAL MODELS OF dσ/dσ INNER-SPHERE ELECTRON-TRANSFER TRANSITION STATES

It has long been assumed that the rates of inner-sphere electron-transfer reactions for transition-metal complexes should be sensitive to the nature of the donor and acceptor orbital symmetries. Efforts to demonstrate these effects have largely focused on the behavior of intramolecular redox systems because intermolecular reactions have relatively uncertain transition-state structures, have rates that are affected by the magnitude of the precursor complex formation constants, and can be limited by reactant diffusion. Unambiguous demonstration of these orbital effects remains elusive, despite the huge amount of effort in intramolecular electron-transfer research, because of the lability of dσ metal centers. Thus, most of the dinuclear systems investigated to date have utilized systems in which dπ electrons are transferred.

A potential solution to this problem is evident in the recent synthesis of some dinuclear Cu(II)$_2$L complexes with bridging halide ligands (*16*). These complexes are based on the ligand

= L

Each tetraazamacrocyclic portion of the ligand can bind a Cu(II) ion, and the flexible central methylene linkages permit the dinuclear complexes to adopt face-to-face conformations. Dinuclear Cu(II)$_2$ complexes with bridging (axial) Cl$^-$ and Br$^-$ ligands have been characterized by X-ray crystallography. UV–vis spectral data indicate that these complexes retain their overall structure in solution. Cyclic voltammetry and differential pulse voltammetry experiments imply that Cu(II) is reversibly oxidized to Cu(III) in these halide-bridged systems. Magnetic susceptibility measurements and ESR spectra show that the electronic coupling in the Cu(II)$_2$ complexes is quite weak. This weak coupling is to be expected, since the unpaired electrons reside in "d$_{x^2-y^2}$" orbitals that are coplanar with the macrocylic rings and hence do not overlap. Likewise, the electrochemical data indicate that the mixed-valent Cu(II)–Cl$^-$–Cu(III) state is not significantly stabilized by electronic coupling. On the other hand, the complexes display very strong LMCT absorption bands, which implies quite strong electronic coupling for these transitions. Thus, the two-center M–X coupling matrix elements are large while the three-center Cu–Cl$^-$–Cu coupling matrix element is small. The implication seems to be that the three-center Cu(II)$_2$Cl matrix elements should be closely related to those for electron transfer in the related Cu(III)–Cl–Cu(II) system, and therefore one might expect slow electron transfer rates. For the case with a bridging Br$^-$ ligand, the presence of two well-resolved

CV waves is taken as evidence that the Cu(III) state is high spin, which complicates the prediction one might make regarding its electron-transfer rate.

V. Electron Transfer with Metalloproteins

A. ELECTRON TRANSFER IN CYTOCHROME c OXIDASE

Cytochrome c oxidase is an enzyme that couples the one-electron oxidation of cytochrome c to the four-electron reduction of O_2 and is thus a crucial component of respiration. Cytochrome c contains the redox-active heme c, while cytochrome c oxidase contains a dinuclear Cu_A redox site in subunit II and three redox-active sites in subunit I: heme a, heme a_3, and Cu_B. It is believed that heme a is an electron-transfer site, while heme a_3 and Cu_B function together at the O_2 reduction site.

Recent advances in measuring the kinetics of the various electron-transfer steps in this system have been achieved by use of flash photolysis of ruthenated derivatives of cytochrome c (Ru–Cc) (17–19). In these studies [Ru(bpy)$_3$]$^{2+}$ is covalently bound to a surface residue at a site that does not interfere with the docking of cytochrome c to cytochrome c oxidase. Solutions are then prepared containing both Ru–Cc and cytochrome c oxidase, and the two proteins associate to form a 1:1 complex. Flash photolysis of the solution leads directly to the excitation of the $Ru^{II}(bpy)_3$ site, which then reduces heme c very rapidly. This method thus provides a convenient means to observe the subsequent intracomplex electron transfer from heme c to cytochrome c oxidase and further stages in the process.

In one of the early papers based on this method, yeast cytochrome c and beef cytochrome c oxidase were used (17). The His[39] residue of cytochrome c was replaced with a cysteine residue by site-directed mutagenesis, and then the cysteine was ruthenated through reaction with [Ru(bpy)$_2$(4-BrCH$_2$-4'-CH$_3$-bpy)]$^{2+}$. This study demonstrated that the reduced heme c is reoxidized by cytochrome c oxidase with a rate constant of $6 \times 10^4 \, s^{-1}$. Detection of a transient bleaching at 830 nm that occurs with the same rate constant was taken as evidence that the dinuclear Cu_A site is the initial electron acceptor in cytochrome c oxidase. Recovery of the 830 nm absorbance occurred with a $1.6 \times 10^4 \, s^{-1}$ rate constant that was the same as for a 604 nm absorbance increase, assigned to reduction of heme a. Thus it was inferred that the Cu_A site in subunit II transfers its electron to the heme a site in subunit I.

A more recent study was based on horse heart cytochrome c and *Rhodobacter sphaeroides* cytochrome c oxidase (*18*). Here, the Lys[55] residue was ruthenated and the cytochrome c oxidase was mutated at several surface sites. A model of the complex between cytochrome c and subunit II of cytochrome c oxidase, the heme c and Cu_A cofactors, and the mutated residues is shown in Fig. 8.

In one series of experiments the cytochrome c oxidase mutations replaced acidic residues by neutral ones, and some of them were thus expected to alter the nature of binding of the protein to cytochrome c. From the pattern of dependence of the heme c to Cu_A electron-transfer rate constant on these mutations it was deduced that the binding of cytochrome c to cytochrome c oxidase is mediated by electrostatic interactions between four specific acidic residues on cytochrome c oxidase and lysines on cytochrome c. In another series of experiments, tryptophan 143 of cytochrome c oxidase was mutated to Phe or Ala. These mutations had an insignificant effect on the binding of the two proteins, but they dramatically reduced the rate constant for electron transfer from heme c to Cu_A. It was concluded that electron transfer from

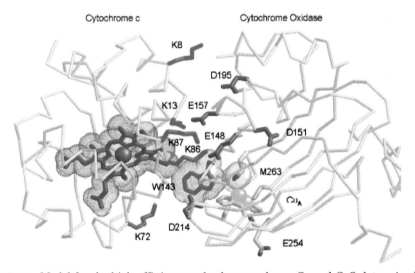

FIG. 8. Model for the high affinity complex between horse Cc and CcO determined by Roberts and Pique (*34*). The backbone of horse Cc and CcO subunit II are shown with the side chains of selected lysines and acidic residues colored blue and red, respectively. The residue numbers on subunit II are for R. *sphaeroides* CcO. Van der Waals surfaces are shown for Cc heme and subunit II Trp[143] and Met[263]. The Cu_A coppers are represented by green Corey–Pauling–Koltun models. Reprinted with permission from Ref. (*18*). Copyright 1999, American Society of Biochemistry and Molecular Biology.

heme c to Cu_A is dependent on the good electronic coupling fostered by Trp^{143}, which appears to bridge the two sites in the two-protein complex.

Most recently this method has been used with cytochrome c oxidase that has been mutated at the Cu_A site (19). One of the mutants had replaced histidine-260 (that was a ligand of one of the Cu atoms) with Asn, while the other replaced methionine-263 (that was a ligand of the other Cu atom) with Leu. These mutations had only a small effect on the rate constant for electron transfer from heme c to Cu_A but the rate constant from Cu_A to heme a was much more strongly inhibited. Both mutations led to a significant increase in the Cu_A redox potential, which would favor the oxidation of heme c by Cu_A but hinder the oxidation of Cu_A by heme a. The small effect of the mutations on the heme c-to-Cu_A rate constant was attributed to an increased reorganizational energy that balanced the more favorable driving force. A complex mutational dependence of the rate constants for electron transfer between Cu_A and heme a was interpreted as an indication that His^{260} is important in establishing electronic coupling between the two sites.

B. Internal Electron Transfer in Sulfite Oxidase

Sulfite oxidase is a dimetallic enzyme that mediates the two-electron oxidation of sulfite by the one-electron reduction of cytochrome c. This reaction is physiologically essential as the terminal step in oxidative degradation of sulfur compounds. The enzyme contains a heme cofactor in the ~ 10 kDa N-terminal domain and a molybdenum center in the ~ 42 kDa C-terminal domain. The catalytic cycle is depicted in Fig. 9.

The two-electron oxidation of sulfite generates a Mo^{IV}–Fe^{III} state, which converts to a Mo^{V}–Fe^{II} state. Cytochrome c then oxidizes this state to Mo^{V}–Fe^{III}, which then undergoes another internal electron transfer (k_3) to form Mo^{VI}–Fe^{II}. A second oxidation by cytochrome c forms Mo^{VI}–Fe^{III}, which completes the catalytic cycle.

A flash photolysis method has been developed that prepares the Mo^{VI}–Fe^{II} state and thus allows the rate constants k_3 and k_{-3} to be measured. Solutions containing 5-deazariboflavin, semicarbazide, and sulfite oxidase are subjected to 555 nm flash photolysis. The deazariboflavin is excited to a triplet state, which is then reduced by semicarbazide to form the 5-deazariboflavin semiquinone radical. This radical is then rapidly oxidized back to its parent species through the one-electron reduction of sulfite oxidase.

In the initial (1993) application of this method a value of $\sim 1000\,s^{-1}$ was obtained for the sum of k_3 and k_{-3}. Subsequent to that work the 1.9 Å X-ray crystal structure of sulfite oxidase was reported (1997), which

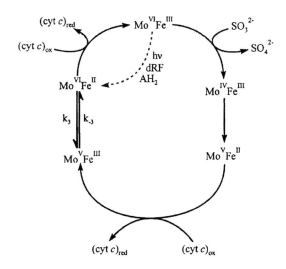

FIG. 9. Postulated oxidation state changes occurring at the Mo and Fe centers of SO during the catalytic oxidation of sulfite, and concomitant reduction of cyt c. The one-electron reduction shown with a *dashed arrow* connecting $Mo^{VI}Fe^{III}$ and $Mo^{VI}Fe^{II}$ can be initiated with a laser pulse, in a solution containing 5-deazariboflavin (dRF) and a sacrificial electron donor (AH_2). Reprinted with permission from Ref. (20). Copyright 1999, Society of Biological Inorganic Chemistry.

shows that the two metal sites are separated by 32 Å, a distance too long to support the observed electron-transfer rates. Confronted by this seeming paradox, members of the original research group recently conducted a reinvestigation of the kinetics (20). The new study verified the original results, used a determination of the electron-transfer equilibrium constant to obtain individual values for k_3 and k_{-3}, and investigated the dependence of the rates on pH and anion concentration. Small anions such as SO_4^{2-} and Cl^- were found to inhibit the rates. A pH dependence is to be expected, since the equilibrium (equilibrium constant, k_3/k_{-3}) involves a net proton gain ($Mo^{VI}O_2 \rightarrow Mo^V O(OH)$). The observed pH dependence was somewhat more complex than might be anticipated, because of the competing inhibition by small anions. Overall, however, the results still led to rate constants that are too large for the 32 Å crystallographic separation between the two redox sites. It was thus proposed that a flexible ~10-residue segment of the peptide chain between the two domains could allow the protein to undergo a large conformational change in solution, bringing the two redox sites much closer together.

The results of a test of this proposal for conformational change were recently published (21). Rates of electron transfer ($k_3 + k_{-3}$) were measured as a function of solution viscosity by adding variable amounts of either polyethylene glycol 400 or sucrose. These additions inhibited the rates, and the degree of inhibition was simply a function of the viscosity rather than the identity of the additive. Moreover, ESR and MCD spectra indicate that neither of the active sites is perturbed by these additives. It is evident that viscosity is the principal factor causing the rate inhibition. This viscosity dependence was taken as evidence that the measured rates are significantly controlled by the rate of conformational change required to bring the two active sites to close proximity.

C. Electron-Transfer Kinetics of Blue Copper Proteins

"Blue" copper proteins, typified by plastocyanin and azurin, contain a single copper atom that is ligated to several of the protein residues and function as electron-transfer agents. The copper cycles between the Cu(II) and Cu(I) oxidation states, and does so quite readily. These proteins have been subjected to intense scrutiny, and a number of significant concepts have developed from these studies. Perhaps most significantly, the concept of the entatic state was heavily grounded on the examples provided by the blue copper proteins. The precise definition of this concept is not easily stated, but in approximate language it asserts that natural proteins achieve their remarkable effectiveness by enforcing specific strained structures. As applied to blue copper proteins it states that their rapid electron-transfer rates are a consequence of the distorted coordination geometry that is conserved in both Cu(I) and Cu(II) oxidation states. The perceived need for a distortion arises from the different geometries "preferred" in the two oxidation states and the belief that to achieve rapid electron transfer requires minimal structural change.

As we have seen in Section II.E, the work of Rorabacher's group has shown that in synthetic small-molecule copper systems, rapid electron transfer can be achieved even when there are gross structural differences between the two oxidation states, such as a change in coordination number or ligand identity. Although the authors of this work interpreted their results as a vindication of the entatic hypothesis (self-exchange rate constants are greater when there is less structural change), an alternative view is that a two-step mechanism (a square scheme as shown in Fig. 5) provides a means to achieve rapid electron transfer in systems showing large structural change. Related insights are also emerging from studies of electron transfer rates in the blue copper proteins themselves (22). The Canters research group has been quite active in this area, and their results have been summarized in a recent Faraday Discussion (22).

Among the relevant points to emerge from these studies is that yes, the structural differences between the oxidized and reduced forms of the blue copper proteins are rather small, but the structures are also quite similar to those of the apoproteins. Thus, the presence of the copper ion, in either oxidation state, does not impose any significant strain on the peptide chain. The amount of strain in the metal–ligand bonds has been debated, but it is clearly not a huge quantity in view of the fact that the affinities of the protein for the metal ions are quite high.

More direct probes are based on assessing the effects on the electron-transfer rates of introducing specific changes (mutations) in the protein structure. For this purpose, self-exchange rate constants are used, because they are independent of changes in the standard potentials that can accompany mutations. It is well established that self-exchange rate constants for the blue copper proteins are highly sensitive to alterations in the ionic medium and in the surface charges, because these factors have a strong influence on the association between the two reactants. Thus, for mutationally altered self-exchange rate constants to provide meaningful insights into questions relating to the entatic hypothesis, it is important that the mutations make no significant perturbations to the protein surface. Among the specific mutations studied, one was the conversion of the azurin methionine ligand to a glutamine; this mutation led to a protein that displayed significant oxidation-state-dependent structural changes yet had a self-exchange rate constant that was only two orders of magnitude less than the wild-type protein. A second series of experiments involved mutations where the basic amicyanin ligand set was preserved but the adjoining peptide regions near the Cu binding site were altered (loop replacement); again, only a two-orders-of-magnitude rate reduction ensued. In yet another example, the Asn[47] residue of azurin, which is strongly conserved in blue copper proteins and forms hydrogen bonds to the copper-binding ligands, was mutated to leucine; this mutation altered $E°$ by 110 mV but it had negligible effect on the self-exchange rate. This body of results seems to imply that the rapid self-exchange rate constants attained by blue copper proteins are not a consequence of metal-induced protein strain, minimal structural change, or a highly specific coordination structure. Although the authors have not ventured to claim that these results disprove the entatic hypothesis, it appears that the evidence seems to be accumulating that the entatic hypothesis is in need of significant reevaluation.

D. "Wired" Cytochrome P450

Cytochrome P450 is a class of mono-heme enzymes that catalyzes the oxidation of a wide spectrum of organic substrates including hydrocarbons by O_2. In the resting state the heme is in the low-spin Fe^{III} state, but other states such as high-spin Fe^{III}, high-spin Fe^{II}, and low-spin Fe^{II} have also been described. Electron-transfer steps are believed to be among the components of the catalytic cycle. The heme site is buried rather deeply within the enzyme (~ 20 Å), so that it is not an efficient electron carrier and is resistant to reaction with conventional electron-transfer reagents. In a fascinating variation on the usual methods of protein surface-residue ruthenation, $Ru(bpy)_3$-based substrates for P450 have been designed that engage readily in electron-transfer transactions with the buried heme site. The key has been to derivatize the $[Ru(bpy)_3]^{2+}$ moiety with a long-chain hydrocarbon tail that is terminated by groups that are conventional substrates for P450 oxidation (23). The substrate groups employed include imidazolyl, adamantyl and ethylphenyl, and the hydrocarbon chains range from seven- to nine-membered alkyls. This design permits the substrate to bind in the P450 heme pocket and have a direct hydrocarbon link to a Ru^{II} located at the enzyme surface. A crystal structure (Fig. 10) nicely demonstrates this binding mode (24).

When solutions of P450 and the sensitizer are subjected to flash photolysis, the Ru^{II} group is promoted to an emissive excited state that is quenched by the protein. Analysis of the quenching kinetics in terms of a Förster energy-transfer mechanism yields a Ru–Fe distance that is consistent with the crystallographic distance in the "wired" enzyme, thus implying that the crystal structure is a good representation of the solution structure (24).

Electron-transfer reactions between the bound Ru and heme centers can be performed by a related flash photolysis method (23). In a typical experiment the C_9-imidazolyl substrate/$P450_{cam}$ complex is photolyzed in the presence of para-methoxy-N,N-dimethylaniline, which leads to rapid reductive quenching of *Ru to form a bound Ru^I-imidazole substrate. Subsequent electron transfer from Ru^I to Fe^{III} occurs with $k = 2 \times 10^4 \, s^{-1}$, as detected by the characteristic shift in the heme Soret band. Similar results were obtained with the ethylphenyl and adamantyl substrate derivatives. The efficiency of this system is highlighted by a comparison with the reduction of P450 by putidaredoxin, a natural redox partner that reacts two orders of magnitude more slowly. A fascinating outcome is the similarity of the rate constants for the various substrate derivatives, even though it is believed that the imidazolyl

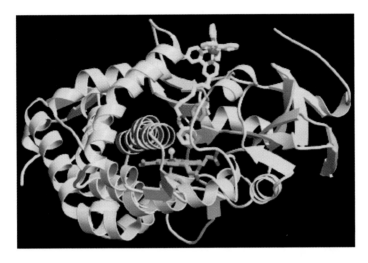

FIG. 10. Crystal structure of the P450$_{cam}$:Ru–C$_9$–Ad complex. The Ru substrate is shown in yellow to highlight docking of Ru(bpy)$_3{}^{2+}$ at the surface of the protein as predicted by computer modeling and energy-transfer experiments. The methylene linker occupies a large channel from the enzyme surface to the heme. A hydrogen bond connects the Ru-substrate amide carbonyl (red atom) to Tyr-96 (orange). The adamantyl moiety (center) resides at the P450 active site above the heme (orange) in the same position as the natural Ad substrate (magenta), shown in superposition from the 4cpp crystal structure. Although both Δ and Λ [Ru(bpy)$_3$]$^{2+}$ enantiomers are present in the complex, only Λ is shown. Reprinted with permission from Ref. (24). Copyright 1999, National Academy of Sciences.

derivative is bound to the iron center while the other two derivatives are not.

In a second type of experiment, *oxidative* quenching is achieved by use of [Co(NH$_3$)$_5$Cl]$^{2+}$ as the quencher. In the one example reported the ethylphenyl derivative of the substrate was used, and the Ru^{III} so generated oxidized the heme with $k = 6 \times 10^3 \, s^{-1}$. From spectroscopic studies it is believed that the heme is oxidized to a porphyrin π-cation radical and has an axial water ligand. One might anticipate the generation of other oxidized states with the use of other substrate derivatives.

VI. Double Electron Transfer

A. INSIGHTS INTO THE MECHANISM OF THE Tl^{3+}/Tl$^+$ SELF-EXCHANGE REACTION

The self-exchange redox reaction between Tl$^+$ and Tl^{3+} has been extensively studied since the 1950s. Several careful determinations of the thermodynamic properties of $Tl^{\bullet 2+}$ are in agreement that this is a quite high-energy species, and it is now clear that the self-exchange reaction cannot proceed through formation of $Tl^{\bullet 2+}$ as in Eq. (18).

$$Tl^{3+} + {}^*Tl + \cancel{\leftrightarrow} \; Tl^{\bullet 2+} + {}^*Tl^{\bullet 2+} \qquad (18)$$

In the presence of complexing anions such as OH^- and Cl^- there is currently a consensus that the reaction proceeds as a "concerted" two-electron transaction through an anion-bridged transition state. There is less consensus regarding the mechanism of the reaction in the absence of such potentially bridging anions. As summarized by Glaser, Taube has discussed the mechanism of this process (25), and he proposes that the reaction proceeds through two sequential one-electron steps without the formation of free $Tl^{\bullet 2+}$. This could occur through the formation of a Tl(II)–Tl(II) bonded intermediate as in Eq. (19).

$$Tl^{3+} + {}^*Tl^+ \longrightarrow (Tl - {}^*Tl)^{4+} \longrightarrow Tl^+ + {}^*Tl^{3+} \qquad (19)$$

Although the proposed Tl_2^{4+} intermediate has never been detected, an analogous species is found in the classical Hg_2^{2+} cation, which also has a metal–metal bond. Further support for this proposal is the observation that low concentrations of Cl^- and Br^- decrease the self-exchange rate: such ligands are expected to stabilize Tl^{3+} more than Tl_2^{4+}.

Quite recently the mechanism of the Tl^{3+}/Tl^+ self-exchange reaction has received detailed theoretical consideration (26), and it considers four specific mechanisms. (1) The first of these mechanisms is the outer-sphere one-electron mechanism as in Eq. (18), and this paper reaffirms that this mechanism can easily be ruled out because of an excessively high activation free energy. (2) The second mechanism to be considered involves simultaneous two-electron outer-sphere electron transfer; if it is assumed that the two reactants approach to a distance of ~ 7–8 Å (contact between the two unperturbed solvation spheres) an activation free energy of $\sim 30 \, \text{kcal mol}^{-1}$ is calculated, which is not that much greater than the apparent experimental value of $23.4 \, \text{kcal mol}^{-1}$. Correction for the effects of nonadiabaticity, however, renders the agreement unsatisfactory. An improvement can be achieved by allowing a closer contact between the ions, with the assumption that the Tl^+ ion is sufficiently weakly solvated that it comes in direct contact with the Tl^{3+} solvation sphere. (3) The third mechanism involves two inner-sphere sequential one-electron transfer steps as in Eq. (19), but it is argued that the $[Tl_2^{II}]^{4+}$ species is "unstable, both thermodynamically and kinetically, and thus can be considered as a virtual intermediate only". From this reader's perspective such a statement must be qualified, as there are many examples of intermediates that are both thermodynamically unstable and kinetically fleeting but nevertheless are quite real species. (4) The last mechanism entails simultaneous two-electron transfer via the "virtual intermediate", $[\text{Tl}_2^{\text{II}}]^{4+}$. It is calculated that an inner-sphere simultaneous two-electron transfer reaction through such a transition state would be adiabatic and have an activation free energy of $\sim 23\,\text{kcal mol}^{-1}$, in good agreement with experiment. Note that the analysis does not rule out the possibility of sequential electron transfer as indicated in Eq. (19). These considerations raise the challenge of either detecting a Tl_2^{4+} species or of setting limits on its lifetime. Further motivation for this work is the suggestion that similar considerations should apply to a whole suite of reactions, including the two-electron oxidation of Hg^0 by Tl^{3+}.

VII. Electrochemical Electron-Transfer Reactions

A. OUTER-SPHERE OXIDATION OF ClO_2^-

A relatively untouched area is the kinetics of outer-sphere electron transfer between an electrode and a main-group species. A significant recent venture in this area is a study of the electrooxidation of aqueous ClO_2^- (27). When this species is oxidized with typical solid electrodes such as platinum, gold, and carbon at 25 °C and pH 7, ClO_2 is generated as a stable product, and hence ClO_2 solutions yield "reversible" cyclic voltammograms at slow scan rates. Such reversibility is almost unique in main-group electrochemistry, largely because of the tendency of the radical species to undergo further chemical reactions.

As the field of electrochemical kinetics may be relatively unfamiliar to some readers, it is important to realize that the rate of an electrochemical process is the current. In transient techniques such as cyclic and pulse voltammetry, the current typically consists of a nonfaradaic component derived from capacitive charging of the ionic medium near the electrode and a "faradaic" component that corresponds to electron transfer between the electrode and the reactant. In a steady-state technique such as rotating-disk voltammetry the current is purely faradaic. The faradaic current is often limited by the rate of diffusion of the reactant to the electrode, but it is also possible that electron transfer between the electrode and the molecules at the surface is the slow step. In this latter case one can define the rate constant as:

$$i = nFAk_{\text{f}}C_{\text{R}}^* \tag{20}$$

Here, i is the faradaic current, n is the number of electrons transferred per molecule, F is the Faraday constant, A is the electrode surface area, k_f is the rate constant, and C_R^* is the bulk concentration of the reactant in units of mol cm^{-3}. In general, the rate constant depends on the applied potential, and an important parameter is k_{el}, the standard rate constant (more typically designated as k^0), which is the forward rate constant when the applied potential equals the formal potential. Since there is zero driving force at the formal potential, the standard rate constant is analogous to the self-exchange rate constant of a homogeneous electron-transfer reaction.

Measurements of the standard rate constant for oxidation of ClO_2^- were performed in three ways. The first was to obtain cyclic voltammograms of ClO_2 solutions at a various scan rates: the peak-to-peak separation, $\Delta E_{p/p}$, is scan-rate dependent when the electrochemical rates are slow (not diffusion limited), and with the use of appropriate working curves one can extract the standard rate constant from the scan rate dependence of $\Delta E_{p/p}$. In the second method rotating-disk electrode (RDE) voltammetry was used. Here the current is measured at a series of fixed applied potentials as a function of the electrode rotation rate. Extrapolation of a plot of $1/i$ vs $1/|\omega$ to infinite rotation rate allows the calculation of the value of k_f at the applied potential. A plot of $\ln k_f$ vs applied potential can then be used to obtain k_{el}. The third method determined k_{el} by phase-sensitive ac voltammetry, a technique that is too complex to be described here. Fine agreement between the three methods was obtained, although it was considered that the accuracy of k_{el} improved from CV through RDE to ac voltammetry.

The results obtained include the values for k_{el} at 25 °C of 0.015 cm s^{-1} at a gold electrode, 0.014 cm s^{-1} at platinum, and 0.0079 cm s^{-1} at glassy carbon. Identical values are to be expected for an outer-sphere mechanism, i.e. where there is no specific bonding between the electrode and the reactant in the transition state for electron transfer. The agreement in k_{el} for Au and Pt is taken in support of an outer-sphere mechanism, and the mildly smaller rate constant at glassy carbon is attributed to the difficulty in preparing a uniform surface on this material. In general, the rate constants can be described as slow (relative to diffusion), and hence there is interest in considering the factors affecting the rate constant. In this endeavor, an estimate of the internal reorganizational energy was made, based on the structures and force constants for both ClO_2^- and ClO_2. The solvent reorganizational energy was estimated by the usual Marcus expression, and the substantial uncertainties introduced by choice of ε_{op} (the optical dielectric constant) were noted. Work terms were also applied, although they were calculated with the unlikely assumption that the potential of zero charge is the same as the potential of the reference electrode. An estimate of the pre-exponential term was performed that included estimates of the nuclear tunneling factor (Γ_n), the nuclear frequency factor (ν_n), the transmission coefficient (κ_{el}°), and the reaction zone thickness (δr). These factors contribute to the rate constant according to

$$k_{el} = \delta r \, \kappa_{el}^{\circ} \nu_n \Gamma_n \exp(-\Delta G^{\ddagger}/RT) \qquad (21)$$

Experimental values of $\Delta G^{\ddagger}$ and the pre-exponential factor were obtained from a plot of $\ln k_{el}$ vs $1/T$ under the assumption that the slope is $-\Delta G^{\ddagger}/R$, and the hidden assumption that $\Delta G^{\ddagger}$ is temperature independent ($\Delta S^{\ddagger}$ is zero). Comparison between the calculated and observed pre-exponential factor was used to infer significant non-adiabaticity, but one may wonder whether inclusion of a nonzero $\Delta S^{\ddagger}$ would alter this conclusion. From an alternative perspective, reasonable agreement was noted for the values of k_{el} and the homogeneous self-exchange rate constant after a standard Marcus-type correction was made for the differing reaction types.

B. ACTIVATION PARAMETERS FOR COUPLED ELECTRON TRANSFER AND SPIN CHANGE

A classic example of a coordination complex that undergoes coupled electron transfer and spin-state change is $[Co(NH_3)_6]^{3+}$, which is a low-spin (diamagnetic) d^6 species. The high-spin d^7 species is produced upon reduction. Three reduction mechanisms may be envisioned: (1) promotion of Co(III) to the high-spin state and production of high-spin Co(II) by subsequent electron transfer; (2) electron transfer to produce low-spin Co(II) followed by relaxation to high-spin Co(II); and (3) concerted spin-state change and electron transfer. The two sequential mechanisms lead to an electrochemical square scheme, while the concerted mechanism corresponds to a diagonal path through the square scheme, as shown in Fig. 11. As spin-state changes for coordination complexes tend to be quite rapid, they are generally treated as equilibria rather than rate-limiting steps. This consideration has made it difficult to distinguish among the three possible pathways for the several systems that have been explored.

In a new twist on this subject, electrochemical activation parameters have been obtained for two series of redox couples that undergo coupled spin-state change and electron transfer (28). One series is $[M(tacn)_2]^{3+/2+}$ where M = Fe, Co, Ni, and Ru, and tacn = 1,4,7-triazacyclononane. The other is $[Fe(pzb)_2]^{+/0}$, where pzb$^-$ = hydrotris(pyrazol-1-yl)borate

$$A\ (E''_{LS},\ k_{th,LS},\ \alpha_{LS})$$

$$LS\text{--}M(III) + e^- \ \rightleftharpoons \ LS\text{--}M(II)$$

$(K_{se,III})\ C\ \updownarrow \qquad\qquad\qquad \updownarrow\ B\ (K_{se,II})$

$$HS\text{--}M(III) + e^- \ \rightleftharpoons \ HS\text{--}M(II)$$

$$D\ (E''_{HS},\ k_{th,HS},\ \alpha_{HS})$$

FIG. 11. Electrochemical square scheme for coupled spin-state change and electron transfer. Reprinted with permission from Ref. (28). Copyright 2002, American Chemical Society.

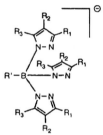

1,4,7-triazacyclononane (tacn)

B(pz)$_4^-$:   R' = pz; R$_1$, R$_2$, R$_3$ = H

HB(pz)$_3^-$:   R', R$_1$, R$_2$, R$_3$ = H

HB(Me$_2$pz)$_3^-$:   R', R$_2$ = H; R$_1$, R$_3$ = CH$_3$

SCHEME 2. Reproduced with permission in part from Ref. (28). Copyright 2002, American Chemical Society.

(HB(pz)$_3^-$), tetrakis(pyrazol-1-yl)borate (B(pz)$_4^-$), and hydrotris(3,5-dimethyl-pyrazol-1-yl)borate (HB(Me$_2$pz)$_3^-$) as shown in Scheme 2.

The cyclic voltammograms of these systems display quasi-reversible behavior, with $\Delta E_{p/p}$ being increased because of slow electrochemical kinetics. Standard electrochemical rate constants, $(k_{s,h})_{obs}$, were obtained from the cyclic voltammograms by matching them with digital simulations. This approach enabled the effects of IR drop (the spatial dependence of potential due to current flow through a resistive solution) to be included in the digital simulation by use of measured solution resistances. These experiments were performed with a non-isothermal cell, in which the reference electrode is maintained at a constant temperature while the temperature at the working electrode is systematically varied. The use of a non-isothermal cell obviates problems associated with re-equilibration of the reference electrode, and it yields absolute activation parameters, i.e. parameters that are independent of ΔH° and ΔS° for the reference electrode. The temperature dependence of $(k_{s,h})_{obs}$ then yields the activation parameters $\Delta H^{\ddagger}_{obs}$ and $\Delta S^{\ddagger}_{obs}$.

The $[Ni(tacn)_2]^{3+/2+}$ and $[Ru(tacn)_2]^{3+/2+}$ couples do not involve spin-state changes and thus serve as points of comparison. The $[Co(tacn)_2]^{3+/2+}$ couple has low-spin Co(III) and high-spin Co(II) and has a much higher value of $\Delta S^{\ddagger}_{obs}$ than do the comparable Ni and Ru couples, as would be expected for a sequential reaction mechanism. The $[Fe(tacn)_2]^{3+/2+}$ couple features a low-spin Fe(III) state and a high-spin/low-spin equilibrium mixture for Fe(II); in this case the intermediate value of $\Delta S^{\ddagger}_{obs}$ supports operation of parallel sequential pathways.

Similar remarks apply to the pyrazolylborate systems. Thus, the $[Fe(B(pz)_4)_2]^{+/0}$ couple is low spin in both oxidation states and serves as a comparison point. The $[Fe(HB(Me_2pz)_3)_2]^{+/0}$ couple has low spin Fe(III) and high spin Fe(II); it shows a large value of $\Delta S^{\ddagger}_{obs}$, consistent with a sequential mechanism. The $[Fe(HB(pz)_3)_2]^{+/0}$ couple has low spin Fe(III) and an equilibrium spin mixture for Fe(II); in this case the spin mixture is largely in the low-spin state, and the value of $\Delta S^{\ddagger}_{obs}$ is similar to that for the low-spin/low-spin analogue.

Overall, the results are taken in support of the sequential (square-scheme) mechanism rather than the concerted alternative. Detailed analysis of the individual systems enabled the authors to distinguish which of the two possible square-scheme pathways was dominant.

C. PHOTOELECTROCHEMICAL CELLS

An area of intense recent research is on photoelectochemical cells. Work in the area has increased exponentially since the 1993 report from Grätzel's group that high-efficiency photoelectrochemical cells could be constructed around dye-sensitized anatase (TiO_2) electrodes (29). The basic design of these cells is to use a photoelectrode fabricated of high-surface-area TiO_2 that is surface-derivatized with compounds related to $[Ru(bpy)_3]^{2+}$. The electrode is bathed in a nonaqueous electrolyte solution such as $I^-/I_2/I_3^-$ in CH_3CN, and the counterelectrode is a simple Pt surface. Upon illumination with visible light, the cell generates current. The basic mechanism of operation is believed to be initiated by photoexcitation of the surface-bound Ru(II) species to *Ru(II), which is followed by rapid transfer of charge to the TiO_2 electrode and the production of Ru(III). The corresponding cathode reaction is the reduction of I_2/I_3^- to

I⁻. Steady-state operation is achieved by reduction of surface-bound Ru(III) back to Ru(II) through the oxidation of I⁻. Thus, chemical reactions mediate the generation of current and the overall process is simply the conversion of light to electrical current. It is believed that these cells hold considerable commercial potential, and since their development entails research in synthetic, electrochemical, photochemical, and inorganic chemistry, the commitment of intensive research to the area is understandable. The dynamics of several types of electron-transfer processes are central in the operation of these cells.

1. Electron-Transfer Dynamics in TiO_2/Sensitizer Systems

The kinetics of three redox processes have been studied for sensitized TiO_2 systems where the sensitizers are [Ru(dicarboxy-bpy)$_2$(CN)$_2$], [Ru(dicarboxy-bpy)$_2$(SCN)$_2$], [Os(dicarboxy-bpy)$_2$(CN)$_2$], and [Os(dicarboxy-bpy)$_2$(SCN)$_2$] (30). The Ru(II) complexes display characteristic excited-state spectra in methanol solution and decay back to the ground state with lifetimes of about 200 ns. For the Os(II) complexes in solution the excited states decay much more rapidly (< 10 ns). On the other hand, when these complexes are adsorbed on TiO_2 excitation leads to the prompt conversion to the M(III) oxidation state, as indicated by transient visible absorption spectra. These results imply that electron injection from all four of the excited sensitizers into the TiO_2 occurs rapidly (< 10 ns).

Efficient operation of these cells requires, *inter alia*, that the M(III) species (obtained by electron injection) be reduced by I⁻ in solution preferentially to recombination with the electrons that were injected into the TiO_2. To probe this competition, both recombination rates and iodide oxidation rates were determined. Recombination rates were investigated by monitoring the recovery of the characteristic ground-state M(II) MLCT bands in the visible region. These recovery rates were found to obey a rate law having two second-order terms, as in Eq. (22).

$$\Delta A = a_1/(1 + a_1 \kappa' t) + a_2/(1 + a_2 \kappa'' t) \qquad (22)$$

Here, the overall absorbance change, ΔA, has two components, a_1 and a_2, and the two second-order rate constants are κ' and κ''. The interpretation of this rate law is that electron injection leads to equal numbers of adsorbed M(III) complexes and injected electrons. Thus, the recombination process is first-order in [M(III)] and [n] where [n] is the concentration of injected electrons. The concentration of M(III) is expressed in molecules cm^{-2} because the M(III) species are surface confined, while the concentration of injected electrons has units of electrons cm^{-3}; these differing concentration dimensions of the two species are unequal but proportional, which leads to an apparent second-order dependence on [M(III)]. This second-order dependence was confirmed by determining the dependence on the excitation intensity. The presence of two terms in the rate law (22) is ascribed to the presence of a distribution of species arising from the heterogeneity of the TiO$_2$ films. An amplitude-weighted average of κ' and κ'' is then computed and converted to a rate constant k_3 with units of cm^3 s^{-1} by correction with a factor of $\Delta\varepsilon d$. A plot of log k_3 as a function of E° for the M(III)/M(II) redox couple, shown in Fig. 12, demonstrates that k_3 decreases smoothly as E° increases, indicating that the recombination reactions are occurring in the Marcus inverted region. Estimates of the driving forces for these reactions can be made based on the assumption that the injected electrons have energies near the edge of the TiO$_2$ conduction band. These driving forces are all greater than 1 eV and are also plotted in Fig. 12. These large driving forces and the small estimated reorganizational energies provide additional support for the conclusion that the recombination reactions lie in the Marcus inverted region. Studies of the dependence on temperature and applied potential were also performed, but need not be discussed here.

For the three M(III) complexes having the largest E° values the recovery of the M(II) signal occurs more rapidly in the presence of iodide in CH$_3$CN, which provides evidence that iodide is oxidized by M(III) competitively with the recombination process. At high iodide concentrations (0.5 M) the recovery obeys first-order kinetics, such that recombination

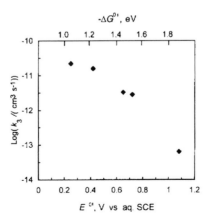

FIG. 12. Dependence of the recombination rate constant k_3 on the ground-state M(III)/M(II) reduction potential of complexes 1–5. Also shown is the estimated driving force $-\Delta G^\circ$ for the charge recombination process. Reprinted with permission from Ref. (*30*). Copyright 2002, American Chemical Society.

can be neglected relative to iodide oxidation. Although the dependence on iodide concentration was not investigated, the pseudo-first-order rate constants for iodide oxidation decreased systematically with decreasing $E°$, thus showing a trend opposite to that of the recombination process.

The above results indicate the great progress being made in understanding the electron-transfer behavior in these sensitizer/semiconductor assemblies. They also highlight the need for further work to establish the full rate laws for iodide oxidation, to obtain better resolution of site heterogeneity in the recombination process, and to define better the nature of the sites for the injected electrons.

2. Electron Transfer with Tripodal Dyes Bound to TiO₂

Classic dye sensitizers in Grätzel cells are compounds of the type [Ru^{II}(R-bpy)$_2$(SCN)$_2$], where R-bpy indicates a derivatized bpy-type ligand that binds specifically to the semiconductor (TiO$_2$) surface. Time-resolved infrared measurements have shown that electron transfer from the *Ru to the TiO$_2$ occurs very rapidly, with $t_{1/2}$ less than 1 ps. In the case where R-bpy is 4,4′-(COOH)$_2$–2,2′-bipyridine the half life is less than 350 fs (31). Despite this very rapid rate of interfacial electron injection, two factors that can limit the net cell quantum yield are recombination (back-electron transfer from the TiO$_2$ to Ru(III)) and excited-state quenching (for example by reduction of I$_2$ by *Ru(II) or by energy-transfer processes). Studies indicate that the rates for these limiting processes are typically several orders of magnitude slower than electron injection, which leads to the suggestion that with appropriate ligand design the rates of electron injection could be slowed without deleterious effects on the overall cell efficiency. These considerations have led to a study on the effects of using dyes that are anchored to the TiO$_2$ with rigid extended spacer groups (32).

The specific rigid dyes have the composition [Ru^{II}(bpy)$_2$L] where L is a bpy or phen-type ligand bearing a tripodal triester moiety as in Scheme 3. The research entailed the design and synthesis of a series of 5 ligands of this type and their Ru(II) complexes (32). Transparent thin films of TiO$_2$ were prepared and soaked in CH$_3$CN solutions of the Ru(II) complexes to prepare the derivatized surfaces. Visible absorption spectra of the soaking solutions and the films were recorded and used to determine the surface coverage; these data were fit with Langmuir isotherms to obtain surface binding constants in the range of 10^6 M^{-1}, which means that 50% coverage is achieved at $\sim 10^{-6}$ M [Ru(II)]. Despite the good fit of the spectroscopic data to Langmuir isotherms,

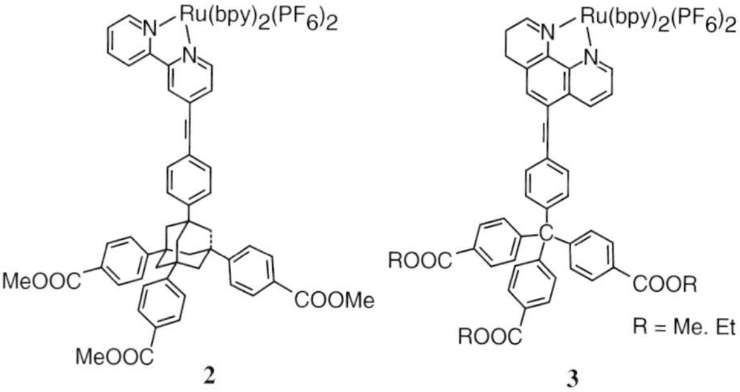

Ru(bpy)$_2$(PF$_6$)$_2$ Ru(bpy)$_2$(PF$_6$)$_2$

MeOOC   COOMe

ROOC   COOR

MeOOC   ROOC   R = Me. Et

2   3

SCHEME 3. Reproduced with permission in part from Ref. (32). Copyright 2002, American Chemical Society.

cyclic voltammetry of the TiO$_2$/Ru(II) films revealed the presence of at least two distinct populations of Ru(II). This heterogeneity was signaled by the magnitude of the integrated CV waves, which showed that only a fraction of the bound Ru(II) was electroactive.

Transient UV–vis absorption spectra showed that the TiO$_2$/Ru(II) films yield prompt electron injection upon photolysis ($k > 10^8 \, s^{-1}$) These same films displayed photoluminescence decays with parallel first- and second-order components, the first-order component having a rate constant of about $1 \times 10^6 \, s^{-1}$. These two sets of results provide further support for the existence of at least two populations of adsorbed Ru(II), one of which injects electrons rapidly and another which does not inject electrons and is thus capable of luminescing on a longer time scale. The second-order component of the luminescence decay is attributed to bimolecular triplet–triplet annihilation of surface-bound *Ru(II). (Note that the "second-order" rate constants reported for luminescence decay have units of s^{-1} because they are actually values for $k_2(\Delta \varepsilon l)$.)

A suggestion for the existence of at least three populations of adsorbed Ru(II) comes from the time evolution of the transient UV–vis absorption spectra. These spectra show that the recovery of the initial Ru(II) spectra occurs with two parallel (fast and slow) second-order components. The rate constants for these two components show remarkably little dependence on the nature of the coordinating ligands. Both of these components are attributed to recombination of the adsorbed Ru(III) with the injected electrons. Thus there is a small luminescent population of Ru(II) that does not engage in electron injection, a non-luminescent population that injects and recombines rapidly, and a third population that injects rapidly and recombines slowly. A detailed picture of the nature of the ligand/semiconductor interaction and how it affects the behavior of these systems awaits further study.

D. METALLOPROTEIN FILM VOLTAMMETRY

It has been shown that a variety of complex redox-linked properties of metalloproteins can be probed by voltammetry of these species adsorbed on the electrodes. Armstrong has been a key investigator in this area, and he has elegantly reviewed salient recent advances (33); the following paragraphs merely summarize some of the highlights of his review.

Armstrong has typically investigated protein films adsorbed on pyrolytic graphite electrodes (PGE). These films are generally monolayers or less, and, for simple electron-transfer proteins such as azurin, they produce reversible cyclic voltammograms indicating that adsorption does not denature the protein or significantly alter its redox properties. Adsorption, however, does eliminate the effects of diffusion and hence can afford considerable insight into complex associated phenomena such as gating and catalytic mechanisms.

One system probed in considerable detail is ferredoxin 1 of *Azotobacter vinelandii*. This protein contains two redox-active groups: a [4Fe–4S] and a [3Fe–4S] cluster. Proton transfer is coupled to electron transfer of the $[3Fe–4S]^{+/0}$ redox couple. Various chemical techniques have indicated that the electron/proton-transfer coupling is due to protonation of a μ-sulfido group in the reduced $[3Fe–4S]^0$ state. Cyclic voltammetry of the [3Fe–4S] site in adsorbed ferredoxin 1 is reversible at low scan rates and shows a pH-dependent redox potential as can be expected. Rapid-scan voltammetry at high pH (pH 8.55) shows displacements of the anodic and cathodic waves, indicative of rate-limiting electron transfer. Quite different behavior is observed at low pH (pH 4.59); at this pH, when the sample is poised at high potential to place the protein in the oxidized $[3Fe–4S]^+$ state, a normal reduction wave is observed at rapid scan rates but the reverse (anodic) wave disappears. This type of irreversibility is an example of gating; it implies that $[3Fe–4S]^+$ is readily reduced to $H^+–[3Fe–4S]^0$, but this species must be deprotonated prior to its oxidation and this proton transfer can be too slow to yield currents at rapid scan rates. Digital simulation of the cyclic voltammograms with a mechanism having rate-limiting proton transfer provided quantitative support for this interpretation.

A mutant form of ferredoxin 1 has been obtained in which the surface aspartate-15 residue is replaced by asparagine. With this mutant protein the voltammograms are similar to the wild type, except that the anodic wave at low pH disappears at a lower scan rate and reappears at very rapid scan rates. "Trumpet plots" of the CV data, as shown in Fig. 13 are quite helpful in displaying these results. The fact that gating occurs at lower scan rates for the mutant than for the wild type implies that

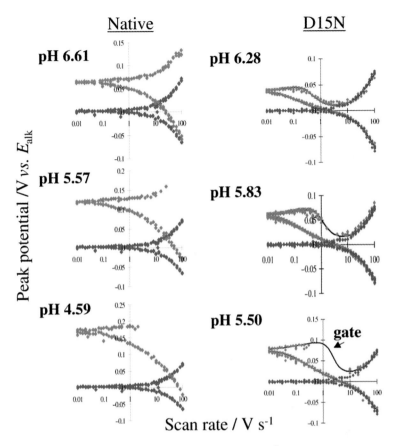

FIG. 13. Representative 'Trumpet Plots' for the $[3Fe–4S]^{+/0}$ couple in native and D15N mutant forms of *Azotobacter vinelandii* ferredoxin I adsorbed on a PGE electrode. The plots for D15N also show the fits based on $k_{off} = 2.5\ s^{-1}$. Note the intermediate region of the plot (pH 5.50) in which an oxidation peak is not observed because ET is gated. Data points shown in red are for the pH values indicated whereas data points shown in blue are for the uncoupled electron-transfer reaction occurring at pH > $pK_{cluster}$. Reproduced from Ref. (*33*) by permission of the Royal Society of Chemistry.

proton transfer in the mutant is slower than in the wild type, and it supports the hypothesis that aspartate-15 functions as a proton relay by swinging its carboxylate proton acceptor group from a solvent-exposed position to one that is proximal to the internal [3Fe–4S] site.

   Studies of protein film electrocatalysis have also been illuminating. For example, succinate dehydrogenase displays an unusual optimal potential for activity. The enzyme contains four redox sites: a flavin, a

[2Fe–2S], a [4Fe–4S] and a [3Fe–4S] center. The enzyme catalyzes the reversible redox conversion of succinate to fumarate. Voltammetry of the enzyme on PGE electrodes in the presence of fumarate shows a catalytic wave for the reduction of fumarate to succinate (much more current than could be accounted for by the stoichiometric reduction of the protein active sites). Typical catalytic waves have a sigmoidal shape at a rotating disk electrode, but in the case of succinate dehydrogenase the catalytic wave shows a definite peak. This window of optimal potential for electrocatalysis seems to be a consequence of having multiple redox sites within the enzyme. Similar results were obtained with DMSO reductase, which contains a Mo-bis(pterin) active site and four [4Fe–4S] centers.

ACKNOWLEDGEMENTS

The NSF (USA) is thanked for support of this work. Dr. Vince Cammarata (Auburn University) is thanked for his comments relating to Section VII.

REFERENCES

1. Becker, R. H.; Bartlett, W. P.; Urbansky, E. T.; Margerum, D. W. *J. Chem. Soc., Dalton Trans.* **2002**, 695–700.
2. Coyle, J. L.; Elias, H.; Herlinger, E.; Lange, J.; Nelson, J. *J. Biol. Inorg. Chem.* **2001**, *6*, 285–291.
3. Rotzinger, F. P. *J. Chem. Soc., Dalton Trans.* **2002**, 719–728.
4. Krishnamurty, K. V.; Wahl, A. C. *J. Am. Chem. Soc.* **1958**, *80*, 5921–5924.
5. Pfeiffer, J.; Kirchner, K.; Wherland, S. *Inorg. Chim. Acta* **2001**, *313*, 37–42.
6. Kandegedara, A.; Krylova, K.; Nelson, T. J.; Schroeder, R. R.; Ochrymowycz, L. A.; Rorarbacher, D. B. *J. Chem. Soc., Dalton Trans.* **2002**, 792–801.
7. Tóth, Z.; Fábián, I. *Inorg. Chem.* **2000**, *39*, 4608–4614.
8. Tóth, Z.; Fábián, I.; Bakač, A. *Inorg. React. Mech.* **2001**, *3*, 147–152.
9. Stanbury, D. M. *Adv. Inorg. Chem.* **1989**, *33*, 69–138.
10. Saha, B.; Stanbury, D. M. *Inorg. Chem.* **2001**, *40*, 5139–5146.
11. Lente, G.; Fábián, I. *J. Chem. Soc., Dalton Trans.* **2002**, 778–784.
12. Sun, J.; Stanbury, D. M. *J. Chem. Soc., Dalton Trans.* **2002**, 785–791.
13. Li, C.; Hoffman, M. Z. *J. Phys. Chem. B* **1999**, *103*, 6653–6656.
14. Li, C.; Hoffman, M. Z. *J. Phys. Chem. A* **2000**, *104*, 5988–6002.
15. Vlcek, A., Jr.; Farrell, I. R.; Liard, D. J.; Matousek, P.; Towrie, M.; Parker, A. W.; Grills, D. C.; George, M. *J. Chem. Soc., Dalton Trans.* **2002**, 701–712.
16. Udugala-Ganehenege, M. Y.; Heeg, M. J.; Hryhorczuk, L. M.; Wenger, L. E.; Endicott, J. F. *Inorg. Chem.* **2001**, *40*, 1614–1625.
17. Geren, L. M.; Beasley, J. R.; Fine, B. R.; Saunders, A. J.; Hibdon, S.; Pielak, G. J.; Durham, B.; Millett, F. *J. Biol. Chem.* **1995**, *270*, 2466–2472.
18. Wang, K.; Zhen, Y.; Sadoski, R.; Grinnell, S.; Geren, L.; Ferguson-Miller, S.; Durham, B.; Millett, F. *J. Biol. Chem.* **1999**, *274*, 38 042–38 050.

19. Wang, K.; Geren, L.; Zhen, Y.; Ma, L.; Ferguson-Miller, S.; Durham, B.; Millett, F. *Biochemistry* **2002**, *41*, 2298–2304.
20. Pacheco, A.; Hazzard, J. T.; Tollin, G.; Enemark, J. H. *J. Biol. Inorg. Chem.* **1999**, *4*, 390–401.
21. Feng, C.; Kedia, R. V.; Hazzard, J. T.; Hurley, J. K.; Tollin, G.; Enemark, J. H. *Biochemistry* **2002**, *41*, 5816–5821.
22. Canters, G. W.; Kolczak, U.; Armstrong, F.; Jeuken, L. J. C.; Camba, R.; Sola, M. *Faraday Discussions* **2000**, *116*, 205–220.
23. Wilker, J. J.; Dmochowski, I. J.; Dawson, J. H.; Winkler, J. R.; Gray, H. B. *Angew. Chem. Int. Ed.* **1999**, *38*, 90–92.
24. Dmochowski, I. J.; Crane, B. R.; Wilker, J. J.; Winkler, J. R.; Gray, H. B. *Proc. Natl. Acad. Sci. USA* **1999**, *96*, 12 987–12 990.
25. Glaser, J. *Adv. Inorg. Chem.* **1995**, *43*, 1–78.
26. Khoshtariya, D. E.; Dolidze, T. D.; Zusman, L. D.; Lindbergh, G.; Glaser, J. *Inorg. Chem.* **2002**, *41*, 1728–1738.
27. Sinkaset, N.; Nishimura, A. M.; Pihl, J. A.; Trogler, W. C. *J. Phys. Chem. B* **1999**, *103*, 10 461–10 469.
28. Turner, J. W.; Schultz, F. A. *J. Phys. Chem. B* **2002**, *106*, 2009–2017.
29. Nazeeruddin, M. K.; Kay, A.; Rodicio, I.; Humphry-Baker, R.; Müller, E.; Liska, P.; Vlachopoulos, N.; Grätzel, M. *J. Am. Chem. Soc.* **1993**, *115*, 6382–6390.
30. Kuciauskas, D.; Freund, M. S.; Gray, H. B.; Winkler, J. R.; Lewis, N. S. *J. Phys. Chem. B* **2001**, *105*, 392–403.
31. Heimer, T. A.; Heilweil, E. J.; Bignozzi, C. A.; Meyer, G. J. *J. Phys. Chem. B* **2000**, *104*, 4256–4262.
32. Galoppini, E.; Guo, W.; Zhang, W.; Hoertz, P. G.; Qu, P.; Meyer, G. J. *J. Am. Chem. Soc.* **2002**, *124*, 7801–7811.
33. Armstrong, F. A. *J. Chem. Soc., Dalton Trans.* **2002**, 661–671.
34. Roberts, V. A.; Pique, M. E. *J. Biol. Chem.* **1999**, *274*, 38 051–38 060.

METAL ION CATALYZED AUTOXIDATION
REACTIONS: KINETICS AND MECHANISMS

ISTVÁN FÁBIÁN and VIKTOR CSORDÁS

University of Debrecen, Department of Inorganic and Analytical Chemistry,
Debrecen, Egyetem tér 1, H-4010, Hungary

I. Introduction

Indisputable biological significance, industrial applications and environmental issues have combined to generate considerable interest in the redox chemistry of dioxygen which offers an inexhaustible pool of scientific problems and has challenged researchers for well over a century. Indeed, considerable effort has been invested into understanding both the basic principles and the fine details of autoxidation processes. The most fundamental properties are discussed extensively in inorganic chemistry text books, and a recent monograph by Sawyer (1) provides a deep insight into oxygen chemistry. Numerous publications deal with preparative, kinetic, mechanistic, theoretical, biochemical, analytical and other aspects of autoxidation reactions and thorough reviews have been compiled from time to time to cover specific areas. Due to the enormous amount of information reported on these reactive

ADVANCES IN INORGANIC CHEMISTRY
VOLUME 54 ISSN 0898-8838

systems, such compilations usually reflect the interest of the authors and were often criticized as somewhat biased. This paper is no exception as we will limit our discussion to recent results on the kinetics and mechanism of metal ion[1] catalyzed autoxidation reactions. While we do not intend to cover older studies in detail, in some cases the most important results from earlier publications will be surveyed in order to put the new studies in context.

Metal ions play an important role as catalysts in many autoxidation reactions and have been considered instrumental in regulating natural as well as industrial processes. In these reactive systems, in particular when the reactions occur under environmental or in vivo biochemical conditions, the metal ions are involved in complicated interactions with the substrate(s) and dioxygen, and the properties of the actual matrix as well as the transport processes also have a pronounced impact on the overall reactions. In most cases, handling and analyzing such a complexity is beyond the capacity of currently available experimental, computational and theoretical methods, and researchers in this field are obliged to use simplified sub-systems to mimic the complex phenomena. When the simplified conditions are properly chosen, these studies provide surprisingly accurate predictions for the 'real' systems. In this paper we review the results obtained in kinetic and mechanistic studies on the model systems, but we do not discuss their broad biological or environmental implications.

A brief overview on why most of the autoxidation reactions develop complicated kinetic patterns is given in Section II. A preliminary survey of the literature revealed that the majority of autoxidation studies were published on a small number of substrates such as L-ascorbic acid, catechols, cysteine and sulfite ions. The results for each of these substrates will be discussed in a separate section. Results on other metal ion mediated autoxidation reactions are collected in Section VII. In recent years, non-linear kinetic features were discovered in some systems containing dioxygen. These reactions form the basis of a new exciting domain of autoxidation chemistry and will be covered in Section VIII.

---

[1]The term "metal ion" will be used in a general sense throughout this paper. Distinction between metal ions and complexes as well as between different isomers of the complexes will be made only when it is required for the clarity of the presentation. Furthermore, the accessible coordination sites of the metal ions are always assumed to be occupied by the corresponding solvent molecules which will be shown only when they are directly involved in a given redox reaction.

II. General Considerations

Dioxygen is a formidable oxidizing agent with relatively high standard redox potential under acidic conditions (2):

$$O_2 + 4H^+ + 4e^- \rightleftharpoons 2H_2O \qquad \varepsilon^0 = 1.229 \text{ V} \qquad (1)$$

The redox potential is strongly pH dependent and it decreases considerably in alkaline solution:

$$O_2 + 2H_2O + 4e^- \rightleftharpoons 4OH^- \qquad \varepsilon^0 = 0.401 \text{ V} \qquad (2)$$

The thermodynamic properties are also affected by the solvent and the composition of the reaction mixture, for example the corresponding values in acetonitrile are $+1.79$ V (1 M $HClO_4$) and -0.53 V (1 M Bu_4NOH), respectively (3).

The reduction of dioxygen to its fully reduced form, H_2O, requires the transfer of 4 electrons, and the transfer may proceed via a series of intermediate oxidation states, such as $O_2^{\bullet-}/HOO^{\bullet}$, $HOO^-/HOOH$, $O^{\bullet-}/OH^{\bullet}$. These reduced forms of oxygen exhibit different redox properties and in the presence of substrate(s) and/or catalyst(s) may open different reaction paths for the electron transfer process. Fast proton transfer reactions between the corresponding acid-base pairs can introduce composite pH dependencies into the kinetic and stoichiometric characteristics of these systems.

Oxidation reactions by dioxygen are often kinetically hindered and do not occur even if the reaction is favored thermodynamically. The simplest explanation for this can be given by assuming that the electron transfer to produce lower oxidation state intermediates occurs step by step. The redox potentials for the formation of $HOO^{\bullet}$ and $O_2^{\bullet-}$ are -0.05 V (2) and -0.325 V (4), respectively. It follows that only relatively strong reducing agents are capable of reducing dioxygen to $O_2^{\bullet-}/HO_2^{\bullet}$ and, in terms of thermodynamics, the limiting step is the formation of the superoxide radical or its acidic form in any stepwise electron transfer reaction of O_2. The superoxide radical is a relatively strong oxidant and it can readily be reduced to hydrogen peroxide. Thus, once $O_2^{\bullet-}$ ($HO_2^{\bullet}$) is formed, thermodynamic constraints are not expected to affect the overall kinetics.

For a more in-depth interpretation of the inertness of dioxygen, the fact that O_2 is a triplet state bi-radical, i.e. it has two unpaired electrons in the $2\pi_g$ orbitals, needs to be considered. It follows that the oxidation of singlet state substrates by the triplet O_2 to form singlet products is spin-forbidden and, as a consequence, relatively slow.

The main role of the metal catalysts is to enhance the $O_2 + e^- \rightarrow O_2^{\bullet-}$ transition or mediate the electron transfer between oxygen and substrate (S) via an alternative path. The following basic reaction schemes can be envisioned:

$$M^{n+} + O_2 \longrightarrow M^{(n+1)+} + O_2^{\bullet-} \tag{3}$$

$$O_2^{\bullet-} + S \longrightarrow I \longrightarrow P \tag{4}$$

$$M^{(n+1)+} + I \longrightarrow M^{n+} + P \tag{5}$$

SCHEME 1.

where I and P stand for one or more intermediates and products, respectively.

In this case, direct electron transfer between the catalyst and O_2 produces the superoxide radical (or other reduced forms of oxygen) which can be involved in a series of subsequent redox reactions. If these reactions are relatively fast, the rate determining step is Eq. (3) and the overall process can be interpreted in terms of relatively simple rate laws.

It should be noted that the metal ion may enhance the oxidation rate of the substrate without being a real catalyst. In this case the oxidation of the substrate becomes faster because of the formation of reactive species in Eqs. (3) and (4), but Eq. (5) does not occur and the metal ion is not recycled to its original oxidation state. Consequently, the 'catalyst' quickly loses its activity over the course of the reaction.

$$M^{n+} + S \longrightarrow MS^{n+} \tag{6}$$

$$MS^{n+} + O_2 \longrightarrow MS \cdot O_2^{n+} \tag{7}$$

$$MS \cdot O_2 \longrightarrow P + M^{n+} \tag{8}$$

SCHEME 2.

In this case, the actual redox step is preceded by the formation of an adduct or a complex between the catalyst, the substrate and dioxygen. The order of these reaction steps is irrelevant as long as the rate determining step is Eq. (8). If Eqs. (6) and (7) are rapidly established pre-equilibria the reaction rate depends on the concentrations of all reactants. In some instances, the rate determining step is the formation of the MS^{n+} complex and the reaction rate is independent of the concentration of dioxygen.

It should be added that $MS \cdot O_2$ is not necessarily a mono-nuclear complex. It could be shown in a few cases that the catalytic activity of the metal ion is due to the formation of dinuclear metal-substrate complexes. Presumably in these species each oxygen atom of dioxygen coordinates to a different metal center. Such systems were extensively used to model the reactivity patterns of various enzymes containing a bimetallic active center.

$$M^{n+} + S_1 \longrightarrow M^{(n-1)+} + I \qquad (9)$$

$$I + O_2 \longrightarrow I' \qquad (10)$$

$$M^{(n-1)+} + O_2 \longrightarrow M^{n+} + O_2^{\bullet -} \qquad (11)$$

$$I', \ O_2^{\bullet -} + S_2 \longrightarrow P \qquad (12)$$

SCHEME 3.

This scheme implies that dioxygen is activated only in a secondary step by an intermediate formed in the initial electron transfer reaction between the metal ion and a co-substrate, S_1. The reduced form of the metal ion is re-oxidized to its original oxidation state by O_2, but such a reaction with a secondary intermediate cannot be excluded. If the rate determining step is Eq. (9), the overall reaction is again zeroth order with respect to dioxygen.

The common element of Schemes 1–3 is that they each postulate direct interaction between the metal center and dioxygen. Although it is not stated explicitly, Eqs. (3) and (11) most likely proceed via an inner-sphere mechanism. Thus, the metal–dioxygen interaction implies spin pairing between the reactants when the metal ion is paramagnetic. As a consequence, the formation of the $M–O_2$ type intermediates circumvents the restriction posed by the triplet to singlet transition which seems to be the major kinetic barrier of autoxidation reactions (5).

It should be emphasized that clear-cut situations described in Schemes 1–3 are uncommon and typically the combination of these models needs to be considered for kinetic and mechanistic description of a real system. However, even when one of the limiting cases prevails, each of these models may predict very different formal kinetic patterns depending on where the rate determining step is located. For the same reason, different schemes may be consistent with the same experimental rate law, i.e. thorough formal kinetic description of a reaction and the analysis of the rate law may not be conclusive with respect to the mechanism of the autoxidation process.

The general features discussed so far can explain the complexity of these reactions alone. However, thermodynamic and kinetic couplings between the redox steps, the complex equilibria of the metal ion and/or the proton transfer reactions of the substrate(s) lead to further complications and composite concentration dependencies of the reaction rate. The speciation in these systems is determined by the absolute concentrations and the concentration ratios of the reactants as well as by the pH which is often controlled separately using appropriately selected buffers. Perhaps, the most intriguing task is to identify the active form of the catalyst which can be a minor, undetectable species. When the protolytic and complex-formation reactions are relatively fast, they can be handled as rapidly established pre-equilibria (thermodynamic coupling), but in any other case kinetic coupling between the redox reactions and other steps needs to be considered in the interpretation of the kinetics and mechanism of the autoxidation process. This may require the use of comprehensive evaluation techniques.

The model shown in Scheme 2 indicates that a change in the formal oxidation state of the metal is not necessarily required during the catalytic reaction. This raises a fundamental question. Does the metal ion have to possess specific redox properties in order to be an efficient catalyst? A definite answer to this question cannot be given. Nevertheless, catalytic autoxidation reactions have been reported almost exclusively with metal ions which are susceptible to redox reactions under ambient conditions. This is a strong indication that intramolecular electron transfer occurs within the MS^{n+} and/or $MS \cdot O_2$ precursor complexes. Partial oxidation or reduction of the metal center obviously alters the electronic structure of the substrate and/or dioxygen. In a few cases, direct spectroscopic or other evidence was reported to prove such an internal charge transfer process. This electronic distortion is most likely necessary to activate the substrate and/or dioxygen before the actual electron transfer takes place. For a few systems where deviations from this pattern were found, the presence of trace amounts of catalytically active impurities are suspected to be the cause. In other words, the catalytic effect is due to the impurity and not to the bulk metal ion in these cases.

Examples of how the general principles apply to specific reactions will be discussed in the following Sections.

III. Autoxidation of L-Ascorbic Acid

Autoxidation reactions of L-ascorbic acid (H_2A) have been the subject of intensive studies for decades. It was shown that some of the most common transition metal ions are very active catalysts for this reaction (6). The majority of earlier studies focused on copper catalysis because of the biological importance of the reactions of ascorbic acid with oxidase enzymes containing copper binding sites. Under slightly acidic or neutral conditions the copper catalyzed reaction was confirmed to produce dehydroascorbic acid (A) and hydrogen peroxide in the initial phase (7):

$$H_2A + O_2 \longrightarrow A + H_2O_2 \tag{13}$$

The oxidation of dehydroascorbic acid to a higher oxidation product is relatively slow and could be neglected for all practical purposes in kinetic and stoichiometric studies.

Earlier results regarding the kinetics and mechanism of the copper(II) catalysis are controversial. Reaction orders for $[O_2]$, $[Cu^{II}]$, $[H_2A]$ and $[H^+]$ were reported in the following respective ranges: 0.5 to 1, 0.5 to 1, 0 to 1, and -2 to $+1$ (8). It is also disputed whether the redox cycling of the catalyst includes oxidation states $+1$ and $+2$ or $+2$ and $+3$. The discrepancies are too marked to be explained only by the differences in the experimental conditions applied.

The first thorough study on the Cu(II) and Fe(III) catalyzed autoxidation of ascorbic acid was reported by Taqui Khan and Martell (6). These authors found evidence for a slow, overall second-order reaction between the HA^- form of ascorbic acid and dioxygen in the absence of added catalyst. The corresponding rate constant was reported to be $0.57 \, M^{-1} s^{-1}$ (aqueous solution, $25\,^\circ$C, $\mu = 0.1 \, M \, KNO_3$). Later studies also supported the existence of such a reaction path (9), which was negligible under catalytic conditions.

In the presence of Cu(II) or Fe(III), a two-term rate law was confirmed in which the two terms correspond to the reaction paths via the mono- and diprotonated forms of ascorbic acid. The reaction was found to be first-order with respect to $[H_2A]_{tot}$, $[O_2]$ and $[M^{n+}]$ with both catalysts:

$$-d[H_2A]/dt = (k_{14}^a [HA^-] + k_{14}^b [H_2A])[M^{n+}][O_2] \tag{14}$$

The third-order rate constants for the Cu(II) and Fe(III) catalyzed reactions are: $k_{14}^a = 6.0 \times 10^7$ and $2.4 \times 10^7 \, M^{-2} s^{-1}$, $k_{14}^b = 3.8 \times 10^5$ and $4.0 \times 10^5 \, M^{-2} s^{-1}$, respectively.

The possibility that the metal ion is reduced by ascorbate ion in a rate determining step could be excluded because in that case the rate would be independent of $[O_2]$. Thus, the following mechanism was proposed for the catalytic oxidation of HA^-:

$$H_2A \rightleftharpoons HA^- + H^+ \tag{15}$$

$$HA^- + M^{n+} \rightleftharpoons [M(HA)]^{(n-1)+} \tag{16}$$

$$[M(HA)]^{(n-1)+} + O_2 \rightleftharpoons [M(HA)(O_2)]^{(n-1)+} \tag{17}$$

$$[M(HA)(O_2)]^{(n-1)+} \longrightarrow [M(HA)^+(O_2^-)]^{(n-1)+} \tag{18}$$

$$[M(HA)^+(O_2^-)]^{(n-1)+} \longrightarrow A^{\bullet-} + M^{n+} + HO_2^\bullet \tag{19}$$

$$A^{\bullet-} + M^{n+} \longrightarrow A + M^{(n-1)+} \tag{20}$$

$$A^{\bullet-} + O_2 \longrightarrow A + O_2^{\bullet-} \tag{21}$$

$$M^{(n-1)+} + HO_2^\bullet + H^+ \longrightarrow M^{n+} + H_2O_2 \tag{22}$$

$$O_2^{\bullet-} + H^+ \rightleftharpoons HO_2^\bullet \tag{23}$$

In this mechanism, the rate-determining step is the intramolecular electron transfer from the ligand to dioxygen, Eq. (18), via the metal center of the $[M(HA)(O_2)]^{(n-1)+}$ complex for which the following structure was proposed:

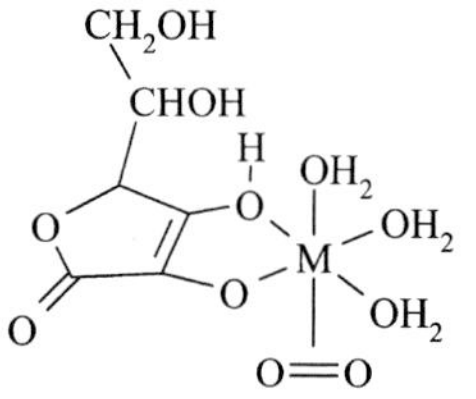

$$MHA(O_2)^{(n-1)+}$$

The intramolecular electron transfer leads to fast formation of semiquinone and the lower oxidation state metal ion. The catalytic cycle is completed by fast reoxidation of the metal ion. Significant deviations from this model were observed at low dioxygen concentrations and it was suggested that another oxidation path becomes operative under such conditions. Although earlier they had been proposed to participate (10), side reactions with dehydroascorbic acid could be excluded.

The same model was applied to the oxidation of the H_2A form of ascorbic acid. In this case iron(III) was found to be a somewhat more active catalyst than copper(II). The difference could be explained by assuming that Fe(III) forms a more stable complex with H_2A than does Cu(II) because of the higher charge of the metal ion.

The formation of the $[M(HA)]^{(n-1)+}$ complex was confirmed in independent pH-metric experiments in the case of copper(II). These studies also provided evidence that ascorbic acid is coordinated to the metal center in its monoprotonated form. Because of relatively fast redox reactions between iron(III) and ascorbic acid, similar studies to confirm the formation of $[Fe(HA)]^{2+}$ were not feasible. However, indirect kinetic evidence also supported the formation of the $[M(HA)]^{(n-1)+}$ complex in both systems (6).

Electron transfer within the $[M(HA)(O_2)]^{(n-1)+}$ complex was envisioned as a two-stage process in which first, a 2p electron of the ascorbate oxygen is transferred to a t_{2g} non-bonding or an e_g antibonding orbital of the metal ion. The subsequent step is the transfer of an electron to the π_y^*2p or π_z^*2p orbital of the oxygen molecule.

Although some of the results reported by Taqui Khan and Martell (6) were challenged in subsequent papers, several elements of their model were included in the interpretation of the results obtained in related studies.

The kinetic results reported by Jameson and Blackburn (11,12) for the copper catalyzed autoxidation of ascorbic acid are substantially different from those of Taqui Khan and Martell (6). The former could not reproduce the spontaneous oxidation in the absence of added catalysts when they used extremely pure reagents. These results imply that ascorbic acid is inert toward oxidation by dioxygen and earlier reports on spontaneous oxidation are artifacts due to catalytic impurities. In support of these considerations, it is worthwhile noting that trace amounts of transition metal ions, in particular Cu(II), may cause irreproducibilities in experimental work with ascorbic acid (13). While this problem can be eliminated by masking the metal ion(s), the masking agent needs to be selected carefully since it could become involved in side reactions in a given system.

Jameson and Blackburn confirmed the following rate law in 0.1 M KNO_3 and at 25 °C (11):

$$-d[O_2]/dt = k_{24}[Cu(II)][HA^-][O_2]^{1/2} \qquad (24)$$

with $k_{24} = 2.2 \times 10^3 \, M^{-3/2} s^{-1}$.

The difference in the rate expressions shown in Eqs. (14) and (24) probably reflects the differences in the experimental conditions of the corresponding studies. It should be noted that the validity of Eq. (14) was confirmed at relatively high $[O_2]$ (6), and the deviations from the rate law at lower oxygen concentrations found in that study seem to be consistent with Eq. (24). The half-order dependence on $[O_2]$ led to the conclusion that a chain mechanism is operative in this reaction, and the following kinetic model was proposed (11):

$$Cu^{2+} + HA^- \rightleftharpoons [Cu(HA)]^+ \qquad\qquad K_{25} \qquad (25)$$

$$2[Cu(HA)]^+ \rightleftharpoons [Cu_2(HA)_2]^{2+} \qquad\qquad K_{26} \qquad (26)$$

$$[Cu_2(HA)_2]^{2+} + O_2 \rightleftharpoons [Cu_2(HA)_2(O_2)]^{2+} \qquad K_{27} \qquad (27)$$

$$[Cu_2(HA)_2(O_2)]^{2+} \longrightarrow [CuA(O_2H)]^\bullet + Cu^{2+} + A^{\bullet-} + H^+ \quad k_{28} \qquad (28)$$

$$[Cu_2(HA)_2(O_2)]^{2+} + A^{\bullet-} \longrightarrow [CuA(O_2H)]^\bullet + [Cu(HA)]^+ + A \quad k_{29} \qquad (29)$$

$$[CuA(O_2H)]^\bullet \longrightarrow Cu^{2+} + A^{\bullet-} + HO_2^- \qquad k_{30} \qquad (30)$$

$$2[CuA(O_2H)]^\bullet \longrightarrow 2Cu^{2+} + HA^- + A + 2O_2^{2-} \qquad k_{31} \qquad (31)$$

Equilibrium studies under anaerobic conditions confirmed that $[Cu(HA)]^+$ is the major species in the Cu(II)–ascorbic acid system. However, the existence of minor polymeric, presumably dimeric, species could also be proven. This lends support to the above kinetic model. Provided that the catalytically active complex is the dimer produced in reaction (26), the chain reaction is initiated by the formation and subsequent decomposition of $[Cu_2(HA)_2(O_2)]^{2+}$ into $[CuA(O_2H)]^\bullet$ and $A^{\bullet-}$. The chain carrier is the semi-quinone radical which is consumed and regenerated in the propagation steps, Eqs. (29) and (30). The chain is terminated in Eq. (31). Applying the steady-state approximation to the concentrations of the radicals, yields a rate law which is fully consistent with the experimental observations:

$$k_{24} = k_{30}(k_{28}/k_{31})^{1/2} (K_{25}K_{26}K_{27})^{1/2} \qquad (32)$$

Chloride ion has a profound effect on the kinetics because the rate is half-order in both ascorbic acid and dioxygen, but first-order in Cu(II) in 0.1 M KCl (12). It was assumed that in this case the catalytically active species is a $[Cu_2A]^{2+}$ type complex which somehow also incorporates chloride ion; however, the exact composition of the complex was not clarified. The role of this dimer is very similar to that of $[Cu_2(HA)_2(O_2)]^{2+}$ in the absence of Cl^-, i.e. after coordinating dioxygen it generates $[CuA(O_2)]^{\bullet-}$ (the conjugate base form of $[CuA(O_2H)]^\bullet$) and $A^{\bullet-}$. These species are involved in propagation steps analogous to those

in Eqs. (29) and (30). An additional reaction sequence was also included
in the model which postulates the formation of Cu(III):

$$[Cu_2A(O_2)]^{2+} \longrightarrow [CuA(O_2)]^{\bullet-} + Cu^{3+} \tag{33}$$

$$Cu^{3+} + H_2O \rightleftharpoons Cu^{2+} + OH^{\bullet} + H^{+} \tag{34}$$

$$OH^{\bullet} + HA^{-} \longrightarrow A^{\bullet-} + H_2O \tag{35}$$

According to these models, oxygen is activated via the formation of the
dinuclear oxygen complexes. It was hypothesized that the binding of
oxygen is perpendicular to the Cu–Cu bond and the π_x and π_y orbitals of
dioxygen interact with the d_{xz} and d_{yz} metal orbitals. Thus, the reduction
of O_2 may occur in a concerted two-electron process. This assumption is
in strict contrast with the stepwise one-electron mechanism proposed
by Taqui Khan and Martell (6). A further difference is that the formation
of a Cu(I) intermediate was excluded from the latter models.

Features of the UV–Vis spectra obtained in the absence of oxygen indicate ligand to metal charge transfer in the Cu(II)–ascorbate complexes
(11). This would suggest that the formation of complexes, formally
copper(I), is favored in this reaction even if the oxidation of ascorbic
acid by copper(II), i.e. the complete electron transfer is thermodynamically unfavorable (14). First, Jameson and Blackburn presented convincing arguments that the electronic structure of the ascorbate
complexes in the absence of oxygen is irrelevant with respect to the
autoxidation reaction and the catalytic reaction proceeds via a Cu(II)–
Cu(III) cycle (11,12). They proposed that the two-electron reduction of
oxygen in $[Cu_2(HA)_2(O_2)]^{2+}$ yields two non-equivalent copper(III)–ascorbate units. While one of them immediately dissociates into Cu^{2+} and the
semi-quinone radical, copper(III) remains bound to ascorbate and
peroxide in the other species. It was noted that the oxidation state of
copper cannot be established firmly in the latter species and a
copper(II)–semiquinone–peroxide arrangement can also be feasible. In
any case, the formation of copper(I) was clearly excluded, and a Cu(III)
intermediate was explicitly included in the kinetic model in the
presence of Cl^{-}.

In a follow-up study, Jameson and Blackburn studied the effect of
chloride ion in more detail and reported the following concentration
dependence of the pseudo-half-order rate constant for the dioxygen
decay (15):

$$k_{obs} = \frac{A + B[Cl^{-}]}{C + D[Cl^{-}] + E[Cl^{-}]^2} \tag{36}$$

where A, B, C, D and E are experimentally determined parameters.

The effect of chloride ion was interpreted in terms of the formation of various ternary complexes between Cu(II)–ascorbate and Cl⁻. It was demonstrated that the involvement of copper(I) is feasible and a corresponding mechanism was presented as an alternative to the Cu(II)/Cu(III) model.

In the absence of chloride ion, the Cu(I)/Cu(II) catalytic cycle could be initiated by an intramolecular redox reaction of the $[Cu_2(HA)_2]^{2+}$ dimer to yield unsymmetrical products:

$$[Cu_2(HA)_2]^{2+} \longrightarrow [Cu(HA)] + Cu^+ + A + H^+ \tag{37}$$

$$Cu^+ + HA^- \longrightarrow [Cu(HA)] \tag{38}$$

In the following steps dioxygen coordinates to the Cu(I) complex and is subsequently reduced to hydrogen peroxide:

$$[Cu(HA)] + O_2 \longrightarrow [Cu(HA)O_2] \tag{39}$$

$$[Cu(HA)O_2] + [Cu(HA)] \longrightarrow [Cu_2(HA)_2]^{2+} + O_2^{2-} \tag{40}$$

$$O_2^{2-} + 2H^+ \longrightarrow H_2O_2 \tag{41}$$

The composite effect of Cl⁻ was interpreted by assuming that the formation of Cu(I) intermediates is enhanced via the $[Cu_2ACl_2]$ complex. However, the relative concentration of this species decreases with increasing [Cl⁻] because of the formation of the $[Cu_2ACl_4]^{2-}$ complex, and ultimately this leads to chloride ion inhibition at higher concentrations of Cl⁻.

Reports by Li and Zuberbühler were in support of the formation of Cu(I) as an intermediate (16). It was confirmed that Cu(I) and Cu(II) show the same catalytic activity and the reaction is first-order in [Cu(I) or (II)] and [O₂] in the presence of 0.6–1.5 M acetonitrile and above pH 2.2. The oxygen consumption deviated from the strictly first-order pattern at lower pH and the corresponding kinetic traces were excluded from the evaluation of the data. The rate law was found to be identical with the one obtained for the autoxidation of Cu(I) in the absence of Cu(II) under similar conditions (17). Thus, the proposed kinetic model is centered around the reduction of Cu(II) by ascorbic acid and reoxidation of Cu(I) to Cu(II) by dioxygen:

$$2Cu^{2+} + HA^- \longrightarrow 2Cu^+ + A + H^+ \tag{42}$$

$$2Cu^+ + O_2 + 2H^+ \longrightarrow 2Cu^{2+} + H_2O_2 \tag{43}$$

The reaction features a complex pH dependence which was not resolved. Nevertheless, it was suggested that the variation of the reaction rate as a function of pH was consistent with an additional term in the rate law which was proportional to $[HA^-]$. The kinetic role of $[HA^-]$ was interpreted in terms of its reaction with CuO_2^+, which is presumably formed as an intermediate during the autoxidation of Cu(I). In the presence of acetonitrile, Cu(I) can be stabilized as $[Cu(MeCN)_2]^+$ and $[Cu(MeCN)_3]^+$. As a consequence, the oxidation of Cu(I) becomes rate determining and the overall rate of the catalytic reaction becomes independent of $[HA^-]$.

The feasibility of the above model rests on the formation of Cu(I) in the copper(II)–ascorbic acid system. A recent study firmly established that Cu(I) can indeed accumulate in the presence of a stabilizing ligand, Cl^-, and in the absence of O_2 (14). The actual form of the rate law is determined by the relative rates of Cu(I) formation and consumption, and further studies should clarify how the stability and reactivity of copper(I) are affected by the presence of various components and the conditions applied.

A non-participating ligand (L) in the coordination sphere of Cu(II) may significantly alter the reactivity patterns discussed above. When bis(histidine)copper(II),[2] CuL_2, is used as a catalyst the initial rate shows saturation as a function of both $[HA^-]$ and $[O_2]$ at physiological pH (18). This observation was interpreted by postulating the formation of a ternary complex which undergoes an intramolecular redox reaction and produces $A^{\bullet-}$ and $O_2^{\bullet-}$ in a rate determining step. These species are involved in fast subsequent reactions:[3]

$$[HA^- - CuL_2 - O_2] \longrightarrow CuL_2 + A^{\bullet-} + O_2^{\bullet-} + H^+ \qquad (44)$$

$$HA^- + O_2^{\bullet-} + H^+ \longrightarrow A^{\bullet-} + H_2O_2 \qquad (45)$$

$$A^{\bullet-} + O_2^{\bullet-} + 2H^+ \longrightarrow A + H_2O_2 \qquad (46)$$

According to ESR measurements, the semiquinone radical forms at nM concentration levels and its steady-state concentration was reported to increase by increasing the total concentration of ascorbic acid. The kinetic role of $O_2^{\bullet-}$ was confirmed by the inhibitory effect of

---

[2]The bis-complex, CuL_2, was marked as a double positive ion in the original paper. However, the carboxylic group of histidine is expected to be deprotonated and we prefer to consider this species as a neutral complex.

[3]For sake of simplicity, the recombination steps of $A^{\bullet-}$ and $O_2^{\bullet-}$ are not shown.

Cu,Zn superoxide dismutase. Essentially, these observations support a stepwise one-electron model again. Interestingly, the oxidation state of copper does not change during the catalytic reaction, i.e. the sole kinetic role of the histidine coordinated metal center is to alter the electronic structures of the substrate and O_2 in order to facilitate the electron transfer process between them.

The coordination mode and structure of a given copper(II) complex are important characteristics in determining how the complex is involved in the autoxidation reaction. For example, the dinuclear $[Cu_2(BPY)_2 (\mu\text{-}OX)]X_2$ complex (BPY = 2,2′-bipyridine, OX = $C_2O_4^{2-}$, X = ClO_4^-; PF_6^-) catalyzes the autoxidation of ascorbic acid via a Cu(I)/Cu(II) cycle. In this case, ascorbic acid is presumably oxidized in a two-electron transfer process after it cleaves the dimer by protonating the bridging ligand (19). This reaction was used to model the functional properties of dopamine β-hydroxylase and was shown to convert benzylamine to benzaldehyde effectively. The Cu(I)/Cu(II) cycle was also proposed in a series of reactions catalyzed by dinuclear copper(II) complexes of Schiff bases (L), Cu_2LCl_3. However, detailed kinetic analysis was not performed in these systems (20).

Iron(III)-catalyzed autoxidation of ascorbic acid has received considerably less attention than the comparable reactions with copper species. Anaerobic studies confirmed that Fe(III) can easily oxidize ascorbic acid to dehydroascorbic acid. Xu and Jordan reported two-stage kinetics for this system in the presence of an excess of the metal ion, and suggested the fast formation of iron(III)–ascorbate complexes which undergo reversible electron transfer steps (21). However, Bänsch and co-workers did not find spectral evidence for the formation of ascorbate complexes in excess ascorbic acid (22). On the basis of a combined pH, temperature and pressure dependence study these authors confirmed that the oxidation by $Fe(H_2O)_6^{3+}$ proceeds via an outer-sphere mechanism, while the reaction with $Fe(H_2O)_5OH^{2+}$ is substitution-controlled and follows an inner-sphere electron transfer path. To some extent, these results may contradict with the model proposed by Taqui Khan and Martell (6), because the oxidation by the metal ion may take place before the ternary oxygen complex is actually formed in Eq. (17).

An interesting, pH-dependent mechanistic changeover was reported in the H_2A–O_2–$[Fe^{III}(TPPS)]$ (TPPS = 5,10,15,20-tetrakis(p-sulfonatophenyl)porphyrinate) system in aqueous solution (23). This water-soluble metalloporphyrin exists as a monomer under slightly acidic conditions and reacts with ascorbic acid on the time-scale of several hours to produce $[Fe^{II}(TPPS)]$ in the absence of dioxygen. The formation of the $[Fe^{III}(TPPS^{\bullet+})]$ radical was also reported at pH 5, but it is not clear how this species could form when no oxidizing agent was present. In the presence of oxygen, the decay of the characteristic Soret band of $[Fe^{III}(TPPS)]$ becomes considerably faster. The kinetic model for this reaction assumes a series of one-electron steps which are initiated by the oxidation of HA^- to $A^{\bullet -}$ by $[Fe^{III}(TPPS)]$. The reactions of the semi-quinone and $[Fe^{II}(TPPS)]$ products with O_2 generate the $O_2^{\bullet -}$ radical which in turn oxidizes the catalyst to $[Fe^{III}(TPPS^{\bullet +})]$. The model predicts redox cycling, but the experimental observations clearly indicate that a substantial amount of the catalyst is lost during the reaction. Apparently, an efficient reaction channel is not available to reduce the $[Fe^{III}(TPPS^{\bullet +})]$ radical back to the original complex. The model offers reasonable qualitative interpretation of the experimental data. However, it requires further refinement in order to explain why the initial $HA^- - [Fe^{III}(TPPS)]$ electron transfer step, which controls the overall reaction, becomes faster when oxygen is added.

In alkaline solution (pH 11), the complex is present as a µ-oxo dimer and ascorbic acid is fully deprotonated. In the absence of oxygen, kinetic traces show the reduction of Fe(III) to Fe(II) with a reaction time on the order of an hour at $[H_2A] = 5 \times 10^{-3}$ M. The product $[Fe^{II}(TPPS)]$ is very sensitive to oxidation and is quickly transformed to Fe(III) when O_2 is added. This leads to a specific induction period in the kinetic traces which increases with increasing $[O_2]$. The net result of the induction period is the catalytic two-electron autoxidation of ascorbic acid in accordance with the following kinetic model (23):

$$[Fe^{III}(TPPS)]_2O + A^{2-} \longrightarrow 2[Fe^{II}(TPPS)] + A + O^{2-} \qquad (47)$$

$$2[Fe^{II}(TPPS)] + O_2 \longrightarrow 2[Fe^{III}(TPPS)] + O_2^{2-} \qquad (48)$$

$$[Fe^{III}(TPPS)] + 2OH^- \longrightarrow [Fe^{III}(TPPS)]_2O + H_2O \qquad (49)$$

This reaction sequence is completed by protolytic reactions of the dianionic radicals.

The kinetic consequence of the non-participating ligand was also noticed in the autoxidation reactions catalyzed by Ru(III) ion, Ru(EDTA) (1:1) and Ru(IMDA) (1:1) (EDTA = ethylenediaminetetraacetate, IMDA = iminodiacetate) (24,25). Each reaction was found to be first order in ascorbic acid and the catalysts and, owing to the protolytic equilibrium between HA^-/H_2A, an inverse concentration dependence was confirmed for $[H^+]$. Only the oxygen dependencies were different as the Ru(III)-catalyzed reaction was half-order in $[O_2]$, whereas the rates of the Ru(III)-chelate-catalyzed reactions were independent of $[O_2]$. In the latter cases, the rate constants were in good agreement with those obtained for the oxidation of H_2A by the same Ru(III) complexes under anaerobic conditions.

The kinetic models for these reactions postulate fast complex-formation equilibria between the HA^- form of ascorbic acid and the catalysts. The noted difference in the rate laws was rationalized by considering that some of the coordination sites remain unoccupied in the $[Ru(HA)Cl_2]$ complex. Thus, O_2 can form a μ-peroxo bridge between two monomer complexes: $[Cl_2(HA)Ru-O-O-Ru(HA)Cl_2]$. The rate determining step is probably the decomposition of this species in an overall four-electron transfer process into A and H_2O_2. Again, this model does not postulate any change in the formal oxidation state of the catalyst during the reaction.

In the case of the Ru(III)-chelates, the crowded coordination sphere around the metal center prevents the coordination of O_2. Thus, the corresponding kinetic model postulates that ascorbic acid is oxidized by Ru(III) in two subsequent redox steps and Ru(II) is reoxidized by O_2:

$$[Ru^{III}L]^{(n-3)-} + HA^- \rightleftharpoons [Ru^{III}L(HA)]^{(n-2)-} + H_2O \qquad K_{50} \qquad (50)$$

$$[Ru^{III}L(HA)]^{(n-2)-} \longrightarrow [Ru^{II}L]^{(n-2)-} + A^{\bullet-} + H^+ \qquad k_{51} \qquad (51)$$

$$[Ru^{III}L]^{(n-3)-} + A^- \longrightarrow [Ru^{II}L]^{(n-2)-} + A \qquad\qquad \text{fast} \qquad (52)$$

$$2[Ru^{II}L]^{(n-2)-} + O_2 + 2H^+ \longrightarrow [Ru^{III}L]^{(n-3)-} + H_2O_2 \qquad \text{fast} \qquad (53)$$

where $n = 2$ (IMDA) or 4 (EDTA). Reported values for $\log K_{50}$ and k_{51} are respectively $4.22\,M^{-1}$ and $6.7 \times 10^{-4}\,s^{-1}$ (IMDA) and $5.23\,M^{-1}$ and $4.4 \times 10^{-4}\,s^{-1}$(EDTA) at $25\,°C$ and $\mu = 0.1\,M$ KNO_3. On the basis of these results, it is not obvious why the Ru(III) > Ru(IMDA) > Ru(EDTA) reactivity order was proposed for the catalysts (24).

The similarities in the rate constants and activation parameters (25) for the reactions with the Ru(III) chelates strongly suggest that the corresponding reactions proceed via the same mechanism. Peculiarly, the EDTA complex binds HA^- about an order of magnitude stronger than the IMDA complex. This trend is in contrast to plausible considerations, because the number of available coordination sites and electrostatic interactions would favor stronger coordination to the IMDA complex. A clear-cut explanation is not readily available for this finding.

An interesting aspect of the results is that catalysis by the IMDA complex proceeds via the one-electron instead of the four-electron path. The three empty coordination sites in this complex would allow simultaneous coordination of HA^- and O_2 and the formation of a dinuclear complex similar to the one formed with the Ru(III) ion.

This may indicate that factors other than the availability of empty coordination sites may also have important mechanistic implications. In the autoxidation kinetics of a series of Fe(II)-aminopolycarboxylato complexes the significance of steric effects was unequivocally confirmed (26). These results may also bear some relevance with respect to the Ru(III)-catalyzed reactions.

The Ru(III)–H_2A–O_2 systems were also used for the oxidation and epoxidation of various organic substrates. These reactions will be discussed in Section VII.

IV. Autoxidation of Catechols and Related Compounds

The kinetics and mechanisms of autoxidation reactions of catechols (H_2C) are similar to those of ascorbic acid in many ways. This is due to the common HO–C–C–OH motif in these molecules, which are transformed into the corresponding di-ketone species in the oxidation process. In addition, most of these reactions proceed via an inner-sphere mechanism and the two neighboring OH groups are the key structural factors in the formation of the precursor complex between the catalyst and the substrate. Beside the similarities, these reactions may also show significant differences in the product distributions. The oxidation of ascorbic acid always leads to the formation of dehydroascorbic acid in the first stage of the reaction. This compound is reasonably stable and resistant to further oxidation which may occur only in very slow subsequent reaction steps. In contrast, the formation of the di-ketonic quinone (Q) is not necessarily preferential over the formation of other products in the oxidation of catechols. These reactions typically yield more than one product which can form either directly from the substrate in parallel reaction steps or through a common intermediate. Metal ions catalyze not only the quinone formation but also intra- and extradiol oxidative cleavage of catechols leading to a variety of aliphatic products such as muconic acid or furanone derivatives, (27–31) and various polymeric species (32,33).

In non-aqueous solution, the copper catalyzed autoxidation of catechol was interpreted in terms of a Cu(I)/Cu(II) redox cycle (34). It was assumed that the formation of a dinuclear copper(II)-catecholate intermediate is followed by an intramolecular two-electron step. The product Cu(I) is quickly reoxidized by dioxygen to Cu(II). A somewhat different model postulated the reversible formation of a substrate-catalyst-dioxygen ternary complex for the Mn(II) and Co(II) catalyzed autoxidations in protic media (35).

In a thorough study by Balla and coworkers, it was shown that in slightly acidic aqueous solution (25 °C, $\mu = 1.0$ KNO$_3$), the mechanism is very similar to that followed in aprotic media (36). Parallel measurement of the O$_2$ consumption and quinone formation confirmed the following stoichiometry up to 35% conversion:

$$+ O_2 \longrightarrow \qquad + H_2O_2 \tag{54}$$

At longer reaction times, the formation of an acidic product, probably cis,cis-muconic acid, and a copper containing precipitate were observed. This latter could be a polynuclear product of quinone and semiquinone fragments (37,38). In agreement with this stoichiometry, about 50% of O$_2$ was regenerated by adding a small amount of catalase to the reaction mixture at relatively short reaction times.

The stability constants of the mono- and bis-complexes between Cu(II) and catecholate were determined under anaerobic conditions and were found to be the same as reported earlier, i.e. $\log \beta_1 = 13.64$ (CuC) and $\log \beta_2 = 24.92$ (CuC$_2^{2+}$) (36,39). A comparison of the speciation and oxidation rate as a function of pH clearly indicated that the mono-catechol complex is the main catalytic species, though the effect of other complexes could not be fully excluded. The rate of the oxidation reaction was half-order in [O$_2$] and showed rather complex concentration dependencies in [H$_2$C]$_0$, [Cu(II)]$_0$ and pH. The experimental data were consistent with the following rate equation:

$$-\mathrm{d}[O_2]/\mathrm{d}t = (k_{55}^a\,[HC^-] + k_{55}^b\,[CuC] + k_{55}^c\,[CuC_2^{2-}])\,([Cu^{2+}][CuC][O_2])^{1/2} \tag{55}$$

with $k_{55}^a = (3.2 \pm 0.2) \times 10^3\,\mathrm{M}^{-3/2}\,\mathrm{s}^{-1}$, $k_{55}^b = 15.5 \pm 0.7\,\mathrm{M}^{-3/2}\,\mathrm{s}^{-1}$ and $k_{55}^c = (3.0 \pm 0.3) \times 10^2\,\mathrm{M}^{-3/2}\,\mathrm{s}^{-1}$.

The possibility of a free-radical type mechanism was excluded on the basis that neither superoxide dismutase (which would remove the superoxide radical, O$_2^-$) nor acrylonitrile (which is a very efficient radical scavenger) had a significant effect on the reaction rate. In contrast, the reaction rate went through a maximum upon adding acetonitrile in increasing concentrations. A similar effect was also observed in copper(II) catalyzed autoxidation of ascorbic acid by Shtamm et al. (40). This observation was interpreted by assuming that the formation of a ternary Cu(II)–acetonitrile-substrate complex enhances the rate of intramolecular electron transfer, but at higher concentrations acetonitrile stabilizes Cu(I) against reoxidation. Though this interpretation is not confirmed thoroughly, the kinetic effect of acetonitrile strongly suggests the formation of copper(I) as an intermediate. This conclusion is also supported by the results reported by Speier for copper(I)-catalyzed autoxidation of 3,5-di-*tert*-butylcatechol (H$_2$DTBC) in non-aqueous solvents *(41)*. In that reaction, the formation of a dimeric species, [Cu$_2$O$_2$]$^{2+}$, was postulated in a rate determining step.

The following non-radical chain mechanism was proposed for the reaction in aqueous solution (for sake of simplicity, fast protolytic and complex-formation reactions are not shown) *(36)*:

Initiation:

$$Cu^{2+} + CuC \rightleftharpoons 2Cu^+ + Q \qquad\qquad k_{56}, \ k_{-56} \quad (56)$$

Chain propagation:

$$Cu^+ + O_2 \rightleftharpoons CuO_2^+ \qquad\qquad K_{57} \qquad (57)$$

$$CuO_2^+ + HC^- \longrightarrow Cu^+ + HO_2^- + Q \qquad k_{58} \qquad (58)$$

$$CuO_2^+ + CuC \longrightarrow Cu^+ + O_2^{2-} + Cu^{2+} + Q \qquad k_{59} \qquad (59)$$

$$CuO_2^+ + CuC_2^{2-} \longrightarrow Cu^+ + O_2^{2-} + CuA + Q \qquad k_{60} \qquad (60)$$

Termination:

$$CuO_2^+ + Cu^+ \longrightarrow 2Cu^{2+} + O_2^{2-} \qquad\qquad k_{61} \qquad (61)$$

It was suggested that initiation proceeds via the same dimer that was proposed in aprotic media *(34)*. Furthermore, the reverse step of this reaction was considered to be very slow. The CuO$_2^+$ intermediate is formed in a reversible step in this model, which is in agreement with the results reported for autoxidation of Cu(I) *(17)*. It should be added that Shtamm *et al.* assumed that this reaction is irreversible *(40)*.

Provided that the rates of the initiation and termination are equal, the model predicts the experimental rate law, and it can be shown that $k_{55}^a = k_{58}(k_{56}K_{57}/k_{61})^{1/2}$, $k_{55}^b = k_{59}(k_{56}K_{57}/k_{61})^{1/2}$ and $k_{55}^c = k_{60}(k_{56}K_{57}/k_{61})^{1/2}$.

Results from subsequent studies were consistent with the model proposed by Balla *et al.* *(36)* in that copper accelerates the autoxidation of catechins[4] through the Cu(I)/Cu(II) redox cycle and complex formation

---

[4]The term catechin is often used for the derivatives of catechol (1,2 dihydroxy benzene).

of the substrate with Cu(II) is an important part of the overall reaction
(42–44). Nevertheless, in some cases one-electron oxidation and the
formation of a semiquinone intermediate was proposed instead of the
two-electron model.

Catalytic autoxidation of four different catechins occurring
naturally in green tea, i.e. (−)-epicatechin, (−)-epicatechin gallate,
(−)-epigallocatechin and (−)-epigallocatechin gallate, was studied by
Mochizuki *et al.* by monitoring the reaction with amperometry, spectrophotometry and a Clark-type oxygen sensing electrochemical system
(44). Special attention was paid to (−)-epigallocatechin gallate which
is particularly prone to oxidation. Typical kinetic traces showed
an induction period (Fig. 1) which was presumably due to the initial
accumulation of a reactive intermediate.

After the addition of Cu(II), absorbance changes were identical under
both anaerobic and aerobic conditions. This was taken as evidence for
semi-quinone (SQ) formation, though it is not clear why the oxidation
would stop at the SQ stage in the presence of excess dioxygen. It should
be noted that the two-electron oxidation would also lead to the same
spectral changes in the presence and absence of dioxygen provided that
the spectrum of the oxidized species corresponds to that of the quinone.
The dependence of the initial steady-state autoxidation rate (defined as

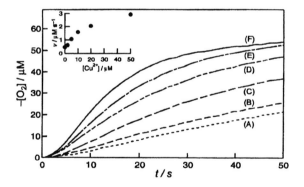

Fɪɢ. 1. Amperometric monitoring of the autoxidation of epigallocatechin gallete in
the presence of (A) 0, (B) 2.0, (C) 5.0, (D) 10, (E) 20, and (F) 50 μm CuCl$_2$. The measurements were performed in 0.1 M Tris buffer (pH 9.0) with a Clark type oxygen electrode
at 28 °C. The epigallocatechin gallate concentration was fixed at 50 μm. The catechin
stock solution was injected into the test solution at $t = 0$. The inset shows the (initial)
steady-state autoxidation rate as a function of Cu^{2+} concentration. Reprinted
from *Biochimica et Biophysica Acta*, vol. 1569, Mochizuki, M.; Yamazaki, S.; Kano, K.;
Ikeda, T., Kinetic analysis and mechanistic aspects of autoxidation of catechins, p. 35,
Copyright (2002), with permission from Elsevier Science.

the rate just after the induction period) on $[Cu^{2+}]$ (cf. Fig. 1, inset) indicated some sort of complex formation between the substrate and catalyst, but this effect was not further evaluated. A one-electron oxidation model, which postulates the formation of the $O_2^{\bullet-}$ intermediate, was proposed:

$$H_2C\text{--}R + Cu^{2+} \longrightarrow SQ\text{--}R^{\bullet-} + Cu^+ + 2H^+ \tag{62}$$

$$Cu^+ + O_2 \longrightarrow Cu^{2+} + O_2^{\bullet-} \tag{63}$$

However, direct evidence was not presented for the formation of the superoxide radical in the presence of Cu(II) and, as indicated above, the reported observations can be interpreted in terms of the two-electron oxidation model equally well.

It was established that Cu(II) promotes the formation of hydrogen peroxide. However, most likely this effect was partly offset by a Fenton-type side reaction with Cu(I):

$$H_2O_2 + Cu^+ \longrightarrow OH^\bullet + OH^- + Cu^{2+} \tag{64}$$

In agreement with literature results (45,46), independent experiments confirmed that the addition of an aliquot of a CuCl acetonitrile solution to an H_2O_2 solution induced the immediate decay of H_2O_2 at pH 9.0 (44). Most likely, the $OH^\bullet$ radical produced is involved in fast oxidation of the H_2C–R, SQ–$R^{\bullet-}$ or even the Q–R forms of the substrate and the stoichiometry shown in Eq. (54) is not valid anymore. The formation of free radicals was excluded under acidic conditions (36) implying that the Fenton-type decomposition of H_2O_2 may gain significance only in alkaline solution.

Mochizuki et al. also demonstrated that the uncatalyzed autoxidation of catechins is suppressed in the presence of a sufficient amount of borate buffer (44). This was interpreted by considering the adduct formation between borate ion and the catechins. The oxidation resumes on addition of Cu(II) in the presence of the borate ions. This may indicate that Cu(II) forms more stable complex(es) with the catechins than the borate ion, but direct oxidation of the catechin–borate complex by Cu(II) cannot be excluded either.

The autoxidation of 3,5-di-tert-butylcatechol (H_2DTBC) was frequently used to test the catalytic activity of various metal complexes. Speier studied the reaction with [Cu(PY)Cl] (PY = pyridine) in CH_2Cl_2 and $CHCl_3$, and reported second-, first- and zeroth-order dependence with respect to Cu(I), O_2 and substrate concentrations, respectively (41). The results are consistent with a kinetic model in which the rate determining oxidation of Cu(I) is followed by fast reduction of Cu(II) by H_2DTBC.

In spite of the significant differences in the catalysts and conditions applied, essentially the same kinetic model was proposed for the catalytic reactions with two pyrazolate-bridged dicopper(II) complexes, $[Cu_2(LEP)_2]^{2+}$ and $[Cu_2(BLEP)(OH)]^{2+}$ in 1:1 methanol–water mixture (47). The product was confirmed to be 3,5-di-*tert*-butyl-1,2-benzoquinone (DTBQ).

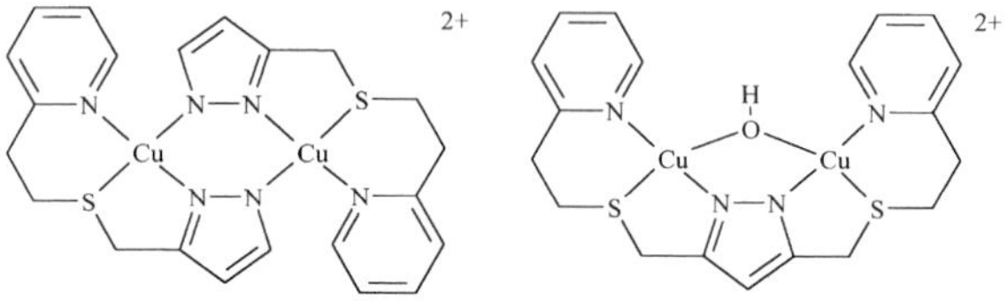

$[Cu_2(LEP)_2]^{2+}$ $[Cu_2(BLEP)(OH)]^{2+}$

In the case of these catalysts, the two metallic redox centers are provided by the same molecule. Therefore, the reaction is first-order in the complex and dioxygen concentrations:

$$-d[O_2]/dt = k_{65}[Cu_2L][O_2] \qquad (65)$$

The metal centers are separated by 370 and 365 pm in the solid state in $[Cu_2(LEP)_2](ClO_4)_2$ and $[Cu_2(BLEP)Cl(H_2O)]Cl$, respectively. In both complexes, two square pyramidal copper ions are connected in the equatorial plane with the apical sites located on the same side of the complex. This cofacial geometry is favorable for simultaneous binding the catechol molecule to the two metal ions. In the case of the reduced forms, O_2 may bind to one apical site but coordination in a bridge position is also feasible. Additional coordination modes can also be envisioned in the case of $[Cu_2(BLEP)(OH)]^{2+}$ by assuming that the OH group is displaced. Thus, the substrate may bind to the Cu(II)$\cdots$Cu(II) complex asymmetrically by replacing the OH group with one oxygen and coordinating to one of the apical sites with the other. In this case, the reduced form of the catalyst may coordinate dioxygen as a μ-peroxo bridge between the two Cu(I) centers.

Alternative kinetic models were considered for this reaction. Both of them predict rapid reduction of Cu(II)$\cdots$Cu(II) to Cu(I)$\cdots$Cu(I) by H_2DTBC and subsequent formation of the $Cu(II)(O_2^{2-})Cu(II)$ intermediate in the reaction of the reduced form with O_2. The first model assumes that the rate determining formation of the intermediate is followed by a fast, acid assisted dissociation into the oxidized form of the catalyst and H_2O_2. In the other model, the rate determining step is the oxidation of H_2DTBC by the intermediate. The two models predict the same rate law and cannot be distinguished. However, it is very likely that the μ-peroxo-bridged intermediate decomposes before any intermolecular redox reaction takes place, such that the second model can be rejected.

Autoxidation became slower when the pH was decreased. Because the substrate is not involved in the rate-determining steps, its protolytic equilibria cannot account for this observation. A straightforward explanation was not readily available for this observation.

At pH 6.40 and 25 °C, the values of k_{65} for $[Cu_2(LEP)_2]^{2+}$ and $[Cu_2(BLEP)(OH)]^{2+}$ are 77.5 $M^{-1}s^{-1}$ and $6.98 \times 10^3 M^{-1} s^{-1}$, respectively (47). The difference in the rate constants indicates that the interaction between the reduced form of the catalyst and O_2 is stronger with the latter complex. This may reflect that the BLEP ligand maintains the dimer structure in the reduced oxidation state whereas the LEP complex may dissociate into monomers. An additional factor could be that the replacement of the OH bridge allows the formation of a μ-peroxo bridge in an equatorial position with the BLEP complex, while O_2 can coordinate only in the apical position(s) in the other case. The catalytic activity of $[Cu_2(BDPDZ)]Cl_2$ (BDPDZ = 3,6-bis-(di-2-pyridylmethyl)pyridazine) in the autoxidation of H_2DTBC was also attributed to the adduct formation between dioxygen and the two Cu(I) centers (48).

A study with a series of $[Cu(R-TMED)X_2]$ complexes (R-TMED = N,N,N'-trimethyl-N'-alkylethylenediamine, X = Cl^-, Br^-, ACAC = acetylacetone, R = CH_3-, $C_{12}H_{25}-$, $C_{16}H_{33}-$) demonstrated that a micellized environment enhances the catalytic activity of copper(II) in the autoxidation of catechols (49). The observations, in the presence of Cu(II) surfactants above their critical micelle concentrations (CMC), can be summarized as follows: (i) in large excess of O_2 and at constant catalyst concentration the reactions were first-order in the substrate concentration; (ii) when the same head group and counter ion were used the surfactant with longer alkyl chain was more active; (iii) when the head group and the alkyl chain were the same, the bromide surfactant was a better catalyst than the chloride analogue; (iv) the $[Cu(R-TMED)(ACAC)X]$ complex was less efficient than the $[Cu(R-TMED)X_2]$ complexes (X = halogenide ion); (v) the reactivity order of the catechols was found to be 3,5-di-tert-butylcatechol > catechol > D- or L-dopamine.

The results were interpreted in terms of the model proposed by Balla and co-workers (36). It is reasonable to assume that the micelle formation produces a somewhat organized pattern of the metal centers and, due to the shortened distance between the copper(II) containing head groups, the coordination of catechol to two metal centers may increase the stability of the catalyst–substrate complex. Perhaps, the same principles apply to the interaction between the reduced form of the catalyst and O_2, and the reaction may also feature the formation of an intermediate with a μ-peroxo bridge between two copper centers. While these conclusions require further confirmation, the effect of added salt on the kinetics could be rationalized on the basis of the above model. It was assumed that the addition of NaCl at low concentration levels increased the oxidation rate because it favored the formation of larger micelles with shorter distances between the metal centers (50). The reaction rate goes through a maximum and finally levels off as a function of [NaCl]. The somewhat smaller rate at higher electrolyte concentration is probably due to the saturation of the cationic head group layer of the micelles with anions which obstruct the complexation with the catechols.

The catalytic activities of Cu(II), Co(II) and Mn(II) are considerably enhanced by sodium dodecyl sulfate (SDS) in the autoxidation of H_2DTBC (51). The maximum catalytic activity was found in the CMC region. It was assumed that the micelles incorporate the catalysts and the short metal–metal distances increase the activity in accordance with the kinetic model discussed above. The concentration of the micelles increases at higher SDS concentrations. Thus, the concentrations of the catalyst and the substrate decrease in the micellar region and, as a consequence, the catalytic reaction becomes slower again.

In the presence of cobalt complexes as catalysts, the following stoichiometry was reported for the autoxidation of H_2DTBC in chloroform (52):

$$H_2DTBC + 1/2O_2 \longrightarrow DTBQ + H_2O \tag{66}$$

The yield of quinone was practically 100% and H_2O_2 accumulated in negligible concentration.

A comparison of the initial rates obtained with various cobalt complexes (Table I) reveals that the chelate complexes of Co(II) are more efficient than the simple salts, the catalytic activity of Co(III) is lower than that of Co(II) and the reaction becomes slower by increasing the number of N atoms in the coordination spheres in both oxidation states. In general, the addition of amine derivatives increased the activity of the catalysts.

The reaction with $Co^{II}(ACAC)_2$ was studied in more detail and the rate law was established. The reaction was found to be first-order with respect to the substrate and the catalyst concentrations, and the partial pressure of O_2. The corresponding kinetic model postulates reversible formation of a $H_2DTBC–Co^{II}(ACAC)_2–O_2$ adduct which undergoes redox decomposition in the rate-determining step. Hydrogen peroxide is also a primary

TABLE I

CATALYTIC ACTIVITY OF Co(II) AND Co(III) COMPLEXES IN AUTOXIDATION OF
DI-*tert*-BUTYLCATECHOL (*52*)[a]

Catalyst	v_0 (M s^{-1})
bis(acetylacetonato)cobalt(II)	1.6×10^{-3}
bis(benzoylacetonato)cobalt(II)	1.6×10^{-3}
bis(ethylbenzoylacetonato)cobalt(II)	1.8×10^{-3}
bis(N-cyclohexylsalicylaldiminato)cobalt(II)	1.2×10^{-3}
bis(N-phenylsalicylaldiminato)cobalt(II)	8.6×10^{-4}
cobaloxime(II)-bis(pyridine)	2.1×10^{-4}
tris(acetylacetonato)cobalt(III)	1.4×10^{-4}
chloro(pyridine)cobaloxime(III)[b]	0
cobalt(II) acetate	4.0×10^{-5}
cobalt(II) nitrate	0
cobalt(II) nitrate-pyridine[c]	2.8×10^{-6}

[a] [H$_2$DTBC]=100 mM, [catalyst]=1.0 mM, 27 °C, in CHCl$_3$.
[b] [catalyst]=3.2 mM.
[c] [PY]=250 mM.

product of this reaction, but it cannot accumulate because of fast decomposition in a cobalt(II) catalyzed reaction:

$$H_2O_2 \longrightarrow H_2O + 1/2 O_2 \qquad (67)$$

The kinetic role of amines was interpreted by assuming that they are capable of promoting the formation of the precursor complex by partially deprotonating the substrate via hydrogen bonding with the OH group.

Semi-quinones are formed as reactive intermediates in the autoxidation of catechols and they are present either bonded to a metal center or as free radicals. Recently, Simándi and Simándi reported ESR spectra for each type of these species in the triphenylphosphine-bis-(dimethylglyoximato)cobalt(II)-catalyzed autoxidation of H$_2$DTBC (catalyst = [Co^{II}(HDMG)$_2$(Ph$_3$P)]) (*53*). An 8-line spectrum with $g = 2.0017$, $a_{Co} = 10.4$ G and coupling constant of 3.3 G is characteristic of DTBSQ$^{\bullet-}$ when it is coordinated to the catalyst as a unidentate ligand (Fig. 2a). At higher resolution, the 18-line spectrum of the free radical was observed at the center of the 8-line signal with $g = 2.0046$ and $a_{2H} = 1.7$ G, a_{18H} 0.35 G (Fig. 2b).

Detailed kinetic studies confirmed a two-stage reaction for the cobaloxime(II)-catalyzed autoxidation of this system in methanol (*54,55*). First, within about 30 s, the reaction reached steady-state conditions via reversible oxygenation of Co(II) to the corresponding

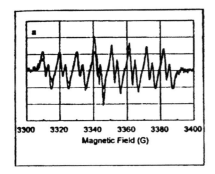 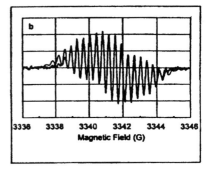

FIG. 2. ESR spectra of free-radical intermediates during the cobaloxime(II) catalyzed oxidation of 3,5-CatH$_2$ by O$_2$ in benzene. Reprinted with permission from Simándi, T. L.; Simándi, L. I. *React. Kin. Catal. Let.* **1998**, *65*, 301.

superoxo complex to which the substrate is bonded via a hydrogen bond (for sake of simplicity, Co^{II} and Co^{III} correspond to the respective [Co^{II}(HDMG)$_2$(Ph$_3$P)] and [Co^{III}(HDMG)$_2$(Ph$_3$P)]$^+$ moieties in the following equations):

$$Co^{II} + O_2 \rightleftharpoons Co^{III}O_2 \qquad\qquad K_{68} \qquad (68)$$

$$Co^{III}O_2 + H_2DTBC \rightleftharpoons [Co^{III}O_2 \cdots H_2DTBC] \qquad K_{69} \qquad (69)$$

It was assumed that the hydrogen bond forms between the superoxo moiety and one of the OH-groups of the substrate. The rate-determining step is the redox decomposition of the hydrogen bonded adduct into hydroperoxocobaloxime and semiquinone. The latter participates in an equilibrium reaction with Co(II):

$$[Co^{III}O_2 \cdots H_2DTBC] \longrightarrow Co^{III}O_2H + DTBSQ^{\bullet -} + H^+ \qquad k_{70} \qquad (70)$$

$$Co^{II} + DTBSQ^{\bullet -} + H^+ \rightleftharpoons Co^{III}(HDTBC) \qquad\qquad K_{71} \qquad (71)$$

The major cobaloxime species, Co^{III}(HDTBC), was present at an approximately constant concentration level in the slow stage, in which the following additional reaction steps were considered:

$$2Co^{III}O_2H \longrightarrow 2Co^{III}OH + O_2 \qquad\qquad (72)$$

$$Co^{III}OH + DTBSQ^{\bullet -} + H^+ \longrightarrow Co^{II} + DTBQ + H_2O \qquad\qquad (73)$$

$$Co^{III}OH + Co^{III}(HDTBC) \longrightarrow Co^{II} + Co^{III}DTBSQ^{\bullet -} + H_2O \qquad (74)$$

$$Co^{III}DTBSQ^{\bullet-} \rightleftharpoons Co^{II} + DTBQ \tag{75}$$

In agreement with the experimental results, this model predicts the following expression for the initial rate of oxygen uptake after steady-state is reached:

$$v_i = k_{70} \frac{aK_{68}K_{69}[O_2][H_2DTBC]_0[Co]_0}{1 + K_{68}K_{69}[O_2][H_2DTBC]_0} \tag{76}$$

where a is the ratio of free cobaloxime(II) compared to the total amount of the catalyst. An analysis of the concentration dependencies of the initial rate gave $K_{68}K_{69} = 6.9 \times 10^3$ and $5.1 \times 10^4\,M^{-2}$ as well as $k_{70} = 7.0 \times 10^{-2}$ and $3.8 \times 10^{-2}\,s^{-1}$ in benzene and methanol, respectively.

It should be noted that the dioxygen complex formed in Eq. (68) would ultimately be converted into cobaloxime(III) which is catalytically inactive. Thus, in order to maintain the catalytic cycle, the cobaloxime(II) species needs to be regenerated relatively quickly.

A very similar kinetic model was proposed for the ferroxime(II)-catalyzed autoxidation of H_2DTBC in methanol (55). The catalyst was added to the reaction mixture in the form of $[Fe(HDMG)_2(MeIM)_2]$ (MeIM = N-methylimidazole). Upon dissolution in methanol, this complex undergoes solvolysis and one of the imidazole ligands is replaced by a solvent molecule. The solvolysis occurs at the same rate under a N_2 or O_2 atmosphere. Again, the reactions between the catalyst, substrate and dioxygen leads to the formation of a hydrogen bonded precursor complex which undergoes relatively slow redox decomposition into $Fe^{III}O_2H$ and semiquinone. (Fe^{II} and Fe^{III} represent the $[Fe(HDMG)_2(MeIM)]$ and $[Fe(HDMG)_2(MeIM)]^+$ complexes, respectively.) Subsequent fast reaction steps are:

$$Fe^{III}O_2H + DTBSQ^{\bullet-} + H^+ \longrightarrow [Fe^{IV}O] + DTBQ + H_2O \tag{77}$$

$$[Fe^{IV}O] + H_2DTBC \longrightarrow Fe^{II} + DTBQ + H_2O \tag{78}$$

The intense blue color of the reaction mixture was assigned to the paramagnetic $[Fe(HDMG)_2(MeIM)(DTBSQ^{\bullet-})]^+$ complex which is characterized by a broad spectral band at $\lambda_{max} \sim 680$ nm and a distinct doublet with $g = 2.00425$ and $a_{1H} = 3.135$ G in the visible and ESR spectra, respectively. This iron(II) species is not involved in a direct redox step and acts only as a reservoir for the semi-quinone radical.

The sole product of the reaction is DTBQ with the dimethylglyoximato complexes. The high chemoselectivity was rationalized by considering that the cleavage of the substrate may occur when the semi-quinone radical is directly attacked by O_2 (presumably this is a slow reaction path) or by a superoxide radical coordinated to a metal center. The most efficient path assumes adjacent coordination of the semi-quinone and superoxide radicals. However, such a coordination mode is not feasible in the presence of HDMG because of the very strong di-equatorial coordination which resembles that of a macrocycle. Consequently, the ring scission is negligible in these reactions if it occurs at all.

An Ir(III) catecholate complex $[(TRIPHOS)Ir(DTBC)]^+$ was reported to catalyze chemoselective oxidation of H_2DTBC to DTBQ via the formation of a dioxygen adduct $[(triphos)Ir(\mu-O_2)(DTBSQ)]^+$ (TRIPHOS = $CH_3C(CH_2PPh_2)_3$) (56). This oxygen complex was prepared by bubbling O_2 through a CH_2Cl_2 solution of the catecholate complex and characterized by using UV/Vis, IR, ^1H and ^{31}P NMR spectroscopies, as well as X-ray analysis (57).

An interesting feature of the reaction is that it produces exclusively DTBQ and H_2O_2 with the Ir(III) catalyst (Scheme 4). Deviations from this pattern were reported only at relatively high partial pressures of O_2. In contrast, other organic products also formed when the rhodium analogue of the catecholate complex is used as a catalyst, although the Ir and Rh complexes were reported to possess the same structure (58,59). This clearly indicates that factors other than the geometry and the overall charge of the catalyst may have an important role in the mechanism.

The rate of the Ir(III) catalyzed reaction was found to be first-order in [Ir] and [H_2DTBC], but independent of O_2 concentration in chloroform (56). The mechanism proposed for the reaction (Scheme 4) postulates that the protonation of the hydroperoxo α-oxygen by the hydroxy group of the bonded catechol in **Int 1** leads to the formation of H_2O_2. The o-quinone ligand of **Int 2** is replaced by the partially coordinated catechol in the next step. In order to comply with the experimental rate law, the rate-determining step needs to be the reaction of the oxygen adduct (**B**) with catechol.

The differences between the iridium and rhodium systems were interpreted by considering that **Int 1** may open an oxygenation path via intramolecular rearrangement. In this case, the scission of a M–O bond between the semi-quinone and metal center would be followed by intradiol insertion of an oxygen and ultimately by the formation of muconic acid and water. The results indicate partial preference of the rhodium complex toward the oxygenation path.

The cleavage of catechols with the incorporation of oxygen is clearly favored in the presence of some of the iron(III) complexes as catalysts. Que and co-workers proposed a substrate activation mechanism for these reactions, wherein the delocalization of the unpaired spin density

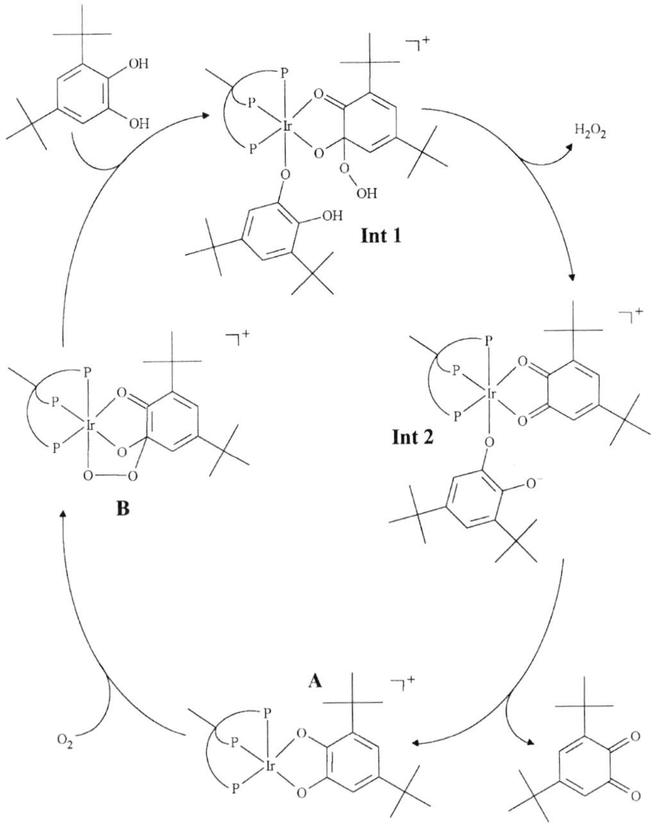

SCHEME 4. Reprinted with permission from Barbaro, P.; Bianchini, C.; Frediani, P.; Meli, A.; Vizza, F. *Inorg. Chem.* **1992**, *31*, 1523. Copyright (2002) American Chemical Society.

from the metal center onto the catecholate in the corresponding complex makes the substrate more susceptible to dioxygen attack (*60*). The non-participating ligand was shown to have a profound effect on the Lewis acidity of the metal center and, as a consequence, on the ligand to metal charge transfer. This effect also manifests itself in the kinetics and product distribution of the autoxidation reactions (*28*). It was shown that oxidative cleavage of catechols becomes preferred over extradiol oxidation to quinone when the Lewis acidity of the ferric center is increased.

In a systematic study, the reaction of [FeL(DTBC)] with dioxygen was studied in various non-aqueous solvents (*29*). In the ternary catecholate complex, L represents the following $(RCH_2)_2N(CH_2R')$ type tripodal ligands: NTA, $R = R' = COO^-$; PDA, $R = COO^-$, $R' = 2$-pyridyl; BPG, $R = 2$-pyridyl, $R' = COO^-$; HDP, $R = 2$-pyridyl, $R' = 2$-hydroxy-3,5-dimethyl-phenyl. These ligands induce relatively pronounced Lewis acidity on the metal center in the following order: BPG > PDA > NTA > HDP. It should be noted that the isotropic shift of the ^1H NMR signal of the coordinated catecholate was considered to be the best spectroscopic parameter as indicator of the Lewis acidity of the ferric center. The oxidation reaction yielded almost exclusively the intradiol scission product in all cases (Table II).

The kinetic experiments were not performed under true catalytic conditions, i.e. the pre-prepared [FeL(DTBC)] complexes were introduced into the reaction mixtures as reactants and excess substrate was not used. Nevertheless, the results are important in exploring the intimate details of the activation mechanisms of the metal ion catalyzed autoxidation reactions of catechols. In excess oxygen the reaction was first-order in the complex concentration and the first-order dependence in dioxygen concentration was also confirmed with the BPG complex. As shown in Table II, the rate constants clearly correlate with the Lewis character of the complex, i.e. the rate of the oxidation reaction increases by increasing the Lewis acidity of the metal center.

The scission product from the diamagnetic [Ga(BPG)(DTBC)] complex is considerably less (4%) than that from the corresponding Fe complex (97%) (29). Gallium(III) is a diamagnetic d^{10} ion with an ionic radius very similar to that of iron(III) and expected to possess similar Lewis properties to that of Fe(III). Thus, the results clearly demonstrate that the ability of the metal center to transfer paramagnetic spin density to

TABLE II

KINETIC PARAMETERS FOR THE REACTIONS OF [FeL(DTBC)] COMPLEXES WITH DIOXYGEN IN DIFFERENT SOLVENTS (29) [a]

Ligand	% Scission [b,c]	$10^2 \times k$ (DMF)	$10^2 \times k$ (CH$_3$OH)	$10^2 \times k$ (CH$_2$Cl$_2$)	$\Delta H^{\ddagger}$ [b] (kJ/mol)	$\Delta S^{\ddagger}$ [b] J/(mol K)
HDP	91	0.33	0.49	0.81		
NTA	84	3.7	1.0		53.5	−92
PDA	95	4.3	5.0	3.7	47.7	−113
BPG	97	18.3	42.6	36.6	42.6	−117

[a] Rate constants are in $M^{-1}s^{-1}$ at 25 °C.
[b] In DMF.
[c] Under 1 bar O_2 in the presence of excess piperidine.

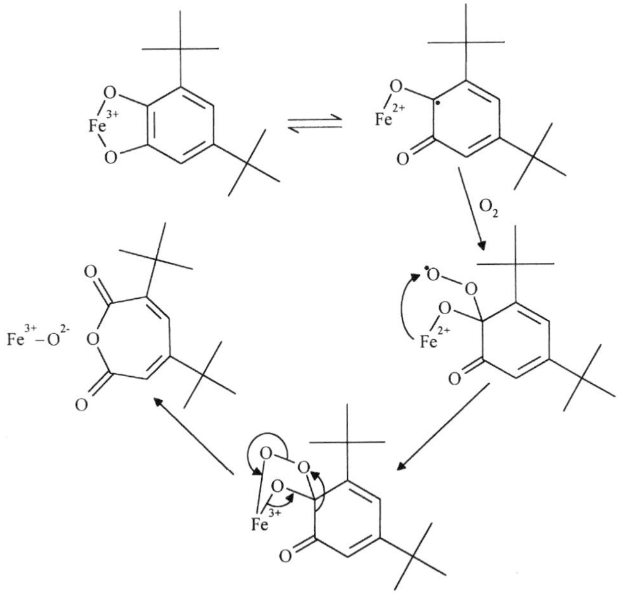

SCHEME 5. Reprinted with permission from Cox, D. D.; Que, L. Jr. *J. Am. Chem. Soc.* **1988**, *110*, 8085. Copyright (2002) American Chemical Society.

the substrate is, besides the Lewis acidity, an essential requirement for the scission reaction.

The mechanism shown in Scheme 5 postulates the formation of a Fe(II)-semi-quinone intermediate. The attack of O_2 on the substrate generates a peroxy radical which is reduced by the Fe(II) center to produce the Fe(III) peroxide complex. The semi-quinone character of the [FeL(DTBC)] complexes is clearly determined by the covalency of the iron(III)–catechol bond which is enhanced by increasing the Lewis acidity of the metal center. Thus, ultimately the non-participating ligand controls the extent of the Fe(II) – semi-quinone formation and the rate of the reaction provided that the rate-determining step is the reaction of O_2 with the semiquinone intermediate. In the final stage, the substrate is oxygenated simultaneously with the release of the $Fe^{III}L$ complex. An alternative model, in which O_2 attacks the Fe(II) center instead of the semi-quinone, cannot be excluded either.

According to a recent study with iron(III) complexes of tripodal ligands, systematic variation of one ligand arm strongly affects the steric shielding of the iron(III) center and the bonding of catechol substrates (*61*). It was shown that the dioxygenation reactions of catechols depend on the redox potentials of the complex and substrate as well as the steric properties of the tripodal ligand. As the redox potential of the catechol was increased the oxidation became slower. It was concluded that the ideal non-participating tripodal ligand needs to induce high Lewis acidity on the iron(III) center and to consist of non-bulky coordinating groups.

The oxidation of adrenaline (H_2C–RH^+: $R = -CH(OH)CH_2NH_2^+CH_3$) to andrenochrome is efficiently catalyzed by VO^{2+} in aqueous solution (62). The following experimental rate law was reported for the oxygen consumption ($\mu = 0.1$ M KNO_3, 25 °C):

$$-\frac{d[O_2]}{dt} = \left(\frac{A[VO]_T[L]_T}{B[H^+]^2 + C[L]_T + D[L]_T} + \frac{E}{[H^+]} + F \right)[O_2] + p \qquad (79)$$

where A, B, C, D, F, p are experimentally determined parameters which could be expressed by combining characteristic equilibrium and rate constants for this system; $[VO]_T$ and $[L]_T$ are the total concentrations of the catalyst and substrate, respectively.

The experimental observations were interpreted by assuming that the redox cycle starts with the formation of a complex between the catalyst and the substrate. This species undergoes intramolecular two-electron transfer and produces vanadium(II) and the quinone form of adrenaline. The organic intermediate rearranges into leucoadrenochrome which is oxidized to the final product also in a two-electron redox step. The +2 oxidation state of vanadium is stabilized by complex formation with the substrate. Subsequent reactions include the autoxidation of the V(II) complex to the product as well as the formation of a VOV^{4+} intermediate which is reoxidized to VO^{2+} by dioxygen. These reactions also produce H_2O_2. The model also takes into account the rapidly established equilibria between different vanadium-substrate complexes which react with O_2 at different rates. The concentration and pH dependencies of the reaction rate provided evidence for the formation of a $V(C-RH)_3^+$ complex in which the formal oxidation state of vanadium is +4.

V. Autoxidation of Cysteine

Earlier studies demonstrated a rich variety of oxidation states, geometries and compositions of the intermediates and products formed in the autoxidation reactions of cysteine (RSH). Owing to the complexity of these systems, only a limited number of detailed kinetic papers were published on this subject and, not surprisingly, some of the results are contradictory. The majority of available data was reported on copper catalysis because of the significance of copper–sulfur interactions in enzymatic electron transfer, oxygen transport and oxygenation reactions (*63*).

Zwart and co-workers confirmed that the catalytic autoxidation produces cystine (RSSR) with 100% selectivity and the actual stoichiometry can be given by the following equation in alkaline solution ([NaOH] = 0.25 M) (*64*):

$$4RS^- + (1 + a)O_2 + 2H_2O \longrightarrow 2RSSR + 2(2 - a)OH^- + 2aHO_2^- \qquad (80)$$

which is the linear combination of Eqs. (81) and (82)

$$4RS^- + O_2 + 2H_2O \longrightarrow 2RSSR + 4OH^- \qquad (81)$$

$$4RS^- + 2O_2 + 2H_2O \longrightarrow 2RSSR + 2OH^- + 2HO_2^- \qquad (82)$$

The stoichiometry strongly suggests that HO_2^- (the basic form of H_2O_2) also oxidizes the substrate.

As shown in Fig. 3, the kinetic traces feature a specific break-point when cysteine is oxidized by excess H_2O_2 under anaerobic conditions, and in the presence of Cu(II). The break-point occurs when all of the

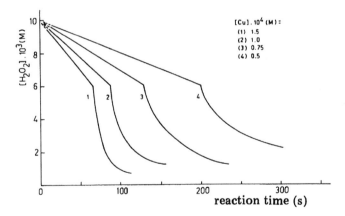

FIG. 3. Decay of the H_2O_2 concentration *versus* time during the anaerobic oxidation reaction with cysteine in the presence of $CuSO_4$. First stage of constant rate (first-order in [Cu]) during the period of oxidation, second stage of increasing rate after completion of the oxidation of cysteine to cystine. Reprinted from *Journal of Molecular catalysis*, vol. 11, Zwart, J.; van Wolput, J. H. M. C.; van der Cammen, J. C. J. M.; Koningsberger, D. C. Accumulation and Reactions of H_2O_2 During the Copper Ion Catalyzed Autoxidation of Cysteine in Alkaline Medium, p. 69, Copyright (2002), with permission from Elsevier Science.

substrate is oxidized. In this part of the reaction, the rate of hydrogen peroxide consumption is practically constant, independent of $[H_2O_2]$ and [RSH] but shows first-order dependence on the concentration of the catalyst. Once cysteine is used up, the reaction starts to generate oxygen in the copper catalyzed decomposition of H_2O_2.

A coherent qualitative interpretation can be given for the observations by assuming that the primary product of the catalytic $RSH–O_2$ reaction is H_2O_2 which also oxidizes the substrate. The latter reaction needs to be considerably slower than the first one because accumulation of significant amounts of H_2O_2 was observed under aerobic conditions. In the absence of added oxygen, catalytic decomposition of H_2O_2 also needs to be considered. The product of this reaction is O_2 which in turn regenerates hydrogen peroxide via the oxidation of the substrate.

Quantitative ESR measurements confirmed that almost all of the total quantity of copper is present as [Cu(RS)] complex during the reaction (65). The kinetic data were consistent with a rate law which is zeroth-order in cysteine concentration:

$$-d[O_2]/dt = k_{83}^a [O_2]^{1/2}[Cu(II)] + k_{83}^b [O_2]^{1/2}[Cu(II)]^2 \qquad (83)$$

$$k_{83}^a = 2.2\,M^{-1/2}\,s^{-1}, \quad k_{83}^b = 1.2 \times 10^4\,M^{-3/2}\,s^{-1}$$

The half-order of the rate with respect to $[O_2]$ and the two-term rate law were taken as evidence for a chain mechanism which involves one-electron transfer steps and proceeds via two different reaction paths. The formation of the dimer $[(RS)_2Cu(\mu\text{-}O_2)Cu(RS)_2]$ complex in the initiation phase is the core of the model, as asymmetric dissociation of this species produces two chain carriers. Earlier literature results were contested by rejecting the feasibility of a free-radical mechanism which would imply a redox shuttle between Cu(II) and Cu(I). It was assumed that the substrate remains bonded to the metal center throughout the whole process and the free thiyl radical, $RS^\bullet$, does not form during the reaction. It was argued that if free $RS^\bullet$ radicals formed they would certainly be involved in an almost diffusion-controlled reaction with dioxygen, and the intermediate peroxo species would open alternative reaction paths to generate products other than cystine. This would clearly contradict the noted high selectivity of the autoxidation reaction.

Kinetic studies in the physiological pH range (6.5 to 7.8) provided consistency with the above results in that the accumulation of H_2O_2 was also observed and the stoichiometry of the reaction depended on the conditions applied (66–69). However, a simple 2:1 stoichiometry was confirmed between cysteine consumed and hydrogen peroxide formed in dilute solution. The reaction followed Michaelis–Menten kinetics with respect to $[O_2]$ which is in clear contradiction with the half-order $[O_2]$ dependence in alkaline solution. The results were interpreted by assuming the formation and re-oxidation of Cu(I) in the catalytic cycle, though the mechanism was not explained in detail.

Ehrenberg and co-workers found Michaelis-Menten type kinetics not only with respect to $[O_2]$ but also with respect to the substrate concentration at pH 7.2 (67). They pointed out that the independence of the reaction rate upon the substrate concentration at high [RSH] and low [Cu(II)] is consistent with zeroth-order kinetics in [RSH]. The reported kinetic trace under such conditions features a slight induction period, but this observation was not addressed in that paper. For comparison, the catalytic effect of iron(III) was also studied. It was concluded that the catalytic activities of the two metal ions are fundamentally different. Copper efficiently enhanced the rate of H_2O_2 formation but practically had no effect on further reduction of hydrogen peroxide. In contrast, H_2O_2 was not detected in the presence of iron(III). The reaction was first-order in [RSH] with freshly prepared catalyst solution; however, conditions favoring the formation of colloidal Fe(III) hydroxide(s) shifted the kinetics toward zeroth-order. This observation seems to be consistent with a mechanistic changeover to surface catalysis in the presence of Fe(III)sol.

Interestingly, the catalytic effects of the two metal ions were not additive; the copper(II) catalyzed reaction became significantly slower on the addition of Fe(III) to the reaction mixture. This observation was interpreted in terms of the model proposed by Zwart and co-workers (64) by assuming that one of the two copper centers is replaced by Fe in the catalytically active dimer (67).

The effect of non-participating ligands on the copper catalyzed autoxidation of cysteine was studied in the presence of glycylglycine-phosphate and catecholamines, $(2\text{-R}-)H_2C$, (epinephrine, $R = CH(OH)-CH_2-NHCH_3$; norepinephrine, $R = CH(OH)-CH_2-NH_2$; dopamine, $R = CH_2-CH_2-NH_2$; dopa, $R = CH_2-CH(COOH)-NH_2$) by Hanaki and co-workers (68,69). Typically, these reactions followed Michaelis-Menten kinetics and the autoxidation rate displayed a bell-shaped curve as a function of pH. The catecholamines had no kinetic effects under anaerobic conditions, but catalyzed the autoxidation of cysteine in the following order of efficiency: epinephrine = norepinephrine $\geq$ dopamine > dopa. The concentration and pH dependencies of the reaction rate were interpreted by assuming that the redox active species is the $[L–Cu^{II}(RS^-)]$ ternary complex which is formed in a very fast reaction between $Cu^{II}L$ and cysteine. Thus, the autoxidation occurs at maximum rate when the conditions are optimal for the formation of this species. At relatively low pH, the ternary complex does not form in sufficient concentration.

Upon increasing the pH, first this species is present in increasing concentrations, but eventually the non-participating ligand is replaced with another cysteine molecule and the redox inactive $Cu(RS)_2$ is formed. Consequently, the reaction rate decreases sharply. The following reaction sequence was proposed for the redox reaction:

$$L-Cu^{II}(RS^-) \longrightarrow L-Cu^I + RS^{\bullet} \tag{84}$$

$$L-Cu^I + RS^- \rightleftharpoons Cu^I(RS^-) + L \tag{85}$$

$$Cu^I(RS^-) + O_2 \rightleftharpoons Cu^I(RS^-) - O_2 \tag{86}$$

$$Cu^I(RS^-)-O_2 + RS^- + 2H^+ \longrightarrow Cu^{II}(RS^-) + RS^{\bullet} + H_2O_2 \tag{87}$$

In the presence of the catecholamines the following two steps also need to be considered:

$$Cu^I(RS^-)-O_2 + (2\text{-}R\text{-})HC^- + 2H^+$$
$$\longrightarrow Cu^{II}(RS^-) + (2\text{-}R\text{-})HSQ^{\bullet} + H_2O_2 \tag{88}$$

$$RS^- + (2\text{-}R\text{-})HSQ^{\bullet} \longrightarrow RS^{\bullet} + (2\text{-}R\text{-})HC^- \tag{89}$$

The product cystine is presumably formed in the recombination of two thiyl radicals. This free-radical model is suitable for formal treatment of the kinetic data; however, it does not account for all possible reactions of the $RS^{\bullet}$ radical (68). The rate constants for the reactions of this species with RS^-, O_2 and $Cu^I L_n$ ($n = 2, 3$) are comparable, and on the order of 10^9–$10^{10}\,M^{-1}\,s^{-1}$ (70–72). Because all of these reaction partners are present in relatively high and competitive concentrations, the recombination of the thiyl radical must be a relatively minor reaction compared to the other reaction paths even though it has a diffusion controlled rate constant. It follows that the $RS^{\bullet}$ radical is most likely involved in a series of side reactions producing various intermediates. In order to comply with the noted chemoselectivity, at some point these transient species should produce a common intermediate leading to the formation of cystine.

In a recent study, spectral evidence was found for the formation of a $[RS-Cu^I-RS]$ dimer when cysteine was added to Cu(II) under anaerobic conditions and at pH 7.4 (73). On exposure to air, the dimer slowly decayed after the cysteine excess was consumed. This may indicate that indeed all the reaction proceeds via an intra-molecular mechanism.

Kachur and co-workers found evidence for a two-stage autoxidation reaction in an excess of dioxygen (74). In the first stage, the formation of hydroxyl radicals could not be detected and 0.25 mole of O_2 was consumed by one mole of cysteine in agreement with the stoichiometry shown in Eq. (81). Further oxygen consumption occurred in the second stage of the reaction in a hydrogen peroxide-mediated Fenton-type process with sulfonate as product. Significant O^-/OH production was also observed. Copper was proposed to catalyze the reaction via the formation of a cysteine complex and the corresponding mechanism postulates simultaneous superoxide- and peroxide-dependent paths for the first stage and a peroxide-dependent path for the second.

Copper(II) retains its catalytic activity when attached to a polyethylenepolyamine base. The synthetic conditions of the sorbents have a significant effect on the catalytic reaction which most likely is related to a change in the coordination sphere around the metal center (75).

Metal ion catalyzed autoxidation reactions of glutathione were found to be very similar to that of cysteine (76,77). In a systematic study, catalytic activity was found with Cu(II), Fe(II) and to a much lesser extent with Cu(I) and Ni(I). The reaction produces hydrogen peroxide, the amount of which strongly depends on the presence of various chelating molecules. It was noted that the catalysis requires some sort of complex formation between the catalyst and substrate. The formation of a radical intermediate was not ruled out, but a radical initiated chain mechanism was not necessary for the interpretation of the results (76).

VI. Autoxidation of Sulfur(IV)

A review by Brandt and van Eldik provides insight into the basic kinetic features and mechanistic details of transition metal-catalyzed autoxidation reactions of sulfur(IV) species on the basis of literature data reported up to the early 1990s (78). Earlier results confirmed that these reactions may occur via non-radical, radical and combinations of non-radical and radical mechanisms. More recent studies have shown evidence mainly for the radical mechanisms, although a non-radical, two-electron decomposition was reported for the $HgSO_3$ complex recently (79). The possiblity of various redox paths combined with protolytic and complex-formation reactions are the sources of manifest complexity in the kinetic characteristics of these systems. Nevertheless, the predominant sulfur containing product is always the sulfate ion. In spite of extensive studies on this topic for well over a century, important aspects of the mechanisms remain to be clarified and the interpretation of some of the reactions is still controversial. Recent studies were

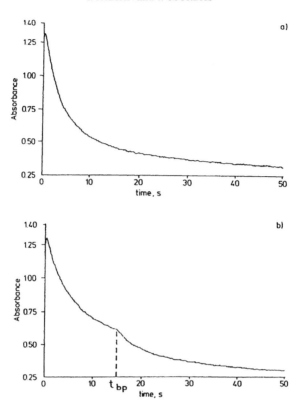

FIG. 4. Absorbance-time traces for the iron(III) catalyzed autoxidation of sulfur(IV) oxides: (a) $[O_2] = 0$ m; (b) $[O_2] = 7.5 \times 10^{-4}$ M. Experimental conditions: $[Fe(III)] = 5.0 \times 10^{-6}$ M; $[S(IV)] = 5.0 \times 10^{-3}$ M; ionic strength = 0.5 M; $T = 25\,^{\circ}$C; pH = 2.5; $\lambda = 390$ nm; absorbance scale is in V (10 V = 1 absorbance unit). Reprinted with permission from Brandt, C.; Fábián, I.; van Eldik, R. *Inorg. Chem.* **1994**, *33*, 687. Copyright (2002) American Chemical Society.

designed to improve previous kinetic models and to understand how specific conditions and environmental factors affect the overall reactions.

Due to the environmental and industrial significance of the reaction, the most thorough studies were reported on the iron(III) catalyzed oxidation of sulfur(IV). As shown in Fig. 4, the kinetic traces are distinctively different in the presence and absence of dioxygen.

Brandt and co-workers proposed a detailed mechanism for this reaction which was validated using kinetic modeling and the most viable values of the rate constants were estimated on the basis of sensitivity analysis (*80*). In this model, the absorbance increase observed at the beginning of the kinetic traces was assigned to the formation of a generalized iron(III) monosulfito complex which undergoes redox decomposition in the rate determining step:

$$Fe^{III}S^{IV} \longrightarrow Fe(II) + SO_3^{\bullet-} \tag{90}$$

Fast reaction of the sulfite ion radical with a second Fe(III) concludes the overall reaction and yields the main final products which are Fe(II), SO_4^{2-}. Small amounts of $S_2O_6^{2-}$ are also formed in the reaction. In the presence of dioxygen, a characteristic break point (at time t_{bp}) was observed in the kinetic traces indicating total consumption of O_2. After the break point, the absorbance decay was the same as in the absence of oxygen. An increase in $[O_2]$ increased t_{bp} while variation of $[Fe(III)]$ had the opposite effect. The formation and decay of the iron(III)-sulfito complex became slower when iron(II) was added initially to the reaction mixture, but t_{bp} was only slightly affected. Hydroquinone, a well known radical scavenger, had no effect on the decomposition rate of the sulfito complex under an argon atmosphere, although it inhibited the oxygen-induced step. This is a clear indication that a free-radical mechanism is operative in the autoxidation stage.

In the presence of oxygen, $SO_3^{\bullet-}$ generates the peroxomonosulfate anion radical (Eq. (91)) in a reaction step with a rate constant close to the diffusion controlled limiting value on the order of 1.0×10^9 to $2.5 \times 10^9\,M^{-1}\,s^{-1}$ (81,82):

$$SO_3^{\bullet-} + O_2 \longrightarrow SO_5^{\bullet-} \tag{91}$$

The $SO_5^{\bullet-}$ radical, being a reasonably powerful and stronger oxidant than dioxygen ($\varepsilon^0 = 1.24\,V$ for the $SO_5^{\bullet-}/HSO_5^-$ couple at pH 2.0 (83)), has a central role in the mechanism, because it can easily oxidize either sulfite ion or Fe(II). Reoxidation of Fe(II) to Fe(III) is essential to establish the catalytic redox cycle, and parallel reactions of $SO_5^{\bullet-}$ with HSO_3^- develop a branch of the reaction chain via the formation of additional reactive intermediates such as $SO_4^{\bullet-}$ and HSO_5^-. The possible reactions of various sulfur oxyanions and oxyanion radicals in the uncatalyzed autoxidation of HSO_3^- were discussed by Connick and co-workers in detail (84), and the rate constants for some of them were reported by Das recently (85). These reactions occur after the rate-determining step, i.e. Eq. (90), of the catalytic cycle, and only the ratios of their rates is relevant with respect to the kinetic model, because the product distribution is determined by the relative kinetic weight of the competing reaction paths. The formation of $S_2O_6^{2-}$ at relatively low concentration levels compared to SO_4^{2-} confirms that the recombination of the $SO_3^{\bullet-}$ radical is of marginal kinetic importance in the catalytic reaction. Warneck and

Ziajka proposed essentially the same model on the basis of kinetic studies in which benzene was used as a radical scavenger (86). They used a steady-state approach for evaluation of the data and reported the following ratios of the second-order rate constants for the reactions of a given intermediate with different reactants: $SO_4^{\bullet-}$ reacting with HSO_3^- and benzene: 0.31; $SO_5^{\bullet-}$ reacting with Fe^{2+} and HSO_3^-: 54.1; HSO_5^- reacting with Fe^{2+} and HSO_3^-: 2.2 and the reaction of $SO_5^{\bullet-}$ with HSO_3^- leading to $SO_3^{\bullet-}$ and $SO_4^{\bullet-}$: ≤ 0.04. Kinetic studies under hydrometallurgical conditions at $80\,^{\circ}C$ were also consistent with the considerations discussed above (87).

The model proposed by Brandt $et\ al.$ is consistent with the experimental observations, reproduces the peculiar shape of the kinetic curves in the absence and presence of dioxygen reasonably well, and predicts the same trends in the concentration dependencies of t_{bp} that were observed experimentally (80). It was concluded that there is no need to assume the participation of oxo-complexes in the mechanism as it has been proposed in the literature (88–90). However, the model provides only a semi-quantitative description of the reaction because it was developed at constant pH by neglecting the acid-base equilibria of the sulfite ion and the reactive intermediates, as well as the possible complex-formation equilibria between various iron(III) species. In spite of the obvious constraints introduced by the simplifications, the results shed light on the general mechanistic features of the reaction and could be used to identify the main tasks for further model development.

The most controversial issue is the number and exact stoichiometries of the iron(III)-sulfito complexes formed under different experimental conditions. Earlier, van Eldik and co-workers reported the formation of a series of $[Fe(SO_3)_n]^{3-2n}$ ($n = 1$ to 3) complexes and the $[Fe(SO_3)(OH)]$ complex ($89,91,92$). The stability constants of these species were determined by evaluating time resolved rapid-scan spectra obtained from the sub-second to several minutes time domain. The cis–$trans$ isomerization of the complexes was also considered, under feasible circumstances. In contrast, Betterton interpreted his results assuming the formation and linkage isomerization of a single complex, $[Fe(SO_3)]^+$ (93). In agreement with the latter results, Conklin and Hoffmann also found evidence only for the formation of a mono-complex (94). However, their results were criticized on the basis that the experiments were made in 1.0 M formic acid/formate buffer where iron(III) existed mainly as formato complex(es). Although these reactions could interfere with the formation of the sulfito complex, they were not considered in the evaluation of the results (95). Finally, van Eldik and co-workers re-examined the complex-formation reactions and presented additional data in support of their earlier results (96–98). The inherent difficulty of obtaining reliable experimental evidence for the complexes formed is the strong overlap between concurrent complex-formation reactions and kinetic coupling between the equilibrium and redox reactions. Nevertheless, the complexity of the spectroscopic data seems to be consistent with the existence of more than one sulfito complex.

A kinetic study was designed by Lente and Fábián to explore the properties of the mono-sulfito complex in the presence of a substantial metal ion excess over sulfite ion concentration. The observations indicated surprisingly complex, multi-stage kinetic behavior (95). In order to separate the overlapping kinetic effects and impede the complex-formation steps, the initial part of the reaction was studied at $10\,^\circ C$ where the overall process lasted for over 20 minutes. Typical stopped-flow traces showed an absorbance increase within the dead-time of the instrument ($\sim 1\,ms$) indicating an instantaneous reaction. It could be shown that this absorbance change corresponds to the formation of the $[Fe(SO_3)]^+$ complex which is much faster than expected on the basis of the rates of other ligand substitution reactions of iron(III). This may indicate that the complex-formation occurs via an oxygen exchange mechanism instead of the usual ligand-exchange path. Such a mechanism was proposed earlier for the formation of other sulfito complexes, (99,100) and supported by the extremely fast complex formation between sulfite ion and the otherwise substitution inert Cr(III) (101,102).

Simple first-order kinetics were observed in the next, well-separated stage which was complete within 200 ms. The main absorbing species at 340 nm is $[Fe_2(OH)_2]^{4+}$ and the absorbance decay at this wavelength indicated that the hydroxo dimer is directly involved in the fast process. An analysis of the concentration dependence of the absorbance change confirmed the formation of a unique dinuclear complex presumably with the following composition: $[(H_2O)_4Fe(\mu\text{-}OH)(\mu\text{-}SO_3)Fe(H_2O)_4]^{4+}$. In follow-up studies the formation of this species was modeled without the interference of redox steps, and the equilibrium and kinetic features of a series of novel di- and tetranuclear iron(III) complexes with simple inorganic oxoanions were described (103–105). The results seem to confirm that the formation of a dinuclear complex from $[Fe_2(OH)_2]^{4+}$ is always feasible when the O–X–O moiety of the entering ligand has the appropriate geometry to replace one of the OH bridges of the hydroxo dimer. These complexes should be regarded as reactive intermediates as they dissociate into the corresponding mono-complexes at a rate comparable with that of the reactions between the monomer and dimeric hydroxo complexes. The combination of these reactions typically yields composite kinetic traces, and the pH dependencies of the individual reaction steps complicate the situation even further.

The kinetic models developed for the formation of dinuclear iron(III) complexes served as a basis for the evaluation of the sulfite ion reaction. Matrix rank analysis of the time-resolved spectra did not indicate the formation of additional absorbing species at longer reaction times, and the common complex-formation model was modified by considering redox reactions only with the monosulfito and the dinuclear complex (106). A series of model calculations assuming different redox paths revealed that the dinuclear species is not involved in direct inner- or outer-sphere electron transfer reactions. Finally, the kinetic model given in Eqs. (92)–(101) was verified by simultaneous evaluation of kinetic traces recorded at 340 and 430 nm under different experimental conditions. Thus, the equilibrium and rate constants were estimated by fitting the kinetic model directly to about 10^5 time-dependent absorbance values. These parameters are given at $25\,^\circ$C and $\mu = 1.0\,$M $NaClO_4$ in the following equations:

$$Fe^{3+} \rightleftharpoons Fe(OH)^{2+} + H^+ \qquad K_{92} = 1.91 \times 10^{-3}\,M,\ \text{fast} \quad (92)$$

$$Fe^{3+} + HSO_3^- \rightleftharpoons Fe(SO_3)^+ + H^+ \qquad K_{93} = 1.35,\ \text{fast} \quad (93)$$

$$H_2O \cdot SO_2 \rightleftharpoons HSO_3^- + H^+ \qquad K_{94} = 1.82 \times 10^{-2}\,M,\ \text{fast} \quad (94)$$

$$HSO_3^- \rightleftharpoons SO_3^{2-} + H^+ \qquad K_{95} = 4.57 \times 10^{-7}\,M,\ \text{fast} \quad (95)$$

$$Fe_2(OH)_2^{4+} + HSO_3^- \rightleftharpoons Fe_2(OH)(SO_3)^{3+} \quad K_{96} = 5.37 \times 10^3\,M$$
$$k_{96} = 3.9 \times 10^5\,M^{-1}\,s^{-1} \quad (96)$$

$$Fe_2(OH)_2^{4+} + SO_3^{2-}(+H^+) \rightleftharpoons Fe_2(OH)(SO_3)^{3+} \quad k_{97} \sim 4 \times 10^9\,M^{-1}\,s^{-1} \quad (97)$$

$$Fe_2(OH)_2^{4+} \rightleftharpoons 2Fe(OH)^{2+} \qquad \begin{aligned} k_{98}^a &= 0.35\,M^{-1}\,s^{-1} \\ k_{98}^b &= 3.5\,s^{-1} \\ k_{98}^c &= 3.6 \times 10^{-3}\,M\,s^{-1} \end{aligned} \quad (98)$$

$$Fe_2(OH)(SO_3)^{3+} \rightleftharpoons Fe(OH)^{2+} + Fe(SO_3)^+ \quad k_{99} = 3.6\,s^{-1} \quad (99)$$

$$Fe(SO_3)^+ \longrightarrow Fe^{2+} + SO_3^{\bullet-} \qquad k_{100} = 0.19\,s^{-1} \quad (100)$$

$$Fe^{3+} + SO_3^{\bullet-} + H_2O \longrightarrow Fe^{2+} + SO_4^{2-} + 2H^+ \qquad \text{fast} \quad (101)$$

The three rate constants for Eq. (98) correspond to the acid-catalyzed, the acid-independent and the hydrolytic paths of the dimer-monomer equilibrium, respectively, and were evaluated independently (*107*). The results clearly demonstrate that the complexity of the kinetic processes is due to the interplay of the hydrolytic and the complex-formation steps and is not a consequence of electron transfer reactions. In fact, the first-order decomposition of the $FeSO_3^+$ complex is the only redox step which contributes to the overall kinetic profiles, because subsequent reactions with the sulfite ion radical and other intermediates are considerably faster. The presence of dioxygen did not affect the kinetic traces when a large excess of the metal ion is present, confirming that either the formation of the $SO_5^{\bullet-}$ radical (Eq. (91)) is suppressed by reaction (101), or the reactions of Fe(II) with $SO_5^{\bullet-}$ and HSO_5^- are preferred over those of HSO_3^- as was predicted by Warneck and Ziajka (*86*). Recently, first-order formation of iron(II) was confirmed in this system (*108*), which supports the first possibility cited, though the other alternative can also be feasible under certain circumstances.

Although the above model was developed under non-catalytic conditions, some of the results may bear significance under natural conditions or in the presence of excess sulfite ions. Thus, the decomposition of the mono-sulfito complex was considered to be the rate-determining step in the catalytic cycle, but only estimates could be given for the rate constant in earlier studies. The comprehensive data treatment used by Lente and Fábián yielded a well established value for this parameter (*106*), which can then be used to improve previous kinetic models. Furthermore, the participation of reactions of the $[Fe_2(OH)(SO_3)]^{3+}$ complex was never considered in kinetic studies where excess sulfite ion was used over low iron(III) concentration in mildly acidic solution (pH 2.5–3.0). The above model predicts that in some cases the formation of the dimeric sulfito complex could make a substantial contribution to the spectral changes and omission of this species could lead to biased conclusions. Reevaluation of data sets reported earlier by including the reactions of $[Fe_2(OH)(SO_3)]^{3+}$ may resolve some of the controversies found in literature results.

Recent studies demonstrated that the composition of the reaction mixture, and in particular the pH have significant effects on the kinetics of iron(III)-catalyzed autoxidation of sulfur(IV) oxides. When the reaction was triggered at pH 6.1, the typical pH profile as a function of time exhibited a distinct induction period after which the pH sharply decreased (*98*). The S-shaped kinetic traces were interpreted by assuming that the buffer capacity of the HSO_3^- / SO_3^{2-} system efficiently reduces the acidifying effect of the oxidation process. The activity of the catalyst was significantly reduced by increasing the pH or aging the reagent iron(III) solutions. These observations were rationalized by considering the formation of polymeric iron(III) hydroxo species which are catalytically less active than the aqua or the monohydroxo iron(III) complex.

In order to model chemical features of aerosols, the effect of oxalate ion on the iron(III) catalyzed autoxidation of sulfur(IV) was reported in related papers by Grgič et al. (109,110) and Wolf et al. (111). Oxalate strongly inhibits the catalytic reaction, and when the catalyst was added as Fe(II) the kinetic traces showed autocatalysis with an induction period that increased upon increasing the oxalate concentrations. The observations were interpreted on the basis of the general kinetic model proposed for iron(III) catalysis. It was assumed that the inhibiting effect is due to the relatively low free iron(III) concentrations arising from its consumption in the formation of complexes with the organic component. According to Grgič and co-workers, the induction period in the presence of Fe(II) is required to reach an equilibrium between Fe(III)-oxalato complexes and free Fe(III) (110).

Manganese catalyzed autoxidation of sulfur(IV) has been studied in different pH regions and under a large variety of experimental conditions. The results show spectacular diversity with respect to the rate equations which may contain one, two or even three terms. Reaction orders of sulfite ion and manganese concentrations that vary between 0 and 2 and from less than 1 to 2, respectively, have been reported. At the same time, there seems to be agreement in that the rate of the catalytic cycle is independent of the oxygen concentration and most of the studies support a free-radical chain mechanism. Recent studies clarified important details of manganese catalysis, but verification of a generally applicable mechanism still appears to be a remote goal (112–115).

Connick and Zhang reported a three-term rate law for the oxygen consumption at constant pH 4.5, 25 °C and $\mu = 0.05$ M (112):

$$-d[O_2]/dt = k_{102}^a [HSO_3^-]^2 + k_{102}^b [HSO_3^-][Mn(II)] + k_{102}^c [Mn(II)]^2 \quad (102)$$

with $k_{102}^a = 3.6 \times 10^{-3}$ M^{-1} s^{-1}, $k_{102}^b = 1.23$ M^{-1} s^{-1}; $k_{102}^c = 98.6$ M^{-1} s^{-1}.

The first term was assigned to the uncatalyzed path which includes the formation of the same reactive radical and non-radical intermediates that were reported in related studies. A clear distinction from other kinetic models is that accumulation of the disulfate radical was also postulated during the autoxidation phase (84). The continued production of H^+ after the total consumption of dioxygen was assigned to the relatively slow decomposition of this species with $k_{104} \sim 0.013 \text{ s}^{-1}$:

$$HSO_5^- + HSO_3^- \longrightarrow S_2O_7^{2-} + H_2O \tag{103}$$

$$S_2O_7^{2-} + H_2O \longrightarrow 2SO_4^{2-} + 2H^+ \tag{104}$$

The obvious challenge in the interpretation of the data is to find a suitable explanation for the independence of the third term of the rate law, Eq. (102), on the concentrations of HSO_3^- and O_2. The rate expression determined experimentally could be modeled quantitatively by combining the following propagation steps with the uncatalyzed reaction mechanism:

$$Mn(II) + SO_5^{\bullet-} \longrightarrow Mn(III) + HSO_5^- \tag{105}$$

$$Mn(III) + HSO_3^- \longrightarrow Mn(II) + SO_3^{\bullet-} \tag{106}$$

The above equations are not balanced for proton budget, and it was noted that a change in the pH may have significant kinetic consequences.

Provided that manganese is present predominantly as manganous ion, the kinetic model predicts a second-order dependence on [Mn(II)] and zeroth-order in the concentration of the other components at high catalyst concentrations, as well as first-order in both [Mn(II)] and [HSO$_3^-$] when the [Mn(II)]/[HSO$_3^-$] ratio is increased and [HSO$_3^-$] is decreased. Furthermore, it provides a coherent interpretation of other kinetic effects such as the inhibition by methanol or catalysis by $S_2O_8^{2-}$. On the basis of the results, the formation of stable sulfito complex(es) could be excluded. This conclusion is in contrast to an earlier study which reported the formation of the stable [Mn(HSO$_3$)]$^+$ complex (113).

Berglund and co-workers (113) observed an induction period in the manganese(II) catalyzed autoxidation of sulfur(IV) followed by strictly first-order decay of HSO$_3^-$ at pH $= 2.4$ and at a considerably lower concentration of sulfur(IV) than used by Connick and Zhang. In the presence of added Mn(III) the induction period was eliminated and the reaction became faster. The validity of the following experimental rate expression was confirmed:

$$k_{\text{obs}} = k_{107} [\text{Mn(II)}] \frac{1 + B[\text{Mn(III)}]_0}{A + [\text{Mn(II)}]} \tag{107}$$

where A and B are constants and [Mn(III)]$_0$ represents the concentration of Mn(III) initially added.

First, the observations were explained on the basis of a kinetic model which was centered around the formation of a stable hydrogensulfito complex with Mn(II). However, in light of the results reported by Connnick and Zhang (*112*), the same authors reevaluated their data and presented evidence for the formation of a mixed-valence dimanganese (II,III) complex, $(OH)Mn^{III}OMn^{II}$ (*114*). An attack by HSO_3^- on the Mn(II) moiety and the formation of a precursor complex were hypothesized and within the latter a bridged electron transfer occurs from S(IV) to Mn(III). The mixed-valence complex was expected to be a more active oxidant than Mn(III) alone. The catalytic cycle is completed by subsequent reactions of the products, Mn(II) and $SO_3^{\bullet-}$. It was concluded that significantly different rate expressions may correspond to the same catalytic cycle, and the differences can be rationalized by considering relocation of the rate-determining step when the experimental conditions are altered.

Connick and Zhang did not specify how exactly the catalytic cycle was initiated, though it can be inferred that the presence of some impurity is critical in triggering the reaction (*112*). Fronaeus and co-workers presented strong arguments that trace amounts of iron(III) are sufficient to initiate the catalytic reaction due to synergism between the two metal ions (*114*). When iron(III) is added to the Mn(II)–sulfite ion system, an equilibrium between the Mn(II)/Mn(III) and Fe(II)/Fe(III) couples is rapidly established. The dimanganese(II,III) complex is capable of oxidizing HSO_3^- much faster than by iron(III); thus, the main kinetic role of Fe(III) is to reoxidize Mn(II) to Mn(III). This reaction produces Fe(II) which is quickly reoxidized to Fe(III) by $SO_5^{\bullet-}$ formed as a secondary radical in the oxidation of HSO_3^-. This qualitative model is supported by the results presented by Grgič and Berčič (*115*). However, other conclusions of these authors are inconsistent with previous results as they propose a simplified kinetic model which postulates the formation of a high stability sulfito complex with manganese(II). As in the case of iron(III) catalysis, the speciation of Mn(II) and Mn(III) is crucial in the mechanism, and further studies should be directed toward exploring the nature of complexes formed under different experimental conditions.

While many important details of the iron- and manganese-catalyzed reactions are yet to be explored, the common features of the corresponding mechanisms are well established and also applicable in the presence of other catalysts. Thus, the formation of the $SO_5^{\bullet-}$, $SO_4^{\bullet-}$ and HSO_5^- intermediates was reported in all of the free-radical type reactions. These species are very reactive oxidants and this explains the apparent paradox of sulfite autoxidation reactions that various substrates undergo oxidation with high yields, in the presence of the strongly reducing sulfur(IV).

The oxidation of two water-soluble Ni(II) complexes with a macrocyclic (CR = 2,12-dimethyl-3,7,11,17-tetraazabicyclo-[11.3.1]heptadeca-1(17),2,11, 13,15-pentaene) and a polydentate (KGH-CONH$_2$ = lysylglycylhistidine-carboxamide) ligand has been reported (116). The reaction with [NiCR]$^{2+}$ shows typical auto-catalytic behavior with an induction period which decreases by increasing the sulfite ion concentration. In this case, the reaction was complete within 200 s in an oxygen-saturated solution ([[NiCR]$^{2+}$] = 1.25×10^{-4} M, [S(IV)] = 1.0–4.0×10^{-3} M, pH = 6.5) and the Ni(III) complex product is stable in solution over an extended period of time. The oxidation of Ni(II) was found to be faster when peroxomonosulfate was used instead of sulfite / O$_2$. The oxidation of the other complex is also auto-catalytic with SO$_3^{2-}$ / O$_2$ and considerably faster with both oxidants than the corresponding reactions with [NiCR]$^{2+}$. The formation of the Ni(III) product was observed only in a limited sulfite ion concentration range and analysis of kinetic traces provided evidence for a back reaction to Ni(II) at longer reaction times. The kinetic model for these reactions postulates a Ni(II) / Ni(III) redox cycle and a series of redox reactions with various oxy-sulfur radicals.

Most of the kinetic models predict that the sulfite ion radical is easily oxidized by O$_2$ and/or the oxidized form of the catalyst, but this species was rarely considered as a potential oxidant. In a recent pulse radiolysis study, the oxidation of Ni(II and I) and Cu(II and I) macrocyclic complexes by SO$_3^{\bullet-}$ was studied under anaerobic conditions (117). In the reactions with Ni(I) and Cu(I) complexes intermediates could not be detected, and the electron transfer was interpreted in terms of a simple outer-sphere mechanism. In contrast, time resolved spectra confirmed the formation of intermediates with a ligand-radical nature in the reactions of the M(II) ions. The formation of a product with a sulfonated macrocycle and another with an addi-tional double bond in the macrocycle were isolated in the reaction with [NiCR]$^{2+}$. These results may require the refinement of the kinetic model proposed by Lepentsiotis for the [NiCR]$^{2+}$–SO$_3^{2-}$–O$_2$ system (116).

The oxidizing power of the catalytic sulfite ion/O$_2$ systems was utilized in oxidative cleavage of DNA (118–121), in an analytical application for the determination of sulfur dioxide in air (122) and in developing a luminescent probe for measuring oxygen uptake (123).

VII. Autoxidation of Miscellaneous Substrates

Copper complexes were used as efficient catalysts for selective autoxidations of flavonols (HFLA) to the corresponding o-benzoyl salicylic acid (o-BSH) and CO in non-aqueous solvents and at elevated temperatures (124–128). The oxidative cleavage of the pyrazone ring is also catalyzed by some cobalt complexes (129–131).

In the $[Cu^{II}(FLA)_2]$-catalyzed reaction, at the beginning of the kinetic traces, a relatively slow induction period was observed; this was assigned to the conversion of the complex into a catalytically active species (127). The actual oxygenation of flavonol was believed to occur in the considerably faster second phase of the reaction which was confirmed to be first-order in the concentrations of the catalyst, the substrate and dioxygen with $k = 2.0 \times 10^3 \, M^{-2} s^{-1}$ at 120 °C in DMF. The kinetic model was developed by assuming that a third flavonolate weakly coordinates to the $[Cu^{II}(FLA)_2]$ complex. It should be noted that the reactive form of the catalyst was not specified, although it must be different from the original bis-flavonalate complex because of the slow transformation process in the induction period.

The third ligand was assumed to be coordinated to the metal center via the deprotonated 3-hydroxy and 4-carbonyl groups. This coordination mode allows delocalization of the electronic structure and intermolecular electron transfer from the ligand to Cu(II). The Cu(I)-flavonoxy radical is in equilibrium with the precursor complex and formed at relatively low concentration levels. This species is attacked by dioxygen presumably at the C2 carbon atom of the flavonoxyl ligand. In principle, such an attack may also occur at the Cu(I) center, but because of the crowded coordination sphere of the metal ion it seems to be less favourable. The reaction is completed by the formation and fast rearrangement of a trioxametallocycle.

Essentially the same mechanism was proposed for the $[Cu^I(FLA)(PPh_3)]$ (127) and $[Cu^{II}(FLA)(IDPA)]^+$ (IDPA = 3,3′-iminobis(N,N-dimethylpropylamine) (128) catalyzed reactions. In the former case, composite kinetic features were found in the initial phase of the reaction indicating a slow transformation of the complex presumably to the same catalytically active species which was postulated in the Cu(FLA)$_2$ system.

For the reaction of $[Cu^{II}(FLA)(IDPA)]^+$ straightforward kinetics were reported, as an induction period was not observed at the beginning of the reaction, and the rate was first-order in the catalyst and O$_2$ concentrations but independent of the substrate concentration, with $k = 4.2 \times 10^{-2} \, M^{-1} s^{-1}$ at 130 °C in DMF. In the absence of flavonol, the catalyst is also oxidized in an overall second-order process and the

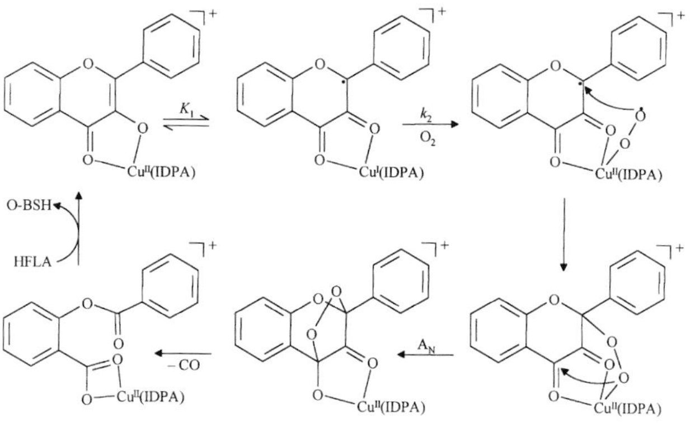

SCHEME 6. Reprinted from *Inorganica Chimica Acta*, vol. 320, Barhács, L.; Kaizer, J.; Pap, J.; Speier, G. Kinetics and mechanism of the stoichiometric oxygenation of (Cu^{II}(fla)(idpa)]ClO$_4$ [fla=flavanolate, idpa=3,3′-imino-bis(*N,N*-dimethylpropylamine)] and the (Cu^{II}(fla)(idpa)]ClO$_4$-catalyzed oxygenation of flavanol, p. 83. Copyright (2002), with permission from Elsevier Science.

corresponding rate constant, $k = 6.1 \times 10^{-3} \, M^{-1} s^{-1}$ (100 °C), 2.9×10^{-2} $M^{-1} s^{-1}$ (130 °C), is in reasonable agreement with that of the catalytic reaction under the same conditions. This suggests that the rate-determining step in the catalytic cycle is the oxidation of the coordinated FLA ligand of the catalyst which is rapidly regenerated as long as the substrate is present in excess. The details of the mechanism proposed for this reaction are shown in Scheme 6.

A comparison of the rate constants for the [Cu^{II}(FLA)(IDPA)]$^+$-catalyzed autoxidation of 4′-substituted derivatives of flavonol revealed a linear free energy relationship (Hammett) between the rate constants and the electronic effects of the *para*-substituents of the substrate (*128*). The logarithm of the rate constants linearly decreased with increasing Hammett σ values, i.e. a higher electron density on the copper center yields a faster oxidation rate.

Catalytic activation of dioxygen by Co(ClO$_4$)$_2$·6H$_2$O, cobalt(II) phthalocyanines (PC = tetrakis(3,5-di-*tert*-butyl-4-hydroxyphenyl)dodecachlorophthalocyaninate) and cobaloxime(II) ([Co^{II}(HDMG)$_2$(L)]$_2$; H$_2$DMG = dimethylglyoxime, L = Ph$_3$P, Ph$_3$As, Ph$_3$Sb) has been studied in oxidation reactions of phenol, thiophenol, and aniline derivatives, primarily, at ambient temperature and in non-aqueous solvents by Simándi and co-workers (*132–141*). A common kinetic feature of these reactions is the accumulation of reactive intermediates in a relatively fast initial phase which persisted for several minutes under typical experimental conditions. The rate laws were evaluated by analyzing the concentration dependencies of the oxygen uptake rates measured after the initial stage, i.e. when the catalytic systems reached steady states. The observations were consistent with a mechanism very similar to that postulated for cobaloxime(II)-catalyzed oxidation of H_2DTBC (*54,55*) (cf. Section IV). Thus, fast reversible formation of a superoxocobalt(III) complex, which is bonded reversibly to the substrate via a hydrogen bond, was proposed. The rate-determining step is the dissociation of this adduct into a hydroperoxocobalt complex and a radical derived from the substrate. Subsequent steps include the recombination of the free radical, its reaction with cobalt(III) and decomposition of the hydroperoxocobalt species. These findings were supported by results from ESR measurements which were used to identify some of the intermediate radical species (*135,139,141*).

With some modifications, the above mechanism is also operative when the substrate is suitable for oxidative attack at more than one position. For example, in the cobaloxime(II)-catalyzed oxidation of 3,3′,5,5′-tetra-*tert*-butyl-4,4′-dihydroxystilbene (H_2StA), oxidative dehydrogenation yielded stilbenequinone (StQ), and simultaneous cleavage of the double bond gave 2,6-di-*tert*-butyl-4-hydroxybenzaldehyde (*141*). When the system reached steady-state, the formation of the two products proceeded at comparable rates presumably via the common $StQ^{\bullet -}$ anion radical. This species can be involved in a further hydrogen abstraction step with $Co^{III}OH$ (Co^{III} represents the $[Co^{III}(HDMG)_2(Ph_3P)]^+$ moiety) or attacked in its mesomeric form at one of the olefinic carbon atoms by the superoxocobalt(III) species forming an alkylperoxocobaloxime(III) intermediate. The aldehyde product is formed in subsequent fast steps. Such a reaction does not occur in the absence of the catalyst and this reaction serves as a rare example of the ability of cobaloxime(II) to catalyze the oxidative cleavage of an olefinic double bond.

In a broad sense, the model developed for the cobaloxime(II)-catalyzed reactions seems to be valid also for the autoxidation of the alkyl mercaptan to disulfides in the presence of cobalt(II) phthalocyanine tetra-sodium sulfonate in reverse micelles (*142*). It was assumed that the rate-determining electron transfer within the catalyst-substrate-dioxygen complex leads to the formation of the final products via the $RS^{\bullet}$ and $O_2^{\bullet -}$ radicals. The yield of the disulfide product was higher in water–oil microemulsions prepared from a cationic surfactant than in the presence of an anionic surfactant. This difference is probably due to the stabilization of the monomeric form of the catalyst in the former environment.

Ruthenium(III) was shown to be a potent catalyst for the autoxidation of various organic substrates by Taqui Khan and co-workers (*143–148*).

TABLE III

CONCENTRATION DEPENDENCIES OF THE EXPERIMENTAL RATE LAW FOR RUTHENIUM CATALYZED AUTOXIDATION OF ORGANIC SUBSTRATES [a,b]

Substrate	Catalyst	H_2A [c]	Ph	n_O	Reference
cyclohexene	Ru(III)	no	2.0	1	(146)
cyclohexene	Ru^{III}(EDTA)	no	1.0–3.0	1/2	(145)
cyclohexene	Ru(III)	yes	2.0	1	(146)
cyclohexene	Ru^{III}(EDTA)	yes	2.5	1	(148)
cyclohexane [d]	Ru^{III}(EDTA)	yes	2.5	1	(148)
cyclohexanol	Ru(III)	no	1.5–3.0	1	(143)
cyclohexanol	Ru^{III}(EDTA)	no	1.5–3.0	0	(143)
cyclohexanol	Ru^{III}(EDTA)	yes	2.5	1	(148)
allyl alcohol	Ru(III)	no	1.0–2.0	0	(144)
		no	2.0–3.0	1/2	(144)
allyl alcohol	Ru^{III}(EDTA)	no	1.0–2.0	0	(147)
		no	2.0–3.0	1	(147)

[a] Typical experimental conditions: T=288–313 K, μ=0.1 M KNO$_3$ or KCl, in water, 1:1 water–ethanol or water-1,4-dioxane.

[b] n_O is the reaction order in O_2. The reaction was first-order in the catalyst and the substrate concentration.

[c] The reaction was first-order in ascorbic acid (H_2A) concentration when it was used as a co-substrate.

[d] Fractional-order was reported for the substrate.

Some of the details of these reactions are summarized in Table III. The corresponding kinetic models were centered around the formation of a complex which included all the reactants, i.e. the catalyst, the substrate, in most cases dioxygen and also ascorbic acid when it was used as a co-substrate. This precursor complex was considered to undergo an internal redox process in a rate-determining step. According to this general model, the rate expression can take different forms depending on how many components are assembled in the precursor complex and where the rate determining-step is located in the reaction sequence.

The oxidation of allyl alcohol shows an interesting pH dependence (144,147). When the pH < 2.0, the main product is acrylaldehyde and the rate is independent of, or inversely proportional to [H$^+$], with the catalysts Ru(III) and Ru(EDTA), respectively.[5] In this case, hydride

---

[5] In these studies Ru(III) and Ru(EDTA) represent [RuCl$_2$(H$_2$O)$_4$]$^+$ and a generalized EDTA complex of Ru(III), respectively. The coordination mode and composition, in particular with respect to possible protonated complexes, were not specified for the EDTA complex in the cited references.

abstraction from the substrate to the metal center within the catalyst-substrate complex was considered to be the rate-determining step which is followed by fast oxidation of the hydrido complex. Above pH 2.0, the reaction becomes independent of pH with both catalysts, and epoxidation of the double bond occurs simultaneously with the oxidation of the alcoholic group. In the case of Ru(III), the formation of a $[S–Ru^{III}–(\mu\text{-}O_2)–Ru^{III}–S]$ (S stands for substrate) dinuclear complex, which decomposes via the cleavage of the O–O bond in a rate determining step, was postulated. The reaction supposedly proceeds via mononuclear complexes when Ru(EDTA) is used as catalyst.

The above models imply that the proton loss of the OH group of the coordinated substrate shifts the mechanism from oxidation to epoxidation with Ru(III). Such a straightforward interpretation of the pH effect was not presented for reactions of the other substrates, i.e. the protolytic reaction, which would act as a switch between the two mechanisms, cannot be identified.

The models proposed for the Ru(III) and Ru(EDTA) catalyzed epoxidation of allyl alcohol replace each other in the corresponding reactions of cyclohexene. Thus, mono-nuclear and di-nuclear paths were reported for the Ru(III) and Ru(EDTA) catalysis, respectively (*145,146*).

The presence of ascorbic acid as a co-substrate enhanced the rate of the Ru(EDTA)-catalyzed autoxidation in the order cyclohexane < cyclohexanol < cyclohexene (*148*). The reactions were always first-order in $[H_2A]$. It was concluded that these reactions occur via a $Ru(EDTA)(H_2A)(S)(O_2)$ adduct, in which ascorbic acid promotes the cleavage of the O_2 unit and, as a consequence, O-transfer to the substrate. While the model seems to be consistent with the experimental observations, it leaves open some very intriguing questions. According to earlier results from the same laboratory (*24,25*), the Ru(EDTA) catalyzed autoxidation of ascorbic acid occurs at a comparable or even a faster rate than the reactions listed in Table III. It follows, that the interference from this side reaction should not be neglected in the detailed kinetic model, in particular because ascorbic acid may be completely consumed before the oxidation of the other substrate takes place.

It is not clear either how the Ru center can accommodate four ligands simultaneously. The crowded coordination sphere around the metal center in the Ru(EDTA)–ascorbate complex is expected to hinder the coordination of other ligands as was proposed earlier (*24,25*). The contradiction between the two sets of results reported in Refs. (*24,25*) and (*148*) is obvious. While the $Ru(EDTA)(H_2A)(O_2)$ complex was not considered in the kinetic model proposed for the oxidation of ascorbic acid, a relatively high equilibrium constant, $\log K = 3.18$, was reported for the formation of the same species from Ru(EDTA)(H$_2$A) and O$_2$ in the latter study. Because differences in the experimental conditions cannot account for the above discrepancies, the need for a coherent re-interpretation of the kinetic data seems to be imperative and inevitable.

The enhanced chemiluminescence associated with the autoxidation of luminol (5-amino-2,3-dihydro-1,4-phthalazinedione) in the presence of trace amounts of iron(II) is being used extensively for selective determination of Fe(II) under natural conditions (149–152). The specificity of the reaction is that iron(II) induces chemiluminescence with O$_2$, but not with H$_2$O$_2$, which was utilized as an oxidizing agent in the determination of other trace metals. The oxidation of luminol by O$_2$ is often referred to as an iron(II)-catalyzed process but it is not a catalytic reaction in reality because iron(II) is not involved in a redox cycle, rather it is oxidized to iron(III). In other words, the lower oxidation state metal ion should be regarded as a co-substrate in this system. Nevertheless, the reaction deserves attention because it is one of the few cases where a metal ion significantly affects the autoxidation kinetics of a substrate without actually forming a complex with it.

The source of chemiluminescence in the oxidation of luminol was explored by Merényi and co-workers in detail (153). The oxidation of luminol yields aminophthalate as a final product and the reaction proceeds via a series of electron transfer steps. The primary oxidation product is the luminol radical which is transformed into either diazaquinone or the α-hydroxide-hydroperoxide intermediate (α-HHP). The latter oxidation step occurs between the deprotonated form of the luminol radical and O$_2^{\bullet-}$. The chemiluminescence is due to the decomposition of the mono-anionic form of α-HHP into the final products:

$$\tag{108}$$

α-HHP

An alternative path includes oxidation, in the absence of light, of the diazaquinone with weak chemiluminescence (154). The effect of iron(II) on the luminescent intensity was interpreted by considering that it can efficiently generate the O$_2^{\bullet-}$ radical in a reaction with O$_2$ and, as a consequence, increase the importance of reaction (108) in the overall process (155).

Trace amounts of Cu(II) were reported to catalyze the oxidation of I^- to I_2 (156) and the phosphinate ion ($H_2PO_2^-$) to peroxodiphosphate ion (PDP), which could be present as $P_2O_8^{4-}$, $HP_2O_8^{3-}$ or $H_2P_2O_8^{2-}$ (157). Individual kinetic traces showed some unusual patterns in these reactions, such as the variation between first- and zeroth-order kinetics with respect to the formation of I_2 under very similar conditions, or an autocatalytic feature in the concentration profiles of PDP, but these events were not studied in detail. The catalytic effect was interpreted in terms of a Cu(II)/Cu(I) redox cycle and the superoxide ion radical, $O_2^{\bullet-}/HO_2^{\bullet}$, was considered as a reactive intermediate in both cases. The addition of radical scavengers strongly retarded the oxidation of the phosphinate ion confirming the radical type mechanism. It was also demonstrated that the reaction ceased when the catalyst was masked with EDTA.

Iron(III) chelate complexes were utilized as catalysts for oxidative removal of H_2S from natural gas. The application of $Fe^{III}(NTA)$ is plagued by the degradation of the ligand to weaker chelating agents such as iminodiacetic acid, glycine, oxalate and eventually CO_2 is generated. According to Chen et al., the degradation occurred on the time-scale of 3–4 days when air and H_2S were included in a reaction mixture originally containing only the catalyst (158). Substantial acidification was observed in unbuffered reaction mixtures, and the pH dropped from 8.6 to 3.5 in a typical experiment. This combined with the fact that the reaction became much slower at constant pH 8.5 indicates the autocatalytic feature of the overall reaction.

It was proposed that the degradation of the catalyst occurs simultaneously with the following catalytic cycle:

$$2Fe(NTA) + H_2S \longrightarrow 2Fe(NTA)^- + 1/8S_8 + 2H^+ \qquad (109)$$

$$2Fe(NTA)^- + 1/2O_2 + H_2O \longrightarrow 2Fe(NTA) + 2OH^- \qquad (110)$$

A free radical mechanism was considered for the reoxidation of the Fe(II) complex involving the formation of H_2O_2. The decomposition of hydrogen peroxide generates the primary $OH^\bullet$ and the secondary $O_2^{\bullet-}$ radicals. While O_2 is not a sufficiently strong oxidant to attack NTA, the free radicals are reactive enough to induce oxidative degradation of the ligand and byproducts. The formation of H_2O_2 was confirmed in experiments with the enzyme catalase and the reaction was inhibited by addition of $Na_2S_2O_3$, a well known free radical scavenger. These observations fully support the proposed model.

VIII. Exotic Kinetic Phenomena

A number of autoxidation reactions exhibit exotic kinetic phenomena under specific experimental conditions. One of the most widely studied systems is the peroxidase-oxidase (PO) oscillator which is the only enzyme reaction showing oscillation in vitro in homogeneous stirred solution. The net reaction is the oxidation of nicotinamide adenine dinucleotide (NADH), a biologically vital coenzyme, by dioxygen in a horseradish peroxidase enzyme (HRP) catalyzed process:

$$NADH + O_2 \longrightarrow 2NAD^+ + 2H_2O \tag{111}$$

Since the first report of oscillation in 1965 (*159*), a variety of other nonlinear kinetic phenomena have been observed in this reaction, such as bi-stability, bi-rhythmicity, complex oscillations, quasi-periodicity, stochastic resonance, period-adding and period-doubling to chaos. Recently, the details and sub-systems of the PO reaction were surveyed and a critical assessment of earlier experiments was given by Scheeline and co-workers (*160*). This reaction is beyond the scope of this chapter and therefore, the mechanistic details will not be discussed here. Nevertheless, it is worthwhile to mention that many studies were designed to explore non-linear autoxidation phenomena in less complicated systems with an ultimate goal of understanding the PO reaction better.

The main features of the copper catalyzed autoxidation of ascorbic acid were summarized in detail in Section III. Recently, Strizhak and coworkers demonstrated that in a continuously stirred tank reactor (CSTR) as well as in a batch reactor, the reaction shows various non-linear phenomena, such as bi-stability, oscillations and stochastic resonance (*161*). The results from the batch experiments can be suitably illustrated with a two-dimensional parameter diagram shown in Fig. 5.

At high [Cu(II)] and low [H$_2$A] initial concentrations, the Pt electrode potential, used to follow the chemical process, increased monotonously. When both species were present at high initial concentrations, a monotonous decrease was observed. Various non-monotonic transient regimes were found at approximate initial concentrations of [Cu(II)] $\sim 10^{-4}$ M and [H$_2$A] $\sim 10^{-4}$ M. Thus, the batch experiments properly illustrate that the system is sensitive to variations of the initial concentrations of ascorbic acid and copper(II) ion, and the observations can be indicative of a transient bi-stability.

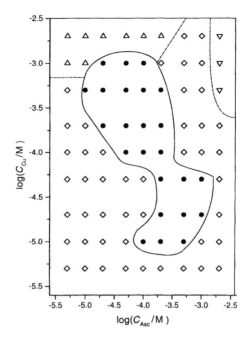

FIG. 5. Two-dimensional parametric diagram of system response at different initial concentrations of reagents in batch: $\triangle$, monotonic growth of Pt potential [Fig. 1(a)]; $\triangledown$, monotonic decrease of Pt potential [Fig. 1(b)]; $\diamond$, Pt electrode potential first decreases and then increases in time [Fig. 1(c)]; $\bullet$, various nonmonotonic transient regimes [Fig. 1(d–f)]. Strizhak, P. E.; Basylchuk, A. B.; Demjanchyk, I.; Fecher, F.; Shcneider, F. W.; Munster, A. F. *Phys. Chem. Chem. Phys.* **2000**, *2*, 4721. Reproduced by permission of The Royal Society of Chemistry on behalf of the PCCP Owner Societies.

The experiments were performed at a constant inflow concentration of ascorbic acid ([H_2A]) in the CSTR. Oscillations were found by changing the flow rate and the inflow concentration of the copper(II) ion systematically. At constant Cu(II) inflow concentration, the electrode potential measured on the Pt electrode showed hysteresis between two stable steady-states when first the flow-rate was increased, and then decreased to its original starting value. The results of the CSTR experiments were summarized in a phase diagram (Fig. 6).

In Fig. 6, separate regions of bi-stability, oscillations and single stable steady-states can be noticed. This "cross-shaped" phase diagram is common for many non-linear chemical systems containing autocatalytic steps, and this was used as an argument to suggest that the Cu(II) ion catalyzed autoxidation of the ascorbic acid is also autocatalytic. The

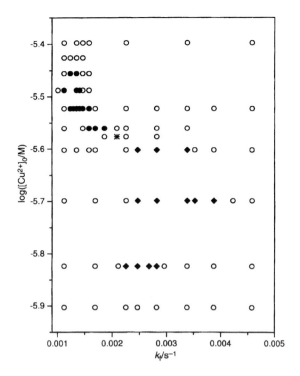

FIG. 6. Dynamical phase diagram of the ascorbic acid/copper(II)/oxygen system in a CSTR in the $k_f - [Cu^{2+}]_0$ plane. Fixed reactor concentrations: $[H_2Asc]_0 = 5.0 \times 10^{-4}$ M; $[H_2SO_4]_0 = 6.0 \times 10^{-5}$ M; $[Na_2SO_4]_0 = 0.04$ M. Symbols: $\bigcirc$, steady state; $\bullet$, oscillations; $\blacklozenge$, bistability. The asterisk (*) marks the Takens-Bogdanov point. Strizhak, P. E.; Basylchuk, A. B.; Demjanchyk, I.; Fecher, F.; Shcneider, F. W.; Munster, A. F. *Phys. Chem. Chem. Phys.* **2000**, *2*, 4721. Reproduced by permission of The Royal Society of Chemistry on behalf of the PCCP Owner Societies.

following reaction scheme was proposed for the interpretation of the results:

$$H_2A + Cu^{2+} \longrightarrow HA^{\bullet} + Cu^{+} + H^{+} \tag{112}$$

$$HA^{\bullet} + O_2 \longrightarrow HO_2^{\bullet} + A \tag{113}$$

$$HO_2^{\bullet} + H_2A \longrightarrow HA^{\bullet} + H_2O_2 \tag{114}$$

$$Cu^{+} + H_2O_2 + H^{+} \longrightarrow OH^{\bullet} + H_2O + Cu^{2+} \tag{115}$$

$$OH^{\bullet} + HA \longrightarrow HA^{\bullet} + H_2O \tag{116}$$

This model is consistent with the results discussed in Section III in that it also postulates the Cu(II)/Cu(I) redox cycle. Otherwise, there are significant differences, such as the removal of H_2O_2 as in Eq. (115), or the formation of the hydroxyl radical, which would require justification. Since calculations were not presented to support the kinetic model, it is open to question as to how well the observations would be reproduced by reactions (*112–116*) or, for that matter, by the alternative models.

The oxidation of benzaldehyde with dioxygen in the presence of Co(II) and bromide ion also shows non-linear kinetic phenomena. The net reaction of the oxidation process is given as follows:

$$2PhCHO + O_2(soln) \longrightarrow 2PhCO_2H \tag{117}$$

Jensen was the first to report in 1983 that the color of the solution oscillated between pink and dark brown in the presence of cobalt(II) and bromide ions when the reaction was carried out in a 90/10 (w/w) acetic acid/water mixture (*162*). This color change was accompanied by a change in the redox potential and the oscillations were observed for over 16 h and 800 cycles. Presumably, the pink color corresponds to a low Co(III)/Co(II) ratio, the dark brownish black to a high Co(III)/Co(II) ratio or to a Co(III)Br complex in this reaction.

In later work, Roelofs and co-workers discovered further details of the reaction by investigating the sub-systems, and they suggested a 21 step chemical model (RWJ model) to explain the observed non-linear kinetic patterns (*163*). According to the experimental observations, the oscillation process can be divided into two distinct alternating stages, the stoichiometries of which can be approximated as follows:

$$PhCHO + O_2(soln) + 2Co(II) + 2H^+ \longrightarrow PhCO_2H + 2Co(III) + H_2O \tag{118}$$

$$PhCHO + 2Co(III) + H_2O \longrightarrow PhCO_2H + 2Co(II) + 2H^+ \tag{119}$$

At the beginning of stage I the concentration of dissolved oxygen is high, and that of Co(III) is low. Equation (118) represents an autocatalytic process, and as the concentration of the product Co(III) increases the rate accelerates, until it reaches a point where the rate of depletion of dissolved oxygen becomes higher than the rate of oxygen transfer from the gas phase to the solution. As a consequence, the concentration of dissolved oxygen falls, and at the end of Stage I dioxygen is totally depleted in the liquid phase. At this point reaction (119) becomes operative, i.e. the Co(III) formed in stage I is consumed in a rapid reaction with benzaldehyde. The model postulates that the reactive Co(III) species is a dimer complex, and a number of radicals ($PhCO^\bullet$, $PhCOO^\bullet$,

PhCOOO$^\bullet$, Co(II)Br$^\bullet$) were assumed to be important intermediates in the reaction. The presence of some of the free radicals was confirmed by EPR spectroscopy, providing additional support for the validity of the RWJ model (164).

The model proposed on the basis of the experimental work of Colussi et al. (CGYN) is not consistent with the RWJ model, in some details (165). The most important difference lies in the interpretation of the role of bromide ion and of bromine that could be formed from bromide ion. Roelofs and co-workers assumed that Br$^-$ is important only in the complex-formation and they did not take into consideration the formation and further reactions of Br$_2$ in their oscillatory model (164). Colussi et al. suggested that Br$_2$ may form at low concentration levels and verified experimentally that bromine added to the system during the oscillation reduces the concentration of Co(III), and makes the periods shorter. A detailed molecular mechanism was proposed in agreement with the observations on the overall reaction and sub-systems (165).

The oxidation of benzaldehyde by molecular oxygen is an autocatalytic process which may feature propagating reaction fronts. Boga et al. investigated this phenomenon in the absence of bromide ion by using a mixture of Co(II) acetate and benzaldehyde solutions saturated with an O$_2$–N$_2$ mixture (166). The wave was initiated in a capillary tube by addition of perbenzoic acid, an intermediate of the reaction. The velocity of front propagation as a function of Co(II), PhCHO, and O$_2$ concentrations was explored. At low concentration levels, Co(II) is a catalyst, but at high concentrations it acts as a scavenger of the PhCOOO$^\bullet$ radical and inhibits the autoxidation of benzaldehyde, Eq. (124). The authors proposed the following skeleton model for interpretation of the results:

$$PhCOOOH + Co(II) \longrightarrow PhCOO^\bullet + Co(III) + OH^-$$
$$k_{120} = 1.9 \times 10^5 \, M^{-1} \, s^{-1} \tag{120}$$

$$Co(III) + PhCHO \longrightarrow Co(II) + PhCO^\bullet + H^+$$
$$k_{121} = 1.6 \times 10^{-4} \, M^{-1} \, s^{-1} \tag{121}$$

$$PhCO^\bullet + O_2 \longrightarrow PhCOOO^\bullet \qquad k_{122} = 5.0 \times 10^8 \, M^{-1} \, s^{-1} \tag{122}$$

$$PhCOOO^\bullet + PhCHO \longrightarrow PhCO^\bullet + PhCOOOH$$
$$k_{123} = 3.9 \times 10^3 \, M^{-1} \, s^{-1} \tag{123}$$

$$PhCOOO^\bullet + Co(II) \longrightarrow Co(III) + PhCOOO^-$$
$$k_{124} = 8.0 \times 10^6 \, M^{-1} \, s^{-1} \tag{124}$$

$$PhCOO^\bullet + PhCHO \longrightarrow PhCO^\bullet + PhCOOH$$
$$k_{125} = 1.0 \times 10^7 \, M^{-1} \, s^{-1} \tag{125}$$

It was shown that an appropriate combination of the above steps, i.e. (120) + (121) + 2(122) + 2(123) + (125), predicts autocatalytic accumulation of the perbenzoic acid:

$$PhCOOOH + 2PhCHO + 2O_2 + 2Co(II)$$
$$\longrightarrow 2PhCOOOH + 2Co(III) + PhCOO^- + OH^- \tag{126}$$

Guslander and co-workers developed another simple skeleton model on the basis of the CGYN model, named Cobaltolator, which consists of only four steps (*167*). Numerical simulations with this model showed that it can reproduce the main features of the oscillation reaction, and it can also simulate the front propagation found by Boga *et al.*

Searching for other oscillatory autoxidation reactions led Druliner and Wasserman to use cyclohexanone as a substrate instead of benzaldehyde (*168*). Unlike the simple stoichiometry found for the benzaldehyde reaction, the ketone gives at least six or more products, and the relative amounts of these vary substantially with the experimental conditions (Scheme 7).

The overall appearance of the oscillation in the cyclohexanone autoxidation is different from that in the corresponding reaction of benzaldehyde, indicating significant differences in the mechanisms. The oscillation can be divided into three stages. In the first stage, organic intermediates, mainly KO_2H and HOK, are formed in a slow autoxidation of cyclohexanone (cf. Scheme 7). During this phase the electrode potential of the system is practically constant, which indicates that the Co(III)/Co(II) ratio is not altered significantly. When enough peroxide is generated the second stage begins, where first HOK reacts with Co(III) to give $RCO^\bullet$, and subsequent steps follow as shown in Scheme 8.

In this stage, the conversion of Co(II) to Co(III) is indicated by the rapid rise in the electrode potential, and fast generation of Co(III) accelerates the formation of HOK. Finally, in the third stage the accumulated Co(III) oxidizes the organic substrates to radical intermediates, while itself is reduced back to Co(II). Because of the noted complexity, the oscillation model for the cyclohexanone reaction was not elaborated upon in detail.

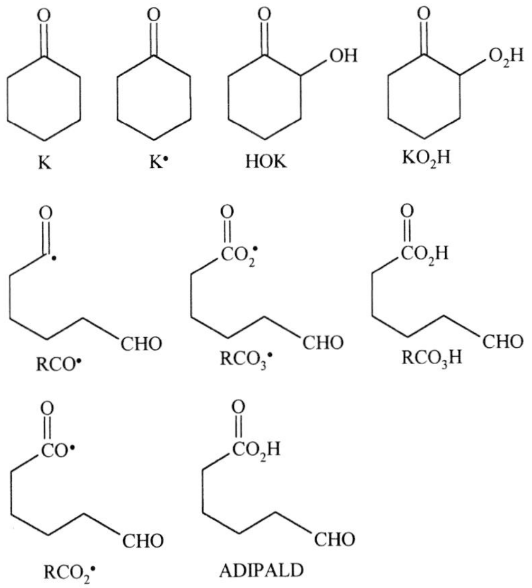

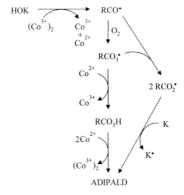

SCHEME 7. Reprinted with permission from Druliner, J. D.; Wasserman, E. *J. Am. Chem. Soc.* **1988**, *110*, 5270. Copyright (2002) American Chemical Society.

SCHEME 8.

IX. Perspectives

The results surveyed in this chapter demonstrate the composite nature of autoxidation reactions and the potential problems involved in exploring the intimate details of the appropriate mechanisms. The kinetic observatiuons have been interpreted to different depths, but there is still plenty of room for improving available kinetic models. Most

certainly, a great number of future studies will be designed to characterize better the reactive intermediates and to identify the dominant
reaction paths in these systems. Recent developments in reaction
kinetics offer sophisticated technical and computational tools to accomplish these objectives. New fast kinetic devices allow the recording of
time resolved spectra on the millisecond time-scale with a reliability
comparable to that of high quality spectrophotometers; the sensitivity,
detection limits and response times of electrochemical instruments
have been improved by orders of magnitude using new technologies. The
availability of micro and nano devices has opened new territories for
time-dependent studies.

The quantity and quality of experimental information determined by
the new techniques call for the use of comprehensive data treatment
and evaluation methods. In earlier literature, quite often kinetic studies
were simplified by using pseudo-first-order conditions, the steady-state
approach or initial rate methods. In some cases, these simplifications
were fully justified but sometimes the approximations led to distorted
results. Autoxidation reactions are particularly vulnerable to this
problem because of strong kinetic coupling between the individual
steps and feed-back reactions. It was demonstrated in many cases, that
these reactions are very sensitive to the conditions applied and their
kinetic profiles and stoichiometries may be significantly altered by
changing the pH, the absolute concentrations and concentration ratios
of the reactants, and also by the presence of trace amounts of impurities
which may act either as catalysts and/or inhibitors.

By lifting the simplifying restrictions, the kinetic observations can be
examined in more detail over much wider concentration ranges of the
reactants than those relevant to pseudo-first-order conditions. It should
be added that sometimes a composite kinetic trace is more revealing
with respect to the mechanism than the conventional concentration
and pH dependencies of the pseudo-first-order rate constants.
Simultaneous evaluation of the kinetic curves obtained with different
experimental methods, and recorded under different conditions, is
based on fitting the proposed kinetic models directly to the primary
data. This method yields more accurate estimates for the rate constants
than conventional procedures. Such an approach has been used sporadi-
cally in previous studies, but it is expected to be applied more widely
and gain significance in the near future.

Reliable mechanisms can serve as the basis for the design of efficient
new catalysts for autoxidation reactions. A systematic analysis of the
effects of the non-participating ligands on the kinetics of the overall
reaction and on the catalytic activity of the metal center(s) could be a

driving force to prepare novel complexes with special functionalities. Furthermore, the transfer of the catalyst to a non-traditional environment may be an exciting field for mechanistic studies. Already there are a few examples where certain catalytic properties are retained when the metal centers are attached to polymeric surfaces; however, the porosity and other properties of the carrier also affect the reactions. Non-conventional solvents, such as supercritical fluids, or biphasic systems can also be used to tune the activity of some of the catalysts.

Finally, kinetic and mechanistic studies of autoxidation reactions will not only lead to a better understanding of these essential reactions but also may trigger extensive studies on various areas of dioxygen chemistry.

ACKNOWLEDGEMENTS

This work was supported by the Hungarian National Research Foundation under grant Nos. OTKA T 029568 and 042755.

REFERENCES

1. Sawyer, D. T. "*Oxygen Chemistry*"; Oxford University Press: New York, 1991.
2. "*Standard Potentials in Aqueous Solution*"; Eds. Bard, A. J.; Parson, R.; Jordan, J.; Marcel Dekker: New York, 1985.
3. Cofré, P.; Sawyer, D. T. *Inorg. Chem.* **1986**, *25*, 2089.
4. Czapski, G. *Annu. Rev. Phys. Chem.* **1971**, *22*, 171.
5. Miller, D. M.; Buettner, G. R.; Aust, S. D. *Free Radic. Biol. Med.* **1990**, *8*, 95.
6. Taqui Khan, M. M.; Martell, A. E. *J. Am. Chem. Soc.* **1967**, *89*, 4176.
7. Gampp, H.; Zuberbühler, A. D. "*Metal Ions in Biological Systems*", vol. 12; Ed. Sigel, H.; Marcel Dekker: New York, 1981, pp. 133–189.
8. Zuberbühler, A. D. "*Copper Coordination Chemistry: Biochemical and Inorganic Perspectives*"; Eds. Karlin, K. D.; Zubieta, J.; Adenine, New York, 1983; pp. 237–258.
9. Shtamm, E. B.; Purmal, A. B.; Skurlatov, I. *Zhur. Fiz. Khim.* **1974**, *48*, 2229.
10. Weissberger, A.; Luvalle, J. E.; Thomas, D. S., Jr. *J. Am. Chem. Soc.* **1943**, *65*, 1934.
11. Jameson, R. F.; Blackburn, N. J. *J. Chem. Soc., Dalton Trans.* **1976**, 534.
12. Jameson, R. F.; Blackburn, N. J. *J. Chem. Soc., Dalton Trans.* **1976**, 1596.
13. Hadady-Kovács, K.; Fábián, I. *J. Pharm. Biomed. Anal.* **1996**, *14*, 1479.
14. Sisley, M. J.; Jordan, R. B. *J. Chem. Soc., Dalton Trans.* **1997**, 3883.
15. Jameson, R. F.; Blackburn, N. J. *J. Chem. Soc., Dalton Trans.* **1982**, 9.
16. Mi, L.; Zuberbühler, A. D. *Helv. Chim. Acta* **1992**, *75*, 1547.
17. Mi, L.; Zuberbühler, A. D. *Helv. Chim. Acta* **1991**, *74*, 1679.
18. Scarpa, M.; Vianello, F.; Signor, L.; Zennaro, L.; Rigo, A. *Inorg. Chem.* **1996**, *35*, 5201.
19. Thomas, A. M.; Mandal, G. C.; Tiwary, S. K.; Rath, R. K.; Chakravarty, A. R. *J. Chem. Soc., Dalton Trans.* **2000**, 1395.
20. Srinivas, B.; Arulsamy, N.; Zacharias, P. S. *Polyhedron* **1991**, *10*, 731.
21. Xu, J. H.; Jordan, R. B. *Inorg. Chem.* **1990**, *29*, 4180.
22. Bänsch, B.; Martinez, P.; Uribe, D.; Zuluaga, J.; van Eldik, R. *Inorg. Chem.* **1991**, *30*, 4555.
23. Lepentsiotis, V.; van Eldik, R. *J. Chem. Soc., Dalton Trans.* **1998**, 999.
24. Taqui Khan, M. M.; Shukla, R. S. *J. Mol. Catal.* **1986**, *34*, 19.

25. Taqui Khan, M. M.; Shukla, R. S. *J. Mol. Catal.* **1986**, *37*, 269.
26. Seibig, S.; van Eldik, R. *Inorg. React. Mech.* **1999**, *1*, 91.
27. Speier, G.; Tyeklár, Z. *J. Chem. Soc., Perkin Trans. 2* **1981**, 1176.
28. Que, L., Jr.; Kolanczyk, R. C.; White, L. S. *J. Am. Chem. Soc.* **1987**, *109*, 5373.
29. Cox, D. D.; Que, L., Jr. *J. Am. Chem. Soc.* **1988**, *110*, 8085.
30. Que, L., Jr.; Ho, R. Y. N. *Chem. Rev.* **1996**, *96*, 2607.
31. Funabiki, T. "*Oxygenases and Model Systems*"; Ed. Funabiki, T.; Kluwer: Boston, 1997, pp. 19–156.
32. Demmin, T. R.; Swerdloff, M. D.; Rogic, M. M. *J. Am. Chem. Soc.* **1981**, *103*, 5795.
33. Bolus, D.; Vigee, G. S. *Inorg. Chim. Acta* **1982**, *67*, 19.
34. Rogic, M. M.; Swerloff, M. D.; Demmim, T. R. "*Copper Coordination Chemistry: Biochemical and Inorganic Perspectives*"; Eds. Karlin, K. D., Zubieta, J.; Adenine, New York, 1983; p. 259.
35. Tyson, C. A.; Martell, A. E. *J. Am. Chem. Soc.* **1972**, *94*, 939.
36. Balla, J.; Kiss, T.; Jameson, R. F. *Inorg. Chem.* **1992**, *31*, 58.
37. Abakumov, G. A.; Lobanov, A. V.; Cherkasov, V. K.; Razuvaev, G. A. *Inorg. Chem. Acta* **1981**, *49*, 135.
38. Pierpont, C. G.; Buchanan, R. M. *Coord. Chem. Rev.* **1981**, *38*, 45.
39. Tyson, C. A.; Martell, A. E. *J. Am. Chem. Soc.* **1968**, *90*, 3379.
40. Shtamm, E. V.; Purmal, A. P.; Skurlatov, Y. I. *Int. J. Chem. Kinet.* **1979**, *11*, 461.
41. Speier, G. *J. Mol. Catal.* **1986**, *37*, 259.
42. Li, Y. B.; Trush, M. A. *Arch. Biochem. Biophys.* **1993**, *300*, 346.
43. Zhang, L. P.; Bandy, B.; Davison, A. J. *Free Radic. Biol. Med.* **1996**, *20*, 495.
44. Mochizuki, M.; Yamazaki, S.; Kano, K.; Ikeda, T. *Biochim. Biophys. Acta* **2002**, *1569*, 35.
45. Hayakawa, F.; Kimura, T.; Maeda, T.; Fujita, M.; Sohmiya, H.; Fujii, M.; Ando, T. *Biochim. Biophys. Acta* **1997**, *1336*, 123.
46. Hayakawa, F.; Kimura, T.; Hoshino, N.; Ando, T. *Biosci. Biotechnol. Biochem.* **1999**, *63*, 1654.
47. Chyn, J. P.; Urbach, F. L. *Inorg. Chim. Acta* **1991**, *189*, 157.
48. Manzur, J.; Garcia, A. M.; Cordova, C.; Pizzaro, O.; Acuna, V.; Spodine, E. *Polyhedron* **2002**, *21*, 181.
49. Lim, Y. Y.; Tan, E. H. L.; Foong, P. C. *J. Mol. Catal.* **1993**, *85*, 173.
50. Mazer, N. A.; Benedek, G. B.; Carey, M. C. *J. Phys. Chem.* **1976**, *80*, 1075.
51. Lim, Y. Y.; Fong, H. Y.; Tan, L. H. *J. Mol. Catal. A* **1995**, *97*, L1.
52. Tsuruya, S.; Yanai, S.; Masai, M. *Inorg. Chem.* **1986**, *25*, 141.
53. Simándi, T. L.; Simándi, L. I. *React. Kin. Catal. Let.* **1998**, *65*, 301.
54. Simándi, L. I.; Simándi, T. L. *J. Chem. Soc., Dalton Trans.* **1998**, 3275.
55. Simándi, T. L.; Simándi, L. I. *J. Chem. Soc., Dalton Trans.* **1999**, 4529.
56. Barbaro, P.; Bianchini, C.; Frediani, P.; Meli, A.; Vizza, F. *Inorg. Chem.* **1992**, *31*, 1523.
57. Barbaro, P.; Bianchini, C.; Mealli, C.; Meli, A. *J. Am. Chem. Soc.* **1991**, *113*, 3181.
58. Bianchini, C.; Frediani, P.; Laschi, F.; Meli, A.; Vizza, F.; Zanello, P. *Inorg. Chem.* **1990**, *29*, 3402.
59. Barbaro, P.; Bianchini, C.; Linn, K.; Mealli, C.; Meli, A.; Vizza, F. *Inorg. Chim. Acta* **1992**, *200*, 31.
60. Que, L., Jr.; Lauffer, R. B.; Lynch, J. B.; Murch, B. P.; Pyrz, J. W. *J. Am. Chem. Soc.* **1987**, *109*, 5381.
61. Pascaly, M.; Duda, M.; Schweppe, F.; Zurlinden, K.; Müller, F. K.; Krebs, B. *J. Chem. Soc., Dalton Trans.* **2001**, 828.
62. Jameson, R. F.; Kiss, T. *J. Chem. Soc., Dalton Trans.* **1986**, 1833.

63. Karlin, K. D.; Tyeklár, Z. *"Advances in Inorganic Biochemistry"*, vol. 9; Eds. Eichhorn, G. L.; Marzilli, L. G.; Prentice Hall: New Jersey, 1994, p. 123.

64. Zwart, J.; van Wolput, J. H. M. C.; van der Cammen, J. C. J. M.; Koningsberger, D. C. *J. Mol. Catal.* **1981**, *11*, 69.

65. Zwart, J.; van Wolput, J. H. M. C.; Koningsberger, D. C. *J. Mol. Catal.* **1981**, *12*, 85.

66. Hanaki, A.; Kamide, H. *Bull. Chem. Soc. Jpn.* **1983**, *56*, 2065.

67. Ehrenberg, L.; Harms-Ringdahl, M.; Fedorcsák, I.; Granath, F. *Acta Chem. Scand.* **1989**, *43*, 177.

68. Hanaki, A. *Bull. Chem. Soc. Jpn.* **1995**, *68*, 831.

69. Hanaki, A.; Kinoshita, T.; Ozawa, T. *Bull. Chem. Soc. Jpn.* **2000**, *73*, 73.

70. Barton, J.; Packer, J. E. *Int. J. Radiat. Phys. Chem.* **1970**, *2*, 159.

71. Hoffman, M. Z.; Hayon, E. *J. Am. Chem. Soc.* **1972**, *94*, 7950.

72. Mezyk, S. P.; Armstrong, D. A. *Can. J. Chem.* **1989**, *67*, 736.

73. Pecci, L.; Montefoschi, G.; Musci, G.; Cavallini, D. *Amino Acids* **1997**, *13*, 355.

74. Kachur, A. V.; Koch, C. J.; Biaglow, J. E. *Free Rad. Res.* **1999**, *31*, 23.

75. Menshikov, S. Y.; Vurasko, A. V.; Molochnikov, L. S.; Kovalyova, E. G.; Efendiev, A. A. *J. Mol. Catal. A* **2000**, *158*, 447.

76. Albro, P. W.; Corbett, J. T.; Shcroeder, J. L. *J. Inorg. Biochem.* **1986**, *27*, 191.

77. Kachur, A. V.; Koch, C. J.; Biaglow, J. E. *Free Rad. Res.* **1998**, *28*, 259.

78. Brandt, C.; van Eldik, R. *Chem. Rev.* **1995**, *95*, 119.

79. Loon, L. V.; Mader, E.; Scott, S. L. *J. Phys. Chem. A* *2000*, *104*, 1621.

80. Brandt, C.; Fábián, I.; van Eldik, R. *Inorg. Chem.* **1994**, *33*, 687.

81. Huie, R. E.; Neta, P. *Atmos. Eviron.* **1987**, *21*, 1743.

82. Buxton, G. V.; McGowan, S.; Salmon, G. A.; Williams, J. E.; Wood, N. D. *Atmos. Environ.* **1996**, *30*, 2483.

83. Das, T. N.; Huie, R. E.; Neta, P. *J. Phys. Chem. A* **1999**, *103*, 3581.

84. Connick, R. E.; Zhang, Y. X.; Lee, S. Y.; Adamic, R.; Chieng, P. *Inorg. Chem.* **1995**, *34*, 4543.

85. Das, T. N. *J. Phys. Chem. A* **2001**, *105*, 9142.

86. Warneck, P.; Ziajka, J. *Ber. Bunsenges. Phys. Chem.* **1995**, *99*, 59.

87. Zhang, W. S.; Singh, P.; Muir, D. *Hydrometallurgy* **2000**, *55*, 229.

88. Boyce, S. D.; Hoffmann, M. R.; Hong, P. A.; Molberly, L. M. *Environ. Sci. Technol.* **1983**, *17*, 602.

89. Kraft, J.; van Eldik, R. *Inorg. Chem.* **1989**, *28*, 2306.

90. Pasiuk-Bronikowska, W.; Bronikowski, T. *Chem. Eng. Sci.* **1989**, *44*, 1361.

91. Kraft, J.; van Eldik, R. *Atmos. Environ.* **1989**, *23*, 2709.

92. Kraft, J.; van Eldik, R. *Inorg. Chem.* **1989**, *28*, 2297.

93. Betterton, E. A. *J. Atmos. Chem.* **1993**, *17*, 307.

94. Conklin, M. H.; Hoffmann, M. R. *Environ. Sci. Technol.* **1988**, *22*, 899.

95. Lente, G.; Fábián, I. *Inorg. Chem.* **1998**, *37*, 4204.

96. Lepentsiotis, V.; Prinsloo, F. F.; van Eldik, R.; Gutberlet, H. *J. Chem. Soc., Dalton. Trans.* **1996**, 2135.

97. Prinsloo, F. F.; Brandt, C.; Lepentsiotis, V.; Pienaar, J. J.; van Eldik, R. *Inorg. Chem.* **1997**, *36*, 119.

98. Brandt, C.; van Eldik, R. *Transition Met. Chem.* **1998**, *23*, 667.

99. Carlyle, D. W.; King, E. L. *Inorg. Chem.* **1970**, *9*, 2333.

100. van Eldik, R.; von Jouanne, J.; Kelm, H. *Inorg. Chem.* **1982**, *21*, 2818.

101. Bazsa, G.; Diebler, H. *React. Kin. Catal. Lett.* **1975**, *3*, 217.

102. Magalhães, M.E.A.; Lente, G.; Fábián, I. *"Abstracts, 34th Int. Conf. Coord. Chem."*; Edinburgh, UK, 2000; p. 277.

103. Lente, G.; Magalhães, M. E. A.; Fábián, I. *Inorg. Chem.* **2000**, *39*, 1950.

104. Lente, G.; Fábián, I. *React. Kin. Catal. Lett.* **2001**, *73*, 117.
105. Lente, G.; Fábián, I. *Inorg. Chem.* **2002**, *41*, 1306.
106. Lente, G.; Fábián, I. *J. Chem. Soc., Dalton Trans.* **2002**, 778.
107. Lente, G.; Fábián, I. *Inorg. Chem.* **1999**, *38*, 603.
108. Hadady, Zs.; Lente, G.; Fábián, I. unpublished results.
109. Grgič, I.; Dovžan, A.; Berčič, G.; Hudnik, V. *J. Atmos. Chem.* **1998**, *29*, 315.
110. Grgič, I.; Poznič, M.; Bizjak, M. *J. Atmos. Chem.* **1999**, *33*, 89.
111. Wolf, A.; Deutsch, F.; Hoffmann, P.; Ortner, H. M. *J. Atmos. Chem.* **2000**, *37*, 125.
112. Connick, R. E.; Zhang, Y. X. *Inorg. Chem.* **1996**, *35*, 4613.
113. Berglund, J.; Fronaeus, S.; Elding, L. I. *Inorg. Chem.* **1993**, *32*, 4527.
114. Fronaeus, S.; Berglund, J.; Elding, L. I. *Inorg. Chem.* **1998**, *37*, 4939.
115. Grgič, I.; Berčič, G. *J. Atmos. Chem.* **2001**, *39*, 155.
116. Lepentsiotis, V.; Domagala, J.; Grgič, I.; van Eldik, R.; Muller, J. G.; Burrows, C. J. *Inorg. Chem.* **1999**, *38*, 3500.
117. Dutta, S. K.; Ferraudi, G. *J. Phys. Chem. A* **2001**, *105*, 4241.
118. Song, Y. H.; Yang, C. M..; Kluger, R. *J. Am. Chem. Soc.* **1993**, *115*, 4365.
119. Muller, J. G.; Hickerson, R. P.; Perez, R. J.; Burrows, C. J. *J. Am. Chem. Soc.* **1997**, *119*, 1501.
120. Hickerson, R. P.; Watkins-Sims, C. D.; Burrows, C. J.; Atkins, J. F.; Gesteland, R. F.; Felden, B. *J. Mol. Biol.* **1998**, *279*, 577.
121. Wietzerbin, K.; Muller, J. G.; Jameton, R. A.; Pratviel, G.; Bernadou, J.; Meunier, B.; Burrows, C. J. *Inorg. Chem.* **1999**, *38*, 4123.
122. Neves, E. A.; Valdes, J.; Klockow, D. *Fresenius J. Anal. Chem.* **1995**, *351*, 544.
123. Lima, S.; Bonifácio, R. L.; Azzellini, G. C.; Coichev, N. *Talanta* **2002**, *56*, 547.
124. Utaka, M.; Hojo, M.; Fujii, Y.; Takeda, A. *Chem. Lett.* **1984**, 635.
125. Utaka, M.; Takeda, A. *J. Chem. Soc., Chem. Comm.* **1985**, 1824.
126. Balogh-Hergovich, É.; Speier, G. *J. Mol. Catal.* **1992**, *71*, 1.
127. Balogh-Hergovich, É.; Kaizer, J.; Speier, G. *J. Mol. Catal.* **2000**, *159*, 215.
128. Barhács, L.; Kaizer, J.; Pap, J.; Speier, G. *Inorg. Chim. Acta* **2001**, *320*, 83.
129. Nishinaga, A.; Tojo, T.; Matsuura, T. *J. Chem. Soc., Chem. Comm.* **1974**, 896.
130. Nishinaga, A.; Numada, N.; Maruyama, K. *Tetrahedron Lett.* **1989**, *30*, 2257.
131. Nishinaga, A.; Kuwashige, T.; Tsutsui, T.; Mashino, T.; Maruyama, K. *J. Chem. Soc., Dalton Trans.* **1994**, 805.
132. Simándi, L. I.; Németh, S.; Rumelis, N. *J. Mol. Catal.* **1987**, *42*, 357.
133. Simándi, L. I.; Fülep-Poszmik, A.; Németh, S. *J. Mol. Catal.* **1988**, *48*, 265.
134. Szeverényi, Z.; Simándi, L. I. *J. Mol. Catal.* **1989**, *51*, 155.
135. Milaeva, E. R.; Szeverényi, Z.; Simándi, L. I. *Inorg. Chim. Acta* **1990**, *167*, 139.
136. Szeverényi, Z.; Simándi, L. I.; Iwanejko, R. *J. Mol. Catal.* **1991**, *64*, L15.
137. Szeverényi, Z.; Milaeva, E. R.; Simándi, L. I. *J. Mol. Catal.* **1991**, *67*, 251.
138. Simándi, L. I.; Barna, T.; Szeverényi, Z.; Németh, S. *Pure. Appl. Chem.* **1992**, *64*, 1511.
139. Simándi, L. I.; Barna, T.; Korecz, L.; Rockenbauer, A. *Tetrahedron Lett.* **1993**, *34*, 717.
140. Simándi, L. I.; Barna, T.; Németh, S. *J. Chem. Soc., Dalton Trans.* **1996**, 473.
141. Simándi, L. I.; Simándi, T. L. *J. Mol. Catal. A* **1997**, *117*, 299.
142. Chauhan, S. M. S.; Gulati, A.; Sahay, A.; Nizar, P. N. H. *J. Mol. Catal. A* **1996**, *105*, 159.
143. Bajaj, H. C.; Taqui Khan, M. M. *React. Kin. Catal. Lett.* **1985**, *28*, 339.
144. Taqui Khan, M. M.; Rao, A. P. *J. Mol. Catal.* **1986**, *35*, 237.
145. Taqui Khan, M. M.; Rao, A. P. *J. Mol. Catal.* **1987**, *39*, 331.
146. Taqui Khan, M. M.; Rao, A. P.; Shukla, R. S. *J. Mol. Catal.* **1989**, *49*, 299.
147. Taqui Khan, M. M.; Rao, A. P.; Shukla, R. S. *J. Mol. Catal.* **1989**, *50*, 45.
148. Taqui Khan, M. M.; Shukla, R. S.; Rao, A. P. *Inorg. Chem.* **1989**, *28*, 452.

149. O'Sullivan, D. W.; Hanson, A. K., Jr.; Kester, D. R. *Mar. Chem.* **1995**, *49*, 65.
150. King, D. W.; Lounsbury, H. A.; Millero, F. J. *Environ. Sci. Technol.* **1995**, *29*, 818.
151. Emmenegger, L.; King, D. W.; Sigg, L.; Sulzberger, B. *Environ. Sci. Technol.* **1998**, *32*, 2990.
152. King, D. W. *Environ. Sci. Technol.* **1998**, *32*, 2997.
153. Merényi, G.; Lind, J.; Eriksen, T. E. *J. Biolumin. Chemilumin.* **1990**, *5*, 53.
154. Merényi, G.; Lind, J.; Eriksen, T. E. J. *J. Am. Chem. Soc.* **1986**, *108*, 7716.
155. Rose, A. L.; Waite, D. T. *Anal. Chem.* **2001**, *73*, 5909.
156. Kimura, M.; Tokuda, M.; Tsukahara, K. *Bull. Chem. Soc. Jpn.* **1994**, *67*, 2731.
157. Kimura, M.; Seki, K.; Horie, H.; Tsukahara, K. *Bull. Chem. Soc. Jpn.* **1996**, *69*, 613.
158. Chen, D. A.; Motekaitis, R. J.; Martell, A. E.; McManus, D. *Can. J. Chem.* **1993**, *71*, 1524.
159. Yokota, K.; Yamazaki, I. *Biochim. Biophys. Acta* **1965**, *105*, 301.
160. Scheeline, A.; Olson, D. L.; Williksen, E. P.; Horras, G. A.; Klein, M. L.; Larter, R. *Chem. Rev.* **1997**, *97*, 739, and references therein.
161. Strizhak, P. E.; Basylchuk, A. B.; Demjanchyk, I.; Fecher, F.; Schneider, F. W.; Munster, A. F. *Phys. Chem. Chem Phys.* **2000**, *2*, 4721.
162. Jensen, J. H. *J. Am. Chem. Soc.* **1983**, *105*, 2639.
163. Roelofs, M. G.; Wasserman, E.; Jensen, J. H. *J. Am. Chem. Soc.* **1987**, *109*, 4207.
164. Roelofs, M. G.; Jensen, J. H. *J. Phys. Chem.* **1987**, *91*, 3380.
165. Colussi, A. J.; Ghibaudi, E.; Yuan, Z.; Noyes, R. M. *J. Am. Chem. Soc.* **1990**, *112*, 8660.
166. Boga, E.; Peintler, G.; Nagypál, I. *J. Am. Chem. Soc.* **1990**, *112*, 151.
167. Guslander, J.; Noyes, R. M.; Colussi, A. J. *J. Phys. Chem.* **1991**, *95*, 4387.
168. Druliner, J. D.; Wasserman, E. *J. Am. Chem. Soc.* **1988**, *110*, 5270.

INDEX

CONTENTS OF PREVIOUS VOLUMES